Recent Advances in Scientific Computing and Applications

CONTEMPORARY MATHEMATICS

586

Recent Advances in Scientific Computing and Applications

Eighth International Conference
on Scientific Computing and Applications
April 1–4, 2012
University of Nevada
Las Vegas, Nevada

Jichun Li
Hongtao Yang
Eric Machorro
Editors

American Mathematical Society
Providence, Rhode Island

2000 *Mathematics Subject Classification.* Primary 65N55, 76M50, 78A45, 81V55, 49N45, 68W25, 35Q61, 78M40, 35B36, 92C15.

Library of Congress Cataloging-in-Publication Data

International Conference on Scientific Computing and Applications (8th : 2012 : Las Vegas, Nev.)
Recent advances in scientific computing and applications : Eighth International Conference on Scientific Computing and Applications, April 1–4, 2012, University of Nevada, Las Vegas, Nevada / Jichun Li, Hongtao Yang, Eric Machorro, editors.
p. cm – (Contemporary Mathematics ; v. 586)
Includes bibliographical references.
ISBN 978-0-8218-8737-0 (alk. paper)
1. Multigrid methods (Numerical analysis)–Congresses. 2. Numerical analysis–Congresses. I. Li, Jichun, editor of compilation. II. Yang, Hongtao, 1962– editor of compilation. III. Machorro, Eric A. (Eric Alexander), 1969– editor of compilation. IV. Title.

QA377.I5678 2012
518′.64–dc23 2012043433

Contemporary Mathematics ISSN: 0271-4132 (print); ISSN: 1098-3627 (online)

∞ The paper used in this book is acid-free and falls within the guidelines
established to ensure permanence and durability.
Visit the AMS home page at `http://www.ams.org/`

10 9 8 7 6 5 4 3 2 1 18 17 16 15 14 13

Contents

Preface

The Eighth International Conference on Scientific Computing and Applications (SCA) was held in University of Nevada Las Vegas (UNLV) during April 1 - 4, 2012. This series of conferences were held in the Pacific Rim region, including Hong Kong (twice), Alberta of Canada (twice), Shanghai in China, Busan in Korea, and Dalian in China. It is the first time this SCA conference was held in USA, and it was the largest of all SCA conferences. It attracted about 180 participants from Australia, Brasil, Canada, China, Cyprus, Czech Republic, France, Germany, Hong Kong, Ireland, Italy, Philippines, Saudi Arabia, Sweden, United Arab Emirates, United Kingdom, and USA. More than 140 papers were presented on various subjects in modern scientific computing and its applications, such as finite element methods, multiscale methods, finite difference methods, spectral methods, collocation methods, adaptive methods, parallel computing, linear solvers, and applications to fluid flow, nano-optics, biofilms, finance, magnetohydrodynamics flow, electromagnetic waves, fluid-structure interaction problem, and stochastic PDEs. This book contains 39 selected papers, which represent some currently active subjects. This book can serve as an excellent reference for graduate students and researchers who work in scientific computing and its applications in various areas of science and engineering.

This conference would not be possible without the support of many organizations and assistance of many people. It received generous support from National Science Foundation (under grant DMS 1139712), National Security Technologies LLC (under Contract No. DE-AC52-06NA25946 with the U.S. Department of Energy and supported by the Site-Directed Research and Development Program), and Department of Mathematical Sciences (DMS) of UNLV. The hard work of the local organizers (Drs. Derrick Dubose, Monica Neda, Pengtao Sun and Yitung Chen) and our graduate students (Sean Breckling, Jiacheng Cai, Xudong Sun, Yuzhou Sun, Jiajia Waters, and Lanxuan Yu) were critical to the success of the conference. We also like to thank Dr. Derrick Dubose (Chair of DMS), Patricia Pablo and Erin McNamara (staff of DMS) for their enthusiastic support. Finally, we would like to thank Mrs. Christine M. Thivierge for her kind assistance in publishing this volume.

Jichun Li, Hongtao Yang, and Eric Machorro (all in Las Vegas)

Multifrequency inverse source problem for elastic waves

Sebastian Acosta, S.-Sum Chow, and Vianey Villamizar

ABSTRACT. We study an inverse source problem involving an elastic wave equation, where an unknown source is to be identified from the knowledge of its radiated wave. We show that with a multi-frequency approach, the source may be uniquely determined. Several theoretical results developed for acoustic waves are extended inverse source problems for elastic waves. We discuss two algorithms for the reconstruction of the source using multi-frequency data. One algorithm is based on an incomplete Fourier transform of the measured data and the other involves the solutions of an adjoint problem corresponding to different frequencies.

1. Introduction

An inverse source problem associated with a wave equation involves the identification of an unknown source from the knowledge of its radiated waves. The waves may be acoustic waves, elastic waves or electromagnetic waves and will give rise to different problems. Inverse source problems arise naturally in many applications such as antenna synthesis and medical imaging.

As an example, the field of photoacoustic tomography (PAT) has received much attention in recent years. PAT is capable of providing high-resolution images with high ultrasonic resolution and strong optical contrast. The mathematics behind PAT is also very interesting, and a very good summary was provided by Kuchment and Kunyansky **[9]**. As indicated by Anastasio *et al* **[3]**, the PAT problem can be posed as an inverse source problem. Most of the model problems for PAT involve the acoustic wave (or Helmholtz) equation, while in reality, these problems are better modeled as wave problems in an elastic medium. Hence it is of interest to study the inverse source problem for elastic waves.

Due to the existence of non-radiating sources (at a given frequency,) inverse source problems in general possess non-unique solutions. For inverse source problems involving acoustic waves, the work by Eller and Valdivia **[6]** and Bao, Lin and Triki **[4]**, and Acosta *et al* **[1]** clearly imply that in order to uniquely identify the source, one must adopt a multi-frequency approach, e.g., by letting the frequency varies within some open interval on the positive real line. In this work we prove that many of the results developed for acoustic waves may be extended to elastic

---

2000 *Mathematics Subject Classification.* Primary 73D50, 65N21.

Supported in part by CHIRP grant from the College of Physical and Mathematical Sciences at Brigham Young University.

waves, and propose some reconstruction algorithms for an inverse source problem involving elastic waves in homogeneous media.

2. Direct and Inverse Source Problems for Elastic Wave Propagation

In this section, we recall how the scalar and vector Helmholtz equations are obtained from the time harmonic wave equation and review their corresponding radiation conditions. We will then describe the direct source problem and inverse source problem using the weak formulation of the Helmholtz equations.

In a homogeneous isotropic medium represented by a bounded domain Ω in $\mathbb{R}^n$, with $n = 2$ or 3, the time harmonic elastic wave equation is obtained by coupling the equation of motion and Hooke's law:

$$\nabla^2 \mathbf{u} + k^2 \mathbf{u} + \gamma \nabla(\nabla \cdot \mathbf{u}) = \mathbf{F} \tag{2.1}$$

where $\mathbf{u}$ is the displacement, $\mathbf{F}$ is related to the external force and k, γ are positive constants.

Let

$$\psi = -k_S^2 \nabla \cdot \mathbf{u} \tag{2.2}$$

where $k_S^2 = k^2/(1+\gamma)$. Taking the divergence of the elastic wave equation (2.1) and substituting in ψ, we have the scalar Helmoltz equation

$$\nabla^2 \psi + k_S \psi = -f_1 \tag{2.3}$$

where

$$f_1 = k^{-2} \nabla \cdot \mathbf{F}$$

Define

$$\mathbf{\Phi} = \mathbf{u} - \nabla \psi - k^{-2} \mathbf{F} \tag{2.4}$$

Using (2.2),(2.3,) (2.4) it is easy to check that $\mathbf{\Phi}$ is divergence free:

$$\nabla \cdot \mathbf{\Phi} = 0$$

and that it satisfies the vector Helmholtz equation

$$k^2 \mathbf{\Phi} + \nabla^2 \mathbf{\Phi} = -\mathbf{G} \tag{2.5}$$

where

$$\mathbf{G} = -k^{-2} \nabla^2 \mathbf{F}$$

In this paper, we assume that Ω contains the origin and that the support of the source $\mathbf{F}$ lies entirely within Ω. As we will need to solve the Helmholtz equations (2.3) and (2.5) in the unbounded domain $\mathbb{R}^n \backslash \Omega$, it is necessary to impose the Kuradze-Sommerfeld radiation conditions.

$$\lim_{r\to\infty} r^{(n-1)/2} \left(\frac{\partial \nabla \psi}{\partial n} - i k_S \nabla \psi \right) = \mathbf{0} \tag{2.6}$$

and

$$\lim_{r\to\infty} r^{(n-1)/2} \left(\frac{\partial \mathbf{\Phi}}{\partial n} - i k \mathbf{\Phi} \right) = \mathbf{0} \tag{2.7}$$

As note in [**5**], it is more convenient to work with ψ than with $\nabla\psi$, and the radiation condition (2.6)may be replace by the usual Sommerfeld radiation condition on ψ, which is a scalar version of (2.7).

For simplicity, we assume that the functions ψ and $\mathbf{\Phi}$ are known on the boundary Γ of Ω and write

$$\psi = \psi_0, \quad \mathbf{\Phi} = \mathbf{\Phi}_0 \text{ on } \Gamma. \tag{2.8}$$

Note that if $\mathbf{u}$ is known on the boundary, the functions ψ_0 and $\mathbf{\Phi}_0$ may be calculated.

In view of the similarity of the scalar and vectorial Helmholtz problems, we will focus on the vector Helmholtz problem. We further assume that the source is separable in the sense that

$$\mathbf{G}(\mathbf{x}, k) = g(k)\mathbf{f}(\mathbf{x}) \tag{2.9}$$

where $g(k)$ is a known function.

In what follows, we use $\mathbf{L}^2(\cdot)$ to denote $[L^2(\cdot)]^n$ and $\mathbf{H}^s(\cdot)$ to denote $[H^s(\cdot)]^n$. Let $\langle\cdot,\cdot\rangle = \langle\cdot,\cdot\rangle_{L^2(\Omega)}$ and $\langle\cdot,\cdot\rangle_{L^2(\Gamma)}$ denote the usual L^2 inner product for vector functions on Ω and on Γ respectively.

The direct source problem is a weak formulation of the Helmholtz problem (2.5) with boundary condition (2.8) and radiation condition (2.7).

DEFINITION 2.1. (**Direct source problem**) For given $\mathbf{f} \in \mathbf{L}^2(\Omega)$, we seek a solution $\mathbf{\Phi} \in \mathbf{H}^1(\Omega)$ satisfying the boundary condition (2.8) and

$$B(\mathbf{\Phi}, \mathbf{v}) = g\,\langle \mathbf{f}, \mathbf{v}\rangle_{L^2(\Omega)} \quad \text{for all } \mathbf{v} \in \mathbf{H}^1(\Omega) \tag{2.10}$$

where

$$B(\mathbf{u}, \mathbf{v}) = \sum_{i=1}^{n} \langle \nabla \mathbf{u}_i, \nabla \mathbf{v}_i\rangle - k^2 \langle \mathbf{u}, \mathbf{v}\rangle - g\,\langle \mathbf{MTu}, \mathbf{Tv}\rangle_{L^2(\Gamma)}$$

with $\mathbf{T} : \mathbf{H}^1(\Omega) \to \mathbf{H}^{1/2}(\Gamma)$ being the trace map and $\mathbf{M} : \mathbf{H}^{1/2}(\Gamma) \to \mathbf{H}^{-1/2}(\Gamma)$ being the Dirichlet to Neumann map.

For the inverse source problem, we do not have information about the source $\mathbf{f}$ but do have some information of the solution on the boundary. More specifically,

DEFINITION 2.2. (**Inverse source problem**) Let $\mathbf{\Phi} \in \mathbf{H}^1(\Omega)$ be the solution of the direct source problem (2.10) for given frequency $k \in \mathcal{K}$ and some unknown source $\mathbf{f} \in \mathbf{L}^2(\Omega)$. Determine the source $\mathbf{f}$ using the trace $\mathbf{T\Phi}$ on Γ as data.

2.1. Summary of theoretical results. In what follows we state several theoretical results for the direct and indirect source problems. Many of these theorems are natural extensions of corresponding results described in [**1**], where a more general scalar inverse source problem is considered.

First we note that the direct source problem is a well-posed problem. This result follows immediately from the existence, uniqueness and stability results of Bramble and Pasciak [**5**] for the exterior problems for the vector Helmholtz equations (2.5) under the boundary and radiation conditions (2.8) and (2.7).

For results related to the inverse source problem, we need some information about non-radiating sources. First recall that a source $\mathbf{f} \in \mathbf{L}^2(\Omega)$ is non-radiating at a frequency k if the solution $\mathbf{\Phi} \in \mathbf{H}^1(\Omega)$ of the direct source problem (2.10) with such source has zero trace on the boundary Γ.

Let $N(\Omega, k)$ denote the set of all non-radiating sources for the vector Helmholtz equation at a frequency k. A source $\mathbf{f} \in \mathbf{L}^2(\Omega)$ is said to be purely-radiating at a frequency k if $\mathbf{f} \in N(\Omega, k)^{\perp}$.

It is possible to characterize $N(\Omega, k)$ in term of the a non-trivial vector space, which has the important implication that the inverse source problem cannot be solved uniquely for given boundary data at a single frequency.

Let

$$\mathcal{N}(\Omega, k) = \{\mathbf{g} \in \mathbf{L}^2(\Omega) : \quad \mathbf{g} = \nabla^2 \mathbf{w} + k^2 \mathbf{w} \text{ for some } \mathbf{w} \in [C_c^\infty(\Omega)]^n\}$$

then we have the following result

THEOREM 2.3. *The set $\mathcal{N}(\Omega, k)$ is dense in $N(\Omega, k)$ in the $\mathbf{L}^2(\Omega)$ norm.*

Furthermore, as it may be shown that $N(\Omega, k)$ is a closed subspace of $\mathbf{L}^2(\Omega)$, we have the orthogonal decomposition

$$\mathbf{L}^2(\Omega) = N(\Omega, k) \oplus N(\Omega, k)^\perp \tag{2.11}$$

Due to the presence of non-radiating source, it is not possible to use one single frequency to resolve the inverse source uniquely. A natural question is whether it is possible to determine the source uniquely using a finite number of distinct frequencies. Unfortunately the answer is also in the negative because there exist nontrivial sources that do not radiate at a finite number of distinct frequencies. We summarize this result in the follow theorem.

THEOREM 2.4. *For any finite set of frequencies $\mathcal{K}_n = \{k_1, k_2, \ldots, k_n\}$, there exists $\mathbf{g} \in \mathbf{L}^2(\Omega)$ such that*

$$\mathbf{g} \in \bigcap_{k \in \mathcal{K}_n} N(\Omega, k), \qquad \textit{and} \qquad \mathbf{g} \neq \mathbf{0}$$

2.2. Multi-frequency uniqueness. In contrast to Theorem 2.4, we now show that the source is uniquely determined if the set of frequencies $\mathcal{K}$ used to identify the source is an open interval.

Suppose the variational direct source problem (2.10) is satisfied by the wave field $\mathbf{\Phi}_k \in \mathbf{H}^1(\Omega)$. By choosing $\mathbf{v} = \mathbf{v}_k$ to be plane waves of the form $\mathbf{v}_k = e^{ik\hat{\mathbf{x}}\cdot\mathbf{z}}\, \mathbf{e}_\ell$ where $\mathbf{e}_\ell$ is the ℓ-th column of the identify matrix, and $\hat{\mathbf{x}}$ is a unit vector, (2.10) may be rewritten as an equation for the Fourier transform of f_ℓ, the ℓ-th component of $\mathbf{f}$:

$$(\mathcal{F} f_\ell)(k\hat{\mathbf{x}}) = (2\pi)^{-3/2} \Big(\langle \mathbf{T\Phi}_k,\ D_\nu \mathbf{v}_k \rangle_{\mathbf{L}^2(\Gamma)} - \langle \mathbf{MT\Phi}_k,\ \mathbf{Tv}_k \rangle_{L^2(\Gamma)} \Big) \tag{2.12}$$

where D_ν denotes the normal derivative on Γ. The above expression gives rise to a reconstruction algorithm based on the inverse Fourier transform to be described later.

The Fourier transform expression (2.12) also allows us to prove the following uniqueness result:

THEOREM 2.5. *Let $\mathcal{K}$ be an open interval on the positive real line $\subset \mathbb{R}_+$. Suppose $\mathbf{f}^1, \mathbf{f}^2 \in \mathbf{L}^2(\Omega)$ are two sources such that their corresponding solutions to the direct problem (2.10) take the same trace value on the surface Γ for all frequencies $k \in \mathcal{K}$. Then $\mathbf{f}^1 = \mathbf{f}^2$.*

In practice, it is of course impossible to make use of infinitely many frequencies to achieve uniqueness as required by the above theorem. Instead we seek a minimum norm solution in the sense that if $\mathbf{f}$ is a solution to the inverse source problem, we take as a solution to the inverse source problem the function $\tilde{\mathbf{f}}$ such that

$$\|\tilde{\mathbf{f}}\|_{\mathbf{L}^2(\Omega)} = \min\{\|\mathbf{f}+\mathbf{g}\|_{L^2(\Omega)}, \quad \mathbf{g} \in N(\Omega,k)\}$$

From the orthogonal decomposition (2.11), it is easy to establish the existence of such minimum norm solution.

COROLLARY 2.6. *Given an arbitrary source* $\mathbf{f} \in \mathbf{L}^2(\Omega)$, *there exist a unique decomposition* $\mathbf{f} = \mathbf{f}_N + \mathbf{f}_P$ *with* $\mathbf{f}_N \in N(\Omega,k)$ *and* $\mathbf{f}_P \in N(\Omega,k)^{\perp}$, *such that* $\mathbf{f}_P$ *is the minimum-norm solution to the inverse source problem associated with* $\mathbf{f}$.

2.3. A characterization of purely-radiating sources. In order to find a characterization of the set $N(\Omega,k)^{\perp}$, we follow the approach of [2] by first defining a variational problem whose solutions are dense in $N(\Omega,k)^{\perp}$:

DEFINITION 2.7. (Adjoint problem) Given boundary data $\boldsymbol{\eta} \in \mathbf{L}^2(\Gamma)$ and fixed frequency k, find a function $\boldsymbol{\psi} \in \mathbf{H}^1(\Omega)$ such that

$$A(\boldsymbol{\psi},\boldsymbol{\phi}) = g\,\langle \boldsymbol{\eta},\ \mathbf{T}\boldsymbol{\phi}\rangle_{L^2(\Gamma)}, \qquad \text{for all } \boldsymbol{\phi} \in \mathbf{H}^1(\Omega), \tag{2.13}$$

where $A : \mathbf{H}^1(\Omega) \times \mathbf{H}^1(\Omega) \to \mathbb{C}$ is given by

$$A(\boldsymbol{\psi},\boldsymbol{\phi}) = \sum_{i=1}^{n} \langle \nabla\boldsymbol{\psi}_i,\ \nabla\boldsymbol{\phi}_i\rangle_{L^2(\Omega)} - k^2\,\langle\boldsymbol{\psi},\ \boldsymbol{\phi}\rangle_{L^2(\Omega)} - \langle \mathbf{M}^*\mathbf{T}\boldsymbol{\psi}, \mathbf{T}\boldsymbol{\phi}\rangle)_{L^2(\Gamma)} \tag{2.14}$$

here $\mathbf{M}^* : \mathbf{H}^{1/2}(\Gamma) \to \mathbf{H}^{-1/2}(\Gamma)$ denotes the adjoint of the DtN map.

Using the result from [1], it is possible to show that the adjoint problem (2.13) is well-posed. Consequently, the following sets are well-defined:

$\mathcal{P}(\Omega,k) = \{\boldsymbol{\psi} \in \mathbf{H}^1(\Omega) \text{ satisfies (2.13) for some } \boldsymbol{\eta} \in \mathbf{L}^2(\Gamma)\}$
$P(\Omega,k) = \text{closure of } \mathcal{P}(\Omega,k) \text{ in the } \mathbf{L}^2(\Omega)\text{-norm}$

It turns out that the solutions of the adjoint problem are closely related to the set of purely-radiating sources:

THEOREM 2.8.

$$N(\Omega,k)^{\perp} = P(\Omega,k), \quad N(\Omega,k) = P(\Omega,k)^{\perp}$$

The above result provides an alternate way to characterize the set of non-radiating sources through the solutions of the adjoint problem; and the set of purely-radiating sources through the subspace orthogonal to the solutions of the adjoint problem. With this result, we now have a computable representation of the set of purely-radiating sources.

3. Reconstruction algorithms

3.1. Reconstruction via Fourier transform. Recall that by using a plane wave as a test function in (2.10), we obtain an expression (2.12) of the Fourier transforms for each component of the source function for all $k \in (0,K)$. In oder to compute an approximation of the Fourier transform, it is necessary to evaluate the DtN map explicitly. This may be done using some of the many existing ideas ([7], [8]) of calculating approximations to the DtN map.

It is usually not possible to perform a simple inversion to recover the unknown source, since the Fourier transform is only known for a finite range of frequencies.

However, a natural and simple algorithm is to define an approximate reconstructed source $f_{\ell K}$ via

$$f_{\ell K} = \mathcal{F}^{-1}\left(\chi_{B(K)}\mathcal{F}\mathbf{f}_\ell\right)$$

where $\chi_{B(K)}$ is the characteristic function of the open ball $B(k) \subset \mathbb{R}^n$ with center at the origin and radius K.

One unfortunate consequence of the approximate and incomplete inverse Fourier transform is that the computed source can no longer be guaranteed to have its support inside Ω. Also, in the presence of noise represented by $\epsilon \in L^2(\mathbb{R}^n)$, we have

$$f_{\ell K,\epsilon} = \mathcal{F}^{-1}\left(\chi_{B(K)}\mathcal{F}\mathbf{f}_\ell + \epsilon\right) \tag{3.1}$$

It is of interest to see the effect of noise to the solution. We summarize the stability result as follows:

THEOREM 3.1. *Let* $\mathbf{f}$ *be the solution of the inverse source problem* (2.10) *and let the components of* $\mathbf{f}_{K,\epsilon}$ *be given by* (3.1). *If* $\mathbf{f} \in \mathbf{H}^s(\mathbb{R}^n)$ *for some* $s > 0$, *then there exists a constant* C *such that*

$$\|\mathbf{f} - \mathbf{f}_{K,\epsilon}\|_{\mathbf{L}^2(\mathbb{R}^n)} \leq \frac{C}{1+K^s}\|\mathbf{f}\|_{\mathbf{H}^s(\mathbb{R}^n)} + \|\epsilon\|_{\mathbf{L}^2(\mathbb{R}^n)}$$

Once the sources for the scalar and vectorial inverse source problems are found, we can proceed to recover the original source $\mathbf{F}$ of the elastic wave equation with some post-processing. Note that if we apply the Helmholtz decomposition theorem and boundary conditions of $\mathbf{F}$ to write $\mathbf{F}$ as the unique sum of a divergence free component $\mathbf{F}_d$ and a curl free component $\mathbf{F}_c$, then

$$f_1 = k^{-2}\nabla\cdot\mathbf{F}_c, \quad \mathbf{G} = -k^{-2}\nabla^2\mathbf{F}_d \tag{3.2}$$

then the divergence free component $\mathbf{u}_d$ and then curl free component $\mathbf{u}_c$ of $\mathbf{u}$ are given by

$$\mathbf{u}_d = \mathbf{\Phi} + k^{-2}\mathbf{F}_d, \quad \mathbf{u}_c = \nabla\psi + k^{-2}\mathbf{F}_c\,.$$

If $\mathbf{f}$ is known, then from (3.2) and (2.9)

$$-k^{-2}\nabla^2\mathbf{F}_d = g(k)\mathbf{f} \quad \text{in } \Omega$$

$$\mathbf{F}_d = \mathbf{0} \quad \text{on } \Gamma$$

and so the divergence free component of $\mathbf{F}$ may be found by solving the above vector Poisson equation.

Likewise, the curl free component of $\mathbf{F}$ may be found by letting $\mathbf{F}_d = \nabla\phi$ and solving the Poisson problem

$$k^{-2}\nabla^2\phi = f_1 \quad \text{in } \Omega$$

$$\phi = 0 \quad \text{on } \Gamma$$

3.2. Reconstruction via adjoint problem. An alternative approach to reconstruct the source is to make use of the adjoint problem (2.13) and the characterization theorem 2.8. As we can only construct purely radiating component of the unknown source, and the set of purely radiating sources $N(\Omega,k)^\perp = P(\Omega,k)$, so the idea is find a least squares approximation to the the unknown source using solutions of the adjoint problems corresponding to different frequencies. A variation of this idea is to construct the Fourier coefficients of the unknown source using a finite subset of an orthogonal basis of $\mathcal{P}(\Omega,k)^\perp$.

Setting $\mathbf{v} = \boldsymbol{\psi}$ in the original problem (2.10), and $\boldsymbol{\phi} = \boldsymbol{\Phi}$ in the corresponding adjoint problem (2.13), we have

$$\langle \mathbf{f}, \boldsymbol{\psi} \rangle_{L^2(\Omega)} = \langle \boldsymbol{\eta}, \ \mathbf{T}\boldsymbol{\Phi} \rangle_{L^2(\Gamma)} \tag{3.3}$$

since

$$B(\boldsymbol{\Phi}, \boldsymbol{\psi}) = A(\boldsymbol{\psi}, \boldsymbol{\Phi})$$

Consider a finite set of frequencies $\mathcal{K}_J = \{k_1 < k_2 < \cdots < k_J\}$. For each frequency k_j, $j = 1, ..., J$, assume that we have the measured boundary data $\boldsymbol{\eta}_j$, of the solution $\boldsymbol{\Phi}_j$ to (2.10). From these data, we can compute a set of solutions $\boldsymbol{\psi}_j$ of the corresponding adjoint problem (2.13) and in turn calculate the coefficients $\langle \mathbf{f}, \boldsymbol{\psi} \rangle_{L^2(\Omega)}$ in (3.3). However, since the solutions $\boldsymbol{\psi}_j$ are in general not orthogonal to one another, the sum $\sum_{j=1}^{J} \langle \mathbf{f}, \boldsymbol{\psi}_j \rangle \boldsymbol{\psi}_j$ is not necessarily the desired approximation to $\mathbf{f}$. Instead, we seek a minimum $\mathbf{L}^2$-norm approximation with the following reconstruction algorithm:

(1) Use the measured boundary data $\boldsymbol{\eta} = \boldsymbol{\eta}_1, \boldsymbol{\eta}_2, \cdots, \boldsymbol{\eta}_j$, to find approximate solutions $\boldsymbol{\psi}_j^h$ of the solutions of the adjoint problems $\boldsymbol{\psi}_j$, $j = 1, ..., J$:

$$A(\boldsymbol{\psi}_j, \boldsymbol{\phi}) = g \left\langle \boldsymbol{\eta}_j, \mathbf{T}\boldsymbol{\psi} \right\rangle_{L^2(\Gamma)}$$

for all $\boldsymbol{\phi} \in \mathbf{H}^1(\Omega)$ and for all the frequencies in $\mathcal{K}_J$.

(2) Find the best approximation to $\mathbf{f}$ by solving the normal equations

$$\sum_{j=1}^{J} \alpha_j \left\langle \boldsymbol{\psi}_i^h, \boldsymbol{\psi}_j^h \right\rangle_{L^2(\Omega)} = \|\boldsymbol{\eta}_i\|^2$$

Alternatively, one may solve an equivalent least squares problem using other stable algorithm.

(3) Form an approximation $\mathbf{f}_J^h$ to the minimum norm solution $\mathbf{f}_J$ of the unknown source $\mathbf{f}$

$$\mathbf{f}_J^h = \sum_{j=1}^{J} \alpha_j \boldsymbol{\psi}_j^h$$

Again, the post processing steps described previously are required to recover the original source of the elastic wave equation. There are some stability issues in the algorithm that need to be addressed. This and the numerical performance of the algorithms will be reported in a separate paper.

4. Conclusion

The multifrequency method discussed in the paper provides a strong theoretical and practical approach to tackle the inverse source problem for elastic waves. The abstract framework gives insight to investigate several related application problems that more closely reflect realistic situations. In particular, the work reported here will be useful in the study of elastic wave problems in non homogeneous media and for inverse source problems involving electromagnetic waves. There are also many interesting numerical and implementational issues such as choice of preconditioners, convergence and error bounds for the reconstruction algorithms that merit further studies.

References

[1] S. Acosta, S. Chow, J. Taylor, and V. Villamizar, *On the multi-frequency inverse source problem in heterogeneous media*, Inverse Problems **28** (2012), 045011.

[2] R. Albanese and P. B. Monk, *The inverse source problem for Maxwell's equations*, Inverse Problems **22** (2006), no. 3, 1023–1035, DOI 10.1088/0266-5611/22/3/018. MR2235652 (2007k:35509)

[3] Mark A. Anastasio, Jin Zhang, Dimple Modgil, and Patrick J. La Rivière, *Application of inverse source concepts to photoacoustic tomography*, Inverse Problems **23** (2007), no. 6, S21–S35, DOI 10.1088/0266-5611/23/6/S03. MR2440996 (2009g:78041)

[4] Gang Bao, Junshan Lin, and Faouzi Triki, *A multi-frequency inverse source problem*, J. Differential Equations **249** (2010), no. 12, 3443–3465, DOI 10.1016/j.jde.2010.08.013. MR2737437 (2012c:35464)

[5] James H. Bramble and Joseph E. Pasciak, *A note on the existence and uniqueness of solutions of frequency domain elastic wave problems: a priori estimates in* $\mathbf{H}^1$, J. Math. Anal. Appl. **345** (2008), no. 1, 396–404, DOI 10.1016/j.jmaa.2008.04.028. MR2422659 (2009j:74053)

[6] Matthias Eller and Nicolas P. Valdivia, *Acoustic source identification using multiple frequency information*, Inverse Problems **25** (2009), no. 11, 115005, 20, DOI 10.1088/0266-5611/25/11/115005. MR2546000

[7] Dan Givoli, *Numerical methods for problems in infinite domains*, Studies in Applied Mechanics, vol. 33, Elsevier Scientific Publishing Co., Amsterdam, 1992. MR1199563 (94j:65003)

[8] Frank Ihlenburg, *Finite element analysis of acoustic scattering*, Applied Mathematical Sciences, vol. 132, Springer-Verlag, New York, 1998. MR1639879 (99g:65114)

[9] P. Kuchment and L. Kunyansky, *Mathematics of thermoacoustic and photoacoustic tomography*, Handbook of Mathematical Methods in Imaging (O. Scherzer, ed.), Springer Verlag, 2010, pp. 817 – 866.

RICE UNIVERSITY, HOUSTON, TEXAS
E-mail address: sebastian.acosta@rice.edu

BRIGHAM YOUNG UNIVERSITY, PROVO, UTAH
E-mail address: schow@math.byu.edu

Multiscale mortar mixed methods for heterogeneous elliptic problems

Todd Arbogast, Zhen Tao, and Hailong Xiao

ABSTRACT. Consider solving a second order elliptic problem when the elliptic coefficient is highly heterogeneous. Generally, a numerical method either uses a very fine computational mesh to resolve the heterogeneities and therefore becomes computationally inefficient, or it performs efficiently on a coarse mesh but gives inaccurate results. Standard nonoverlapping domain decomposition using mortar spaces to couple together the subdomains efficiently handles these equations in parallel, but the issue of heterogeneity is not directly addressed. We define new mortar spaces that incorporate fine scale information obtained from local cell problems, using the theory of homogenization as a heuristic guide to limit the number of degrees of freedom in the mortar space. This gives computational efficiency in parallel, even when the subdomain problems are fully resolved on a fine mesh. In the case of an elliptic coefficient satisfying the two-scale separation assumption, the method is provably accurate with respect to the heterogeneity. Formally first and second order mortar space approximations are constructed explicitly in two dimensions. Numerical tests are presented for one medium with the two-scale separation assumption and two without it. The results show that these new homogenization based mortar spaces work much better than simple polynomial based mortar spaces, and that generally the second order spaces work better than the first order ones.

1. Introduction

We consider a second order elliptic problem with a heterogeneous coefficient (i.e., one that is highly variable or oscillatory in space) that models, for example, the single phase flow of fluid in the Earth's subsurface according to Darcy's Law. In mixed form [**BF91**, **RT91**, **BS94**], the problem is

$$\mathbf{u} = -a_\varepsilon \nabla p \quad \text{in } \Omega, \tag{1.1}$$

$$\nabla \cdot \mathbf{u} = f \quad \text{in } \Omega, \tag{1.2}$$

$$\mathbf{u} \cdot \nu = 0 \quad \text{on } \partial\Omega, \tag{1.3}$$

1991 *Mathematics Subject Classification.* Primary 65N55, 76M50; Secondary 65N30, 76S05.

Key words and phrases. Nonoverlapping domain decomposition, mixed method, heterogeneous, multiscale mortar, homogenization, convergence.

This work was supported as part of the Center for Frontiers of Subsurface Energy Security, an Energy Frontier Research Center funded by the U.S. Department of Energy, Office of Science, Office of Basic Energy Sciences under Award Number DE-SC0001114.

where $\Omega \subset \mathbb{R}^d$, $d = 2$ or 3, is the problem domain, ν is the outer unit normal, $a_\varepsilon(x)$ is a symmetric, uniformly positive definite tensor coefficient with $L^\infty(\Omega)$ components representing the *permeability*, $f \in L^2(\Omega)$ is the source or sink term, and the unknowns are *pressure* $p(x)$ and *velocity* $\mathbf{u}(x)$. The homogeneous Neumann boundary condition is considered for simplicity.

Domain decomposition for mixed methods [**GW88**] has been developed as a divide and conquer strategy to increase parallelism in computations and to handle interdomain multiphysics. In a nonoverlapping domain decomposition approach, a relatively small mortar finite element space [**BMP94, ACWY00, APWY07**] can be introduced to reduce coupling between subdomains. Let Ω_i, $i = 1, 2, \ldots, n$, be nonoverlapping subdomains of Ω, let p_i and $\mathbf{u}_i$ be the pressure p and velocity $\mathbf{u}$ restricted to Ω_i, and let ν_i be the outer unit normal to $\partial\Omega_i$. We rewrite (1.1)–(1.3) in a domain decomposition setting as

$$\mathbf{u} = -a_\varepsilon \nabla p \quad \text{in } \Omega_i, \tag{1.4}$$

$$\nabla \cdot \mathbf{u} = f \quad \text{in } \Omega_i, \tag{1.5}$$

$$p_i = p_j \quad \text{on } \partial\Omega_i \cap \partial\Omega_j \equiv \Gamma_{ij}, \tag{1.6}$$

$$\mathbf{u}_i \cdot \nu_i + \mathbf{u}_j \cdot \nu_j = 0 \quad \text{on } \Gamma_{ij}, \tag{1.7}$$

$$\mathbf{u} \cdot \nu = 0 \quad \text{on } \partial\Omega. \tag{1.8}$$

Accurate approximation of (1.1)–(1.3) or (1.4)–(1.8) is difficult when the permeability coefficient a_ε is a highly varying function, where ε is some measure of the correlation length of the medium. When $\varepsilon \ll 1$, the medium is highly heterogeneous, and its resolution requires a fine computational mesh (see, e.g., the error estimate Theorem 2.1 and (2.7)–(2.8)). This is reasonable for the subdomain problems, since they can be computed independently in parallel without the need for communication. The mortar interface problem is not so easily solved in parallel, but it can be made computationally efficient if it is small in size.

Recently, one of the current authors in [**Arb11a, Arb11b**] suggested a new multiscale finite element space based on the homogenization microstructure theorem (see Theorem 3.1) to handle the heterogeneity. More recently, two of the current authors in [**AX12**] adapted the idea to define a new multiscale mortar space and numerically tested a formally first order mortar space approximation with only three degrees of freedom on each subdomain interface. In this paper, we extend the idea to give a formally second order multiscale mortar space approximation with five degrees of freedom on each subdomain interface.

Briefly, the idea is to efficiently sample the microstructure by solving local cell problems. Heuristically, homogenization theory tells us that these local solutions can be used implicitly to reconstruct the pressure p in terms of a fixed operator and a smooth homogenized function p_0 (see (3.6)). Rather than approximating p directly, we approximate p_0 by a polynomial (see (4.7)), which gives an efficient multiscale mortar space with only a few degrees of freedom per subdomain interface.

We close this introduction by outlining the paper. We first give a brief review of the domain decomposition mortar method and homogenization theory in Sections 2 and 3, respectively. Then in Section 4, we define the first and second order mortar space approximations based on solutions to a localized cell problem as in homogenization theory. We also note that in the case of an elliptic coefficient satisfying the two-scale separation assumption of periodic homogenization [**BLP78, JKO94**], the method is provably accurate with respect to the heterogeneity, and we show that our new mortar method can be viewed as an implicitly defined multiscale finite element method. In Section 5, three numerical examples are given. Although our new mortar space was designed based on homogenization theory, which requires a locally periodic coefficient, our numerical tests on nonperiodic permeability fields show that the new method performs well for problems with general heterogeneities.

2. Mortar domain decomposition mixed method

Throughout, let $\Gamma = \bigcup_{i,j} \Gamma_{ij}$ and $\Gamma_i = \partial\Omega_i \bigcap \Gamma$ denote interior subdomain interfaces. For any $\omega \subset \Omega$ and $\gamma \subset \Gamma$, let $(\cdot,\cdot)_\omega$ and $\langle \cdot,\cdot \rangle_\gamma$ denote the $L^2(\omega)$ and $L^2(\gamma)$ inner products, respectively.

2.1. The variational form. Define the function spaces

$$\mathbf{V}_i = \{\mathbf{v} \in H(\mathrm{div};\Omega_i) : \mathbf{v}\cdot\nu|_{\partial\Omega\cap\partial\Omega_i} = 0\}, \qquad \mathbf{V} = \bigoplus_{i=1}^{n} \mathbf{V}_i,$$

$$W_i = L^2(\Omega_i), \qquad W = \left\{ w \in L^2(\Omega) : \int_\Omega w\, dx = 0 \right\},$$

$$M = H^{1/2}(\Gamma).$$

The variational form of (1.4)–(1.8) is: Find $\mathbf{u} \in \mathbf{V}$, $p \in W$, and $\lambda = p \in M$ such that for $1 \le i \le n$,

$$(a_\varepsilon^{-1}\mathbf{u},\mathbf{v})_{\Omega_i} - (p, \nabla\cdot\mathbf{v})_{\Omega_i} + \langle \lambda, \mathbf{v}\cdot\nu_i\rangle_{\Gamma_i} = 0 \qquad \forall \mathbf{v} \in \mathbf{V}_i, \tag{2.1}$$

$$(\nabla\cdot\mathbf{u}, w)_{\Omega_i} = (f,w)_{\Omega_i} \qquad \forall w \in W_i, \tag{2.2}$$

$$\sum_{i=1}^{n} \langle \mathbf{u}\cdot\nu_i, \mu\rangle_{\Gamma_i} = 0 \qquad \forall \mu \in M. \tag{2.3}$$

2.2. The Finite element approximation. Let $\mathcal{T}_{h,i}$ be a conforming, quasi-uniform, finite element partition of Ω_i with maximum element diameter h_i. Let $h = \max_i h_i$ and $\mathcal{T}_h = \bigcup_{i=1}^n \mathcal{T}_{h,i}$ be the finite element partition over the entire domain Ω. Let $\mathbf{V}_{h,i} \times W_{h,i} \subset \mathbf{V}_i \times W_i$ be any of the usual inf-sup stable mixed finite element spaces [**BF91**, **RT91**, **BS94**] defined over $\mathcal{T}_h$, and set $\mathbf{V}_h = \bigoplus_{i=1}^n \mathbf{V}_{h,i}$ and $W_h = \bigoplus_{i=1}^n W_{h,i}/\mathbb{R}$. Denote by $\mathcal{T}_{H,ij}$ a quasi-uniform finite element partition of Γ_{ij}, with maximal diameter of H_{ij} and $H = \max_{1\le i,j\le n} H_{ij}$. Let $M_{H,ij} \subset L^2(\Gamma_{ij})$ be the local mortar finite element space we will define later, and let $M_H = \bigoplus_{i\ne j} M_{H,ij}$.

In mixed finite element approximation of (2.1)–(2.3), we find $\mathbf{u}_h \in \mathbf{V}_h$, $p_h \in W_h$, and $\lambda_H \in M_H$ such that for $1 \le i \le n$,

$$(a_\varepsilon^{-1}\mathbf{u}_h,\mathbf{v})_{\Omega_i} - (p_h, \nabla\cdot\mathbf{v})_{\Omega_i} + \langle \lambda_H, \mathbf{v}\cdot\nu_i\rangle_{\Gamma_i} = 0 \qquad \forall \mathbf{v} \in \mathbf{V}_{h,i}, \tag{2.4}$$

$$(\nabla\cdot\mathbf{u}_h, w)_{\Omega_i} = (f,w)_{\Omega_i} \qquad \forall w \in W_{h,i}, \tag{2.5}$$

$$\sum_{i=1}^{n} \langle \mathbf{u}_h\cdot\nu_i, \mu\rangle_{\Gamma_i} = 0 \qquad \forall \mu \in M_H. \tag{2.6}$$

If our mixed finite element spaces give approximation of order $\mathcal{O}(h^k)$ for $\mathbf{u}$ and $\mathcal{O}(h^\ell)$ for p, and if we use a mortar space M_H of piecewise continuous or discontinuous polynomials of degree $m-1$ over each $\mathcal{T}_{H,ij}$, then from [**APWY07**] we have the following a-priori estimates.

THEOREM 2.1. *There exists C, independent of h and H, such that for $1 \le r \le k$, $0 \le s \le \ell$, and $0 < t \le m$,*

$$\begin{aligned}
&\|\nabla\cdot(\mathbf{u}-\mathbf{u}_h)\|_0 \le C\|f\|_s h^s,\\
&\qquad\|\mathbf{u}-\mathbf{u}_h\|_0 \le C\big\{\|\mathbf{u}\|_r h^r + \|p\|_{t+1/2}H^{t-1/2} + \|\mathbf{u}\|_{r+1/2}h^r H^{1/2}\big\},\\
&\qquad\|p-p_h\|_0 \le C\big\{\|p\|_s h^s + \|p\|_{t+1/2}H^{t+1/2}\\
&\qquad\qquad + \|f\|_s h^s H + \|\mathbf{u}\|_r h^r H + \|\mathbf{u}\|_{r+1/2}h^r H^{3/2}\big\}.
\end{aligned}$$

To be computationally feasible, we usually assume that $h < \varepsilon < H$. However, recall that the gradients of the solution $(\mathbf{u}, p)$ also depend on ε, i.e.,

$$\|\nabla p\|_0 = \mathcal{O}(\varepsilon^{-1}) \text{ and } \|D^k p\|_0 = \mathcal{O}(\varepsilon^{-k}),$$

and similar for $\mathbf{u}$. Thus Theorem 2.1 implies

$$\|\mathbf{u}-\mathbf{u}_h\|_0 \le C\big\{(h/\varepsilon)^r + (H/\varepsilon)^{t-1/2}/\varepsilon + (h/\varepsilon)^r(H/\varepsilon)^{1/2}\big\}, \tag{2.7}$$

$$\|p-p_h\|_0 \le C\big\{(h/\varepsilon)^s[1+H] + (H/\varepsilon)^{t+1/2} + (h/\varepsilon)^r[1+(H/\varepsilon)^{1/2}]H\big\}. \tag{2.8}$$

The approximation is poor when $h < \varepsilon < H$, so multiscale techniques are required.

3. Resolving heterogeneities using homogenization theory

Homogenization is a classic mathematical theory to resolve heterogeneities in porous media [**BLP78, JKO94**]. The key assumption in periodic homogenization theory is the two-scale separation of $a_\varepsilon(x)$, that is,

$$a_\varepsilon(x) = a(x, x/\varepsilon), \tag{3.1}$$

where $a(x,y)$ is periodic in y in the unit cell $\mathbf{Y} = [0,1]^d$. Now $a(x,y)$ is assumed to vary slowly in $x \in \Omega$, and these variations can be resolved by H, but as $\varepsilon \to 0$, $y = x/\varepsilon$ varies more and more rapidly (i.e., a_ε becomes more heterogeneous).

The homogenized problem is formulated as

$$\mathbf{u}_0 = -a_0\nabla p_0 \quad \text{in } \Omega, \tag{3.2}$$

$$\nabla\cdot\mathbf{u}_0 = f \quad \text{in } \Omega, \tag{3.3}$$

$$\mathbf{u}_0\cdot\nu = 0 \quad \text{on } \partial\Omega. \tag{3.4}$$

The true solution $(\mathbf{u}, p)$ of (1.1)–(1.3) converges to the homogenized solution $(\mathbf{u}_0, p_0)$ as $\varepsilon \to 0$. Here, the homogenized coefficient tensor $a_0(x)$ is given by

$$a_{0,ij}(x) = \int_{\mathbf{Y}} a(x,y)\Big(\delta_{ij} + \frac{\partial\omega_j(x,y)}{\partial y_i}\Big)\,dy, \quad i,j = 1,\dots,d,$$

where $\omega_j(x,y)$, for each fixed $x \in \Omega$, is the y-periodic solution of the cell problem

$$-\nabla_y\cdot\big[a(x,y)\big(\nabla_y\omega_j(x,y) + \mathbf{e}_j\big)\big] = 0 \quad \text{in } \Omega\times\mathbf{Y},\ j = 1,\dots,d, \tag{3.5}$$

with $\mathbf{e}_j \in \mathbb{R}^d$ being the jth Cartesian unit vector. We can further correct the homogenization solution $(\mathbf{u}_0, p_0)$ to first order expansion [**JKO94, MV97, AB06**].

THEOREM 3.1. *Let $\boldsymbol{\omega} = (\omega_1,\dots,\omega_d)^T$ and define the* first order corrector *by*

$$p_\varepsilon^1(x) = p_0(x) + \varepsilon\boldsymbol{\omega}(x, x/\varepsilon)\cdot\nabla p_0(x). \tag{3.6}$$

If $p_0 \in H^2(\Omega)$, then there is some constant C, depending on the solutions to the cell problems but not on ε, such that

$$\|p - p_\varepsilon^1\|_0 \le C\varepsilon\|p_0\|_2. \tag{3.7}$$

Moreover, if $p_0 \in H^2(\Omega) \cap W^{1,\infty}(\Omega)$, *then*

$$\|\nabla(p - p_\varepsilon^1)\|_0 \le C\{\varepsilon\|\nabla p_0\|_1 + \sqrt{\varepsilon}\,\|\nabla p_0\|_{0,\infty}\}. \tag{3.8}$$

4. A multiscale mortar space based on homogenization

We remark that the two-scale assumption (3.1) is used above for theoretical analysis and error estimation. It is not used in this section to define our multiscale mortar space (which follows the construction in [**AX12**]).

4.1. Interface error in the mortar method. Let the weakly continuous velocities [**ACWY00**] be

$$\mathbf{V}_{h,0} = \Big\{\mathbf{v} \in \mathbf{V}_h : \sum_{i=1}^{n} \langle \mathbf{v}|_{\Omega_i} \cdot \nu_i, \mu\rangle_{\Gamma_i} = 0\ \forall\, \mu \in M_H\Big\}, \tag{4.1}$$

and reformulate (2.4)–(2.6) as: Find $\mathbf{u}_h \in \mathbf{V}_{h,0}$ and $p \in W_h$ such that

$$(a_\varepsilon^{-1}\mathbf{u}_h, \mathbf{v}) - \sum_{i=1}^{n}(p_h, \nabla\cdot\mathbf{v})_{\Omega_i} = 0 \qquad \forall\, \mathbf{v} \in \mathbf{V}_{h,0}, \tag{4.2}$$

$$\sum_{i=1}^{n}(\nabla\cdot\mathbf{u}_h, w)_{\Omega_i} = (f, w) \quad \forall\, w \in W_h. \tag{4.3}$$

Subtracting (4.2)–(4.3) from (2.1)–(2.2), we obtain equations for the error (recalling $p = \lambda$ on Γ)

$$(a_\varepsilon^{-1}(\mathbf{u} - \mathbf{u}_h), \mathbf{v})_\Omega - \sum_{i=1}^{n}\big[(p - p_h, \nabla\cdot\mathbf{v})_{\Omega_i} - \langle p, \mathbf{v}\cdot\nu\rangle_{\Gamma_i}\big] = 0 \quad \forall\, \mathbf{v} \in \mathbf{V}_{h,0}, \tag{4.4}$$

$$\sum_{i=1}^{n}(\nabla\cdot(\mathbf{u} - \mathbf{u}_h), w)_{\Omega_i} = 0 \quad \forall\, w \in W_h. \tag{4.5}$$

The non-conforming error term $\langle p, \mathbf{v}\cdot\nu\rangle_{\Gamma_i}$ arises because although p is continuous, it is not *weakly* continuous. However, $\mathbf{v}$ is in the weakly continuous space, so

$$\langle p, \mathbf{v}\cdot\nu\rangle_{\Gamma_i} = \langle p - \mu, \mathbf{v}\cdot\nu\rangle_{\Gamma_i} \quad \forall\, \mu \in M_H, \tag{4.6}$$

leads to coarse H-level approximation error. We next use results from the homogenization theory heuristically as a guide to improve the approximation of p in M_H.

4.2. Formal first and second order approximations. From Theorem 3.1, we should expect that although the solution p of (2.1)–(2.3) is not smooth, it is a fixed operator of a smooth function p_0. Thus we should approximate

$$\begin{aligned}\lambda(x) = p(x) \approx p_\varepsilon^1(x) &= \big(1 + \varepsilon\boldsymbol{\omega}(x, x/\varepsilon)\cdot\nabla\big)p_0(x)\\ &\approx \big(1 + \varepsilon\boldsymbol{\omega}(x, x/\varepsilon)\cdot\nabla\big)q(x),\end{aligned} \tag{4.7}$$

where $q(x)$ is a piecewise polynomial.

Since we may not in general have a local period Y for $a_\varepsilon(x)$, we may also have no cell problem (3.5) defining $\varepsilon\boldsymbol{\omega}(x, x/\varepsilon)$. We approximate the local microstructure near each Γ_{ij} on $\overline{\Omega_i \cup \Omega_j}$ by finding the periodic solution to

$$-\nabla\cdot\big[a_\varepsilon(x)\big(\nabla\omega_k^{\Gamma_{ij}}(x,y) + \mathbf{e}_k\big)\big] = 0 \quad \text{in } \overline{\Omega_i \cup \Omega_j},\ k = 1, \dots, d. \tag{4.8}$$

Let Γ_{ij}^* to be an extension of Γ_{ij} in the normal direction into $\overline{\Omega_i \cup \Omega_j}$, and let $\mathbb{P}_{m-1}(\mathcal{T}_{H,ij}^*)$ to be the piecewise (continuous or discontinuous) polynomials of degree

$m-1$ defined over the interface mesh $\mathcal{T}_{H,ij}$ and extended in the normal direction of the same degree. Then we define [**AX12**]

$$M_H = \left\{\lambda \in L^2(\Gamma) : \lambda\big|_e = \left(1 + \boldsymbol{\omega}^{\Gamma_{ij}} \cdot \nabla\right) q\big|_e,\ q \in \mathbb{P}_{m-1}(\mathcal{T}^*_{H,ij}),\ e \in \mathcal{T}_{H,ij}\right\},$$

wherein the extended polynomials were restricted back to Γ.

In a two dimensional example, suppose we use only a single finite element over each interface Γ_{ij}. We linearly map an interface Γ_{ij} and its neighboring strip in both normal directions to a master rectangle $[-\eta, \eta] \times [0, H]$, where $x = 0$ gives the restriction to Γ_{ij}. If we choose q to be a linear polynomial as in [**AX12**],

$$q(x, y) = a + bx + cy,$$

then

$$\begin{aligned} p(x) &\approx (1 + \boldsymbol{\omega}^{\Gamma_{ij}}(x, y) \cdot \nabla)(a + bx + cy) \\ &= a + b\big[x + \omega_1^{\Gamma_{ij}}(x, y)\big] + c\big[y + \omega_2^{\Gamma_{ij}}(x, y)\big], \end{aligned}$$

and the formally first order mortar approximation on $\{0\} \times [0, H]$ is

$$\lambda_H(y) = a + b\,\omega_1^{\Gamma_{ij}}(0, y) + c\big[y + \omega_2^{\Gamma_{ij}}(0, y)\big].$$

Similarly, we could choose q to be a quadratic polynomial,

$$q(x, y) = a + bx + cy + dx^2 + exy + fy^2,$$

and then we have the formally second order mortar approximation

$$\begin{aligned} \lambda_H(y) = a &+ b\,\omega_1^{\Gamma_{ij}}(0, y) + c\big[y + \omega_2^{\Gamma_{ij}}(0, y)\big] \\ &+ e\,y\,\omega_1^{\Gamma_{ij}}(0, y) + f\big[y^2 + 2y\,\omega_2^{\Gamma_{ij}}(0, y)\big]. \end{aligned}$$

Notice that we have three, not two, degrees of freedom for first order approximation on the one dimensional interface. Similarly, we have five, not three, degrees of freedom for second order approximation.

4.3. Implicitly defined multiscale finite elements. Define the bi-linear form $d_H : M_H \times M_H \to \mathbb{R}$ and linear functional $g_H : M_H \to \mathbb{R}$ by

$$d_H(\lambda, \mu) = -\sum_{i=1}^{n} \langle \mathbf{u}_h^*(\lambda) \cdot \nu_i, \mu\rangle_{\Gamma_i},$$

$$g_H(\mu) = \sum_{i=1}^{n} \langle \bar{\mathbf{u}}_h \cdot \nu_i, \mu\rangle_{\Gamma_i},$$

where $(\mathbf{u}_h^*(\lambda), p_h^*(\lambda)) \in \mathbf{V}_h \times W_h$ solves (wherein λ is given, $f = 0$)

$$\begin{aligned} (a_\varepsilon^{-1}\mathbf{u}_h^*(\lambda), \mathbf{v})_{\Omega_i} - (p_h^*(\lambda), \nabla \cdot \mathbf{v})_{\Omega_i} &= -\langle \lambda, \mathbf{v} \cdot \nu_i\rangle_{\Gamma_i} \quad && \forall\, \mathbf{v} \in \mathbf{V}_{h,i}, \\ (\nabla \cdot \mathbf{u}_h^*(\lambda), w)_{\Omega_i} &= 0 && \forall\, w \in W_{h,i}, \end{aligned}$$

and $(\bar{\mathbf{u}}_h, \bar{p}_h) \in \mathbf{V}_h \times W_h$ solves (wherein $\lambda = 0$, f is given)

$$\begin{aligned} (a_\varepsilon^{-1}\bar{\mathbf{u}}_h, \mathbf{v})_{\Omega_i} - (\bar{p}_h, \nabla \cdot \mathbf{v})_{\Omega_i} &= 0 \quad && \forall\, \mathbf{v} \in \mathbf{V}_{h,i}, \\ (\nabla \cdot \bar{\mathbf{u}}_h, w)_{\Omega_i} &= (f, w)_{\Omega_i} && \forall\, w \in W_{h,i}. \end{aligned}$$

The equivalent coarse variational problem is [**GW88**]: Find $\lambda_H \in M_H$ such that

$$d_H(\lambda_H, \mu) = g_h(\mu) \quad \forall\, \mu \in M_H. \tag{4.9}$$

Let $\{\mu_\ell\}$ be a basis for $M_H = \text{span}\{\mu_\ell\}$. Define $\mathbf{v}_\ell = \mathbf{u}_h^*(\mu_\ell)$, $w_\ell = p_h^*(\mu_\ell)$, and

$$N_{h,H} = \text{span}\{(\mathbf{v}_\ell, w_\ell)\} = \text{span}\{(\mathbf{u}_h^*(\mu_\ell), p_h^*(\mu_\ell))\} \subset \mathbf{V}_h \times W_h.$$

It is easy to show that $d_H(\lambda, \mu) = (a_\varepsilon^{-1}\mathbf{u}_h^*(\lambda), \mathbf{u}_h^*(\mu))$. Then we can reformulate the coarse variational problem (4.9) as: Find $(\mathbf{u}_h, p_h) \in N_{h,H} + (\bar{\mathbf{u}}_h, \bar{p}_h)$ such that

$$(a_\varepsilon^{-1}\mathbf{u}_h, \mathbf{v}) = (f, w) \qquad \forall (\mathbf{v}, w) \in N_{h,H}.$$

The discrete space $N_{h,H}$ incorporates fine-scale information, and is thus a multiscale finite element space [**EH09**]. In this sense, the multiscale mortar method can be viewed as a multiscale finite element method, with the subdomains being coarse elements [**Arb11b, AX12**]. This is a very unusual multiscale mixed finite element, in that each basis function has weakly zero flow, but not zero flow, on all of its element edges, and pressures and velocities are intrinsically coupled together.

4.4. A-priori error estimates. Under certain technical conditions [**AX12**], we have the following bounds on the velocity and pressure errors.

THEOREM 4.1. *Suppose the two-scale separation assumption* (3.1) *holds. Then there exists a constant C, independent of h, H, L (the maximal diameter of the subdomains), and ε, such that for $1 \le r \le k$, $0 \le s \le \ell$, and $0 < t \le m$,*

$$\|\nabla \cdot (\mathbf{u} - \mathbf{u}_h)\|_0 \le C\|f\|_s h^s, \tag{4.10}$$

$$\|p - p_h\|_0 \le \|\hat{p} - p_h\|_0 + C\|p\|_s h^s, \tag{4.11}$$

$$\begin{aligned}\|\mathbf{u} - \mathbf{u}_h\|_0 + \|\hat{p} - p_h\|_0 \le C\big\{&\big[\|\mathbf{u}\|_r + \|\mathbf{u}\|_{r+1/2}((H+\varepsilon)/L)^{1/2}\big]h^r \\ &+ H^{t-1}(H+\varepsilon)(Lh)^{-1/2}\|p_0\|_{t+1/2} + \varepsilon\|p_0\|_2 + \varepsilon^{1/2}\|\nabla p_0\|_{0,\infty}\big\},\end{aligned} \tag{4.12}$$

$$\begin{aligned}\|\mathbf{u} - \mathbf{u}_h\|_0 + \|\hat{p} - p_h\|_0 \le C\big\{&\|\mathbf{u}\|_r((H+\varepsilon)/L)^{1/2}h^{r-1/2} \\ &+ H^{t-1}(H+\varepsilon)(Lh)^{-1/2}\|p_0\|_{t+1/2} + \varepsilon\|p_0\|_2 + \varepsilon^{1/2}\|\nabla p_0\|_{0,\infty}\big\}.\end{aligned} \tag{4.13}$$

Here we can see that the error is small whenever $h < \varepsilon < H \le L$.

5. Numerical results

In the previous section, we noted theoretically that our new mortar method works well under the two-scale separation assumption (3.1). Here we first verify the theory with a test using a streaked permeability field with a locally periodic microstructure. We then present numerical results for permeability fields a_ε that do not possess an obvious two-scale structure.

For simplicity, all of our numerical tests are conducted on rectangular grids with rectangular subdomains. The subdomain problems are approximated in $\mathbf{V}_{h,i} \times W_{h,i}$, which we take to be the lowest order Raviart-Thomas space RT0 [**RT77**], for which $k = \ell = 1$. We take one element per interface $\Gamma_{ij} = \bar{\Omega}_i \cap \bar{\Omega}_j$ (so $H = L$). Four mortar spaces M_H are tested for each example, they are:

(1) P1M, linear polynomial mortars with 2 degrees of freedom per edge;
(2) P2M, quadratic polynomial mortars with 3 degrees of freedom per edge;
(3) MS1, formally first order multiscale mortars based on homogenization with 3 degrees of freedom per edge;
(4) MS2, formally second order multiscale mortars based on homogenization with 5 degrees of freedom per edge.

All test examples use a rectangular domain with a quarter five-spot well pattern for f, that is, injection in the lower left corner and extraction (or production) in the upper right corner. For each interface Γ_{ij}, the cell problem is defined over the region $\mathbf{Y} = \Omega_i \cup \Gamma_{ij} \cup \Omega_j$, as described above and in [**Arb11a, AX12**]. We compare our numerical results with the reference fine-scale RT0 solution, since the true solution is not known analytically.

5.1. A streaked permeability. We first test a streaked permeability field as shown in Fig. 1, where a locally periodic structure can be observed. Also notice that this is a strongly anisotropic permeability field. From Table 1, we observe that by increasing the order of the polynomial space from P1M to P2M, the performance does not improve at all. On the other hand, we obtain a immediate improvement when turning to the homogenization side by using MS1, and we can further improve the performance by applying MS2. A similar performance can be found in the multiscale finite element method [**Arb11a**], where anisotropic problems are better handled with a homogenization-based element.

TABLE 1. Streaked permeability. Relative errors in the pressure and velocity for the mortar spaces relative to the 20×20 reference RT0 solution, using a 2×2 coarse grid and 10×10 subgrid.

	Pressure error		Velocity error	
Method	ℓ^2	ℓ^∞	ℓ^2	ℓ^∞
P1M	0.5964	0.1741	0.6357	0.7889
P2M	0.5615	0.1588	0.6656	1.0684
MS1	0.1755	0.0792	0.4095	0.3595
MS2	0.0305	0.0195	0.1491	0.2264

5.2. A moderately heterogeneous permeability. The permeability field of our second example is moderately heterogeneous, being locally isotropic and geostatistically mildly correlated. It is depicted in Fig. 2 on a logarithmic scale ranging over four orders of magnitude. The domain is 40 meters square and the fine grid is uniformly 40×40.

From Fig. 2, we can see that generally the MS1 and MS2 velocities are closer to the fine-scale RT0 velocity than P1M and P2M. Recall that P2M and MS1 use mortar spaces with the same number of degrees of freedom. Therefore, we can reduce the relative ℓ^2-velocity error from 25.6% to 10.7% without increasing the complexity of the interface problems by using our new mortar space. Moreover, although two more degrees of freedom per edge are introduced in MS2, we can get a 0.15% ℓ^2-pressure error and a 4.1% ℓ^2-velocity error in return, which is quite accurate.

5.3. A channelized permeability from SPE10. Finally, we test the 80th layer of the Tenth Society of Petroleum Engineers Comparative Solution Project (SPE10) [**Chr01**], which is shown in Fig. 3. Obviously, the permeability does not fulfill the two-scale separation assumption (3.1). In Fig. 3, one can see that the velocities of P1M and P2M exhibit extreme inaccuracies that resemble points of singularity, making these methods perform poorly (see Table 3). On the other hand, MS1 and MS2 control the ℓ^2-velocity error within a reasonable range. Although

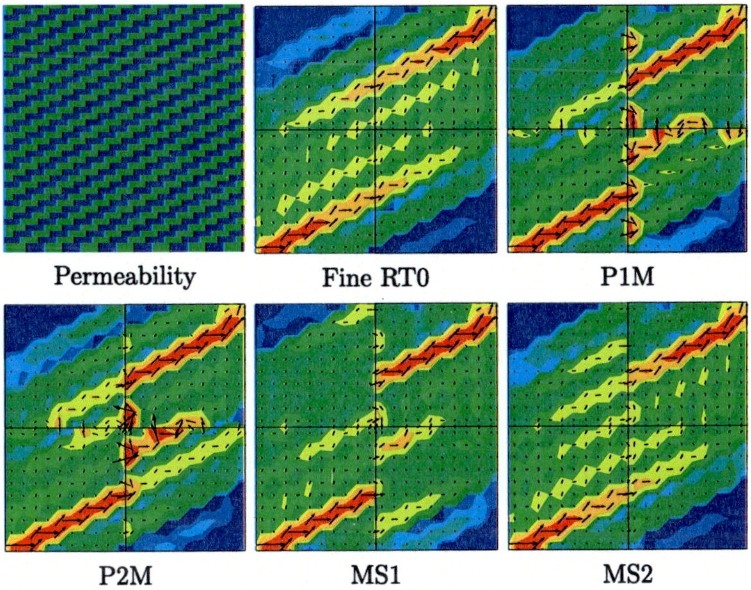

FIGURE 1. Streaked permeability. The permeability, on a 20×20 grid, has only two values 1 and 200. Velocities are computed by RT0 on the fine grid, and by the mortar methods on a 2×2 coarse grid of subdomains with a 10×10 subgrid. Color depicts speed on a log scale from 0.001 (blue) to 1 (red). Arrows show velocities.

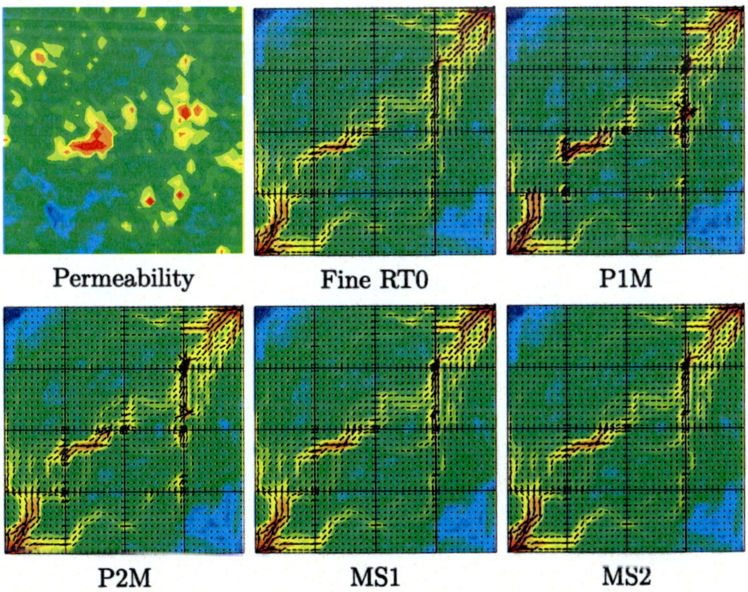

FIGURE 2. Moderate heterogeneity. The 40×40 permeability is shown on a log scale from about 0.32 to 3200 millidarcy. Velocities are computed by RT0 on the fine grid and by mortars on a 4×4 grid of subdomains with a 10×10 subgrid. Color depicts speed, on a log scale from 0.6 to 0.0006. Arrows show velocities.

TABLE 2. Moderate heterogeneity test. Relative errors in the pressure and velocity for the mortar spaces relative to the 40×40 reference RT0 solution, using a 4×4 coarse grid and 10×10 subgrid.

	Pressure error		Velocity error	
Method	ℓ^2	ℓ^∞	ℓ^2	ℓ^∞
P1M	0.1989	0.1452	0.4157	0.8042
P2M	0.0431	0.0353	0.2564	0.5267
MS1	0.0111	0.0137	0.1072	0.1432
MS2	0.0015	0.0020	0.0410	0.0688

TABLE 3. SPE10-80 test. Relative errors in the pressure and velocity for the mortar spaces relative to the 60×220 reference RT0 solution, using a 3×11 coarse grid and 20×20 subgrid.

	Pressure error		Velocity error	
Method	ℓ^2	ℓ^∞	ℓ^2	ℓ^∞
P1M	0.0846	0.0452	0.6584	2.0868
P2M	0.0437	0.0204	0.5287	2.0156
MS1	0.0127	0.0090	0.1459	0.4860
MS2	0.0093	0.0066	0.1143	0.5985

MS2 gives a better ℓ^2-error than MS1, it is a marginal improvement; moreover, MS2 shows somewhat greater ℓ^∞-velocity error.

6. Conclusions

Nonoverlapping domain decomposition using mortar spaces to couple together the subdomains is an efficient way to numerically solve second order elliptic problems (1.1)–(1.3) in parallel. Heterogeneity in the elliptic coefficient a_ε can be handled within the mortar space by using ideas from homogenization theory. Local cell problem can be solved, which then allow implicit reconstruction of the pressure p in terms of a fixed operator and a smooth homogenized function p_0 through (3.6). Approximation of p_0 by a polynomial, as in (4.7), gives a multiscale mortar space with only a few degrees of freedom per subdomain interface, resulting in computational efficiency in parallel. In the two-scale separation case, we have good approximation properties (Theorem 4.1).

In two space dimensions, formally first order mortar spaces were constructed in [**AX12**] and reviewed here, and the formally second order mortar spaces were constructed explicitly here. We can generally expect more accurate numerical results when using homogenization based mortar spaces in domain decomposition, even without a two-scale microstructure. Usually the formally second order mortar approximation based on homogenization (MS2) give a better result than the first order mortars (MS1), and both generally perform much better the simple polynomial mortar space approximations (P1M and P2M).

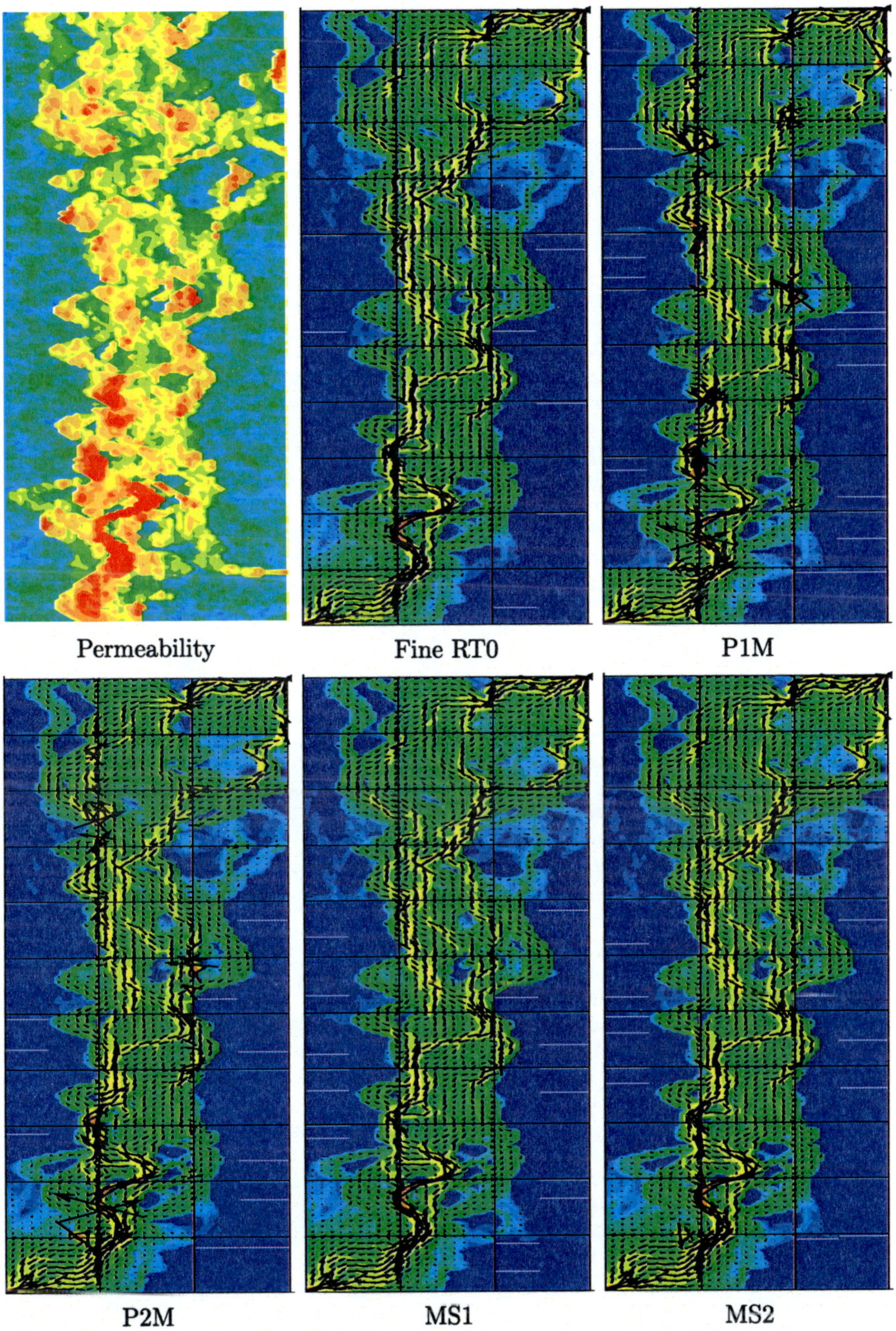

FIGURE 3. SPE10-80 test. The permeability is given on a 60×220 grid plotted using a log scale from 1.9e-11 (red) to 1.0e-18 (blue) m^2. The fine-scale RT0 speed and velocity are plotted on a log scale from 1.3 (red) to 1.0e-3 (blue). The mortar results use a 3×11 coarse grid with a 20×20 subgrid.

References

[AB06] Todd Arbogast and Kirsten J. Boyd, *Subgrid upscaling and mixed multiscale finite elements*, SIAM J. Numer. Anal. **44** (2006), no. 3, 1150–1171 (electronic), DOI 10.1137/050631811. MR2231859 (2007k:65165)

[ACWY00] Todd Arbogast, Lawrence C. Cowsar, Mary F. Wheeler, and Ivan Yotov, *Mixed finite element methods on nonmatching multiblock grids*, SIAM J. Numer. Anal. **37** (2000), no. 4, 1295–1315, DOI 10.1137/S0036142996308447. MR1756426 (2001h:65140)

[APWY07] Todd Arbogast, Gergina Pencheva, Mary F. Wheeler, and Ivan Yotov, *A multiscale mortar mixed finite element method*, Multiscale Model. Simul. **6** (2007), no. 1, 319–346, DOI 10.1137/060662587. MR2306414 (2008k:65234)

[Arb11a] Todd Arbogast, *Homogenization-based mixed multiscale finite elements for problems with anisotropy*, Multiscale Model. Simul. **9** (2011), no. 2, 624–653, DOI 10.1137/100788677. MR2818414 (2012h:65267)

[Arb11b] Todd Arbogast, *Mixed multiscale methods for heterogeneous elliptic problems*, Numerical Analysis of Multiscale Problems (I. G. Graham, Th. Y. Hou, O. Lakkis, and R. Scheichl, eds.), Lecture Notes in Computational Science and Engineering, vol. 83, Springer, 2011, pp. 243–283.

[AX12] T. Arbogast and Hailong Xiao, *A multiscale mortar mixed space based on homogenization for heterogeneous elliptic problems*, SIAM J. Numer. Anal. (2013).

[BF91] Franco Brezzi and Michel Fortin, *Mixed and hybrid finite element methods*, Springer Series in Computational Mathematics, vol. 15, Springer-Verlag, New York, 1991. MR1115205 (92d:65187)

[BLP78] Alain Bensoussan, Jacques-Louis Lions, and George Papanicolaou, *Asymptotic analysis for periodic structures*, Studies in Mathematics and its Applications, vol. 5, North-Holland Publishing Co., Amsterdam, 1978. MR503330 (82h:35001)

[BMP94] C. Bernardi, Y. Maday, and A. T. Patera, *A new nonconforming approach to domain decomposition: the mortar element method*, Collège de France Seminar, Vol. XI (Paris, 1989), Pitman Res. Notes Math. Ser., vol. 299, Longman Sci. Tech., Harlow, 1994, pp. 13–51. MR1268898 (95a:65201)

[BS94] Susanne C. Brenner and L. Ridgway Scott, *The mathematical theory of finite element methods*, Texts in Applied Mathematics, vol. 15, Springer-Verlag, New York, 1994. MR1278258 (95f:65001)

[Chr01] M. A. Christie, *Tenth SPE comparative solution project: A comparison of upscaling techniques*, SPE Reservoir Evaluation & Engineering **4** (2001), no. 4, 308–317, Paper no. SPE72469-PA.

[EH09] Yalchin Efendiev and Thomas Y. Hou, *Multiscale finite element methods*, Surveys and Tutorials in the Applied Mathematical Sciences, vol. 4, Springer, New York, 2009. Theory and applications. MR2477579 (2010h:65224)

[GW88] Roland Glowinski and Mary Fanett Wheeler, *Domain decomposition and mixed finite element methods for elliptic problems*, Partial Differential Equations (Paris, 1987), SIAM, Philadelphia, PA, 1988, pp. 144–172. MR972516 (90a:65237)

[JKO94] V. V. Jikov, S. M. Kozlov, and O. A. Oleĭnik, *Homogenization of differential operators and integral functionals*, Springer-Verlag, Berlin, 1994. Translated from the Russian by G. A. Yosifian [G. A. Iosif′yan]. MR1329546 (96h:35003b)

[MV97] Shari Moskow and Michael Vogelius, *First-order corrections to the homogenised eigenvalues of a periodic composite medium. A convergence proof*, Proc. Roy. Soc. Edinburgh Sect. A **127** (1997), no. 6, 1263–1299, DOI 10.1017/S0308210500027050. MR1489436 (99g:35018)

[RT77] P.-A. Raviart and J. M. Thomas, *A mixed finite element method for 2nd order elliptic problems*, Naz. delle Ricerche (C.N.R.), Rome, 1975), Springer, Berlin, 1977, pp. 292–315. Lecture Notes in Math., Vol. 606. MR0483555 (58 #3547)

[RT91] J. E. Roberts and J.-M. Thomas, *Mixed and hybrid methods*, Handbook of numerical analysis, Vol. II, Handb. Numer. Anal., II, North-Holland, Amsterdam, 1991, pp. 523–639. MR1115239

University of Texas, Institute for Computational Engineering and Sciences, 201 East 24th Street, Stop C0200, Austin, Texas 78712-1229
E-mail address: arbogast@ices.utexas.edu

University of Texas, Institute for Computational Engineering and Sciences, 201 East 24th Street, Stop C0200, Austin, Texas 78712-1229
E-mail address: taozhen@ices.utexas.edu

University of Texas, Institute for Computational Engineering and Sciences, 201 East 24th Street, Stop C0200, Austin, Texas 78712-1229
E-mail address: xialhl@ices.utexas.edu

A parallel hp-adaptive finite element method

Randolph E. Bank and Hieu Nguyen

ABSTRACT. This paper describes a combination of automatic hp-adaptive finite elements and domain decomposition. The combination is based on the Bank-Holst parallel adaptive meshing paradigm. The hp-adaptivity is based on a derivative recovery technique while the domain decomposition method is formulated using a mortar-like formulation with Dirac δ functions for the mortar element space. Numerical results show that the approach can scale to hundreds of processors and the convergence of the domain decomposition method is independent of the number of processors as well as the distribution of element sizes and element degrees among the local meshes.

1. Introduction

hp-adaptive finite elements and domain decomposition are popular methods for numerically solving PDEs. The former is known for its fast rate of convergence and the latter is recognized as an efficient way to solve large problems using parallel computation. In the software package $PLTMG$ **[Ban11]**, we combine these two methods to take advantages of their strong points. In this paper we describe some of the challenges in forming an effective combination, based on extensions of the Bank-Holst paradigm **[BH00, BH03, Ban06b]**.

In Bank-Holst paradigm, the domain is partitioned according to a posteriori error estimates obtained on a coarse mesh of the whole domain solved on a single processor. In this partition, each subdomain has approximately equal error even though they can greatly vary in size, number of elements and degrees of freedom. At the conclusion of the adaptive hp refinement, each subdomain will have (approximately) the same number of degrees of freedom. This is achieved by giving each processor the same target number of degrees of freedom and allowing them to perform adaptive meshing independently (with focus on their own subdomain). The domain decomposition solver is employed only when adaptive enrichment has concluded.

1991 *Mathematics Subject Classification.* Primary 65N50, 65N55.

Key words and phrases. hp adaptivity, hp-FEM, domain decomposition, Bank-Holst paradigm.

The work of this author was supported by the National Science Foundation under contract DMS-0915220.

The work of this author was supported by the National Science Foundation under contract DMS-0915220 and a grant from the Vietnam Education Foundation (VEF).

The *hp*-finite element method in this work is based on gradient/derivative recovery technique introduced in [**BX03a, BX03b, BXZ07**]. High order derivatives of the exact solution are recovered by superconvergent approximations. These approximations are used to formulate a posteriori error estimates that guide the adaptive meshing. The approach is also able to address a crucial question in *hp*-adaptivity; that is whether it is more advantageous to refine a given element into several child elements (h-refinement) or increase its degree. A superconvergent result is used as a consistency check of the a posteriori error estimates. In the region where the exact solution is smooth enough, the result should be consistent. In the regions near singularities and low regularity, we anticipate that the recovery scheme will have difficulties approximating high order derivatives of the exact solution and the superconvergent result is no longer valid. Consequently, the consistency check can be used to identify regions near singularities regions with rapid changes, and signal the adaptive procedure to use *h*-refinement in these regions and *p*-refinement elsewhere.

2. *hp*-Finite Element

2.1. Basis functions. In our study, we use nodal basis functions, rather than a more traditional hierarchical family of functions as in [**MM11**]. For a standard element of degree p, basis functions are defined by their values at nodal points of degree p (equals 1 at the associated nodal point and 0 at all the others) as illustrated in Figure 1 (left).

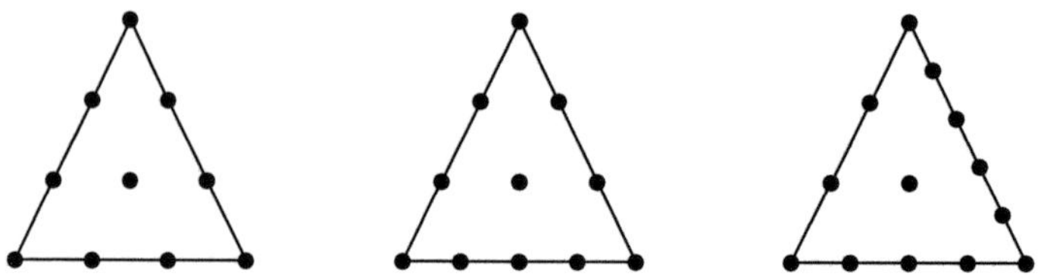

FIGURE 1. A standard cubic element (left), a cubic element with one quartic edge (middle) and a cubic element with one quartic and one quintic edge (right).

Along edges shared by elements of different degrees, the element of the lower degree inherits the degrees of freedom of the higher degree element. This results in elements of degree p with one or two *transition edges* of higher degree. Some typical cases are illustrated in Figure 1. Special treatment is utilized for basis functions associated with the transition edges. A transition basis function of degree $p+k$ is sought in the form of a linear combination of the standard basis functions of degree p associated with the transition edge and special polynomials of degree $p+1, p+2, \ldots, p+k$. These special polynomials can be chosen to be ones that equal zero at all of nodal points of degree p (see [**Ngu10, BN11b**])

This construction of basis functions guarantees there are no hanging nodes and the global finite element space is continuous. In addition, the construction can be generalized to three dimensions (see the Appendix in [**BN11b**]).

2.2. Error estimates. Let Ω be the domain of the PDE and $\mathcal{T}_h$ be a shape regular triangulation of Ω of size h. Denote by $\mathcal{V}_h^{(p)}$ the space of continuous piecewise polynomials of degree p associated with $\mathcal{T}_h$. Define Q be the L^2 projection from the space of discontinuous piecewise constants functions associated with $\mathcal{T}_h$ into the space of continuous piecewise linear polynomials $\mathcal{V}_h^{(1)}$. Finally, let S be a multigrid smoother for the Laplace operator and m be a small integer, typically one or two.

Our error estimate is inspired by the work on gradient/derivative recovery techniques of Bank, Xu, and Zheng ([**BX03a**, **BX03b**, **BXZ07**]), where it is shown that $\|\partial^p(u-S^mQu_{hp})\|_\Omega$ is an asymptotically exact approximation of $\|\partial^p(u-u_{hp})\|_\Omega$ with better than first order of accuracy. In addition, $\partial(S^mQ(\partial^p u_{hp}))$ is proved to be a superconvergent approximation of $\partial^{p+1}u$. Here u is the exact solution and u_{hp} is the finite element solution of the PDE.

To describe our a posteriori estimate for the case of element τ of degree p, let $\mathcal{P}_p(\tau)$ be the space of polynomials of degree p defined on τ. We write

$$\mathcal{P}_{p+1}(\tau) = \mathcal{P}_p(\tau) \oplus \mathcal{E}_{p+1}(\tau)$$

where the hierarchical extension $\mathcal{E}_{p+1}(\tau)$ consists of those polynomials in $\mathcal{P}_{p+1}(\tau)$ that are zero at all degrees of freedom associated with $\mathcal{P}_p(\tau)$. In the case of two dimensions, this is a subspace of dimension $p+2$, with a convenient basis given by

$$\psi_{p+1,k} = \prod_{j=0}^{k-1}(c_1 - j/p)\prod_{m=0}^{p-k}(c_2 - m/p) \quad 0 \le k \le p+1.$$

where c_i is the i-th barycentric coordinate function. Using this basis, we approximate $u - u_{hp}$ on element τ as

$$u - u_{hp}|_\tau \approx \epsilon_\tau = \alpha_\tau \sum_{k=0}^{p+1} \frac{\partial_{c_1}^k \partial_{c_2}^{p+1-k}\hat{u}}{k!(p+1-k)!}\psi_{p+1,k} \tag{2.1}$$

where $\hat{u}$ is a hierarchical extension of degree $p+1$ of u_{hp}.

The partial derivatives of order $p+1$ appearing in (2.1) are formally $O(h_t^{p+1})$ when expressed in terms of ∂_x and ∂_y. The derivative $\partial_x^k\partial_y^{p+1-k}\hat{u}$ is constant on element τ, computed by differentiating the recovered p-th derivatives of u_{hp}, which are linear polynomials on element τ.

The constant α_τ is chosen such that

$$\|\partial^p(u - S^mQu_{hp})\|_{0,\tau} = \|\partial^p(\epsilon_\tau)\|_{0,\tau}.$$

Normally, one should expect $\alpha_\tau \approx 1$, except for elements where the true solution u is not smooth enough to support p derivatives. Therefore, the size of α_τ can be used to formulate *hp*-refinement indicator that decide whether element τ should be refined in h or in p. Then our *local error indicator* is defined by $\eta_\tau = |\epsilon_\tau|_{1,\tau}$. For more details on our formulation of a posteriori error estimates and *hp* refinement indicator, see [**BN11b**].

2.3. *hp* Refinement Procedure. The *hp*-Refinement procedure is outlined in Figure 2. In this procedure, the target number of degrees of freedom for the new mesh, denoted by $NDTRGT_0$, is given by

$$NDTRGT_0 = \min(NDTRGT, NDF \times 4^{1/p_{ave}}). \tag{2.2}$$

Here *NDTRGT* is the target number given by user and the average degree of elements in the current mesh is estimated by the following formula

$$p_{ave} = \sqrt{\frac{NDF}{NVF}} \tag{2.3}$$

where *NDF* and *NVF* are the number of degrees of freedom and the number of vertices of the current mesh respectively .

The use of (2.2) tries to force a geometric increase in the number of degrees of freedom in each refinement step. This is an empirical formula based on the observation that the dimensions of the subspaces (i.e. *NDF*) must grow more slowly for higher order elements or the higher rates of convergence promised by these elements might not be achieved.

R1 Create a heap with respect to η_τ with the largest error estimate $\eta_{\tau_{\max}}$ at the root;
R2 If $NDF \approx NDTRGT_0$ or $\eta^2_{\tau_{max}} \leq \eta^2_{ave}/2$, then **exit**.
R3 Decide to refine $\tau_{\max}$ in h or p based on $\alpha_{\tau_{\max}}$
R4 Refine element $\tau_{\max}$, and possibly others as required.
R5 Update error indicators for affected elements.
Add new elements as needed.
Remake the heap. Go to **R2**.

FIGURE 2. *hp*-refinement procedure

While we normally expect the refinement loop to exit when the target number of degrees of freedom is approximately achieved, we can also exit if the largest error in the current mesh is sufficiently small. In particular,

$$\eta^2_{\tau_{\max}} \leq \frac{\eta^2_{ave}}{2} = \frac{1}{2N} \sum_{t \in \Omega_I} \eta_t^2 \tag{2.4}$$

where Ω_I is the fine subregion associated with processor I in the case of parallel computation, and $\Omega_I \equiv \Omega$ otherwise; N is the number of triangles in Ω_I.

In *hp*-refinement procedure, the most interesting test is to decide between h-refinement and p-refinement for element $\tau_{\max}$. In our implementation, we use h-refinement if

$$\alpha_{\tau_{\max}} > 2\alpha_{ave} \tag{2.5}$$

and use p-refinement otherwise. Here α_{ave} is the average of all of α_τ in the mesh before the refinement. If the scaling factor $\alpha_{\tau_{\max}} \approx 1$, then the recovered derivatives and the error estimate are consistent, and we assume that the solution is locally smooth, which in turn justifies p-refinement. Large values of $\alpha_{\tau_{\max}}$ empirically correspond to locally non-smooth behavior of the solution, and this in turn suggests h-refinement.

While (2.5) is the main test for *hp*-refinement, we also check for round-off error problems, and for the maximum degree (limited by the available quadrature rules), and change the decision suggested by (2.5) if necessary. Finally, for very coarse meshes we always choose h-refinement. In particular, we choose h-refinement until the relative error

$$RELERP \leq \frac{1}{5} \tag{2.6}$$

is satisfied. Condition (2.6) tries to insure that the mesh contains enough elements that *hp*-refinement is a viable option. We have also observed that our error estimates are sometimes unreliable on extremely coarse meshes, perhaps due to data oscillations, or more generally that h is not sufficiently small for the asymptotic error behavior underlying our derivative recovery procedures to hold.

3. Parallel *hp*-Adaptive Paradigm

3.1. Bank-Holst paradigm. In this work, we use the Bank-Holst paradigm [**BH00, BH03, Ban06b, BO07, Ova04, Mit98a, Mit98b**], that addresses the load balancing problem in a new way, requiring far less communication. Another important point is that our approach allows serial adaptive finite element codes to run in a parallel environment without a large investment in additional coding. This approach has three main components:

Step 1: A small (nonlinear) problem is solved on an initial coarse mesh, and *a posteriori* error estimates are computed for the coarse grid solution. The triangulation is partitioned such that each subdomain has approximately equal *error* (although they can significantly differ in size, numbers of elements and degrees of freedom).

Step 2: Each processor is provided the complete coarse mesh and solution, and instructed to solve the *entire* (nonlinear) problem, with the stipulation that its adaptive refinement should be limited largely to its own partition. Load balancing is achieved by instructing each processor to create a refined mesh with the same number of degrees of freedom.

Step 3: A final mesh is computed using the union of the refined partitions provided by each processor. This mesh is reconciled such that the (virtual) mesh made up of the union of the refined subregions would be conforming. A final solution is computed, using a domain decomposition method. An initial guess is provided by the local solutions.

The above approach has several interesting features. First, the load balancing problem (Step 1) is reduced to the numerical solution of a small problem on a single processor. It can be done using any serial adaptive finite element code without any modifications. Second, the adaptive mesh generation calculation (Step 2) takes place independently on each processor, and can also be performed with no communication.

The only parts of the calculation requiring communication are

(1) the initial fan-out of the mesh distribution to the processors, once the decomposition is determined by the error estimator.
(2) the mesh regularization, requiring communication to produce a global conforming mesh.
(3) the final solution phase. Note that a good initial guess for Step 3 is provided in Step 2 by taking the solution from each subregion restricted to its partition.

3.2. Bank-Holst paradigm and *hp*-refinement. Even though *hp*-refinement can be used directly with Bank-Holst paradigm without any restriction, a careful use of *hp*-refinement can make the combination smoother and more effective. The following are the practices that we employ

(1) In Step 1, only h-refinement is used. In this step, having a good approximate solution and accurate a posteriori error estimate is not the first priority. It is more critical to create a mesh with a sufficient number of elements so that a good partition of the domain can be obtained. In addition, avoiding p- and hp-refinement in this step is also appropriate as these strategies are not effective on extremely coarse meshes.

(2) At the beginning of Step 2 on each processor, if the number of elements in the local subdomain is small, adaptive h-refinement or local uniform h-refinement should be used first. After that, automatic hp-refinement can be used.

Since the adaptive enrichment on each processor (Step 2) is completely independent of what happens on other processors, the global refined mesh, constructed from the meshes associated with the refined regions on each of the processors, is initially non-conforming along the interface system. With the use of hp-refinement, the task to reconcile the local meshes can be challenging. The meshes are unstructured in geometry (in h) and have variable degree (variable p). In addition, there is no refinement tree, and nonconformity exists in both h and p. Thus, we need to efficiently identify and resolve these nonconformities, and ultimately to establish links between degrees of freedom on the fine mesh interface system on a given processor and the corresponding degrees of freedom on other processors which share its interface. For detail descriptions of this task, we refer to [**BNar**].

4. Domain Decomposition Solver

In this section, we describe the domain decomposition algorithm implemented in *PLTMG* for Step 3 of the Bank-Holst paradigm presented in Subsection 3.1. This algorithm is described in detail in [**BN11a, BL03, Ban06a, Lu04, BV08**].

For simplicity in our discussion, we restrict attention to the case of just two subdomains. In our scheme, each subregion contributes equations corresponding all fine degrees of freedom, including its interface. Thus in general there will be multiple unknowns and equations in the global system corresponding to the interface degrees of freedom. This is handled by equality constraints that impose continuity at all degrees of freedom on the interface. The result is a mortar-element like formulation, using Dirac δ functions for the mortar element space. In any event, with a proper ordering of unknowns, the global system of equations has the block 5×5 form

$$\begin{pmatrix} A_{11} & A_{1\gamma} & & & \\ A_{\gamma 1} & A_{\gamma\gamma} & & & I \\ & & A_{\nu\nu} & A_{\nu 2} & -I \\ & & A_{2\nu} & A_{22} & \\ & I & -I & & \end{pmatrix} \begin{pmatrix} \delta U_1 \\ \delta U_\gamma \\ \delta U_\nu \\ \delta U_2 \\ \Lambda \end{pmatrix} = \begin{pmatrix} R_1 \\ R_\gamma \\ R_\nu \\ R_2 \\ U_\nu - U_\gamma \end{pmatrix}. \tag{4.1}$$

Here A_{11} and A_{22} correspond to the fine degrees of freedom on processors 1 and 2, respectively, that are not on the interface, while $A_{\gamma\gamma}$ and $A_{\nu\nu}$ correspond to interface points. The fifth block equation imposes continuity, and its corresponding Lagrange multiplier is Λ. The identity matrix appears because the global fine mesh is conforming. The introduction of the Lagrange multiplier and the saddle point formulation (4.1) are only for expository purposes; indeed, Λ is never computed or updated.

On processor 1, we develop a similar but "local" saddle point formulation. That is, the fine mesh subregion on processor 1 is "mortared" to the remaining coarse mesh on processor 1. This leads to a linear system of the form

$$\begin{pmatrix} A_{11} & A_{1\gamma} & & & \\ A_{\gamma 1} & A_{\gamma\gamma} & & & I \\ & & \bar{A}_{\nu\nu} & \bar{A}_{\nu 2} & -I \\ & & \bar{A}_{2\nu} & \bar{A}_{22} & \\ & I & -I & & \end{pmatrix} \begin{pmatrix} \delta U_1 \\ \delta U_\gamma \\ \delta \bar{U}_\nu \\ \delta \bar{U}_2 \\ \Lambda \end{pmatrix} = \begin{pmatrix} R_1 \\ R_\gamma \\ R_\nu \\ 0 \\ U_\nu - U_\gamma \end{pmatrix}, \tag{4.2}$$

where quantities with a bar (e.g., $\bar{A}_{22}$) refer to the coarse mesh. A system similar to (4.2) can be derived for processor 2. With respect to the right hand side of (4.2), the interior residual R_1 and the interface residual R_γ are locally computed on processor 1. We obtain the boundary residual R_ν, and boundary solution U_ν from processor 2; processor 2 in turn must be sent R_γ and U_γ. The residual for the coarse grid interior points is set to zero. This avoids the need to obtain R_2 via communication, and to implement a procedure to restrict R_2 to the coarse mesh on processor 1. Given our initial guess, we expect $R_1 \approx 0$ and $R_2 \approx 0$ at all iteration steps. R_γ and R_ν are not generally small, but $R_\gamma + R_\nu \to 0$ at convergence.

As with the global formulation (4.1), equation (4.2) is introduced mainly for exposition. The goal of the calculation on processor 1 is to compute the updates δU_1 and δU_γ, which contribute to the global conforming solution. To this end, we formally reorder (4.2) as

$$\begin{pmatrix} & -I & & I & \\ -I & \bar{A}_{\nu\nu} & & & \bar{A}_{\nu 2} \\ & & A_{11} & A_{1\gamma} & \\ I & & A_{\gamma 1} & A_{\gamma\gamma} & \\ & \bar{A}_{2\nu} & & & \bar{A}_{22} \end{pmatrix} \begin{pmatrix} \Lambda \\ \delta \bar{U}_\nu \\ \delta U_1 \\ \delta U_\gamma \\ \delta \bar{U}_2 \end{pmatrix} = \begin{pmatrix} U_\nu - U_\gamma \\ R_\nu \\ R_1 \\ R_\gamma \\ 0 \end{pmatrix}. \tag{4.3}$$

Block elimination of the Lagrange multiplier Λ and $\delta\bar{U}_\nu$ in (4.3) leads to the block 3×3 Schur complement system

$$\begin{pmatrix} A_{11} & A_{1\gamma} & \\ A_{\gamma 1} & A_{\gamma\gamma} + \bar{A}_{\nu\nu} & \bar{A}_{\nu 2} \\ & \bar{A}_{2\nu} & \bar{A}_{22} \end{pmatrix} \begin{pmatrix} \delta U_1 \\ \delta U_\gamma \\ \delta \bar{U}_2 \end{pmatrix} = \begin{pmatrix} R_1 \\ R_\gamma + R_\nu + \bar{A}_{\nu\nu}(U_\nu - U_\gamma) \\ \bar{A}_{2\nu}(U_\nu - U_\gamma) \end{pmatrix}. \tag{4.4}$$

The system matrix in (4.4) corresponds to the final adaptive refinement step on processor 1, with possible modifications due to global fine mesh regularization. It is exactly the matrix used in the preliminary local solve to generate the initial guess for the global domain decomposition iteration. In the solution of (4.4), the components δU_1 and δU_γ contribute to the global solution update, while $\delta\bar{U}_2$ is discarded. We remark that the global iteration matrix corresponding to this formulation is not symmetric, even if all local system matrices are symmetric.

The domain decomposition algorithm is incorporated as the solver for the approximate Newton method which is used to solve the discretized equations [1]. In particular, only one domain decomposition iteration (a so-called *inner iteration*) is used in each approximate Newton step. Thus, loosely speaking, each solve of (4.4) alternates with a line search step in which the global solution is updated. The

[1] Even when the original problem is linear, we formally apply the approximate Newton method and the linearity is realized after one Newton step.

Newton line search procedure requires global communication to form some norms and inner products, as well as the boundary exchange described above.

In this paper, the convergence of the domain decomposition algorithm is determined using either one of the the following criteria:

$$\frac{\|\delta \mathcal{U}^k\|_G}{\|\mathcal{U}^k\|_G} \le \max\left(\frac{\|\delta \mathcal{U}^0\|_G}{\|\mathcal{U}^0\|_G}, \frac{\|\nabla e_h\|_{0,\Omega}}{\|\nabla u_h\|_{0,\Omega}}\right) \times 10^{-1}. \tag{4.5}$$

or

$$\frac{\|\delta \mathcal{U}^k\|_G}{\|\mathcal{U}^k\|_G} \le \frac{\|\delta \mathcal{U}^0\|_G}{\|\mathcal{U}^0\|_G} \times 10^{-4}. \tag{4.6}$$

Here $\|\delta \mathcal{U}^k\|_G$ and $\|\mathcal{U}^k\|_G$ are the discrete global norm of the approximate solution and the update respectively, at iteration k, while $\|\nabla e_h\|_{0,\Omega}$ and $\|\nabla u_h\|_{0,\Omega}$ are the a posteriori error estimate and the norm of the initial solution. Normally, (4.5) is sufficient for the purposes of computing an approximation to the solution of the partial differential equation. However, we also use the more stringent criterion (4.6) to illustrate the behavior of the domain decomposition solver as an iterative method for solving linear systems of equations.

5. Numerical Results

In this section, we present some numerical results. Our examples were run on a LINUX-based Beowulf cluster, consisting of 38 nodes, each with two quad core Xeon processors (2.33GHz) and 16GB of memory. The communication network is a gigabit Ethernet switch. This cluster runs the NPACI ROCKS version of LINUX and employs MPICH2 as its MPI implementation. The computational kernels of *PLTMG* [**Ban11**] are written in FORTRAN; the *gfortran* compiler was used in these experiments, invoked using the script *mpif90* and optimization flag *-O*.

In our experiments, we used *PLTMG* to solve the boundary value problem

$$\begin{aligned} -\Delta u &= 1 \quad \text{in } \Omega, \\ u &= 0 \quad \text{on } \partial\Omega, \end{aligned}$$

where Ω is a domain surrounding an airfoil-shaped object (see Figure 3, on the left).

At the beginning, an adaptive mesh of size $N_c = 50K$ was created on one processor. All elements on this mesh were linear elements. This mesh was then partitioned into P subregions, $P = 2^k$, $1 \le k \le 8$. This coarse mesh was broadcast to P processors (simulated as needed) and each processor continued the adaptive process, first in h and then automatic hp, creating a mesh of size N_P. In this experiment, N_P was chosen to be $400K, 600K$, and $800K$. This resulted in global meshes varying in size from approximately $750K$ to $161M$. These global meshes were regularized to be h-conforming and p-conforming by applying appropriate refinement and unrefinement in both h and p to the local meshes. For the case $N_P = 800K$, $P = 32$, the solution and the load balance is shown in Figure 3. The mesh density and degree density of the global mesh and one local mesh are shown in Figure 4 and Figure 5. As expected, both the mesh density and the degree density are high in the local subdomain and much lower elsewhere in the local mesh.

After the global meshes are regularized, a global DD solve was made to obtain the global solution. The results are summarized in Table 1. For the case $N_P =$

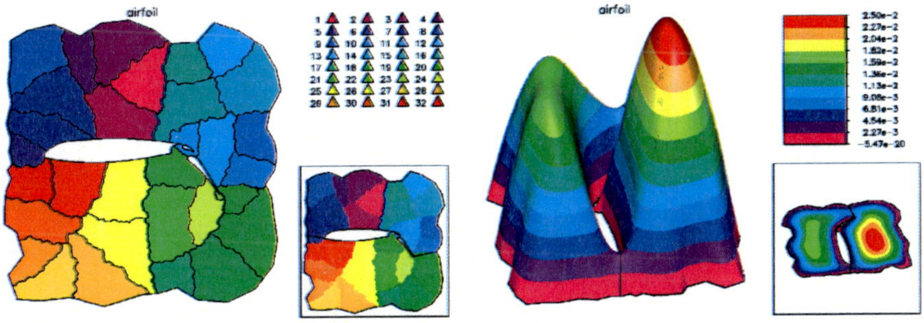

FIGURE 3. The load balance (left) and solution (right) in the case $N_P = 800K$, $P = 32$.

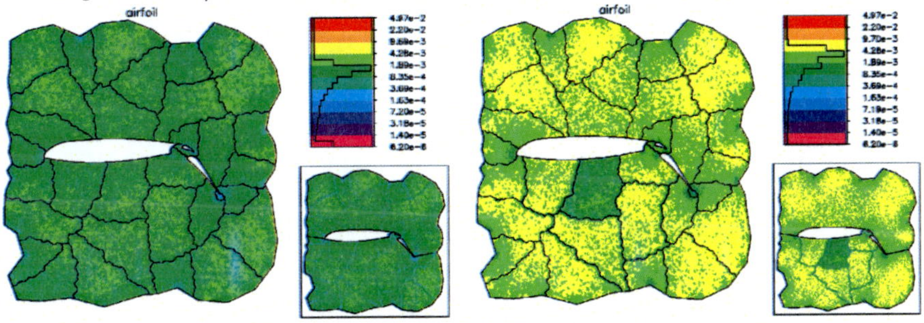

FIGURE 4. The mesh density for the global mesh (left) and for one of the local meshes (right) in the case $N_P = 800K$, $P = 32$.

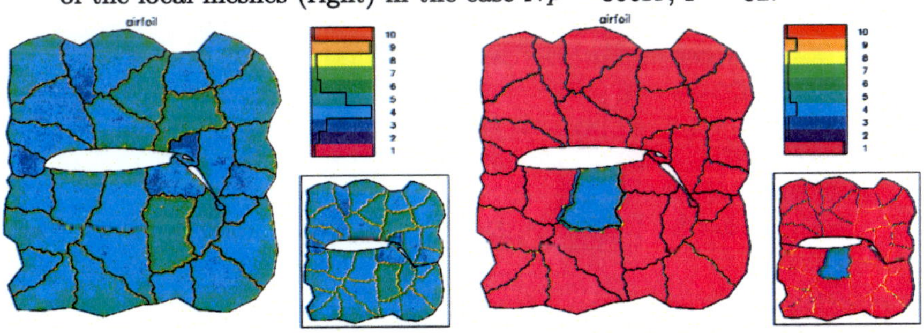

FIGURE 5. The degree density for the global mesh (left) and for one of the local meshes (right) in the case $N_P = 800K$, $P = 32$.

$800K, P = 256$, the convergence history when using the strict criterion (4.5) is shown in Figure 6.

For this approach of domain decomposition, the number of degrees of freedom of the global mesh is predicted by

$$N \approx PN_P - (P-1)N_c. \tag{5.1}$$

Equation (5.1) only predicts an upper bound, as it does not account for refinement outside of Ω_i, needed to keep the mesh conforming and for other reasons. For example, for N_c=50K, N_P=800K, $P = 256$, (5.1) predicts $N \approx 192050000$ when actually $N = 161009439$.

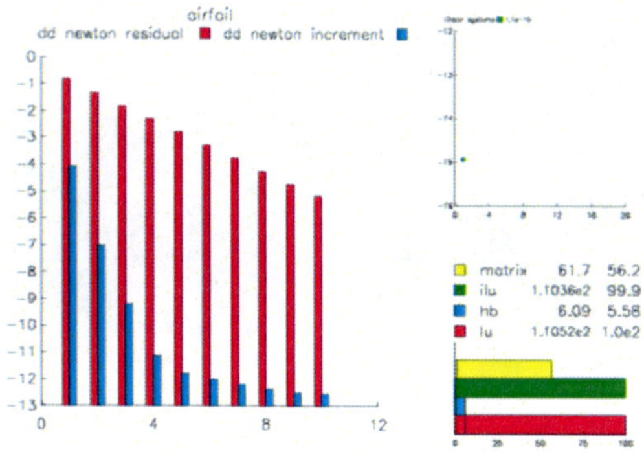

FIGURE 6. Convergence history of the domain decomposition solver for the case $N_P = 800K, P = 256$ when the criterion (4.6) is used.

TABLE 1. Convergence Results. Numbers of iterations needed to satisfy convergence criteria are given in the column labeled DD. The numbers in parentheses are the number of iterations required to satisfy (4.5) which is the default convergence criterion used by *PLTMG*.

	$N_P = 400K$		$N_P = 600K$		$N_P = 800K$	
P	N	DD	N	DD	N	DD
2	750247	14 (4)	1150487	14 (4)	1550418	14 (4)
4	1450884	13 (4)	2252526	19 (5)	3050963	20 (5)
8	2851662	19 (6)	4442935	19 (5)	6101079	20 (5)
16	5670458	20 (4)	8975651	20 (3)	12269476	20 (3)
32	11315140	20 (4)	17721124	20 (4)	24019421	20 (4)
64	20260417	20 (4)	31612049	19 (4)	44327632	18 (4)
128	34750517	10 (3)	56984391	8 (3)	83764874	8 (3)
256	61949578	13 (3)	108554486	10 (3)	161009439	10 (4)

References

[Ban06a] Randolph E. Bank, *A domain decomposition solver for a parallel adaptive meshing paradigm*, Domain decomposition methods in science and engineering XVI, Lect. Notes Comput. Sci. Eng., vol. 55, Springer, Berlin, 2007, pp. 3–14, DOI 10.1007/978-3-540-34469-8_1. MR2334087 (2008c:65364)

[Ban06b] Randolph E. Bank, *Some variants of the Bank-Holst parallel adaptive meshing paradigm*, Comput. Vis. Sci. **9** (2006), no. 3, 133–144, DOI 10.1007/s00791-006-0029-6. MR2271791

[Ban11] Randolph E. Bank, *PLTMG: A software package for solving elliptic partial differential equations, users' guide 11.0*, Tech. report, Department of Mathematics, University of California at San Diego, 2011.

[BH00] Randolph E. Bank and Michael Holst, *A new paradigm for parallel adaptive meshing algorithms*, SIAM J. Sci. Comput. **22** (2000), no. 4, 1411–1443, DOI 10.1137/S1064827599353701. MR1797889 (2002g:65117)

[BH03] Randolph E. Bank and Michael Holst, *A new paradigm for parallel adaptive meshing algorithms*, SIAM Rev. **45** (2003), no. 2, 291–323, DOI 10.1137/S003614450342061. Reprinted from SIAM J. Sci. Comput. **22** (2000), no. 4, 1411–1443 [MR1797889]. MR2010380

[BL03] Randolph E. Bank and Shaoying Lu, *A domain decomposition solver for a parallel adaptive meshing paradigm*, SIAM J. Sci. Comput. **26** (2004), no. 1, 105–127 (electronic), DOI 10.1137/S1064827503428096. MR2114336 (2005i:65199)

[BN11a] Randolph E. Bank and Hieu Nguyen, *Domain decomposition and hp-adaptive finite elements*, Domain decomposition methods in science and engineering XIX, Lect. Notes Comput. Sci. Eng., vol. 78, Springer, Heidelberg, 2011, pp. 3–13, DOI 10.1007/978-3-642-11304-8_1. MR2867640

[BN11b] Randolph E. Bank and Hieu Nguyen, *hp Adaptive finite elements based on derivative recovery and superconvergence*, Comput. Vis. Sci. **14** (2012), 287–299.

[BNar] Randolph E. Bank and Hieu Nguyen, *Mesh regularization in Bank-Holst parallel hp-adaptive meshing*, Domain Decomposition Methods in Science and Engineering XX, Lecture Notes in Computational Science and Engineering, Springer-Verlag, to appear.

[BO07] Randolph E. Bank and Jeffrey S. Ovall, *Dual functions for a parallel adaptive method*, SIAM J. Sci. Comput. **29** (2007), no. 4, 1511–1524 (electronic), DOI 10.1137/060668304. MR2341799 (2009a:65305)

[BV08] Randolph E. Bank and Panayot S. Vassilevski, *Convergence analysis of a domain decomposition paradigm*, Comput. Vis. Sci. **11** (2008), no. 4-6, 333–350, DOI 10.1007/s00791-008-0103-3. MR2425500 (2009h:65202)

[BX03a] Randolph E. Bank and Jinchao Xu, *Asymptotically exact a posteriori error estimators. I. Grids with superconvergence*, SIAM J. Numer. Anal. **41** (2003), no. 6, 2294–2312 (electronic), DOI 10.1137/S003614290139874X. MR2034616 (2004k:65194)

[BX03b] Randolph E. Bank and Jinchao Xu, *Asymptotically exact a posteriori error estimators. II. General unstructured grids*, SIAM J. Numer. Anal. **41** (2003), no. 6, 2313–2332 (electronic), DOI 10.1137/S0036142901398751. MR2034617 (2004m:65212)

[BXZ07] Randolph E. Bank, Jinchao Xu, and Bin Zheng, *Superconvergent derivative recovery for Lagrange triangular elements of degree p on unstructured grids*, SIAM J. Numer. Anal. **45** (2007), no. 5, 2032–2046 (electronic), DOI 10.1137/060675174. MR2346369 (2009b:65293)

[Lu04] Shaoying Lu, *Parallel adaptive multigrid algorithms*, Ph.D. thesis, Department of Mathematics, University of California at San Diego, 2004.

[Mit98a] William F. Mitchell, *The full domain partition approach to distributing adaptive grids*, Proceedings of the International Centre for Mathematical Sciences Conference on Grid Adaptation in Computational PDEs: Theory and Applications (Edinburgh, 1996), 1998, pp. 265–275, DOI 10.1016/S0168-9274(97)00095-0. MR1602805

[Mit98b] William F. Mitchell, *A parallel multigrid method using the full domain partition*, Electron. Trans. Numer. Anal. **6** (1997), no. Dec., 224–233 (electronic). Special issue on multilevel methods (Copper Mountain, CO, 1997). MR1615167

[MM11] W.F. Mitchell and M.A. McClain, *A survey of hp-adaptive strategies for elliptic partial differential equations*, Recent Advances in Computational and Applied Mathematics (2011), 227–258.

[Ngu10] Hieu Nguyen, *p- and fully automatic hp- adaptive finite element methods for elliptic partial differential equations methods*, Ph.D. thesis, University of California, San Diego, 2010.

[Ova04] Jeffrey S. Ovall, *Duality-based adaptive refinement for elliptic PDEs*, ProQuest LLC, Ann Arbor, MI, 2004. Thesis (Ph.D.)–University of California, San Diego. MR2705835

Department of Mathematics, University of California, San Diego, La Jolla, California 92093-0112

E-mail address: rbank@ucsd.edu

Department of Computer Science, University of California, Davis, Davis, California 95616

E-mail address: htrnguyen@ucdavis.edu

A posteriori error estimation via nonlinear error transport with application to shallow water

J. W. Banks, J. A. F. Hittinger, J. M. Connors, and C. S. Woodward

ABSTRACT. Numerical error estimation for time dependent hyperbolic problems is challenging for theoretical and practical reasons. In these systems, error can propagate long distances and produce effects far from the point of generation. In addition, nonlinear interactions of error, as well as discretization nonlinearities can play important roles and must be addressed. In this work, we investigate the use of error transport equations for *a posteriori* error estimation. We discuss the inclusion of nonlinearities in the error equations, which are particularly important for situations where local errors become large, such as near shocks.

1. Introduction

Error estimation for numerical approximations to partial differential equations is an important topic of research within the scientific computing community. Estimates of error can be useful in many ways, including adaptive gridding, model refinement, resource allocation decisions, and uncertainty quantification. For time dependent hyperbolic problems, errors can be generated locally and propagate long distances. For such systems, error transport has been developed as an effective way to incorporate the effect of error propagation [**8**, **11**, **15**]. In recent work [**5**], nonlinear error propagation was shown to be an important extension to the more traditional linear transport. Indeed, incorporating the effect of nonlinear error interactions was found to be critical to developing an accurate estimate of the error for some cases. In the current work, we extend the ideas of nonlinear error transport presented in [**5**] and include results for the shallow water equations in two space dimensions.

The remainder of this paper is organized as follows. In Section 2, we present the basic concepts of nonlinear error transport. Section 3 presents the shallow water equations in 2D. The error equations for the shallow water example are introduced in Section 4, and their discretization is described in Section 5. Section 6 shows

2010 *Mathematics Subject Classification*. Primary 65M12, 65M08; Secondary 76M12, 35L40.

Key words and phrases. A posteriori error estimation, hyperbolic equations, finite volume methods, finite difference methods, weak solutions.

This work was performed under the auspices of the U.S. Department of Energy by Lawrence Livermore National Laboratory under Contract DE-AC52-07NA27344 and was funded by the Uncertainty Quantification Strategic Initiative Laboratory Directed Research and Development Project at LLNL under project tracking code 10-SI-013.

numerical results, including a verification study and results for a problem whose solution transitions from smooth to discontinuous. Conclusions are presented in section 7.

2. Nonlinear Error Transport

The basic idea of error transport is to formulate an auxiliary PDE describing the evolution of the error. Numerical approximations to the original PDEs and error equations are then advanced in tandem. Traditionally the error equations have been linearized [**8**, **11**, **15**], but more recent work [**5**] investigated the use of the full nonlinear error equations.

To introduce the method, consider an evolution equation for $u(x,t)$ of the form

$$\partial_t u + \mathcal{F}(u) = 0, \tag{2.1}$$

where $\mathcal{F}(u)$ is some differential operator on u with initial conditions $u(x,t=0) = g(x)$ for $x \in \mathbb{R}$. Assuming $\tilde{u}(x,t)$ is an approximation to the true solution, the error is defined to be

$$e(x,t) = u(x,t) - \tilde{u}(x,t). \tag{2.2}$$

The error equations are found by substituting (2.2) into (2.1) to yield

$$\partial_t e + \mathcal{F}(e + \tilde{u}) - \mathcal{F}(\tilde{u}) = -\left(\partial_t \tilde{u} + \mathcal{F}(\tilde{u})\right). \tag{2.3}$$

Here $\mathcal{F}(\tilde{u})$ has been subtracted from both sides to show that in the linear case the error and primal solutions evolve by the same differential operator.

In linear error transport, Equation (2.3) is expanded in a Taylor series for small e, and terms that are nonlinear in e are discarded. This approach is valid when e is small. However, for many cases of practical interest, such as when weak solutions containing discontinuities are sought, the local error in a vicinity of the jump may be as large as the solution itself. For these cases, the expansion breaks down, and the resulting linear error equations are not an accurate description of the true error. Nonlinear error transport [**5**] was developed to address these concerns and uses the full nonlinear error equations (2.3).

2.1. A simple example. In order to discuss the potential importance of error nonlinearity consider the inviscid 1-D Burgers' equation,

$$\partial_t u + \partial_x \left(\frac{1}{2}u^2\right) = 0, \qquad x \in \mathbb{R}, \quad t > 0. \tag{2.4}$$

The error equation is

$$\partial_t e + \partial_x \left(\frac{1}{2}e^2 + \tilde{u}e\right) = -\partial_t \tilde{u} - \partial_x \left(\frac{1}{2}\tilde{u}^2\right). \tag{2.5}$$

We approach discretization using a method-of-lines formulation and the standard 4-stage RK-4 time integrator [**1**]. Note that single step schemes are also possible using a space-time, or modified equation, time-stepper (refer to [**5**] for details). Let $u_i \approx u(x_i,t)$ and $e_i \approx e(x_i,t) = u(x_i,t) - u_i$. Furthermore, let D_0, D_+, and D_- represent the usual centered, forward, and backward difference operators defined

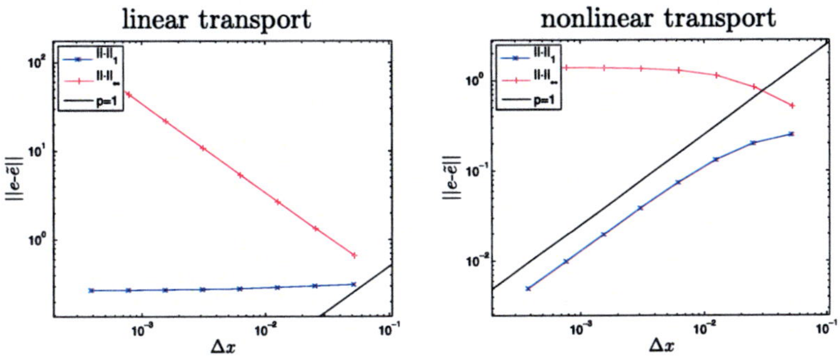

FIGURE 1. Convergence of the approximate error using linear error transport (left), and approximate error using nonlinear error transport (right). Results are presented at $t = 0.5$ for Burgers' equation with initial data $u(x, t = 0) = -\sin(\pi x)$.

by

$$\begin{aligned} 2\Delta x D_0(u_i) &= (u_{i+1} - u_{i-1}), \\ \Delta x D_+(u_i) &= (u_{i+1} - u_i), \\ \Delta x D_-(u_i) &= (u_i - u_{i-1}). \end{aligned}$$

Discretization of (2.4) and (2.5) can be accomplished using the conservative formulation

$$\partial_t u_i = -D_+ \left(\frac{1}{2} (u_{i-1/2})^2 \right), \tag{2.6}$$

and

$$\partial_t e_i = -D_+ \left(\frac{1}{2} (e_{i-1/2})^2 + \hat{u}_{i-1/2} e_{i-1/2} \right) - \mathcal{R}_i, \tag{2.7}$$

where

$$\mathcal{R}_i = D_+ \left(\frac{1}{2} (u_{i-1/2})^2 \right) - D_0 \left(1 - \frac{\Delta x^2}{6} D_+ D_- \right) \left(\frac{1}{2} (u_i)^2 \right).$$

The quantities $u_{i-1/2}$ and $e_{i-1/2}$ are given by the solution to local Riemann problems at cell faces, and $\hat{u}_{i-1/2} = \frac{1}{2}(u_{i-1} + u_i)$ is a second-order accurate approximation to u at the cell face.

We perform a convergence test for the problem $u(x, t = 0) = -\sin(x)$ using both linear and nonlinear transport. Note that linear transport can be achieved by simply omitting terms in (2.5) that are nonlinear in e. Figure 1 shows these results at $t = 0.5$ after the formation of a shock in the exact solution. This figure shows the importance of error nonlinearity in that nonlinear transport converges as expected, while linear transport fails to converge in the L_1 norm and diverges in the max-norm.

3. Shallow Water Equations

An interesting example of nonlinear systems in two space dimension are the shallow water equations. This system sits at the core of climate codes, and estimating the error in simulation results can be important in climate studies. The basic equations can be written as

$$\partial_t \mathbf{u} + \partial_x \mathbf{f}(\mathbf{u}) + \partial_y \mathbf{g}(\mathbf{u}) = 0, \tag{3.1}$$

where

$$\mathbf{u} = \begin{bmatrix} \eta \\ \eta u \\ \eta v \end{bmatrix}, \quad \mathbf{f}(\mathbf{u}) = \begin{bmatrix} \eta u \\ \frac{1}{2} g\eta^2 + \eta u^2 \\ \eta uv \end{bmatrix}, \quad \mathbf{g}(\mathbf{u}) = \begin{bmatrix} \eta v \\ \eta uv \\ \frac{1}{2} g\eta^2 + \eta v^2 \end{bmatrix}.$$

Here η represents the height of the water surface above a given bathymetry (flat in this case), (u, v) represents the velocity, and g is a given constant acceleration due to gravity. The eigenstructure of the flux Jacobian is given by

$$\frac{\partial \mathbf{f}}{\partial \mathbf{u}} = \begin{bmatrix} 1 & 0 & 1 \\ u-c & 0 & u+c \\ v & 1 & v \end{bmatrix} \begin{bmatrix} u-c & 0 & 0 \\ 0 & u & 0 \\ 0 & 0 & u+c \end{bmatrix} \begin{bmatrix} \frac{u+c}{2c} & -\frac{1}{2c} & 0 \\ -v & 0 & 1 \\ -\frac{u-c}{2c} & \frac{1}{2c} & 0 \end{bmatrix},$$

where $c = \sqrt{g\eta}$. A similar result is found for the y-direction. Therefore the equations are hyperbolic when $g\eta > 0$. In addition, the system is seen to support a linear wave traveling at the local velocity and nonlinear "acoustic" waves traveling at the local velocity plus/minus c. More details on the equations and characteristic analysis can be found in [**14**], for example.

4. Error Equations for Shallow Water

In order to derive the error equations for the shallow water equations, we make the familiar *ansatz* that $\mathbf{u} = \tilde{\mathbf{u}} + \mathbf{e}$. Here $\mathbf{e}$ is a vector of errors in the conserved variables. Substitution into (3.1) yields the error equation

$$\partial_t \mathbf{e} + \partial_x \mathbf{F}(\mathbf{e}, \tilde{\mathbf{u}}) + \partial_y \mathbf{G}(\mathbf{e}, \tilde{\mathbf{u}}) = -\partial_t \tilde{\mathbf{u}} - \partial_x \mathbf{f}(\tilde{\mathbf{u}}) - \partial_y \mathbf{g}(\tilde{\mathbf{u}}) \tag{4.1}$$

$$\mathbf{F}(\mathbf{e}, \tilde{\mathbf{u}}) = \mathbf{f}(\mathbf{e} + \tilde{\mathbf{u}}) - \mathbf{f}(\tilde{\mathbf{u}}), \tag{4.2}$$

$$\mathbf{G}(\mathbf{e}, \tilde{\mathbf{u}}) = \mathbf{g}(\mathbf{e} + \tilde{\mathbf{u}}) - \mathbf{g}(\tilde{\mathbf{u}}).$$

Equation (4.1) is a forced nonlinear evolution equation describing the evolution of the error in relation to the given approximation $\tilde{\mathbf{u}}$.

5. Discretization

Discretization of the governing and error equations is accomplished using conservative finite differences in space, and the standard four-stage RK-4 time integration. Succinctly the semi-discrete form is

$$\partial_t \mathbf{u}_{i,j} = -D_{+x} \mathbf{f}(\mathbf{u}_{i-1/2,j}) - D_{+y} \mathbf{g}(\mathbf{u}_{i,j-1/2}), \tag{5.1}$$

$$\partial_t \mathbf{e}_{i,j} = -D_{+x} \mathbf{F}(\mathbf{e}_{i-1/2,j}, \hat{\mathbf{u}}_{i-1/2,j}) - D_{+y} \mathbf{G}(\mathbf{e}_{i,j-1/2}, \hat{\mathbf{u}}_{i,j-1/2}) - \mathcal{T}_{i,j}, \tag{5.2}$$

where D_{+x} and D_{+y} are the divided differences in the x- and y-coordinate directions, respectively, $\hat{\mathbf{u}}_{i-1/2,j} = \frac{1}{2}(\mathbf{u}_{i-1,j} + \mathbf{u}_{i,j})$, $\hat{\mathbf{u}}_{i,j-1/2} = \frac{1}{2}(\mathbf{u}_{i,j-1} + \mathbf{u}_{i,j})$, and $\mathcal{T}_{i,j}$ is an approximation to the residual derived in a similar manner as in Section 2:

$$\begin{aligned} \mathcal{T}_{i,j} = & D_{+x} \mathbf{f}(\mathbf{u}_{i-1/2,j}) - D_{0x}\left(1 - \frac{\Delta x^2}{6} D_{+x} D_{-x}\right) \mathbf{f}(\mathbf{u}_{i,j}) + \\ & D_{+y} \mathbf{g}(\mathbf{u}_{i,j-1/2}) - D_{0y}\left(1 - \frac{\Delta y^2}{6} D_{+y} D_{-y}\right) \mathbf{g}(\mathbf{u}_{i,j}). \end{aligned} \tag{5.3}$$

The values at the half points are determined through the solution of local Riemann problems. For example, consider $\mathbf{u}_{i-1/2,j}$ and $\mathbf{e}_{i-1/2,j}$ (all others are similarly derived). The left and right states for the primal Riemann problem are denoted $\mathbf{u}_L = \mathbf{u}_{i-1+,j}$ and $\mathbf{u}_R = \mathbf{u}_{i-,j}$, respectively, while for the error equation they are $\mathbf{e}_L = \mathbf{e}_{i-1+,j}$ and $\mathbf{e}_R = \mathbf{e}_{i-,j}$, respectively. These states are defined using

$$\mathbf{u}_{i\pm,j} = \mathbf{u}_{i,j} \pm \frac{1}{2} R_{i,j}\Lambda_{i,j}\psi\left(R_{i,j}^{-1}\Delta x D_+ u_{i,j}, R_{i,j}^{-1}\Delta x D_- u_{i,j}\right), \tag{5.4}$$

$$\mathbf{e}_{i\pm,j} = \mathbf{e}_{i,j} \pm \frac{1}{2} R_{i,j}\Lambda_{i,j}\psi\left(R_{i,j}^{-1}\Delta x D_+ e_{i,j}, R_{i,j}^{-1}\Delta x D_- e_{i,j}\right), \tag{5.5}$$

where the function ψ can be used to vary the algorithm between first-order, second-order, and a nonlinear TVD method through the choice of ψ:

$$\begin{aligned} \psi_1(a,b) &= 0, \\ \psi_2(a,b) &= \frac{1}{2}(a,b), \\ \psi_{MM}(a,b) &= \text{minmod}\,(a,b)\,. \end{aligned}$$

See [**2**] for additional details of this approach. Note that $R_{i,j}\Lambda_{i,j}R_{i,j}^{-1} = \frac{\partial \mathbf{f}}{\partial \mathbf{u}}(\mathbf{u}_{i,j})$ depend only on $\mathbf{u}_{i,j}$ and not on the error. This choice has no affect on the formal accuracy of the method and is made because the eigenvalues are real only when η is positive. In practice it is often the case that the discrete solution $\mathbf{u}$ adheres to these bounds, but this may not be the case for $\mathbf{u} + \mathbf{e}$.

Our approximate solutions to local Riemann problems make use of a Roe averaged linearization [**12, 13**]. The Roe averaged state, $\mathbf{u}_*$, is given by

$$\begin{aligned} \eta_* &= \frac{1}{2}\left(u_L + u_R\right), \\ u_* &= \frac{\sqrt{\eta_L}u_L + \sqrt{\eta_R}u_R}{\sqrt{\eta_L} + \sqrt{\eta_R}}, \\ v_* &= \frac{\sqrt{\eta_L}v_L + \sqrt{\eta_R}v_R}{\sqrt{\eta_L} + \sqrt{\eta_R}}. \end{aligned}$$

Define characteristic quantities $\mathbf{w}_L = R_*^{-1}\mathbf{u}_L$, $\mathbf{w}_R = R_*^{-1}\mathbf{u}_R$, $\mathbf{q}_L = R_*^{-1}\mathbf{e}_L$, and $\mathbf{q}_R = R_*^{-1}\mathbf{e}_R$ where $R_*\Lambda_*R_*^{-1} = \frac{\partial \mathbf{f}}{\partial \mathbf{u}}(\mathbf{u}_*)$. The solution to the primal Riemann problem is given by

$$\hat{\mathbf{w}}^{(k)} = \begin{cases} \mathbf{w}_L^{(k)} & \text{if} \quad \Lambda_*^{(k,k)} > 0, \\ \mathbf{w}_R^{(k)} & \text{if} \quad \Lambda_*^{(k,k)} \le 0, \end{cases} \tag{5.6}$$

where the superscripts k are used to denote component number. The solution to the error Riemann problem is given by

$$\hat{\mathbf{q}}^{(k)} = \begin{cases} \mathbf{q}_L^{(k)} & \text{if} \quad \Lambda_*^{(k,k)} > 0, \\ \mathbf{q}_R^{(k)} & \text{if} \quad \Lambda_*^{(k,k)} \le 0. \end{cases} \tag{5.7}$$

The final update is then completely prescribed by assigning $\mathbf{u}_{i-1/2,j} = R_*\hat{\mathbf{w}}$ and $\mathbf{e}_{i-1/2,j} = R_*\hat{\mathbf{q}}$.

6. Results

To demonstrate the validity of our approach to error estimation, we now present results from a number of numerical experiments. We begin with a verification study

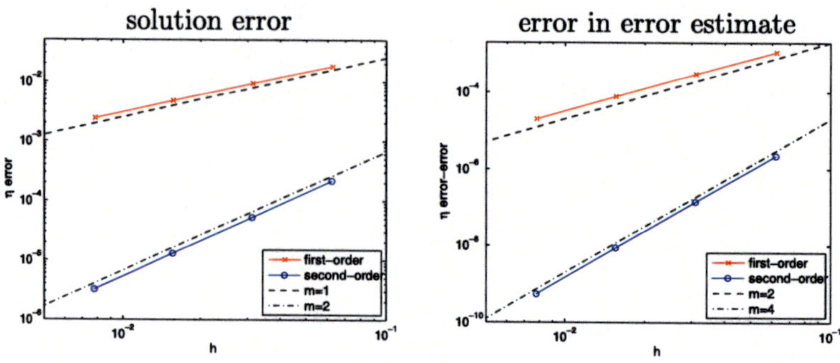

FIGURE 2. Verification using MMS. Results are shown using the first-order and second-order schemes. The error in the primal approximation converges at the expected rate, and the error in the error estimate doubles that.

using the method of manufactured solutions (MMS) and then proceed to a somewhat more physically relevant test case that evolves from smooth to discontinuous as a result of nonlinearities in the operator.

6.1. Verification. In order to verify the overall implementation, we use the MMS, sometimes called twlilight zone [**7**, **9**], whereby a known smooth solution is assumed as the exact answer. The source term necessary to yield this solution is easily derived and added the the governing equations. We make the following choice

$$\eta = 1 + \frac{1}{10}\cos(x)\sin\left(y - \frac{1}{2}\right)\sin(t),$$
$$u = 2\cos\left(x - \frac{1}{2}\right)\sin(y)\sin(2t),$$
$$v = \cos\left(x - \frac{1}{4}\right)\sin\left(y - \frac{1}{4}\right)\sin(3t),$$

where the offset in η is chosen to bound η away from zero so that the governing equations remain well-posed. The periodic simulation domain is chosen to be $(x, y) \in (-\pi, \pi) \times (-\pi, \pi)$, and the exact solution is imposed as an initial condition. As is typical in the literature, limiting is not used for MMS verification tests because nonlinearities in the method produce locally reduced rates that can be difficult to interpret [**4**, **6**]. Convergence results using the first- and second-order schemes and the L_∞ norm are presented in Figure 2. The primal approximations from the first- and second-order schemes are seen to converge at $O(h)$ and $O(h^2)$ respectively, while the error estimates are seen to converge at $O(h^2)$ and $O(h^4)$ respectively. This is the expected behavior as shown in [**5**] using Taylor series analysis.

6.2. Gaussian hump. Consider now the unforced shallow water equations with zero initial velocity and a Gaussian hump for vertical displacement,

$$\eta(x, y, 0) = \frac{1}{10}\left(\exp\left(-\frac{25}{8}\left(x^2 + y^2\right)\right) + \frac{1}{10}\right).$$

Due to nonlinearities in the governing equations, the solution to this problem develops a discontinuity at finite time. We do not know the exact solution to this problem and instead choose to use Richardson extrapolation to indicate the efficacy of our error estimation procedure. As described in [4, 10], Richardson extrapolation can be used to estimate the error in a numerical approximation given three resolutions. This estimated error can then be compared against the error estimate produced using evolution.

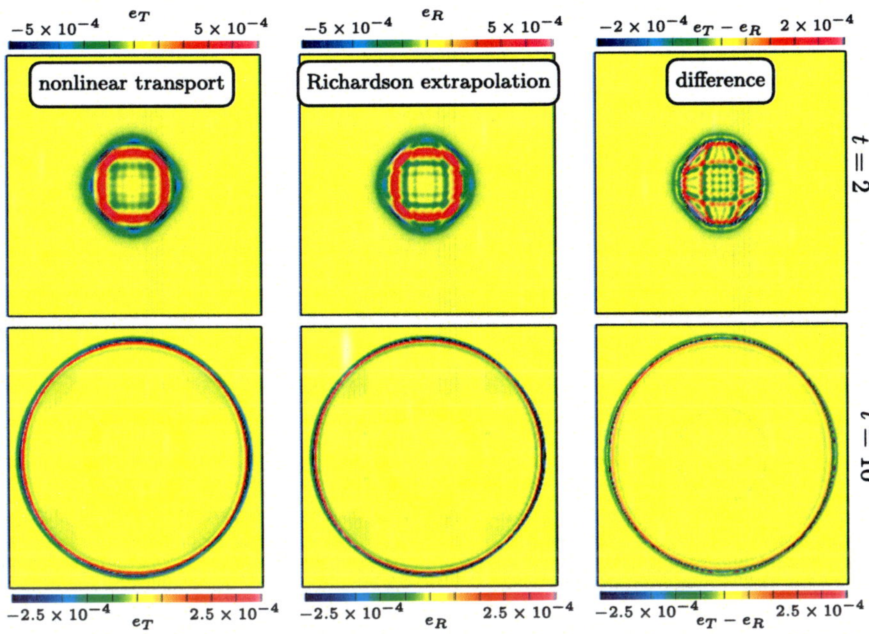

FIGURE 3. Comparison of error estimates for η before shock formation at $t = 2$ (top), and after shock formation at $t = 10$ (bottom). Shown are results for nonlinear error transport (left), Richardson extrapolation (center), and their difference (right).

Figure 3 shows such a comparison on a domain $(x, y) \in (-\pi, \pi) \times (-\pi, \pi)$. Results from nonlinear error transport are shown using 200 points in each direction, and this is also the base resolution for the Richardson estimate. The subsequent two resolutions for Richardson use 400 and 800 points, respectively. We show only results for the nonlinear high-resolution scheme. For many applications, such a scheme is the most practical. In addition, the inherent nonsmooth nature of the errors in such a scheme leads to interesting results for error estimators. For early times ($t = 2$), the solution to this test problem is smooth, and Figure 3 shows good agreement between the two techniques despite the difficulties imposed by discontinuous limiting. At later times ($t = 10$), the solution develops a discontinuity. Despite the discontinuous nature of the solution, the nonlinear error transport approach is shown to estimate the magnitude and location of the error near the shock. Note that in applying Richardson extrapolation for discontinuous solutions great care must be exercised, and we follow the prescription in [3].

7. Conclusions

We have discussed the use of nonlinear error transport as an error estimation technique. The need to include nonlinear error interactions was motivated by the Burgers' equation in 1D. Extension to the 2D shallow water equations was then discussed. The approach was verified using MMS for a smooth problem. In addition, we presented results for a more challenging test problem with no known solution. Here the exact solution transitioned from smooth to discontinuous, and the scheme was shown to still provide reasonable error estimates.

References

[1] Uri M. Ascher and Linda R. Petzold, *Computer methods for ordinary differential equations and differential-algebraic equations*, Society for Industrial and Applied Mathematics (SIAM), Philadelphia, PA, 1998. MR1638643 (99k:65052)

[2] B. van Leer, *Towards the ultimate conservative difference scheme, V. A second-order sequel to Godunov's method*, J. Comput. Phys. **32** (1979), 101–136.

[3] J. W. Banks and T. D. Aslam, *Richardson extrapolation for linearly degenerate discontinuities*, J. Sci. Comput. submitted. arXiv:1205.6504v1.

[4] J. W. Banks, W. D. Henshaw, and J. N. Shadid, *An evaluation of the FCT method for high-speed flows on structured overlapping grids*, J. Comput. Phys. **228** (2009), no. 15, 5349–5369, DOI 10.1016/j.jcp.2009.04.033. MR2541457 (2010h:76108)

[5] J. W. Banks, J. A. F. Hittinger, J. M. Connors, and C. S. Woodward, *Numerical error estimation for nonlinear hyperbolic PDEs via nonlinear error transport*, Comput. Methods Appl. Mech. Engrg. **213/216** (2012), 1–15, DOI 10.1016/j.cma.2011.11.021. MR2880499

[6] J. W. Banks, D. W. Schwendeman, A. K. Kapila, and W. D. Henshaw, *A high-resolution Godunov method for compressible multi-material flow on overlapping grids*, J. Comput. Phys. **223** (2007), no. 1, 262–297, DOI 10.1016/j.jcp.2006.09.014. MR2314392 (2008a:76102)

[7] G. Chesshire and W. D. Hanshaw, *Composite overlapping meshes for the solution of partial differential equations*, J. Comput. Phys. **90** (1990), no. 1, 1–64, DOI 10.1016/0021-9991(90)90196-8. MR1070471 (91f:76043)

[8] A. Hay and M. Visonneau, *Error estimation using the error transport equation for finite-volume methods and arbitrary meshes*, Int. J. Comput. Fluid Dyn. **20** (2006), no. 7, 463–479, DOI 10.1080/10618560600835934. MR2286288

[9] William D. Henshaw, *A fourth-order accurate method for the incompressible Navier-Stokes equations on overlapping grids*, J. Comput. Phys. **113** (1994), no. 1, 13–25, DOI 10.1006/jcph.1994.1114. MR1278187 (95b:76066)

[10] William D. Henshaw and Donald W. Schwendeman, *Parallel computation of three-dimensional flows using overlapping grids with adaptive mesh refinement*, J. Comput. Phys. **227** (2008), no. 16, 7469–7502, DOI 10.1016/j.jcp.2008.04.033. MR2437579 (2009f:76124)

[11] Y. Qin and T. I-P. Shih, *A method for estimating grid-induced errors in finite-difference and finite-volume methods*, AIAA Paper 2003-0845, Reno, Nevada, 2003.

[12] P. L. Roe, *Approximate Riemann solvers, parameter vectors, and difference schemes*, J. Comput. Phys. **43** (1981), no. 2, 357–372, DOI 10.1016/0021-9991(81)90128-5. MR640362 (82k:65055)

[13] Eleuterio F. Toro, *Riemann solvers and numerical methods for fluid dynamics*, 2nd ed., Springer-Verlag, Berlin, 1999. A practical introduction. MR1717819 (2000f:76091)

[14] G. B. Whitham, *Linear and nonlinear waves*, Wiley-Interscience [John Wiley & Sons], New York, 1974. Pure and Applied Mathematics. MR0483954 (58 #3905)

[15] X. D. Zhang, J.-Y. Trépanier, and R. Camarero, *A posteriori error estimation for finite-volume solutions of hyperbolic conservation laws*, Comput. Methods Appl. Mech. Engrg. **185** (2000), no. 1, 1–19, DOI 10.1016/S0045-7825(99)00099-7. MR1755732 (2000m:65124)

Center for Applied Scientific Computing, Lawrence Livermore National Laboratory, Livermore, California 94551

E-mail address: `banks20@llnl.gov`

Multi-physical modeling and multi-scale computation of nano-optical responses

Gang Bao, Guanghui Hu, Di Liu, and Songting Luo

ABSTRACT. Our recent study of multi-physical modeling and multi-scale computation of nano-optical responses is presented in this paper. The semi-classical theory treats the evolution of the electromagnetic (EM) field and the motion of the charged particles concurently by coupling Maxwell equations with Quantum Mechanics. A new efficient computational framework is proposed (in Bao, Liu, and Luo, *A multiscale method for optical responses of nano structures*; and Bao, Liu, and Luo, *Multi-scale modeling and computation of nano-optical responses*) by integrating the Time Dependent Current Density Functional Theory (TD-CDFT), which leads to the coupled Maxwell-Kohn-Sham equations determining the EM field as well as the current and electron densities simultaneously. In the regime of linear responses, a self-consistent multi-scale method is proposed to deal with the well separated space scales. Related recent research on developing adaptive Finite Element Methods for the Kohn-Sham equation (Bao, Hu, and Liu (2012); Bao, Hu, and Liu (to appear)) is also discussed.

1. Multi-Physical Modeling and Nano-Optics

The study of optical responses of nano structures has generated a lot of interest in the development of modern physics. When the optical device is of nano scale, the macroscopic theory for the electromagnetic (EM) field based on constitutive relations can not faithfully capture the microscopic and nonlocal characteristics of the light-matter interaction. In this case, it is necessary to consider the quantum mechanical description of the current and charge densities. Quantum Electrodynamics (QED) [**5**] is able to give a complete description of the interactions between photons and electrons. However, the high computational expense prohibits QED from applications. The semi-classical theories [**6, 7, 8**] combine the classical treatment of the EM field and the first principle approach for the charged particles. Different from QED, in a semi-classical theory, the EM field is not quantized and its time

1991 *Mathematics Subject Classification.* Primary 78A45, 81-08, 81V55.

Key words and phrases. Optical response, nanostructures, multiscale methods, adaptive methods, density functional theory.

The research is supported in part by the NSF FRG DMS-0968360. G. Bao's research was also supported in part by the NSF grants DMS-0908325, CCF-0830161, EAR-0724527, DMS-1211292, the ONR grant N00014-12-1-0319, a Key Project of the Major Research Plan of NSFC (No. 91130004), and a special research grant from Zhejiang University. D. Liu's research was also supported partially by NSF Career Award DMS-0845061.

evolution is described classically by the Maxwell equations. In the meantime, the motion of charged particles is determined quantum mechanically by Schrödinger equations. To avoid solving the high dimensional many body Schrödinger equation involved in the semi-classical theory, in [**1, 2**], we adopted the Time Dependent Current Density Functional Theory (TD-CDFT) [**9, 10**] to further simplify the model and its computation.

1.1. The semi-classical theory for nano-optics. The semi-classical theory for nano-optical responses combines classical treatment of the EM field and quantum mechanical description of the charged particles. The evolution of the EM field can be determined by Maxwell equations. In terms of the vector potential $\mathbf{A}$ and scalar potential ϕ, under the Coulomb gauge $\nabla \cdot \mathbf{A} = 0$, the Maxwell equations have the form:

$$\begin{aligned} \frac{1}{c^2}\frac{\partial^2 \mathbf{A}}{\partial t^2} - \nabla^2 \mathbf{A} + \frac{1}{c}\frac{\partial(\nabla\phi)}{\partial t} &= \frac{4\pi}{c}\mathbf{j}, \\ -\nabla^2\phi &= 4\pi\rho, \end{aligned} \tag{1.1}$$

where c is the speed of light in vacuum, and $\mathbf{j}$ and ρ are the current density and charge density related by the continuity equation:

$$\nabla \cdot \mathbf{j} + \frac{\partial \rho}{\partial t} = 0. \tag{1.2}$$

The electric and magnetic fields, $\mathbf{E}$ and $\mathbf{B}$, can be evaluated by

$$\mathbf{E} = -\nabla\phi - \frac{1}{c}\frac{\partial \mathbf{A}}{\partial t}, \qquad \mathbf{B} = \nabla \times \mathbf{A}. \tag{1.3}$$

Notice that in the Maxwell equations, ρ and $\mathbf{j}$ serve as input to compute the EM field. In a classical macroscopic model, they would be determined by the so called constitutive relations as local functions in terms of $\mathbf{E}$ and $\mathbf{B}$. When the size of the sample is of nano scale such that the spatial structure of the resonant states is comparable to or even larger than the wavelength of the light, a microscopic nonlocal treatment must be considered.

Quantum mechanically, the motion of the charged particles is governed by the Schrödinger equation. For a system consisting of N electrons moving under the influence of a given transverse EM field represented by $\mathbf{A}$, the general nonrelativistic Hamiltonian takes the form [**5, 8**]:

$$H_M = \frac{1}{2}\sum_l \left[\mathbf{p}_l + \frac{1}{c}\mathbf{A}(\mathbf{r}_l)\right]^2 + \sum_l v(\mathbf{r}_l) + U, \tag{1.4}$$

where $\mathbf{r}_l$ and $\mathbf{p}_l$ are the coordinate and conjugate momentum of the lth electron, $v(\mathbf{r})$ is the single particle external potential due to the nuclei, and $U = \frac{1}{2}\sum_{l\neq l'} \frac{1}{|\mathbf{r}_l - \mathbf{r}_{l'}|}$ is the mutual Coulomb interaction among electrons. For simplicity, we will assume the Born-Oppenheimer approximation to separate the electronic motion and the nuclear motion for the molecular structures under consideration.

After the incident light is applied, the system will evolve according to the time dependent Schrödinger equation

$$i\frac{\partial\psi(\mathbf{r}_1, ..., \mathbf{r}_N, t)}{\partial t} = H_M\psi(\mathbf{r}_1, ..., \mathbf{r}_N, t). \tag{1.5}$$

The current density $\mathbf{j}(\mathbf{r},t)$ and electron density $\rho(\mathbf{r},t)$ can be computed through solutions of (1.5) using

$$\mathbf{j}(\mathbf{r},t) =< \psi|\hat{\mathbf{j}}|\psi > \quad \text{and} \quad \rho(\mathbf{r},t) =< \psi|\hat{\rho}|\psi >, \tag{1.6}$$

with the current density operator $\hat{\mathbf{j}}$ and electron density operator $\hat{\rho}$ being given respectively by

$$\begin{aligned} \hat{\mathbf{j}} &= -\frac{1}{2}\sum_l \Big[\mathbf{p}_l\delta(\mathbf{r}-\mathbf{r}_l) + \delta(\mathbf{r}-\mathbf{r}_l)\mathbf{p}_l\Big] - \frac{1}{c}\sum_l \mathbf{A}(\mathbf{r}_l,t), \\ \hat{\rho} &= -\sum_l \delta(\mathbf{r}-\mathbf{r}_l). \end{aligned} \tag{1.7}$$

Notice that in Schrödinger equation (1.5), $\mathbf{A}$ acts as parameters for computing the wavefunction ψ, which will give all physical observables including current and electron densities.

In the semi-classical model, the system is completely described by $(\mathbf{A},\phi)$ and ψ, which affect each other through the coupled Maxwell equations (1.1) and Schrödinger equation (1.5). Therefore they must be determined self-consistently so that equations (1.1) and (1.5) are solved concurrently, which will give rise to the evolution of the EM field and the motion of electrons simultaneously.

1.2. Time dependent current density functional theory. Although the semi-classical theory greatly simplifies the modeling of light-matter interactions at the nano scale, it still poses a significant numerical challenge to solve the high dimensional many body Schrödinger equation involved. Notice that solving the Maxwell equations (1.1) only requires the input of much simpler quantities of the current density $\mathbf{j}$ and electron density ρ. One efficient way to obtain numerical approximations of $(\mathbf{j},\rho)$ is the Time Dependent Current Density Functional Theory (TD-CDFT). The advantage of TD-CDFT is that by restricting to the current and electron densities that are functions of only 3D spatial variables, the computational cost can be greatly reduced.

In TD-CDFT, a system of time dependent Kohn-Sham (KS) equations can be derived in the following form:

$$i\frac{\partial\varphi_l(t)}{\partial t} = H_M^{KS}(t)\varphi_l(t), \tag{1.8}$$

with the following Hamiltonian:

$$H_M^{KS}(t) = \frac{1}{2}\Big[\mathbf{p} + \mathbf{A}_{KS}(\mathbf{r},t)\Big]^2 + v_{KS}(\mathbf{r},t). \tag{1.9}$$

The time dependent KS potential in the above Hamiltonian is given by

$$v_{KS}(\mathbf{r},t) = v(\mathbf{r},t) + v_H(\mathbf{r},t) + v_{xc}(\mathbf{r},t), \tag{1.10}$$

with $v_{xc}(\mathbf{r},t)$ representing the time dependent scalar xc-potential, and

$$\mathbf{A}_{KS}(\mathbf{r},t) = \frac{1}{c}\mathbf{A}(\mathbf{r},t) + \mathbf{A}_{xc}(\mathbf{r},t), \tag{1.11}$$

where $\mathbf{A}$ is the vector potential for the external EM field and $\mathbf{A}_{xc}(\mathbf{r},t)$ is the vector xc-potential. The electron density and the current density can be given by

$$\rho(\mathbf{r}) = \sum_l f_l|\varphi_l(\mathbf{r})|^2, \tag{1.12}$$

and

$$\begin{aligned}\mathbf{j}(\mathbf{r},t) = -\frac{i}{2}\sum_l f_l\Big[\varphi_l^*(\mathbf{r},t)\nabla\varphi_l(\mathbf{r},t) - \varphi_l(\mathbf{r},t)\nabla\varphi_l^*(\mathbf{r},t)\Big] \\ + \sum_l f_l|\varphi_l(\mathbf{r})|^2\mathbf{A}_{KS}(\mathbf{r},t),\end{aligned} \tag{1.13}$$

where f_l is the occupation number.

1.3. The Maxwell-KS system. We can incorporate the TD-CDFT into the semi-classical theory by replacing the current density and the electron density (1.6) given by solutions of the Schrödinger equation with those obtained by TD-CDFT using (1.12) and (1.13). Therefore the Maxwell equations (1.1) and the time dependent Kohn-Sham equations (1.8) form a coupled system for the EM field ($\mathbf{A}$, ϕ) and the current and electron densities ($\mathbf{j}$, ρ) as they are functionals of each other, i.e.,

$$\begin{cases} (\mathbf{A},\phi) & = & \mathcal{M}\big[\mathbf{j},\rho\big] ; \\ (\mathbf{j},\rho) & = & \mathcal{T}\big[\mathbf{A},\phi\big] , \end{cases} \tag{1.14}$$

which suggests that they must be determined self-consistently. From now on, we will refer the above equations as the Maxwell-Kohn-Sham (Maxwell-KS) equations for nano-optics.

For most applications in nano-optics, the induced EM field is varying on a much larger scale than the induced current and electron densities, when the wavelength of the induced EM field is comparable to or larger than the size of the nano structure. Numerically, the mesh size required for the accuracy and stability of solving the Maxwell equations is much larger than the domain we need to handle with TD-CDFT. As a consequence, the coupled Maxwell-KS system can be very ill-conditioned after direct space discretization.

2. Linear Response Theory

The linear response of the Maxwell-KS system (1.14) will further facilitate the computation. In the regime of linear responses, the self-consistent calculation of $(\mathbf{A},\phi)$ and $(\mathbf{j},\rho)$ results in a simple linear system of equations that will allow us to work in the frequency domain.

2.1. Linearized Maxwell-KS system. Rewriting Maxwell equations in the integral form through Green functions shows that the EM field is actually linear functionals of the electron and current densities. On the other hand, the linear response theory of TD-CDFT [**11, 12, 13**] describes the linear relation between the input of the EM field and the output of the microscopic quantities. Combining both theories will lead to the following linear system for the induced EM field ($\delta\mathbf{A}$,

$\delta\phi$) and the induced electron and current densities ($\delta\mathbf{j}$, $\delta\rho$):

$$\text{(2.1)}\qquad \begin{cases} \delta\mathbf{A}(\mathbf{r},\omega) &= -\dfrac{1}{c}\displaystyle\int \mathbf{G}(\mathbf{r}-\mathbf{r}')\delta\mathbf{j}(\mathbf{r}')d\mathbf{r}', \\ \delta\phi(\mathbf{r},\omega) &= -\displaystyle\int \dfrac{\delta\rho(\mathbf{r}')}{|\mathbf{r}-\mathbf{r}'|}d\mathbf{r}', \\ \delta\mathbf{j}(\mathbf{r},\omega) &= \displaystyle\int \Big(\chi_{jj}(\mathbf{r},\mathbf{r}',\omega)-\chi_{jj}(\mathbf{r},\mathbf{r}',0)\Big)\cdot\delta\mathbf{A}_{KS}(\mathbf{r}',\omega)d\mathbf{r}' \\ & \qquad + \displaystyle\int \chi_{j\rho}(\mathbf{r},\mathbf{r}',\omega)\delta v_{KS}(\mathbf{r}',\omega)d\mathbf{r}', \\ \delta\rho(\mathbf{r},\omega) &= \displaystyle\int \chi_{\rho j}(\mathbf{r},\mathbf{r}',\omega)\cdot\delta\mathbf{A}_{KS}(\mathbf{r}',\omega)d\mathbf{r}' \\ & \qquad + \displaystyle\int \chi_{\rho\rho}(\mathbf{r},\mathbf{r}',\omega)\delta v_{KS}(\mathbf{r}',\omega)d\mathbf{r}', \end{cases}$$

where the vector and scalar potentials, $\delta\mathbf{A}_{KS}$ and δv_{KS}, are linear functionals in terms of $\delta\mathbf{A}$, $\delta\mathbf{j}$ and $\delta\rho$ such that

$$\text{(2.2)}\qquad \delta\mathbf{A}_{KS}(\mathbf{r},\omega) = \frac{1}{c}(\mathbf{A}_0(\mathbf{r},\omega)+\delta\mathbf{A}(\mathbf{r},\omega)) + \int \mathbf{f}_{xc}(\mathbf{r},\mathbf{r}',\omega)\delta\mathbf{j}(\mathbf{r}',\omega)d\mathbf{r}',$$

and

$$\text{(2.3)}\qquad \delta v_{KS}(\mathbf{r},\omega) = \int \frac{\delta\rho(\mathbf{r}',\omega)}{|\mathbf{r}-\mathbf{r}'|}d\mathbf{r}' + \int f_{xc}(\mathbf{r},\mathbf{r}',\omega)\delta\rho(\mathbf{r}',\omega)d\mathbf{r}',$$

with $\mathbf{f}_{xc} = \delta\mathbf{A}_{xc}/\delta\mathbf{j}$ and $f_{xc} = \delta v_{xc}/\delta\rho$ being the tensor and scalar xc-kernels respectively. Note that ($\delta\mathbf{j}$, $\delta\rho$) in the above equation satisfy the the continuity equation in the frequency domain such that

$$\text{(2.4)}\qquad \delta\rho = \frac{1}{i\omega}\nabla\cdot\delta\mathbf{j}\ .$$

The tensorial Green function $\mathbf{G}$ can be given as

$$\text{(2.5)}\qquad \mathbf{G}(\mathbf{r}-\mathbf{r}') = \frac{e^{iq|\mathbf{r}-\mathbf{r}'|}}{|\mathbf{r}-\mathbf{r}'|}\mathbf{I} + \frac{1}{q^2}\left(\frac{e^{iq|\mathbf{r}-\mathbf{r}'|}}{|\mathbf{r}-\mathbf{r}'|} - \frac{1}{|\mathbf{r}-\mathbf{r}'|}\right)\nabla'\nabla',$$

where $q = \omega/c$ is the wavenumber in vacuum. The linear response function is given by

$$\text{(2.6)}\qquad \begin{aligned} &\chi_{\alpha\beta}(\mathbf{r},\mathbf{r}',\omega) \\ &= \sum_{ia} f_i \left[\left(\frac{\psi_i(\mathbf{r})\alpha\psi_a(\mathbf{r})\psi_a(\mathbf{r}')\beta\psi_i(\mathbf{r}')}{\epsilon_i-\epsilon_a+\omega}\right) - \left(\frac{\psi_i(\mathbf{r})\alpha\psi_a(\mathbf{r})\psi_a(\mathbf{r}')\beta\psi_i(\mathbf{r}')}{\epsilon_a-\epsilon_i+\omega}\right)^*\right], \end{aligned}$$

where i and a run over the occupied and unoccupied KS orbitals, respectively. The electron density operator $\boldsymbol{\rho} = 1$ and the following paramagnetic current density operator should be substituted for α and β in (2.6):

$$\text{(2.7)}\qquad \mathbf{j}_p = -i(\nabla - \nabla^\dagger)/2,$$

with $\nabla^\dagger$ acting on all terms to the left.

2.2. P-matrix formulation. To further simply the notations, we choose the following spectral representations for the current and electron densities:

$$\begin{cases} \delta\mathbf{j}(\mathbf{r},\omega) = \sum_{ia} f_i \frac{-\omega}{\epsilon_i - \epsilon_a} \psi_i(\mathbf{r})\mathbf{j}_p\psi_a(\mathbf{r})\big[P_{ai}(\omega) - P_{ia}(\omega)\big], \\ \delta\rho(\mathbf{r},\omega) = \sum_{ia} f_i\psi_i(\mathbf{r})\psi_a(\mathbf{r})\big[P_{ai}(\omega) - P_{ia}(\omega)\big], \end{cases} \tag{2.8}$$

with the P-matrix elements defined to be

$$\begin{aligned} &P_{ml}(\omega) \\ &= \frac{\frac{-\omega}{\epsilon_l - \epsilon_m}\int \psi_m(\mathbf{r})\mathbf{j}_p\psi_l(\mathbf{r})\cdot\delta\mathbf{A}_{KS}(\mathbf{r},\omega)d\mathbf{r} + \int \psi_m(\mathbf{r})\psi_l(\mathbf{r})\delta v_{KS}(\mathbf{r},\omega)d\mathbf{r}}{\epsilon_l - \epsilon_m + \omega}, \end{aligned} \tag{2.9}$$

for $\{m,l\} = \{i,a\}$ or $\{a,i\}$. Substituting (2.8) into the first two equations in (2.1) leads to

$$\begin{cases} \delta\mathbf{A}(\mathbf{r},\omega) = \sum_{ia} f_i \frac{\omega}{c(\epsilon_i - \epsilon_a)} \\ \qquad\qquad \times \int \mathbf{G}(\mathbf{r}-\mathbf{r}')\psi_i(\mathbf{r}')\mathbf{j}_p\psi_a(\mathbf{r}')\mathbf{r}'\big[P_{ai}(\omega) - P_{ia}(\omega)\big], \\ \delta\phi(\mathbf{r},\omega) = \sum_{ia} f_i \int \frac{\psi_i(\mathbf{r}')\psi_a(\mathbf{r}')}{|\mathbf{r}-\mathbf{r}'|}d\mathbf{r}'\big[P_{ai}(\omega) - P_{ia}(\omega)\big]. \end{cases} \tag{2.10}$$

By eliminating ($\mathbf{A}_{KS}$, v_{KS}) and ($\delta\mathbf{j}$, $\delta\rho$) in (2.2)-(2.3), (2.8) and (2.9), we arrive at a linear equation satisfied by the elements of the P-matrix such that for n and n' running over occupied and unoccupied orbitals, we have

$$\begin{cases} P_{nn'}(\omega) - \sum_{ia} \frac{K_{nn',ia}(\omega) + M_{nn',ai}(\omega)}{\epsilon_{n'} - \epsilon_n + \omega} P_{ai}(\omega) \\ \qquad + \sum_{ia} \frac{K_{nn',ia}(\omega) + M_{n'n,ia}(\omega)}{\epsilon_{n'} - \epsilon_n + \omega} P_{ia}(\omega) \\ \quad = \frac{-q}{(\epsilon_{n'} - \epsilon_n + \omega)(\epsilon_{n'} - \epsilon_n)} \int \psi_n(\mathbf{r})\mathbf{j}_p\psi_{n'}(\mathbf{r})\mathbf{A}_0(\mathbf{r},\omega)d\mathbf{r}, \\ P_{n'n}(\omega) - \sum_{ia} \frac{K_{n'n,ai}(\omega) + M_{n'n,ia}(\omega)}{\epsilon_n - \epsilon_{n'} + \omega} P_{ai}(\omega) \\ \qquad + \sum_{ia} \frac{K_{n'n,ai}(\omega) + M_{nn',ai}(\omega)}{\epsilon_n - \epsilon_{n'} + \omega} P_{ia}(\omega). \\ \quad = \frac{-q}{(\epsilon_n - \epsilon_{n'} + \omega)(\epsilon_n - \epsilon_{n'})} \int \psi_{n'}(\mathbf{r})\mathbf{j}_p\psi_n(\mathbf{r})\mathbf{A}_0(\mathbf{r},\omega)d\mathbf{r}, \end{cases} \tag{2.11}$$

where the coupling matrix $K_{nn',ia}$ is given as

$$\begin{aligned} &K_{nn',ia}(\omega) \\ &= \frac{\omega^2 f_i}{(\epsilon_{n'} - \epsilon_n)(\epsilon_i - \epsilon_a)} \int \psi_n(\mathbf{r})\mathbf{j}_p\psi_{n'}(\mathbf{r})\mathbf{f}_{xc}(\mathbf{r},\mathbf{r}',\omega)\psi_i(\mathbf{r}')\mathbf{j}_p\psi_a(\mathbf{r}')d\mathbf{r}d\mathbf{r}' \\ &\quad + f_i \int \psi_n(\mathbf{r})\psi_{n'}(\mathbf{r})\left(\frac{1}{|\mathbf{r}-\mathbf{r}'|} + f_{xc}(\mathbf{r},\mathbf{r}',\omega)\right)\psi_i(\mathbf{r}')\psi_a(\mathbf{r}')d\mathbf{r}d\mathbf{r}', \end{aligned} \tag{2.12}$$

and the radiative correction $M_{nn',ia}$ has the form:

$$M_{nn',ia}(\omega) = \frac{-\omega^2 f_i}{c^2(\epsilon_n - \epsilon_{n'})(\epsilon_i - \epsilon_a)} \int \psi_n(\mathbf{r})\mathbf{j}_p\psi_{n'}(\mathbf{r})\mathbf{G}(\mathbf{r}-\mathbf{r}')\psi_i(\mathbf{r}')\mathbf{j}_p\psi_a(\mathbf{r}')d\mathbf{r}d\mathbf{r}'. \tag{2.13}$$

The above formulations (2.11) can be put in a compact form for the P-matrix elements:

$$\left[\begin{pmatrix} \mathbf{S} & \mathbf{T} \\ \mathbf{T} & \mathbf{S} \end{pmatrix} - \omega \begin{pmatrix} \mathbf{I} & \mathbf{0} \\ \mathbf{0} & \mathbf{-I} \end{pmatrix}\right] \begin{pmatrix} \mathbf{P} \\ \mathbf{P'} \end{pmatrix} = \begin{pmatrix} \mathbf{F} \\ \mathbf{-F} \end{pmatrix}, \tag{2.14}$$

with

$$\begin{aligned} &\mathbf{P}_{nn'} = P_{nn'}, \quad \mathbf{P'}_{n'n} = P_{n'n}, \\ &\mathbf{S}_{nn',ia} = \delta_{in}\delta_{an'}(\epsilon_i - \epsilon_a) - K_{nn',ia} - M_{nn',ia}, \\ &\mathbf{T}_{nn',ia} = K_{nn',ia} + M_{nn',ia}, \\ &\mathbf{F}_{nn'} = \frac{q}{\epsilon_{n'} - \epsilon_n} \int \psi_n(\mathbf{r})\mathbf{j}_p\psi_{n'}(\mathbf{r})\mathbf{A}_0(\mathbf{r},\omega)d\mathbf{r}. \end{aligned}$$

Furthermore, if we denote $\mathcal{P}_{nn'} = P_{nn'} - P_{n'n}$, then from (2.14) by addition and subtraction, we can get a linear system on $\mathcal{P}_{nn'}$ such that

$$(\mathcal{S} - \omega^2\mathbf{I})\mathcal{P} = \mathcal{F}, \tag{2.15}$$

with

$$\begin{aligned} &\mathcal{S}_{nn',ia} = \delta_{in}\delta_{an'}(\epsilon_i - \epsilon_a)^2 - 2(\epsilon_n - \epsilon_{n'})(K_{nn',ia} + M_{nn',ia}), \\ &\mathcal{F}_{nn'} = 2(\epsilon_n - \epsilon_{n'})\mathbf{F}_{nn'}. \end{aligned} \tag{2.16}$$

The radiative correction $M_{nn',ia}$ is a consequence of the coupling of Maxwell equations and the linear response theory of TD-CDFT. Without the first two equations in (2.1), there will be no $M_{nn',ia}$ in (2.14), which will be reduced to the standard linear response within TD-CDFT [**13**].

2.3. Resonance conditions. Besides the self-consistent determination of the induced EM field and current density, the linear system (2.14) (or equivalently (2.15)) also enables us to determine the resonant eigenmodes of the nano-optical structure. Resonant eigenmodes exist for particular frequencies such that the matrix in (2.14) or (2.15) is degenerate, which are called self-sustaining (SS) modes [**8**]. The resonant structure of optical spectra in general can be determined by the SS modes. Therefore, we can solve

$$\det\left(\begin{pmatrix} \mathbf{S} & \mathbf{T} \\ \mathbf{T} & \mathbf{S} \end{pmatrix} - \omega \begin{pmatrix} \mathbf{I} & \mathbf{0} \\ \mathbf{0} & \mathbf{-I} \end{pmatrix}\right) = 0, \quad \text{or} \quad \det(\mathcal{S} - \omega^2\mathbf{I}) = 0, \tag{2.17}$$

to determine the eigenfrequencies ω. In particular, we can treat it as an eigenvalue problem to determine the eigenfrequencies ω for the above matrix to have zero eigenvalues.

3. The Self-Consistent Multiscale Method

The Maxwell equations are solved on a much larger domain with a coarse grid compared with the smaller domain and a finer grid for TD-CDFT. In order to deal with the multiscale challenge, we propose a multiscale scheme which consists of two solvers: TD-CDFT serving as a **micro solver** $\mathcal{T}_l$ for the current density and the electron density and a **macro solver** $\mathcal{M}_d$ for the Maxwell equations. A self-consistent iteration is adopted to find the solution of the coupled system (2.1), which will lead to the following procedure:

(1) **Micro solver**: at each step indexed by k, with inputs $(\delta\mathbf{A}_k, \delta\phi_k)$, update the induced current and electron densities through the linear response of TD-CDFT, i.e.,

$$(\delta\mathbf{j}_{k+1}, \delta\rho_{k+1}) = \mathcal{T}_l(\delta\mathbf{A}_k, \delta\phi_k),$$

(2) **Macro solver**: with $(\delta\mathbf{j}_{k+1}, \delta\rho_{k+1})$ as fixed parameters, solve the Maxwell equations to update the EM field such that

$$(\delta\mathbf{A}_{k+1}, \delta\phi_{k+1}) = \mathcal{M}_d(\mathbf{j}_{k+1}, \rho_{k+1}),$$

(3) Repeat until a self-consistent solution is reached.

The micro-solver $\mathcal{T}_l$ can be designed to first solve the equation for the P-matrix (2.15) then obtain the current and electron densities through (2.8). The matrix-vector product $\mathcal{S}_k \cdot \mathcal{P}$ can be obtained for any vector $\mathcal{P}$ as the following:

(3.1)

$$\begin{aligned}[\mathcal{S}_k \cdot \mathcal{P}]_{nn'} =& (\epsilon_n - \epsilon_{n'})^2 \mathcal{P}_{nn'} + (\epsilon_n - \epsilon_{n'}) \int \psi_{n'}(\mathbf{r})\psi_n(\mathbf{r})\delta v_{KS}(\mathbf{r})d\mathbf{r} \\ &+ 2\omega\Big(\int \psi_{n'}(\mathbf{r})\mathbf{j}_p\psi_n(\mathbf{r})\delta\mathbf{A}(\mathbf{r})d\mathbf{r} + \int \psi_{n'}(\mathbf{r})\mathbf{j}_p\psi_n(\mathbf{r})\delta\mathbf{A}_{xc}(\mathbf{r})d\mathbf{r}\Big),\end{aligned}$$

where the right hand side is determined by δA_k and $\delta\phi_k$ through (2.2)-(2.3). Due to the self-consistent structure of the above algorithm, we do not have to pursue an exact solution of (2.15). Instead, a Krylov subspace method will be used to solve (2.15) approximately. For the macro-solver, we can choose a standard scheme such as Finite Difference Method, Finite Element Method, Fast Multipole Method, etc. At each iteration, linear interpolation is used to provide the missing data due to the mismatch between the macro and micro meshes, which essentially allows communications between the macro variable of the EM field and the micro variable of the current and electron densities. The initial EM field $(\delta\mathbf{A}_0, \delta\phi_0)$ can be chosen to be the incident light.

4. Numerical Examples

We present a model calculation of resonant Scanning Near-Field Optical Microscopy (SNOM) as in Figure 1. A substrate supporting the samples is modeled by a semi-infinite local dielectric which occupies the half-space $z < 0$. For our numerical experiments, we choose both the samples and the probe as Copper(I) chloride (CuCl). The Maxwell equations are solved with a locally adaptively refined triangular mesh. The ground state occupied and unoccupied KS orbitals are computed with the OCTOPUS package [**14**]. The local density approximation (LDA) and adiabatic local density approximation (ALDA) are used for v_{xc} for ground state and

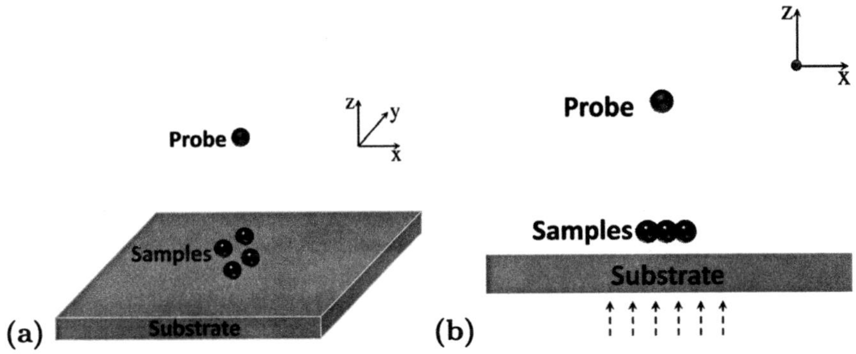

FIGURE 1. SNOM model. (a) model; and (b) example of collection mode: dashed arrows indicate the direction of incident light.

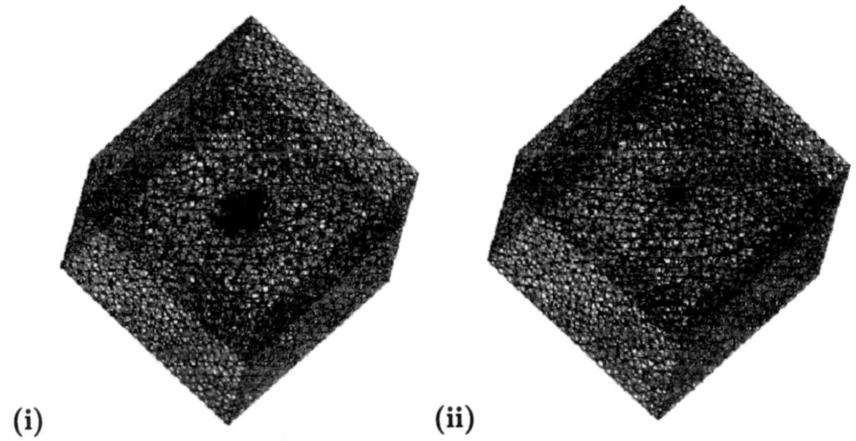

FIGURE 2. Tetrahedral meshes for solving Maxwell equations: meshes are refined near samples (i) and near probe (ii).

time dependent cases respectively [**15, 16**]. For simplicity, the vector xc-potential $\mathbf{A}_{xc}$ is ignored here.

We first compute the resonant conditions for the model corresponding to different positions of the probe, which is performed by solving the eigenvalue proble as in last section. Table 1 shows the computed lowest eigenvalues. The results show that the position of the probe have very small impact on the resonant conditions. Next we verify our computation of the lowest eigenvalues corresponding to different positions of the probe. The incident field is chosen to be

$$\mathbf{A}_0(\mathbf{r},\omega) = -ic\mathbf{p}\exp\left(i\omega/c\mathbf{d}\cdot\mathbf{r}\right)/\omega, \tag{4.1}$$

with polarization $\mathbf{p} = (p_x, p_y, p_z)$ and incident direction $\mathbf{d} = (d_x, d_y, d_z)$ such that $||\mathbf{p}|| = 1, ||\mathbf{d}|| = 1$ and $\mathbf{p}\cdot\mathbf{d} = 0$. With this incident field and given frequencies, we solve (2.15) to get the induced current density and the induced EM field. Then we compute the induced dipole moment given as

$$\delta\mu(\omega) = \frac{i}{\omega}\int \delta\mathbf{j}(\mathbf{r},\omega)d\mathbf{r}. \tag{4.2}$$

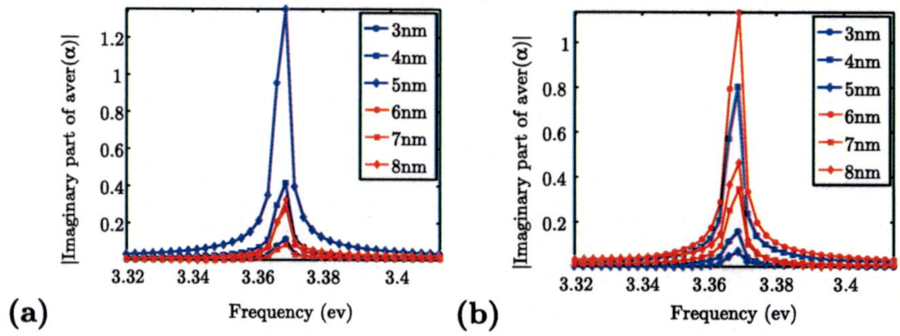

FIGURE 3. Imaginary part of aver(α) corresponding to different position of the probe: (a) $s_x = 0nm$, $s_y = 2.6nm$ and (b) $s_x = 0nm$, $s_y = 0nm$. s_z is indicated in the figure.

The induced dipole moment and the EM field are related through the linear polarizability $\vec{\alpha}$ as

$$\delta\mu = \vec{\alpha}\delta\mathbf{E}; \quad \vec{\alpha} = \begin{pmatrix} \alpha_{xx} & \alpha_{xy} & \alpha_{xz} \\ \alpha_{yx} & \alpha_{yy} & \alpha_{yz} \\ \alpha_{zx} & \alpha_{zy} & \alpha_{zz} \end{pmatrix}. \tag{4.3}$$

Hence we can compute $\vec{\alpha} = \frac{\delta\mu}{\delta\mathbf{E}}$. In particular, we compute aver$(\alpha) \equiv (\alpha_{xx} + \alpha_{yy} + \alpha_{zz})/3$ at the probe. Figure 3 shows the imaginary part of aver(α). We observe a peek at $\omega \approx 3.368(ev)$ which confirms that it is a resonant mode, and the result coincides with the calculation for the eigenvalue value problem (2.17).

5. Adaptive Methods for the Kohn-Sham Equation

A lot of work has been devoted to developing numerical methods for DFT and TDDFT . So far, the plane-wave expansion method (PWE) [**17**] is the most popular. Despite its successes, the PWE method still has limitations. For example, it is nontrivial to deal with the problem with non-periodic boundary condition, or to implement a parallel version because of the scaling problem. These limitations motivate the development of the real-space methods for solving the Kohn-Sham equation such as the finite difference method (FDM) [**18, 19**], the finite volume method (FVM) [**20**], finite element method (FEM)[**21**], discontinuous Galerkin method (DGM)[**22**], and mesh-free method (MFM)[**23**].

TABLE 1. Computed lowest eigenvalues corresponding to different positions of the probe: $s_x = 0nm$, $s_y = 0nm$ or $s_y = 2.6nm$, and $s_z = (2, 3, 4, 5, 6, 7)nm$.

s_z	$3nm$	$4nm$	$5nm$	$6nm$	$7nm$
			$s_y = 2.6nm$		
$\omega(ev)$	3.36761629	3.36759775	3.36761463	3.36757544	3.36761282
			$s_y = 0nm$		
$\omega(ev)$	3.36755793	3.36761002	3.36761394	3.36760513	3.36758880

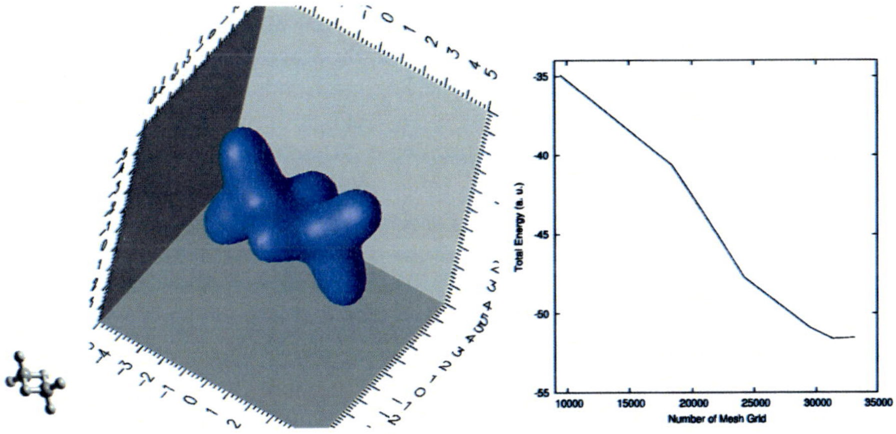

FIGURE 4. Left: The isosurface for a diborane molecule. Right: The convergence curve. In this simulation, the h-adaptive finite element method is used.

In [**3, 24**], the h-adaptive finite element method is introduced for solving the Kohn-Sham equation. Different from other adaptive methods, the mesh topology is changed after the refinement and/or coarsening. An efficient method is necessary to manage the mesh data, which is done by using a specific data structure for the mesh grid. For example, a hierarchical geometry tree (HGT) is utilized for this purpose in [**3**]. With HGT, the mesh refinement and/or coarsening can be easily organized, and an efficient interpolation mechanism between two different meshes can also be obtained. A numerical example is presented in Figure 4, in which a diborane molecule is simulated with the total energy successfully converging to the reference data (-52.628 a.u.). In [**25**], an hp-adaptive method is proposed for solving the Kohn-Sham equation, i.e., besides locally refining and/or coarsening the mesh, the order of the approximate polynomial is also locally enriched.

The r-adaptive method for the Kohn-Sham equation can be found in [**4**]. Different from the h-adaptive method which changes the mesh topology, the r-adaptive method optimizes the distribution of the grid points in the mesh, while keeping the mesh topology unchanged. The general idea is to use a geometry transformation which maps a regular mesh on a domain to a nonuniform mesh on a different domain. An early attempt of the r-adaptive method for the Kohn-Sham equation can be found in [**26**] based on the curvilinear coordinate method. The strategy proposed in [**4**] is to use a harmonic map to optimize the distribution of the mesh grids in the vicinities of the atoms. Compared with the curvilinear coordinate method, the scheme with the harmonic maps can totally separate solving PDEs from redistributing the mesh grids, which makes the code reusable for the mesh redistribution.

6. Concluding Remarks

To study the response of a system under the influence of a weak perturbation, it has been shown that the linear response theory is a quite efficient computational framework. However, this is not the situation when the perturbation of a system is significantly large. In this case, the high order terms in the response function must be taken into account, and a time propagation method should be adopted. A remarkable challenge for the time propagation method is that it is very demanding computationally. Preliminary results on adaptive methods for TDDFT [**27**] have turned out to be promising. A even greater challenge is to overcome the time scale separation between the EM field and the charged particles.

References

[1] G. Bao, D. Liu and S. Luo, *A multiscale method for optical responses of nano structures*, SIAM J. Appl. Math., to appear.

[2] G. Bao, D. Liu and S. Luo, *Multi-scale modeling and computation of nano-optical responses*, submitted.

[3] G. Bao, G. Hu and D. Liu, *An h-adaptive finite element solver for the calculations of the electronic structures*, Journal of Computational Physics, 231, 4967-4979, 2012.

[4] G. Bao, G. Hu and D. Liu, *Numerical solution of the Kohn-Sham equation by finite element methods with an adaptive mesh redistribution technique*, J. Sci. Comput., DOI 10.1007/s10915-012-9636-1, 2012.

[5] C. Cohen-Tannoudji and J. Dupont-Roc and G. Grynberg, *Photons and Atoms: Introduction to Quantum Electrodynamics*, Wiley, New York, 1989.

[6] A. Stahl and I. Balslev, *Electrodynamics of the Semiconductor Band Edge*, Springer Tract in Mod. Phys. 110, Springer-Verlag, New York, 1987.

[7] O. Keller, *Local fields in the electrodynamics of mesoscopic media*, Phys. Rep., 268, 85-262, 1996.

[8] K. Cho, *Optical Response of Nanostructures: Microscopic Nonlocal Theory*, Springer, New York, 2003.

[9] E. Runge and E. K. U. Gross, *Density-functional theory for time-dependent systems*, Phys. Rev. Lett., 52, 997-1000, 1984.

[10] S. K. Ghosh and A. K. Dhara, *Density-functional theory of many-electron systems subjected to time-dependent electric and magnetic fields*, Phys. Rev. A, 38, 1149-1158, 1988.

[11] M. E. Casida, *Time-dependent density functional response theory for molecules*, In Recent advances in density functional methods; D. P. Chong, Ed. World Scientific, Singapore, 155-193, 1995.

[12] G. Vignale, *Current-dependent exchange-correlation potential for dynamical linear response theory*, Phys. Rev. Lett., 77, 2037-2040, 1996.

[13] M. van Faassen and P. L. de Boeij and R. van Leeuwen and J. A. Berger and J. G. Snijders, *Application of time-dependent current-density-functional theory to nonlocal exchange-correlation effects in polymers*, J. Chem. Phys., 118, 1044-1053, 2003.

[14] A. Castro and H. Appel and M. Oliveira and C. A. Rozzi and X. Andrade and F. Lorenzen, and M. A. L. Marques, *Octopus: a tool for the application of time-dependent density functional theory*, Phys. Stat. Sol. B, 243, 2465-2488, 2006.

[15] C. Fiolhais and F. Nogueira and M. A. L. Marques (Editors), *A Primer in Density Dunctional Theory*, Lect. Notes Phys. 620, Springer-Verlag, New York, 2003.

[16] M. A. L. Marques and C. A. Ullrich and F. Nogueira and A. Rubio and K. Burke and E. K. U. Gross (Editors), *Time-Dependent Density Functional Theory*, Lect. Notes Phys. 706, Springer, Heidelberg, 2006. MR2387299 (2009c:81003)

[17] C. Yang, J. C. Meza, B. Lee, and L. W. Wang, *KSSOLV: A Matlab toolbox for solving the Kohn-Sham equations*, ACM Trans. Math. Software, 36, 1-35, 2009. MR2738191

[18] J. R. Chelikowsky, N. Troullier, and Y. Saad, *Finite-difference-pseudopotential method: Electronic structure calculations without a basis*, Phys. Rev. Lett., 72, 1240-1243, 1994.

[19] J. L. Fattebert, M. B. Nardelli, *Finite difference methods for ab initio electronic structure and quantum transport calculations of nanostructures*, Handbook of Numerical Analysis, 10, 571-612, 2003 MR2008392

[20] X. Y. Dai, X. G. Gong, Z. Yang, D. E. Zhang, and A. H. Zhou, *Finite volume discretizations for eigenvalue problems with applications to electronic structure calculations*, Multiscale Modeling & Simulation, 9, 208-240, 2011. MR2769995 (2012a:81310)

[21] J. E. Pask, and P. A. Sterne, *Finite element methods in ab initio electronic structure calculations*, Modelling Simul. Mater. Sci. Eng., 13, R71, 2005.

[22] L. Lin, J. F. Lu, L. X. Ying, and W. E, *Adaptive local basis set for Kohn-Sham density functional theory in a discontinuous Galerkin framework I: Total energy calculation*, J. Comput. Phys., 231, 2140-2154, 2012.

[23] P. Suryanarayana, K. Bhattacharya, and M. Ortiz, *A mesh-free convex approximation scheme for Kohn-Sham density functional theory*, J. Comput. Phys., 230, 5226-5238, 2011. MR2799508 (2012b:81276)

[24] D. E. Zhang, L. H. Shen, A. H. Zhou, and X. G. Gong, *Finite element method for solving Kohn-Sham equations based on self-adaptive tetrahedral mesh*, Phys. Lett. A, 372, 5071-5076, 2008.

[25] T. Torsti, T. Eirola, J. Enkovaara, T. Hakala, P. Havu, V. Havu, T. Höynälänmaa, J. Ignatius, M. Lyly, I. Makkonen, T. T. Rantala, J. Ruokolainen, K. Ruotsalainen, E. Räs"anen, H. Saarikoski, and M. J. Puska. *Three real-space discretization techniques in electronic structure calculations*. Physica Status Solidi (b), 243, 1016-1053, 2006.

[26] E. Tsuchida, and M. Tsukada, *Adaptive finite-element method for electronic-structure calculations*, Phys. Rev. B, 54, 7602-7605, 1996.

[27] G. Bao, G. Hu and D. Liu, *An h-adaptive finite element solver to the calculations of the electronic structures: extension to the time-dependent case*, in preparation.

Department of Mathematics, Zhejiang University, Hangzhou 310027, China, and Department of Mathematics, Michigan State University, East Lansing, Michigan 48824
E-mail address: bao@math.msu.edu

Department of Mathematics, University of Macau, Macau, China
E-mail address: garyhu@umac.mo

Department of Mathematics, Michigan State University, East Lansing, Michigan 48824
E-mail address: richardl@math.msu.edu

Department of Mathematics, Iowa State University, Ames, Iowa 50011
E-mail address: luos@iastate.edu

A Lagged Diffusivity Method for Computing Total Variation Regularized Fluid Flow

Ranil Basnayake, Aaron Luttman, and Erik Bollt

ABSTRACT. There is a great deal of recent work using optical flow methods for analyzing the dynamics of fluids, and much attention has been paid to developing regularization schemes for variational approaches that are consistent with the physics of fluid flow. In this work we show that using total variation to regularize two different kinds of optical flow functionals leads to very good flow field reconstructions for the kinds of dynamical structures that appear in fluid flow. The first optical flow functional is the classical component-based conservation of intensity, and the second approach is to reconstruct the potential of the flow, rather than the flow components. In the two cases, total variation regularization corresponds to imposing different scientific priors on the solution, which we compute using a variation of the Lagged Diffusivity Fixed Point Iteration. Numerical details are presented, and the results are demonstrated on synthetic data and on a data-driven oceanic flow model.

1. Introduction

Optical flow is the term used to describe the computation of the apparent flow field between two images of a particular scene. The original approach, introduced by Horn and Schunck in [**7**], imposed the assumptions that the flow be smooth and that intensity is locally conserved. Given image data $I(x,y,t)$, where $x, y \in \Omega \subset \mathbb{R}^2$ and $t \in [0,T]$, the flow field between two adjacent time instances $\langle u, v\rangle$ was computed as a minimizer of

$$E(u,v) = \int_\Omega (I_t + I_x u + I_y v)^2 \, d\Omega + \alpha \int_\Omega u_x^2 + u_y^2 + v_x^2 + v_y^2 \, d\Omega, \tag{1.1}$$

where $\alpha > 0$ is a regularization parameter. This method has been extended in many directions, introducing new data fidelities [**1, 4**] and regularization terms [**12, 14**], as well as numerical methods to enhance the accuracy and convergence of the corresponding algorithms [**5**]. In the case of imaging fluid dynamics, the flow is generally not smooth, and the turbulent structures are those of greatest interest.

1991 *Mathematics Subject Classification.* Primary 49N45, 49M99; Secondary 65P10, 65T45.

Key words and phrases. Optical flow, fluid dynamics, dynamical systems, total variation regularization.

This research was supported by the United States Office of Naval Research under Grant #N00014-09-1-0647. This manuscript has been authored by National Security Technologies, LLC, under Contract No. DE-AC52-06NA25946 with the U.S. Department of Energy. The United States Government retains and the publisher, by accepting the article for publication, acknowledges that the United States Government retains a non-exclusive, paid-up, irrevocable, world-wide license to publish or reproduce the published form of this manuscript, or allow others to do so, for United States Government purposes. DOE/NV/25946–1553.

In [8, 9], it was shown that, rather than reconstructing the components $\langle u, v\rangle$ of the flow, it is possible to reconstruct the potential or stream function associated with the flow instead, when the flow is the gradient of a potential or the symplectic gradient of a stream. In either case, in order to more accurately capture dynamical structures in the flow, the optical flow functional can be regularized via total variation. It is well known that this results in signals that are approximately piecewise constant, and, in the case of regularizing with the total variation of the flow components, a flow whose components are piecewise constant will be favored. This is especially appropriate for laminar flows or flows that are approximately parallel to the coordinate axes. In the case of regularizing with the total variation of the potential (stream) function, the potential (stream) will be piecewise constant, which is appropriate for sparse flows. In this work we show that TV regularization also leads to excellent reconstructions of other flow structures such as vortices, hyperbolic fixed points, and sources and sinks, in both the component-based and potential-based formulations.

In order to minimize the TV-regularized optical flow functional, we linearize the associated Euler-Lagrange equations by adapting the Lagged Diffusivity Fixed Point Iteration (LDFPI) from [13]. In Section 2, we present the TV optical flow formulation and the LDFPI method. This is followed in Section 3 by results for synthetic data representing the dynamical structures of greatest interest for fluid flows, as well as to data generated from a data-driven ocean model of sea surface temperature data off the coast of Oregon, U.S.A.

2. Total Variation Regularized Optical Flow and LDFPI

The data fidelity term in (1.1) imposes the local conservation of intensity, and the regularization term enforces the smoothness of the flow field. If the flow is the gradient of a potential (or the symplectic gradient of a stream) function ψ, then (1.1) can be written

$$E(\psi) = \int_\Omega (I_t + \nabla I \cdot \nabla\psi)^2 \, d\Omega + \alpha \int_\Omega \psi_{xx}^2 + 2\psi_{xy}^2 + \psi_{yy}^2 \, d\Omega,$$

where ∇I is the spatial gradient of I for fixed t. In the case of a stream flow – which is incompressible – we use the symplectic gradient, $\nabla_H \psi = \langle -\psi_y, \psi_x\rangle$.

Rather than assuming the flow is smooth, we instead regularize the flow using the total variation of the flow components or the potential function, resulting in the alternative optical flow functionals

$$E(u,v) = \int_\Omega (I_t + I_x u + I_y v)^2 \, d\Omega + \alpha \int_\Omega (|\nabla u| + |\nabla v|) \, d\Omega, \tag{2.1}$$

$$E(\psi) = \int_\Omega (I_t + \nabla I \cdot \nabla\psi)^2 \, d\Omega + \alpha \int_\Omega |\nabla\psi| \, d\Omega. \tag{2.2}$$

The corresponding Euler-Lagrange equations for (2.1) are

$$\begin{aligned} I_x(I_t + I_x u + I_y v) + \alpha\nabla \cdot \left(\frac{\nabla u}{|\nabla u|}\right) &= 0 \\ I_y(I_t + I_x u + I_y v) + \alpha\nabla \cdot \left(\frac{\nabla v}{|\nabla v|}\right) &= 0, \end{aligned} \tag{2.3}$$

and, setting $K = -\nabla I \cdot \nabla$, the Euler-Lagrange equation for (2.2) is

$$K^* (I_t - K\psi) + \alpha \nabla \cdot \left(\frac{\nabla \psi}{|\nabla \psi|} \right) = 0, \tag{2.4}$$

where K^* is the operator adjoint of K.

The most straightforward approach to solving the system in (2.3) or the equation (2.4) is to use an explicit-time finite-difference discretization, but the convergence is too slow to be practical in applications. In order to speed up the convergence, we linearize and solve using the Lagged Diffusivity Fixed Point Iteration.

The optical flow functionals (2.1) and (2.2) are least-squares, total-variation functionals of the form

$$E(f) = \|K(f) - h\|^2_{L^2(\Omega)} + \alpha TV(f), \tag{2.5}$$

where $f = \psi$ or $f = \langle u, v \rangle$, $K(u, v) = -I_x u - I_y v$ or $K\psi = -\nabla I \cdot \nabla \psi$, $h = I_t$, and $TV(f)$ is the total variation of ψ or the sum of the total variations of u and v. We use the standard approximation to the total variation of f, given by

$$TV(f) = \int_\Omega |\nabla f| \, d\Omega \approx \int_\Omega \sqrt{f_x^2 + f_y^2 + \beta} \, d\Omega$$

for a fixed numerical regularization $\beta > 0$.

Firstly, suppose that we wish to solve the Euler-Lagrange equations in (2.3), assuming the optical flow functional in (2.1). The flow components u and v are written as column vectors, and we define

$$P_u = \frac{1}{\sqrt{u_x^2 + u_y^2 + \beta}} \quad \text{and} \quad P_v = \frac{1}{\sqrt{v_x^2 + v_y^2 + \beta}},$$

where the division is computed pointwise. The gradient of the approximate TV is then

$$L_n = \begin{bmatrix} D_x^T P_u D_x + D_y^T P_u D_y & 0 \\ 0 & D_x^T P_v D_x + D_y^T P_v D_y \end{bmatrix};$$

the data fidelity operator K is

$$K = \begin{bmatrix} -I_x & 0 \\ 0 & -I_y \end{bmatrix};$$

and the LDFPI algorithm is given by the iteration

(1) Compute gradient direction: $g_n = K^T(K[u_n, v_n]^T - [h, h]^T) + \alpha L_n [u_n, v_n]^T$
(2) Approximate Hessian: $H_n = K^T K + \alpha L_n$
(3) Quasi-Newton Step: $H_n [w_n, y_n]^T = -g_n$
(4) Solution update: $[u_{n+1}, v_{n+1}] = [u_n, v_n] + [w_n, y_n]$,

On the other hand, if we wish to solve the Euler-Lagrange equation in (2.4), assuming the optical flow functional in (2.2), then

$$P = \frac{1}{\sqrt{\psi_x^2 + \psi_y^2 + \beta}}, \qquad L = D_x^T P D_x + D_y^T P D_y, \qquad \text{and} \qquad K = -I_x D_x - I_y D_y.$$

In this case, the iteration follows the same procedure, with $g_n = K^T(K\psi_n - h) + \alpha L_n \psi_n$, $H_n = K^T K + \alpha L_n$, $H_n s_n = -g_n$, and $\psi_{n+1} = \psi_n + s_n$. In either case, one must select an appropriate initial condition, $[u_0^T, v_0^T]$ or ψ_0, which, in the case of optical flow, is usually chosen as the null flow or null potential. In the Quasi-Newton step, we use an LU factorization and Gaussian elimination. For the results

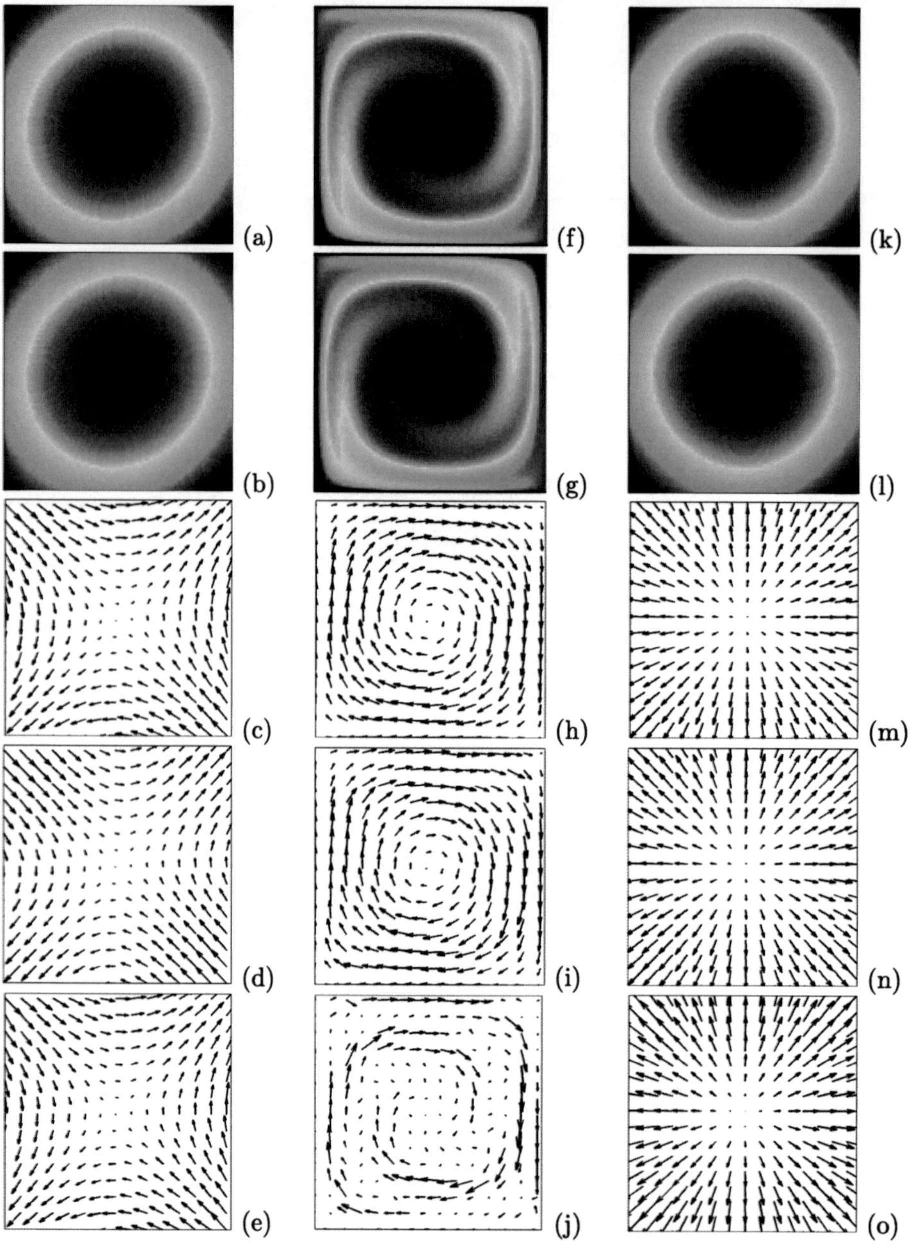

FIGURE 1. Flow Computations – Images (a),(b) show two time instances of the hyperbolic flow. Image (c) shows the true flow field between the images, and images (d),(e) present the computed flows for the u-v and ψ formulations, respectively. The second and third columns show the results for the gyre and diffusive flows.

chosen here, the iteration is stopped when the relative change from one iteration to the next is sufficiently small.

3. Results of Optical Flow Calculations

In this section we present the results of our TV-regularized optical flow calculation on synthetic flows representing three different kinds of dynamical structures that are important in fluid dynamics, as well as results of the flow computation applied to data from a satellite data-driven ocean model.

For the first three sets, the data was obtained by choosing an initial density $I(x,y)$ and evolving it forward in time according to the continuity equation $\frac{dI}{dt} = -(I_x u + I_y v + I u_x + I v_y)$. The first example flow is about a hyperbolic fixed point, given by the velocity field $\langle u,v\rangle = \langle 2y, 2x\rangle$ on the domain $[-0.5, 0.5] \times [-0.5, 0.5]$. Two time instances of the density evolution can be seen in images (a) and (b) of Fig. 1. The true flow is shown in image (c), and the computed flows are shown in images (d) and (e), for the u-v and potential formulations, respectively. At least qualitatively, the images of the true and computed flows are quite similar. One method for measuring error in optical flow is the Mean Angular Error (MAE) [**10**], and in this case the resulting MAE is 2.24° for the u-v and 0.99° for the potential function total variation formulations. Both of these methods outperform the classical Horn-Schunck method, which has an MAE of 5.31°.

The second data set is a so-called "gyre" flow about a vortex, given by the velocity field $\langle u,v\rangle = \langle -\pi\sin(\pi x)\cos(\pi y), \pi\cos(\pi x)\sin(\pi y)\rangle$ on $[0,1]\times[0,1]$. Note that this velocity is not the gradient of a potential function, but it is the symplectic gradient of a stream function, i.e. $\langle u,v\rangle = \langle -\psi_y, \psi_x\rangle$ for $\psi(x,y) = \sin(\pi x)\sin(\pi y)$. Two time instances of the density evolution, the true flow, and the computed flows are shown in Fig. 1 (images (f)-(j)). In this case, the u-v formulation of total variation regularized flow is very close to the true flow–with an MAE of 2.58°– but the stream function formulation is not able to accurately reconstruct the gyre. This is due to the regularization, as a smoothness-based regularization is capable of reconstructing this flow within the stream function framework. In this case, the Horn-Schunck algorithm outperforms the TV-regularized u-v optical flow, but the two mean angular errors are quite close.

Two time instances of the flow about a source are shown in images (k) and (l) of Fig. 1, an evolution given by the velocity field $\langle u,v\rangle = \langle \cos x\cos y, -\sin x\sin y\rangle$ on $[\frac{1}{4}\pi, \frac{3}{4}\pi]\times[\frac{3}{4}\pi, \frac{5}{4}\pi]$. The true flow is shown in image (m) and the computed flows in (n) and (o). The mean angular errors in this case are 1.67° and 1.45° for the u-v and potential TV optical flows, whereas the MAE for the Horn-Schunck method is 22.06°. In all three cases, the computed flows with TV in the u-v formulation capture the dynamical behaviors of the true flows, which is the fundamental goal when using an approach like this to analyze a fluid dynamical system.

For each of these computations, the regularization parameters α and β must be selected. There are numerous methods for choosing the regularization parameter α, such as the discrepancy principle, generalized cross-validation, or the L-curve method, but in these computations we have performed an exhaustive search to approximately choose α to minimize the mean angular error. The actual parameters chosen for the hyperbolic, gyre and diffusive data sets are $\alpha = 10^{-14}$, 10^{-14} and 3.16×10^{-8}, respectively, for the u-v formulation and $\alpha = 8.48\times10^{-14}$, 5.99×10^{-9} and 2.68×10^{-7}, respectively, for the potential/stream formulation. In each case, the minimum mean angular error corresponded to $\beta = 10^{-2}$.

TABLE 1. Mean Angular Errors for the Horn-Schunck method and the total-variation regularized, u-v and potential function formulations for hyperbolic, gyre, and diffusive synthetic flows.

Flow/Method	u-v TV	Potential TV	Horn-Schunck
Hyperbolic	2.24°	0.99°	5.31°
Gyre	2.58°	25.58°	2.26°
Diffusive	1.67°	1.45°	22.06°

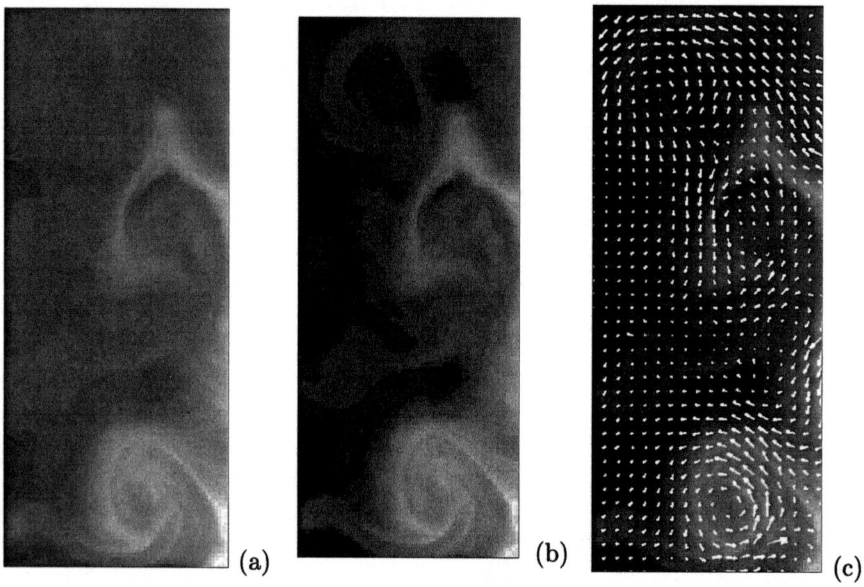

FIGURE 2. Sea Surface Temperature Flow Computations – Images (a) and (b) represent sea surface temperature off the coast of Oregon in August 2002. The lighter gray regions correspond to warmer surface temperatures and the darker regions to cooler temperatures. The computed flow with $\alpha = 10^{-5}$ is shown in (c).

3.1. Flow for oceanic data. Next we apply our method to compute the flow between time instances of the virtual flow of sea surface temperature off the coast of Oregon, U.S.A. This data set was generated from a 3-D ocean model using data obtained from the Geostationary Operational Environmental Satellite (GOES) as the initial condition. In Fig. 2 (a) and (b) represent two time adjacent images on August 1, 2002, representing sea surface temperature. In these images, light gray regions correspond to warmer surface temperatures, and the darker gray regions to cooler surface temperatures. It is clear that there are several vortices about which waters of like temperature are flowing but not mixing with the surrounding waters. These kinds of dynamical structures–and the resulting barriers and pseudo-barriers to transport–are important to study to understand the behavior of large-scale fluid dynamical systems, and the total variation, u-v formulation does a good job of capturing the vortices in the flow. The laminar flows in between the vortices are not captured as well, but it is the voriticial structures that are of the most interest when studying these dynamics.

4. Convergence Analysis of LDFPI for Optical Flow

The primary advantage to the LDFPI for optimizing total variation regularized problems is that the convergence of the scheme is very fast, especially when a good initial condition is chosen. As noted above, for optical flow, the initial condition is usually chosen to be a null flow, and it turns out that this also results in very fast convergence of the algorithm, where numerical convergence is defined by thresholding the relative change in the solution from one iteration to the next.

Fig. 3 shows the mean angular error vs. iteration number for each of the three synthetic data sets analyzed above. The null flow results in an initial MAE of greater than 60°, and, in each case, the algorithm reduced that error in a very few iterations. Even though each iteration of the fixed point method is computationally more intensive than an iteration for an explicit-time scheme, in each case here–the TV-regularized u-v formulation for each of the 3 data sets and the TV-potential function formulation for the diffusive and hyperbolic data sets–the LDFPI method converges in fewer than 10 iterations.

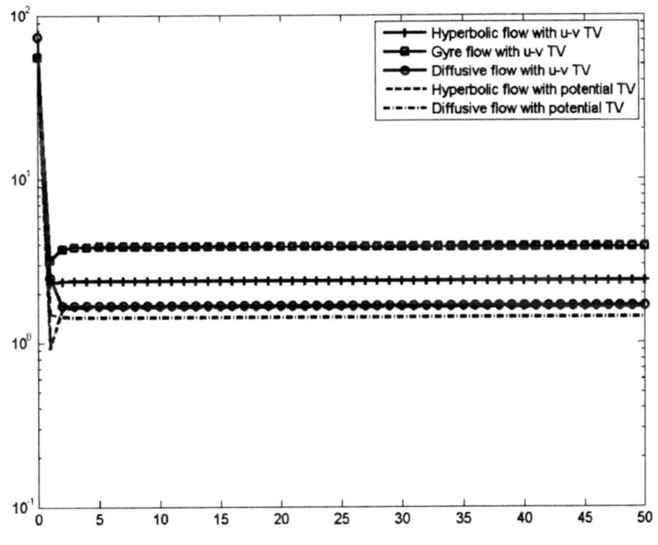

FIGURE 3. Mean Angular Error vs. Iteration Number – The computed mean angular error for the first 50 iterations with a fixed α for the hyperbolic, single gyre, and diffusive data sets.

5. Conclusions

In order to analyze the dynamics of fluid flows measured by imaging systems or evolved as densities by ocean or other flow models, optical flow methods can be used to compute vector fields describing the flow. These vector fields can then be used to compute properties of the dynamics. In this work we have presented a total-variation regularized approach to optical flow and have presented the lagged diffusivity fixed point iteration as a highly efficient algorithm for computing the resulting flows. This approach gives accurate flow reconstruction results for the kinds of dynamics that are of interest in fluid flows, such as flows about hyperbolic and elliptic fixed points and flows out of a source.

Acknowledgements

The authors thank N. Tuffillaro for helpful comments and suggestions on the manuscript and working with the oceanic data and J. Osborne and A. Kurapov for providing access to data from their Regional Ocean Modeling System (ROMS).

References

1. C. Brune, H. Maurer, and M. Wagner, "Detection of Intensity and Motion Edges within Optical Flow via Multidimensional Control," SIAM J. Imaging Science, **2** (2009), no. 4, 1190–1210. MR2559164 (2010j:49053)
2. T. Chang and P. Mulet, "On the convergence of the lagged diffusivity fixed point method in total variation image restoration," SIAM J. Numer. Anal. **36** (1999), no. 2, 354–367. 1997. MR1668254 (2000c:65008)
3. P. Charbonnier, L. Blanc-fraud, G. Aubert and M. Barlaud, "Deterministic edge-preserving regularization in computed imaging," IEEE Trans. Image Processing, **6** (1997), 298–311.
4. T. Corpetti, É. Mémin and P. Pérez, "Adaptation of Standard Optic Flow Methods to Fluid Motion," 9th Int. Symp. Flow Visualisation, (2000), 1–10.
5. A. Doshi and A. G. Bors, "Navier-Stokes formulation for modeling turbulent optical flow," BMVC07, 2007, 10 pages.
6. D. Geman and C. Yang, "Nonlinear image recovery with half-quadratic regularization," IEEE Trans. Image Processing **4** (1995), no. 7, 932–946.
7. B. K. P. Horn and B. G. Schunck, "Determining Optical Flow," Artificial Intelligence **17** (1981), 185–203.
8. T. Kohlberger, É. Mémin, and C. Schnörr, "Variational dense motion estimation using the Helmholtz decomposition," in Scale Space '03, Vol. 2695, (2003), 432–448.
9. A. Luttman, E. Bollt, R. Basnayake, and S. Kramer, "A Stream Function Approach to Optical Flow with Applications to Fluid Transport Dynamics," Proc. Appl. Math. Mechanics, **11** (2011), no. 1, 855–856.
10. B. McCane, K. Novins, D. Crannitch, and B. Galvin, "On Benchmarking Optical Flow," Computer Vision and Image Understanding **84** (2001), 126–143.
11. L. Rudin and S. Osher and E. Fatemi, "Nonlinear total-variation based noise removal algorithms," Physica D: Nonlinear Phenomena **60** (1992), 259–268.
12. D. Suter, "Motion estimation and vector splines," in Proc. Comp. Vision Pattern Rec., 1994, 939–942.
13. C. R. Vogel, *Computational Methods for Inverse Problems*, SIAM, 2002. MR1928831 (2003i:65004)
14. J. Weickert, A. Bruhn, N. Papenberg and T. Brox, "Variational Optic Flow Computation: From Continuous Models to Algorithms," Int. Workshop Computer Vision Image Anal., 2003.

Department of Mathematics, Clarkson University, Potsdam, New York 13699
E-mail address: basnayrk@clarkson.edu

National Security Technologies, LLC, Las Vegas, Nevada 89193-8521
E-mail address: LuttmaAB@nv.doe.gov

Department of Mathematics, Clarkson University, Potsdam, New York 13699
E-mail address: bolltem@clarkson.edu

Estimating the bias of local polynomial approximation methods using the Peano kernel

Jerome Blair, Eric Machorro, and Aaron Luttman

ABSTRACT. Determining the uncertainty in a model estimate of a signal requires both the variance and the bias of the estimate, and, in the case of local polynomial approximation (LPA), calculating the variance of estimates is straightforward. We present a method, using a generalized form of the Peano Kernel Theorem, to estimate the bias of LPA estimates and show how this can be used to optimize the LPA parameters in terms of the bias-variance tradeoff. Figures of merit are derived and values calculated for several common methods. The results in the literature are expanded by giving bias error bounds that are valid for all lengths of the smoothing interval, generalizing the currently available asymptotic results that are only valid in the limit as the length of this interval goes to zero.

1. Introduction

It is standard to model measurements of a continuous signal as samples, often corrupted with noise, giving the discrete model

$$y_i = x(i\Delta t) + \eta_i, \tag{1.1}$$

where x is the signal being measured, y_i is the ith sample, Δt is the fixed time interval between samples, and the η_i are independent random variables with expectation

$$E\left[\eta_i^2\right] = \sigma^2/\Delta t. \tag{1.2}$$

It is desired to estimate $x^{(\nu)}(t_0)$, the νth derivative of the signal, for arbitrary t_0, not necessarily one of the times at which at a sample is taken. In most real-world applications the magnitude of the noise is large enough that data smoothing is required to obtain adequate results, and here we consider local polynomial approximation (LAP), in which a weighted least-squares fit of a polynomial to the data in an interval surrounding t_0 is constructed. The desired value of the function or its derivatives is taken as that of the approximating polynomial at t_0. The whole

---

2010 *Mathematics Subject Classification.* Primary 68W25, 65B99; Secondary 65G99, 65T99.

Key words and phrases. Error analysis, Peano Kernel Theorem, bias error.

This manuscript has been authored by National Security Technologies, LLC, under Contract No. DE-AC52-06NA25946 with the U.S. Department of Energy and supported by the Site-Directed Research and Development Program. The United States Government retains and the publisher, by accepting the article for publication, acknowledges that the United States Government retains a non-exclusive, paid-up, irrevocable, world-wide license to publish or reproduce the published form of this manuscript, or allow others to do so, for United States Government purposes. DOE/NV/25946–1551.

procedure is repeated for each value of t_0 of interest, a common approach covered in [**3–5, 10, 13**], among others.

In order to analyze the quality of the LPA, it is necessary to quantify both how the random noise in the signal is incarnated in the approximation (variance) and the systematic errors in the approximation (bias). The variance In this work, we use Fourier analysis and the Peano Kernel Theorem (PKT) [**2, 7–9, 11**] to provide significantly improved results for the bias error in LAP. The Peano Kernel Theorem is a result for computing estimates in numerical quadrature, and a modern formulation of the theorem from [**6**] is

THEOREM 1.1 (Peano Kernel Theorem). *If J is a linear functional on $C^n([a,b])$ such that $J(p)=0$ for all polynomials p up to degree $n-1$, then there exists a function K_n such that*

$$J(f)=\int_a^b K_n(t)f^n(t)\,dt$$

for all $f\in C^n([a,b])$.

The important points are that the Peano Kernels, K_n depend only on n and that a formula for the kernels is also given. In [**1**], this result is extended to a larger class of linear functionals and provides a formula for the Fourier transform of the Peano Kernels.

In Section 2, we give the mathematical formulation of the local polynomial approximation problem, and we show that the bias error satisfies the hypotheses of the PKT, allowing its use for estimating the bias. The actual formulas for the error bounds are derived in Section 2.1, and the bounds are actually given in Section 2.2. Section 3 gives a comparison between our results and the results previously given in the literature, and conclusions are presented in Section 4.

2. Problem Formulation and Derivation of Error Estimates

In this section we derive the formulas that will be used for computing the error formulas and bounds. The continuous formulation of (1.1) and (1.2) is given by

$$y(t)=x(t)+\eta(t),$$

where $\eta(t)$ is a white noise process satisfying

$$E\left[\left(\int_{\mathbb{R}} f(t)\eta(t)\,dt\right)^2\right]=\sigma^2\int_{\mathbb{R}} f(t)^2\,dt. \tag{2.1}$$

From (1.2) it follows that for any continuous f,

$$E\left[f(i\Delta t)\eta_i\Delta t\right]^2=\frac{\sigma^2}{\Delta t}\sum_i f(i\Delta t)^2\Delta t^2=\sigma^2\sum_i f(i\Delta t)^2\Delta t. \tag{2.2}$$

So, for functions f for which the sum in (2.2) is well approximated by the integral in (2.1), calculations using the continuous time problem will well approximate the analogous calculations of original problem. Without loss of generality, t_0 – the time at which the estimate is to be made – can be taken to be 0. Let $p_{k,W}(t)$ be the $(k+1)^{\text{th}}$ orthonormal polynomial with weight $W(t)$, i.e. $p_{k,W}(t)$ is of degree k and

$$(p_{k,W},p_{j,W})_W\triangleq\int_{\mathbb{R}} p_{k,W}(t)p_{j,W}(t)W(t)\,dt=\delta_{k,j},$$

where $\delta_{k,j}$ is the Kronecker delta. For a positive number, T, let $p_{k,W,T}(t)$ be the $(k+1)^{\text{th}}$ orthonormal polynomial with weight $W(t/T)$. Then

$$p_{k,W,T}(t) = T^{-1/2} p_{k,W}(t/T). \tag{2.3}$$

The local polynomial approximation of order n (degree $n-1$) to the data is then given by

$$P_{n,W,T}(t) = \sum_{k=0}^{n-1} (p_{k,W,T}, y)_W \, p_{k,W,T}(t),$$

which implicitly leads to the estimate for $x^{(\nu)}(0)$ of

$$\begin{aligned} x(\nu, W, n, T) &= P^{(\nu)}_{n,W,T}(0) = \int_{\mathbb{R}} g(\nu, W, n, T; t) y(t)\, dt \qquad \text{with} \\ g(\nu, W, n, T; t) &= \sum_{k=0}^{n-1} p^{(\nu)}_{k,W,T}(0) p_{k,W,T}(t) W(t/T). \end{aligned} \tag{2.4}$$

It is clear from this that local polynomial smoothing is a special case of kernel smoothing (**[3–5, 10]**) with the kernel given by g in (2.4). It is convenient to express the kernel in terms of the polynomials, $p_{k,W}$, that do not depend on T. Applying (2.3) gives

$$g(\nu, W, n, T; t) = \frac{1}{T^{1+\nu}} \sum_{k=0}^{n-1} p^{(\nu)}_{W,k}(0) p_{W,k}(t/T) W(t/T),$$

which leads to the scaling relations

$$\begin{aligned} g(\nu, W, n, T; t) &= \frac{1}{T^{1+\nu}} g(\nu, W, n, 1; T) \qquad \text{and} \\ \hat{g}(\nu, W, n, T; t) &= \frac{1}{T^{\nu}} \hat{g}(\nu, W, n, 1; \omega) \end{aligned} \tag{2.5}$$

for the kernel function and its Fourier transform, respectively.

2.1. Error Formulas. The error in the approximation (2.4) is given by

$$\begin{aligned} e_{\nu,W,n} &= x(\nu, W, n, T) - x^{(\nu)}(0) \\ &= \left(-x^{(\nu)}(0) + \int g(\nu, W, n, T; t) x(t)\, dt\right) + \int g(\nu, W, n, T; t) \eta(t)\, dt \\ &= \qquad b_{\nu,W,n} \qquad + \qquad r_{\nu,W,n}. \end{aligned} \tag{2.6}$$

The first term above is the bias error, the primary concern of this paper. It does not depend on the noise. The second term is a random variable that does not depend on the signal. From (2.1), its variance is given by

$$\sigma^2_{\nu,W,n} = \sigma^2 \int g(\nu, W, n, T; t)^2\, dt = \frac{\sigma^2}{T^{2\nu+1}} \int g(\nu, W, n, 1, t)^2\, dt \triangleq \frac{\sigma^2}{T_{eq}^{2\nu+1}},$$

where T_{eq} is defined by

$$T_{eq} = \frac{T}{[\int g(\nu, W, n, 1, t)^2\, dt]^{1/2\nu+1}} = \frac{2\pi T}{[\int |\hat{g}(\nu, W, n, 1; \omega)|^2\, d\omega]^{1/2\nu+1}}. \tag{2.7}$$

This is the reciprocal of the equivalent noise bandwidth **[12]**. The comparison of the bias errors of two different methods (different n, W, and/or T) should be made for the same value of T_{eq}.

As can be seen in (2.6), the bias error is a linear functional of the signal, $x(t)$, and by construction it is zero if the signal is a polynomial of degree of degree $n-1$

or less, leading to a method for estimating its magnitude via a Corollary to the Peano Kernel Theorem 1.1 and its extension in [**1**].

COROLLARY 2.1 (Peano Kernel Theorem for LAP Bias). *If W is compactly supported and the signal $x(t)$ is m-times differentiable for $\nu \le m \le n$, then*

$$b_{\nu,W,n} = \int_{\mathbb{R}} k(\nu, W, n, m, T; t) x^{(m)}(t)\, dt, \tag{2.8}$$

where the Fourier transform of the Peano Kernel (PK) is given by

$$\hat{k}(\nu, W, n, m, T; \omega) = \frac{\hat{g}(\nu, W, n, T; \omega) - (i\omega)^{\nu}}{(i\omega)^{m}}. \tag{2.9}$$

This result is proven in [**1**] when W is compactly supported, but it can be shown that the formula is valid whenever W is bounded by $\exp(-|at|)$ for some $a \neq 0$ for sufficiently large $|t|$. This condition guarantees that the numerator is an analytic function of ω, while the fact that the bias error vanishes on polynomials of degree less than n guarantees that the numerator has a zero of order $n - \nu$ at the origin. Thus, $\hat{k}$ is a bounded, analytic function of ω. Like g and $\hat{g}$, the PK and its Fourier transform have scaling rules analogous to (2.5), give by

$$\begin{aligned} \hat{k}(\nu, W, n, m, T; \omega) &= T^{m-\nu} \hat{k}(\nu, W, n, m, 1; \omega T) \qquad \text{and} \\ k(\nu, W, n, m, T; t) &= T^{m-\nu-1} k(\nu, W, n, m, 1; t/T). \end{aligned}$$

2.2. Error Bounds. Let V be a normed space to which the PK belongs as a function of t, and let V' be its dual space. Then, from (2.8),

$$\begin{aligned} |b_{\nu,W,n}| &\le \|k(\nu, W, n, m, T; \cdot)\|_V \left\|x^{(m)}\right\|_{V'} \\ &= T^{m-\nu-1} \|k_T(\nu, W, n, m, 1; \cdot)\|_V \left\|x^{(m)}\right\|_{V'}, \end{aligned} \tag{2.10}$$

where $f_T(t) \triangleq f(t/T)$. Norms of particular interest are

$$\|f\|_{L^p} = \left(\int_{\mathbb{R}} |f(t)|^p\, dt\right)^{1/p} \quad \text{and} \quad \|f\|_{F^p} = \left(\frac{1}{2\pi}\int_{\mathbb{R}} \left|\hat{f}(\omega)\right|^q d\omega\right)^{1/q}$$

with $1/p + 1/q = 1$. The definitions in (2.10) are valid for $1 \le p, q < \infty$, with the well-known modification for either variable being ∞. Also, the dual norm of L^p (F^p) is L^q (F^q). A straightforward calculation shows that

$$\|f_T\|_{G^p} = T^{1/p} \|f\|_{G^p}.$$

In this and future expressions, whenever G appears it can be replaced by either L of F. Thus (2.10) gives

$$|b_{\nu,W,n}| \le T^{m-\nu-1+1/p} \|k(\nu, W, n, m, 1; \cdot)\|_{G^p} \left\|x^{(m)}\right\|_{G^q}.$$

Of particular interest is the bound with $p = 1$ and $q = \infty$, so we define the constants

$$\gamma_{\nu,W,n,m,V} = \|k(\nu, W, n, m, 1; \cdot)\|_V \, \|g(\nu, W, n, 1; \cdot)\|_{L^2}^{\frac{2(m-\nu)}{2\nu+1}}. \tag{2.11}$$

Then (2.11) and (2.7) yield

$$|b_{\nu,W,n}| \le \gamma_{\nu,W,n,m,G^p} T_{eq}^{m-\nu-1+1/p} \left\|x^{(m)}\right\|_{G^q}. \tag{2.12}$$

This gives a bound for the bias error proportional to a bound on a norm of the mth derivative of the unknown function. Instead of a bound on a norm of the mth derivative, one could assume that the mth derivative is a sample function of a stochastic process. In this case $b_{\nu,W,n}$ is a random variable. It was shown in [1] that if $x^{(m)}$ is any stationary random process satisfying $E\left[x^{(m)}(t)^2\right] < \infty$, then

$$\sqrt{E\left[b^2_{\nu,W,n,T}\right]} \le T_{eq}^{m-\nu}\gamma_{\nu,W,n,m,F^1}\sqrt{E\left[x^{(m)}(t)^2\right]}. \tag{2.13}$$

It is only necessary that $x^{(m)}$ can be approximated as stationary in the interval in which the PK is nonzero. Since this uses the same constants that appear in (2.12), it does not need special treatment.

These bounds are valid for any value of T. In [3], asymptotic bias error estimates are given for the limit as T goes to zero. This result, in contrast, says that if $x^{(n)}(t)$ is continuous then

$$b_{\nu,W,n,n} = \beta_{\nu,W,n}T^{n-\nu}x^{(n)}(0) + o(T^{n-\nu}), \tag{2.14}$$

where the constants involved are given in [3]. Note that the asymptotic results are only given for $m = n$, and it will be demonstrated later that it almost never desirable to use $n = m$. From (2.8) and (2.10) it can be readily derived that the result (2.14) holds with

$$\beta_{\nu,W,n} = \hat{k}(\nu, W, n, n, 1; 0). \tag{2.15}$$

3. Some Values for the Constants and Comparison With Other Results

In this section numerical values are given for some of the constants for three different weighting functions. The values of the constants are useful for specific error bounds, but the pattern in their sizes is more important.

Three weighting functions that appear regularly in the literature are the Uniform, Epanechnikov and Gaussian weights, given by

$$U(t) = \begin{cases} 1 & |t| < 1 \\ 0 & \text{otherwise} \end{cases}, \qquad E(t) = \begin{cases} 1 - t^2 & |t| < 1 \\ 0 & \text{otherwise} \end{cases}, \qquad G(t) = e^{-t^2}, \tag{3.1}$$

respectively. The values in Table 1 are for $\nu = 0$, $V = F^1$, and various values of m and n. All constants were obtained by numerical calculation of the Fourier transform of the PK. Because in all cases W is an even function, the result of smoothing of order n and $n-1$ are identical for even n and even ν. For each pair (m, n) the smallest constant is given in bold. For $m \le n/2$ the smallest constant is for the Gaussian weighting function. For $m > n/2$ the smallest constants are for the Epanechnikov window, which is shown in [3] to be optimal for $m = n$. There are no situations in the table for which the uniform window is better than either of its alternatives.

The nature of the unknown signal being analyzed determines appropriate values for m, but the value for n is a design choice to minimize the value of the error constant. For all the cases in the table, the error is minimized for $n = 2(m-1)$. In [1] an explanation is given for why to expect that the optimum order should be approximately $2m$. The values in italic (at the extreme right of the table) are the absolute values of the asymptotic constants derived from (2.14) by converting it from an expression involving T to one involving T_{eq}. The values in the table are calculated as the supremum of the absolute value of the Fourier transform of the PK, while those in (2.14) are from the value at $\omega = 0$. For the 12 cases in the table the supremum was at $\omega = 0$.

TABLE 1. Values of Bias Error Constants γ_{0,W,n,m,F^1} for Uniform, Epanechnikov, and Gaussian weighting functions. The derivative being approximated is $\nu = 0$. The order of the polynomial approximation is n (on the vertical axis), and m (horizontal axis) satisfies $0 \leq m \leq n$.

n/m	2	3	4	5	6	7	8
U_2	*4.2e-2*						
E_2	***3.6e-2***						
G_2	*4.0e-2*						
U_4	4.3e-2	1.1e-2	*5.7e-3*				
E_4	3.7e-2	**.95e-2**	***4.8e-3***				
G_4	**3.4e-2**	1.0e-2	*6.3e-3*				
U_6	5.3e-2	1.3e-2	3.5e-3	1.3e-3	*8.8e-4*		
E_6	4.5e-2	1.1e-2	**3.0e-3**	**1.1e-3**	***7.4e-4***		
G_6	**3.8e-2**	**1.0e-2**	3.2e-3	1.3e-3	*11e-4*		
U_8	6.0e-2	1.4e-2	3.8e-3	1.1e-3	$3.8e-4$	1.7e-4	*1.4e-4*
E_8	5.2e-2	1.2e-2	3.2e-3	**.94e-3**	**3.2e-4**	**1.4e-4**	***1.2e-4***
G_8	**4.2e-2**	**1.6e-2**	**3.1e-3**	$1.0e-3$	3.9e-4	2.0e-4	*2.0e-4*

Note that, for each m, either the Epanechnikov window has the smallest constant, or its constant is very close to the smallest one. This window, having bounded support, involves less computation than the Gaussian window. In [**1**], a similar table is given for some common filters that are not based on local polynomial approximation. Generally the constants for the super Gaussian filters are smaller than the values for the Epanechnikov window for $m \geq 5$, and the values are very close to the same (within 10%) for $m \leq 4$.

4. Conclusion

The Peano Kernel Theorem yields bounds for the bias error of local polynomial approximation smoothing that are superior in several ways to what is in the current literature. It provides bounds that are valid for all lengths of the smoothing interval rather than just asymptotic estimates, which are only valid in the limit as the length of the smoothing interval goes to zero. Further, the bounds supplied here for n^{th} order polynomial approximation apply when the unknown function has less than n derivatives. It was demonstrated that errors are minimized by using a polynomial order roughly twice the number of bounded derivatives of the unknown function.

Acknowledgements

The authors would like to thank J. Li and H. Yang for organizing the 8th *International Conference on Scientific Computing and Applications* and to P. Kachroo for co-organizing the minisymposium "Uncertainty Quantification for Signal Processing and Inverse Problems."

References

[1] J. Blair, *Filtering error estimates and order of accuracy via the Peano Kernel Theorem*, Computer Standards and Interfaces **33** (2011), 122–127.

[2] Philip J. Davis, *Interpolation and approximation*, Dover Publications Inc., New York, 1975. Republication, with minor corrections, of the 1963 original, with a new preface and bibliography. MR0380189 (52 #1089)

[3] J. Fan and I. Gijbels, *Local polynomial modelling and its applications*, Monographs on Statistics and Applied Probability, vol. 66, Chapman & Hall, London, 1996. MR1383587 (97f:62063)

[4] Trevor Hastie, Robert Tibshirani, and Jerome Friedman, *The elements of statistical learning*, Springer Series in Statistics, Springer-Verlag, New York, 2001. Data mining, inference, and prediction. MR1851606 (2002k:62048)

[5] V. Katkovnik, K. Egiazarian, and J. Astola, *Local approximation techniques in signal and image processing*, SPIE, 2006.

[6] David Kincaid and Ward Cheney, *Numerical analysis*, 2nd ed., Brooks/Cole Publishing Co., Pacific Grove, CA, 1996. Mathematics of scientific computing. MR1388777 (97g:65003)

[7] G. Peano, *Resto delle formule di quadratura espresso con un integrale definito*, Atti Reale Accademia dei Lincei, Rend. **22** (1913), 562–569.

[8] ———, *Residuo in formulas de quadratura*, Mathesis **4** (1914), 5–10.

[9] Anthony Ralston and Philip Rabinowitz, *A first course in numerical analysis*, 2nd ed., McGraw-Hill Book Co., New York, 1978. International Series in Pure and Applied Mathematics. MR0494814 (58 #13599)

[10] J. O. Ramsay and B. W. Silverman, *Functional data analysis*, 2nd ed., Springer Series in Statistics, Springer, New York, 2005. MR2168993

[11] Arthur Sard, *Linear approximation*, American Mathematical Society, Providence, R.I., 1963. MR0158203 (28 #1429)

[12] H. Van Trees, *Detection, estimation and modulation theory*, John Wiley and Sons, New York, 1968.

[13] Holger Wendland, *Local polynomial reproduction and moving least squares approximation*, IMA J. Numer. Anal. **21** (2001), no. 1, 285–300, DOI 10.1093/imanum/21.1.285. MR1812276 (2002a:65025)

NATIONAL SECURITY TECHNOLOGIES, LLC, P.O. BOX 98521, M/S NLV071, LAS VEGAS, NEVADA 89193-8521

E-mail address: blairjj@embarqmail.com

NATIONAL SECURITY TECHNOLOGIES, LLC, P.O. BOX 98521, M/S NLV078, LAS VEGAS, NEVADA 89193-8521

E-mail address: MachorEA@nv.doe.gov

NATIONAL SECURITY TECHNOLOGIES, LLC, P.O. BOX 98521, M/S NLV078, LAS VEGAS, NEVADA 89193-8521

E-mail address: LuttmaAB@nv.doe.gov

Stability and dispersion analysis of high order FDTD methods for Maxwell's equations in dispersive media

V. A. Bokil and N. L. Gibson

ABSTRACT. Based on recent work (Bokil, Gibson, 2012) that derived a concise formula for the symbol of spatial high order finite difference approximations of first order differential operators, we analyze the stability and dispersion properties of second order accurate in time and $2M, M > 1$ order accurate in space finite difference schemes for Maxwell's equations in dispersive media of Debye type in two dimensions.

1. Introduction

Computational methods for electromagnetic interrogation problems must be highly efficient, consistent and stable. Simulations are used in critical applications such as cancer detection and explosives characterization [**1**]. Many important materials exhibit dielectric dispersion. Thus, an appropriate discretization method should have a numerical dispersion that matches the model dispersion as closely as possible. Common models of dispersive materials include the Debye model for orientational polarization [**5**] and the Lorentz model for electronic polarization [**1**].

We derive stability conditions and demonstrate numerical dispersion error for arbitrary (even) order in space and second order in time finite difference time domain (FDTD) schemes for Maxwell's equations in two dimensions coupled to Debye polarization models via an auxilliary differential equation (ADE) approach [**4**, **7**]. The work here is an extension of the effort in [**3**] where results were obtained for dispersive Maxwell's equations reduced to one spatial dimension based on work done in [**2**, **7**] for the second order Yee FDTD scheme [**8**]. Stability and dispersion analysis for the Yee FDTD scheme applied to Maxwell's equations coupled with different dispersive models such as Debye and Lorentz polarization models, with the ADE approach, in two and three dimensions can be found in [**2**, **7**].

The key result required to perform the stability and dispersion analyses for arbitrary (even) order $2M, M \in \mathbb{N}$ spatial discretizations is the equivalence of the symbol of the $2M$ order finite difference approximation of the first order derivative operator $\partial/\partial z$ with the truncation of an appropriate series expansion of the symbol of $\partial/\partial z$ [**3**].

2010 *Mathematics Subject Classification.* Primary 65M12, 65M06, 35Q61.
The first author was supported in part by NSF grant # DMS-0811223.

2. Model Formulation

Making the assumption that no fields exhibit variation in the z direction, we consider the Maxwell curl equations in two dimensions (the TE mode), which govern the electric field $\mathbf{E}$, and the magnetic field $\mathbf{H}$ in a domain Ω with no free charges in the time interval $(0, T)$, given as

$$\frac{\partial B}{\partial t} = -\text{curl } \mathbf{E}, \tag{2.1a}$$

$$\frac{\partial \mathbf{D}}{\partial t} = \frac{1}{\mu_0}\mathbf{curl}\ B, \tag{2.1b}$$

where $\text{curl } \mathbf{U} = \frac{\partial U_y}{\partial x} - \frac{\partial U_x}{\partial y}$ is the scalar curl operator and $\mathbf{curl}\ U = \left(\frac{\partial U}{\partial y}, -\frac{\partial U}{\partial x}\right)^T$ is the vector curl operator in two dimensions [6]. The fields of interest are $B = B_z$, $\mathbf{E} = (E_x, E_y)^T$, and $\mathbf{D} = (D_x, D_y)^T$. The fields $\mathbf{D}, \mathbf{B}$ are the electric, and magnetic flux densities, respectively. All the fields in (2.1) are functions of position $\mathbf{x} = (x, y)$ and time t. We neglect the effects of boundary conditions and initial conditions.

We will consider the case of a dispersive dielectric medium in which magnetic effects are negligible. The dispersive dielectric is modeled as a single-pole Debye medium exhibiting orientational polarization [4, 5]. Thus, within the dielectric medium we have constitutive relations that relate the flux densities $\mathbf{D}, \mathbf{B}$ to the electric, and magnetic fields, respectively, as

$$\tau\frac{\partial \mathbf{D}}{\partial t} + \mathbf{D} = \epsilon_0\epsilon_\infty\tau\frac{\partial \mathbf{E}}{\partial t} + \epsilon_0\epsilon_s\mathbf{E}, \tag{2.2a}$$

$$\mathbf{B} = \mu_0\mathbf{H}. \tag{2.2b}$$

The parameters ϵ_0, and μ_0, are the permittivity, and permeability, respectively, of free space. In equation (2.2a), the parameter ϵ_s is the static relative permittivity and τ is the relaxation time of the Debye medium. The presence of instantaneous polarization is accounted for by the coefficient ϵ_∞, the infinite frequency permittivity, in the Debye model (2.2a) [1]. We will call the Maxwell's equations (2.1) coupled with the constitutive laws (2.2) as the *Maxwell-Debye* model.

3. High Order Numerical Methods for Dispersive Media

In this section we construct a family of finite difference schemes for the Maxwell-Debye model (2.1)-(2.2). These schemes are based on the discrete higher order ($2M$, $M \in \mathbb{N}$) approximations $\mathcal{D}_{\Delta w}^{(2M)}$, to the first order spatial differential operator $\partial/\partial w, w = x, y$, or z, that were constructed in [3]. For the time discretization we employ the standard leap-frog scheme which is second order accurate in time. We will denote the resulting schemes as $(2, 2M)$ schemes. When $M = 1$, the corresponding $(2, 2)$ schemes are extensions of the famous Yee scheme [8] or FDTD scheme for Maxwell's equations to dispersive media.

Let us denote the time step by $\Delta t > 0$ and the spatial mesh step sizes in the x and y directions by $\Delta x > 0$, and $\Delta y > 0$, respectively. The high order FDTD schemes described here utilize, like the Yee scheme, staggering in space and time of the components of the electric field and flux density with the magnetic field and flux density. We define $t^n = n\Delta t$, $x_\ell = \ell\Delta x$ and $y_j = j\Delta y$, for $n, \ell, j \in \mathbb{Z}$. We also define staggered nodes in the time direction and the x and y direction, respectively as $t^{n+\frac{1}{2}} = t^n + \frac{1}{2}\Delta t$, $x_{\ell+\frac{1}{2}} = x_\ell + \frac{1}{2}\Delta x$, and $y_{j+\frac{1}{2}} = y_j + \frac{1}{2}\Delta y$. The components E_x, D_x of the electric field and electric flux density are discretized at nodes $(t^n, x_{\ell+\frac{1}{2}}, y_j)$, whereas the components E_y, D_y are discretized at nodes $(t^n, x_\ell, y_{j+\frac{1}{2}})$ in the space-time mesh. Finally, the component B of the magnetic flux density is discretized at nodes $(t^{n+\frac{1}{2}}, x_{\ell+\frac{1}{2}}, y_{j+\frac{1}{2}})$. For any field variable $V(t,x,y)$, we denote the approximation of $V(t^n, x_\ell, y_j)$ by $V^n_{\ell,j}$ on the space-time mesh. With the above notation, the $(2,2M)$ discretized schemes for the two dimensional Maxwell-Debye system given in (2.1)-(2.2) are

(3.1a)
$$\frac{B^{n+\frac{1}{2}}_{\ell+\frac{1}{2},j+\frac{1}{2}} - B^{n-\frac{1}{2}}_{\ell+\frac{1}{2},j+\frac{1}{2}}}{\Delta t} = \sum_{p=1}^{M} \frac{\lambda^{2M}_{2p-1}}{(2p-1)\Delta y}\left(E^n_{x_{\ell+\frac{1}{2},j+p}} - E^n_{x_{\ell+\frac{1}{2},j-p+1}}\right)$$
$$- \sum_{p=1}^{M} \frac{\lambda^{2M}_{2p-1}}{(2p-1)\Delta x}\left(E^n_{y_{\ell+p,j+\frac{1}{2}}} - E^n_{y_{\ell-p+1,j+\frac{1}{2}}}\right),$$

(3.1b)
$$\frac{D^{n+1}_{x_{\ell+\frac{1}{2},j}} - D^n_{x_{\ell+\frac{1}{2},j}}}{\Delta t} = \frac{1}{\mu_0}\sum_{p=1}^{M} \frac{\lambda^{2M}_{2p-1}}{(2p-1)\Delta y}\left(B^{n+\frac{1}{2}}_{\ell+\frac{1}{2},j+p-\frac{1}{2}} - B^{n+\frac{1}{2}}_{\ell+\frac{1}{2},j-p+\frac{1}{2}}\right),$$

(3.1c)
$$\frac{D^n_{y_{\ell,j+\frac{1}{2}}} - D^n_{y_{\ell,j+\frac{1}{2}}}}{\Delta t} = -\frac{1}{\mu_0}\sum_{p=1}^{M} \frac{\lambda^{2M}_{2p-1}}{(2p-1)\Delta x}\left(B^{n+\frac{1}{2}}_{\ell+p-\frac{1}{2},j+\frac{1}{2}} - B^{n+\frac{1}{2}}_{\ell-p+\frac{1}{2},j+\frac{1}{2}}\right),$$

where
$$\lambda^{2M}_{2p-1} = \frac{2(-1)^{p-1}[(2M-1)!!]^2}{(2M+2p-2)!!(2M-2p)!!(2p-1)},$$
as given in [3], and the double factorial is defined as
$$n!! = \begin{cases} n\cdot(n-2)\cdot(n-4)\ldots 5\cdot 3\cdot 1 & n>0, \text{ odd} \\ n\cdot(n-2)\cdot(n-4)\ldots 6\cdot 4\cdot 2 & n>0, \text{ even} \\ 1, & n=-1,0. \end{cases}$$

For a Debye media we add the discretized (second order finite difference in time) version of the equation (2.2a) to the discrete scheme defined in (3.1). The discretization of equation (2.2a) in scalar form is

(3.2a)
$$\epsilon_0\epsilon_\infty\tau\frac{E^{n+1}_{x_{\ell+\frac{1}{2},j}} - E^n_{x_{\ell+\frac{1}{2},j}}}{\Delta t} + \epsilon_0\epsilon_s\frac{E^{n+1}_{x_{\ell+\frac{1}{2},j}} + E^n_{x_{\ell+\frac{1}{2},j}}}{2}$$
$$= \tau\frac{D^{n+1}_{x_{\ell+\frac{1}{2},j}} - D^n_{x_{\ell+\frac{1}{2},j}}}{\Delta t} + \frac{D^{n+1}_{x_{\ell+\frac{1}{2},j}} + D^n_{x_{\ell+\frac{1}{2},j}}}{2},$$

(3.2b)
$$\epsilon_0\epsilon_\infty\tau\frac{E^{n+1}_{y_{\ell,j+\frac{1}{2}}} - E^n_{y_{\ell,j+\frac{1}{2}}}}{\Delta t} + \epsilon_0\epsilon_s\frac{E^{n+1}_{y_{\ell,j+\frac{1}{2}}} + E^n_{y_{\ell,j+\frac{1}{2}}}}{2}$$
$$= \tau\frac{D^{n+1}_{y_{\ell,j+\frac{1}{2}}} - D^n_{y_{\ell,j+\frac{1}{2}}}}{\Delta t} + \frac{D^{n+1}_{y_{\ell,j+\frac{1}{2}}} + D^n_{y_{\ell,j+\frac{1}{2}}}}{2},$$

4. Stability Analysis

To determine stability conditions we use von Neumann analysis which allows us to localize roots of certain classes of polynomials [**2**]. We follow the approach in [**2**] in which the author derives stability conditions for the $(2,2)$ (Yee) schemes applied to Debye and Lorentz dispersive media. This analysis is based on properties of Schur and von Neumann polynomials.

Stability conditions for the general $(2,2M)$ schemes are made possible by the results presented in [**3**], in which $2M$ order finite difference approximations $\mathcal{D}^{(2M)}_{\Delta w}$, of the first order derivative operator $\partial/\partial w, w = x, y, z$, are described in terms of a truncation of a series expansion of the symbol of this operator

$$\mathcal{F}\left(\mathcal{D}^{(2M)}_{\Delta w}\right) = \frac{2i}{\Delta w}\sum_{p=1}^{M} \gamma_{2p-1} \sin^{2p-1}(k_w \Delta w/2), \tag{4.1}$$

with $\gamma_{2p-1} = \dfrac{[(2p-3)!!]^2}{(2p-1)!}$, and k_w is the component of the wave vector in the $w = x, y$ or z dimension. In performing the von Neumann analysis for the $(2,2M)$ schemes we show that the resulting amplification matrices retain the same structure as in the $(2,2)$ schemes in [**2**], albeit with a generalized definition of the parameter q in [**2**] that appears in this matrix.

4.1. Stability Analysis for $(2,2M)$ Schemes for Debye Media. We consider the $(2,2M)$ scheme for discretizing Maxwell's equations coupled with the Debye polarization model presented in the form of equations (3.1) and (3.2a).

We assume a spatial dependence of the following form in the field quantities

$$\begin{aligned}
B^{n-\frac{1}{2}}_{\ell+\frac{1}{2},j+\frac{1}{2}} &= \hat{B}^{n-\frac{1}{2}}(k_x,k_y)e^{ik_x x_{\ell+\frac{1}{2}}+ik_y y_{j+\frac{1}{2}}};\\
E^n_{x_{\ell+\frac{1}{2},j}} &= \hat{E}^n_x(k_x,k_y)e^{ik_x x_{\ell+\frac{1}{2}}+ik_y y_j}; \quad E^n_{y_{\ell,j+\frac{1}{2}}} = \hat{E}^n_y(k_x,k_y)e^{ik_x x_\ell+ik_y y_{j+\frac{1}{2}}};\\
D^n_{x_{\ell+\frac{1}{2},j}} &= \hat{D}^n_x(k_x,k_y)e^{ik_x x_{\ell+\frac{1}{2}}+ik_y y_j}; \quad D^n_{y_{\ell,j+\frac{1}{2}}} = \hat{D}^n_y(k_x,k_y)e^{ik_x x_\ell+ik_y y_{j+\frac{1}{2}}};
\end{aligned} \tag{4.2}$$

with $k_w, w = x, y$, the component of the wave vector $\mathbf{k}$ in the w dimension, i.e. $\mathbf{k} = (k_x, k_y)^T$. The wave number is $k = \sqrt{k_x^2 + k_y^2}$. We define the vector

$$\mathbf{U}^n := [c_\infty \hat{B}^{n-\frac{1}{2}},\ \hat{E}^n_x,\ \frac{1}{\epsilon_0\epsilon_\infty}\hat{D}^n_x,\ \hat{E}^n_y,\ \frac{1}{\epsilon_0\epsilon_\infty}\hat{D}^n_y]^T. \tag{4.3}$$

Substituting the forms (4.2) into the higher order schemes (3.1) and (3.2a), and canceling out common terms we obtain the system $\mathbf{U}^{n+1} = \mathcal{A}\mathbf{U}^n$, where the amplification matrix $\mathcal{A}$ is

$$\begin{bmatrix}
1 & \sigma_y & 0 & -\sigma_x & 0\\
-\sigma_y^*\theta^+ & \tilde{\theta} - q_y\theta^+ & \theta^+ - \theta^- & \sigma_y^*\sigma_x\theta^+ & 0\\
-\sigma_y^* & -q_y & 1 & \sigma_y^*\,\sigma_x & 0\\
\sigma_x^*\theta^+ & \sigma_x^*\sigma_y\theta^+ & 0 & \tilde{\theta} - q_x\theta^+ & \theta^+ - \theta^-\\
\sigma_x^* & \sigma_x^*\sigma_y & 0 & -q_x & 1
\end{bmatrix}.$$

In the above, we have used the following simplifying notation

$$\tilde{\theta} := \frac{2 - h_\tau \epsilon_q}{2 + h_\tau \epsilon_q}, \quad \theta^+ := \frac{2 + h_\tau}{2 + h_\tau \epsilon_q}, \quad \theta^- := \frac{2 - h_\tau}{2 + h_\tau \epsilon_q}.$$

For $w = x, y$, we define the parameter $\eta_w := (c_\infty \Delta t)/\Delta w$, where the parameter $c_\infty^2 := 1/(\epsilon_0 \mu_0 \epsilon_\infty) = c_0^2/\epsilon_\infty$. The speed of light in vacuum is donoted by c_0, and c_∞ is the maximum speed of light in the Debye medium. We note that each parameter, η_x and η_y, is a Courant (stability) number. Other parameters are defined as $h_\tau := \Delta t/\tau$ and $\epsilon_q := \epsilon_s/\epsilon_\infty$ with $\epsilon_s > \epsilon_\infty$ (i.e., $\epsilon_q > 1$) and $\tau > 0$. In the above, the parameter $q_w, w = x, y$ is defined to be $q_w := \sigma_w \sigma_w^*$ [**3**] with

$$\sigma_w := -\eta_w \Delta w \mathcal{F}\left(\mathcal{D}_{\Delta w}^{(2M)}\right) \tag{4.4}$$

and $\sigma_w^* = -\sigma_w$ is the complex conjugate of σ_w. We utilize the description of the symbol of the discrete operator $\mathcal{F}\left(\mathcal{D}_{\Delta w}^{(2M)}\right)$ given in (4.1) to evaluate in terms of k_w and Δw only.

Now, using the results of the von Neumann stability analysis performed in [**2**], we can generalize the stability analysis of the Yee scheme to the $(2, 2M)$ schemes. From the assumption $\epsilon_s > \epsilon_\infty$, a necessary and sufficient stability condition for the $(2, 2M)$ schemes in (3.1) and (3.2a) is that $q := q_x + q_y \in (0, 4)$, for all wave vector components, k_x and k_y [**2**], i.e.,

$$4\eta_x^2 \left(\sum_{p=1}^{M} \gamma_{2p-1} \sin^{2p-1}\left(\frac{k_x \Delta x}{2}\right)\right)^2 + 4\eta_y^2 \left(\sum_{p=1}^{M} \gamma_{2p-1} \sin^{2p-1}\left(\frac{k_y \Delta y}{2}\right)\right)^2 < 4, \ \forall k_w, \tag{4.5}$$

$w = x, y$, which implies that

$$\Delta t < \frac{1}{\left(\sum_{p=1}^{M} \frac{[(2p-3)!!]^2}{(2p-1)!}\right) c_\infty \sqrt{\frac{1}{\Delta x^2} + \frac{1}{\Delta y^2}}}. \tag{4.6}$$

In the limiting case (as $M \to \infty$), we may evaluate the infinite series using results in [**3**]. The positivity of the coefficients γ_{2p-1} implies that the following constraint guarantees stability for all orders in two dimensions

$$\Delta t < \frac{2}{\pi c_\infty \sqrt{\frac{1}{\Delta x^2} + \frac{1}{\Delta y^2}}}. \tag{4.7}$$

5. Dispersion Analysis

As mentioned in the introduction, the models for dispersive media have actual physical dispersion which needs to be modeled correctly. In this section we construct the numerical dispersion relations for the $(2, 2M)$ schemes presented in (3.1)-(3.2). We plot the phase error using representative values for all the parameters of the model.

A plane wave solution of the continuous Maxwell-Debye model (2.1)-(2.2) gives us the following (exact) dispersion relation

$$k_{\mathrm{EX}}^{\mathrm{D}}(\omega) = \frac{\omega}{c}\sqrt{\epsilon_r^{\mathrm{D}}(\omega)}; \quad \epsilon_r^{\mathrm{D}}(\omega) := \frac{\epsilon_s \lambda - \mathrm{i}\omega\epsilon_\infty}{\lambda - \mathrm{i}\omega}. \tag{5.1}$$

In the above, $\epsilon_r^{\mathrm{D}}(\omega)$ is the relative complex permittivity of the Debye medium, $\lambda := 1/\tau$ and ω is the angular frequency of the plane wave, and $k_{\mathrm{EX}}^{\mathrm{D}}(\omega)$ is the corresponding wave number.

By considering plane wave solutions for all the discrete variables in the $(2, 2M)$ finite difference schemes for the Maxwell-Debye system given in (3.1)-(3.2), we can derive the numerical dispersion relation of this scheme. The numerical dispersion relations of the $(2, 2M)$ schemes (3.1)-(3.2), for the Maxwell-Debye model, are given implicitly by

$$\left|\mathcal{F}\left(\mathcal{D}_{\Delta x}^{(2M)}\right)\right|^2 + \left|\mathcal{F}\left(\mathcal{D}_{\Delta y}^{(2M)}\right)\right|^2 = \frac{\omega_\Delta^2}{c^2}\left(\frac{\epsilon_s\lambda_\Delta - \mathrm{i}\omega_\Delta\epsilon_{\infty,\Delta}}{\lambda_\Delta - \mathrm{i}\omega_\Delta}\right),$$

where the parameters

$$\epsilon_{\infty,\Delta} := \epsilon_\infty; \quad \lambda_\Delta := \lambda\cos(\omega\Delta t/2), \tag{5.2}$$

are discrete representations of the corresponding continuous model parameters. In addition the parameter ω_Δ, which is a discrete representation of the frequency, is defined as

$$\omega_\Delta := \omega\frac{\sin(\omega\Delta t/2)}{\omega\Delta t/2}. \tag{5.3}$$

We define the phase error Φ for a method applied to a particular model to be

$$\Phi = \left|\frac{k_{\mathrm{EX}} - k_{\Delta,M}}{k_{\mathrm{EX}}}\right|, \tag{5.4}$$

where the numerical wave number $k_{\Delta,M}$ is implicitly determined by the corresponding dispersion relation and k_{EX} is the exact wave number. The components of the wave vector are $k_x = \mathbf{k}\cos(\theta)$ and $k_y = \mathbf{k}\sin(\theta)$, where θ is the angle made by the incident plane wave with the horizontal.

To generate the plots below we have assumed the following values of the physical parameters:

$$\epsilon_\infty = 1; \quad \epsilon_s = 78.2; \quad \tau = 8.1 \times 10^{-12} \text{ sec.} \tag{5.5}$$

These are appropriate constants for modeling water and are representative of a large class of Debye type materials [**1**]. In order to resolve all the time scales, the time step is determined by the choice of h_τ via $\Delta t = h_\tau\tau$, and consequently the spatial step $\Delta := \Delta x = \Delta y$ is chosen based on the CFL number $\eta := \eta_x = \eta_y = c_\infty\Delta t/\Delta$.

In the plots of Figure 1 we depict graphs of the phase error Φ, versus frequency ω, for the $(2, 2M)$th order finite difference methods applied to the Maxwell-Debye model in two dimensions, as given in equations (3.1) and (3.2a), for (spatial) orders $2M = 2, 4, 6, 8$ and the limiting $(M = \infty)$ case. The temporal refinement factor, $h_\tau = \Delta t/\tau$, is fixed at 0.1. The plots use values of η set to the maximum stable value for the order, as given in (4.6). We see that increasing θ to $\pi/4$ decreases the phase error, but that the effect is lessened for smaller frequencies and increasing orders. The graphs corresponding to each angle converge to nearly the one dimensional result for the given h_τ and order (it converges nearby and not to the 1D result because the CFL condition for 2D is more restrictive and we are using the highest stable CFL number). With $h_\tau = 0.01$ the graphs corresponding to various angles converge at order 4 (plot not shown). Regardless of angle, for the frequencies of interest (i.e., those near $\omega\tau = 1$), the higher order methods exhibit a gradual improvement over the second order method.

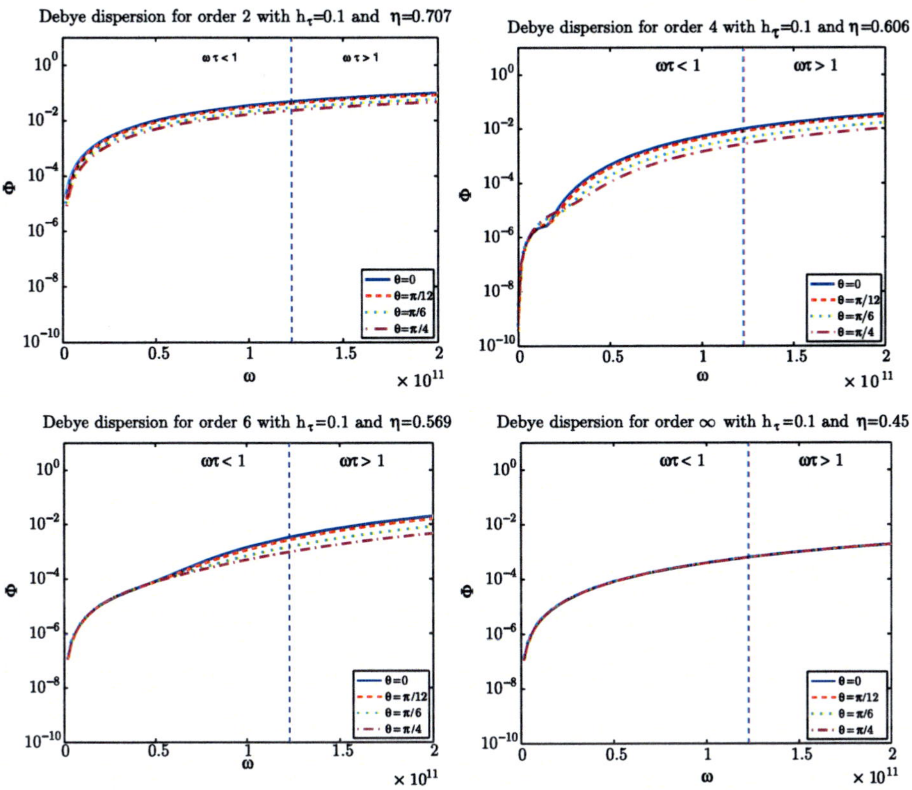

FIGURE 1. Semi-log plots of phase error versus frequency ω for various orders using the largest stable CFL number (η). The vertical line depicts $\omega\tau = 1$. Several angles (θ) are demonstrated.

In Figure 2 we show polar plots of the phase error on a log axis (i.e., the exponent is labeled in the radial dimension). In these plots $\omega = 1/\tau$ and $h_\tau = 0.1$ (left) and $h_\tau = 0.01$ (right). We again see the lessening effect of angle (θ) as the order increases. Further, as in the one dimensional case [**3**], we observe orders of magnitude decreases in Φ as h_τ is decreased. The right plot in particular supports the argument that fourth order is sufficient for minimizing phase error. Any additional decrease in error due to the accuracy of the method is countered by the stability constraint requiring a relatively larger spatial step (Δ).

6. Conclusions

Necessary and sufficient stability conditions, explicitly dependent on the material parameters and the order of the discrete scheme, are presented for Maxwell's equations in Debye dispersive media in two spatial dimensions. Bounds for stability for all orders are obtained by computing the limiting (infinite order) case. We have derived a concise representation of the numerical dispersion relation for each scheme of arbitrary order, which allows an efficient method for predicting the numerical characteristics of a simulation of electromagnetic wave propagation in a dispersive material. This analysis can be extended to higher order discrete schemes

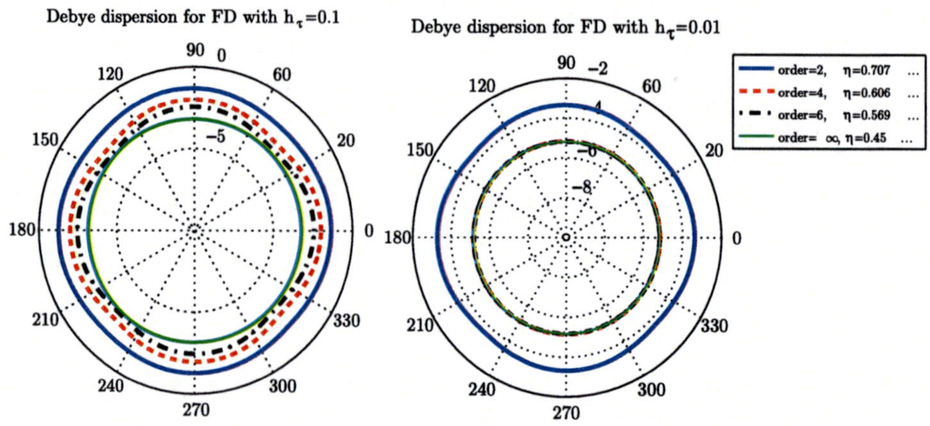

FIGURE 2. Polar semi-log plot of dispersion error versus angle θ for various orders with $h_\tau = 0.1$ (left) and $h_\tau = 0.01$ (right).

for other types of dispersive materials such as Lorentz or Drude media. One can similarly extend the results to three dimensions, however the details are tedious [**2**].

Acknowledgments

This work was partially supported by NSF Computational Mathematics program, grant number DMS-0811223.

References

[1] H. T. Banks, M. W. Buksas, and T. Lin, *Electromagnetic material interrogation using conductive interfaces and acoustic wavefronts*, Frontiers in Applied Mathematics, vol. 21, Society for Industrial and Applied Mathematics (SIAM), Philadelphia, PA, 2000. MR1787981 (2001k:78036)

[2] Brigitte Bidégaray-Fesquet, *Stability of FD-TD schemes for Maxwell-Debye and Maxwell-Lorentz equations*, SIAM J. Numer. Anal. **46** (2008), no. 5, 2551–2566, DOI 10.1137/060671255. MR2421047 (2009c:65183)

[3] V. A. Bokil and N. L. Gibson, *Analysis of Spatial High-Order Finite Difference Methods for Maxwell's Equations in Dispersive Media*, IMA J. Numer. Anal., doi:10.1093/imanum/drr001 **32** (2012), no. 3, 926–956. MR2954735.

[4] R. M. Joseph, S. C. Hagness, and A. Taflove, *Direct time integration of Maxwell's equations in linear dispersive media with absorption for scattering and propagation of femtosecond electromagnetic pulses*, Optics Lett. **16** (1991), no. 18, 1412–1414.

[5] T. Kashiwa, N. Yoshida, and I. Fukai, *A treatment by the finite-difference time domain method of the dispersive characteristics associated with orientational polarization*, IEEE Trans. IEICE **73** (1990), no. 8, 1326–1328.

[6] Peter Monk, *A comparison of three mixed methods for the time-dependent Maxwell's equations*, SIAM J. Sci. Statist. Comput. **13** (1992), no. 5, 1097–1122, DOI 10.1137/0913064. MR1177800 (93j:65184)

[7] P. G. Petropoulos, *Stability and Phase Error Analysis of FDTD in Dispersive Dielectrics*, Antennas and Propagation, IEEE Transactions on **42** (1994), no. 1, 62–69.

[8] K. Yee, *Numerical Solution of Initial Boundary Value Problems Involving Maxwell's Equations in Isotropic Media*, Antennas and Propagation, IEEE Trans. on **14** (1966), no. 3, 302–307.

Department of Mathematics, Oregon State University
E-mail address: bokilv@math.oregonstate.edu

Department of Mathematics, Oregon State University
E-mail address: gibsonn@math.oregonstate.edu

Numerical approximation of a multiscale Leray model for incompressible, viscous flow

Abigail L. Bowers

ABSTRACT. This article develops, analyzes and tests a finite element method for approximating solutions to the Leray-α model, where the filtering is performed on a finer mesh than the momentum/mass system. Typical simulations with Leray-α can oversmooth solutions, and our algorithm is proposed to help avoid this potential drawback. We prove the proposed scheme is unconditionally stable and optimally convergent, with choice of element order and fine mesh width guided by the analysis. A complete and rigorous numerical analysis of the scheme is given, along with experiments that demonstrate its effectiveness.

1. Introduction

The Leray-α regularization of the Navier-Stokes equations (NSE) is given by

$$u_t + \bar{u} \cdot \nabla u + \nabla p - \nu \Delta u = f, \tag{1.1}$$

$$\nabla \cdot u = 0, \tag{1.2}$$

$$-\alpha^2 \Delta \bar{u} + \bar{u} + \nabla \lambda = u, \tag{1.3}$$

$$\nabla \cdot \bar{u} = 0, \tag{1.4}$$

where u denotes the velocity, p the pressure, f a body force, and ν the kinematic viscosity. This model was originally proposed by Leray in 1934 [**10**], although then with the Gaussian filter, but has recently gained the interest of the fluid dynamics community due to its numerous attractive mathematical and physical properties [**3**],[**10**], and its ability to be efficiently computed [**3**]. Some of the more noteworthy results for Leray-α include that it can be fully resolved in 3D with $O(Re^{3/2})$ degrees of freedom (compared to $O(Re^{9/4})$ for the NSE [**3**]), is well-posed for α positive [**10**], conserves energy and 2d enstrophy, and follows the energy cascade of the NSE up to a filtering radius-dependent wavelength [**3**]. Computationally, the Leray-α model admits a natural linearization that is unconditionally stable; such a result is not known for the NS-α regularization model.

A potential problem with the Leray-α model is its tendency to over-regularize solutions, and over-smooth or even remove critical flow structures [**1**],[**6**],[**7**]. This has led to variations of the model, which use different types of regularization. One method is van Cittert deconvolution, however the theory requires very smooth NSE

2000 *Mathematics Subject Classification.* Primary 35Q30, 65M12, 65N30, 76D05, 76F65.

Partially supported by National Science Foundation grants DMS0914478 and DMS1112593.

solutions in order for the model to achieve high accuracy [9]. Another variation is using adaptive nonlinear filtering [2], but this is phenomenological, and currently has no rigorous theory behind it.

We propose herein a method for the Leray-α model, which uses fine mesh filtering to localize more the regularization, which will lessen the over-regularization without significantly increasing the computational resolution. Similar approaches have been found effective in the conservation equation [5], and the idea is related to multiscale deconvolution regularization introduced by Dunca in [4]. In our scheme, the filter system is decoupled from the mass/momentum system, and even though it is solved on a finer mesh, the filter solve can be considered as equivalent to a well conditioned Stokes problem, as we are solving the same matrix at every timestep. Moreover, the filter scheme can use lower order elements than the mass/momentum system solve without any loss of accuracy.

The article is arranged as follows. Section 2 gives notation and presents preliminary results used throughout the paper. Section 3 presents our algorithm, and proves well-posedness and unconditional stability. In section 4, we prove the proposed scheme's solutions converge to NSE solutions, and based on this analysis, give guidance for choice of element and ratio of the fine mesh width to the coarse mesh width. Section 5 presents two numerical experiments which illustrate the effectiveness of our method.

2. Notation and Preliminaries

Let $\Omega \subset \mathbb{R}^d$ (d=2 or 3) be a convex polygon or polyhedra. The $L^2(\Omega)$ norm and inner product will be denoted as $\|\cdot\|$ and $(\cdot,\cdot)$.

For wall-bounded flows, the natural velocity and pressure spaces are defined as

$$X := H_0^1(\Omega) = \{v \in H^1(\Omega), v = 0 \text{ on } \partial\Omega\}, \quad Q := L_0^2(\Omega) = \{q \in L^2(\Omega), \int_\Omega q = 0\}.$$

Denote τ_h and $\tau_{\hat{h}}$ regular conforming meshes, $h > \hat{h}$. We assume conforming velocity-pressure finite element (FE) spaces $X_{\hat{h}}, X_h \subset X$ and $Q_{\hat{h}}, Q_h \subset Q$ corresponding to τ_h and $\tau_{\hat{h}}$ satisfy the LBB condition. The discretely divergence-free subspace of X_h is defined by

$$V_h = \{v_h \in X_h, (\nabla \cdot v_h, q_h) = 0 \ \ \forall q_h \in Q_h\}. \tag{2.1}$$

We use the skew-symmetric trilinear form to ensure stability of the numerical method:

$$b^*(u,v,w) := \frac{1}{2}(u \cdot \nabla v, w) - \frac{1}{2}(u \cdot \nabla w, v). \tag{2.2}$$

DEFINITION 2.1 (Continuous α-filter). For incompressible $v \in L^2(\Omega)$ and $\alpha > 0$ fixed, denote the filtering operation on v by $\overline{v}$, where $(\overline{v}, \lambda) \in (X, Q)$ is the unique solution of

$$-\alpha^2 \Delta \overline{v} + \overline{v} - \nabla \lambda = v \tag{2.3}$$

$$\nabla \cdot \overline{v} = 0 \tag{2.4}$$

DEFINITION 2.2 (Discrete α-Filter). Given incompressible $v \in L^2(\Omega)$, for a given filtering radius $\alpha > 0$, $\overline{v}^{\hat{h}} := F_{\hat{h}} v$ is the unique solution in $X_{\hat{h}}$ of

$$\begin{aligned} \alpha^2(\nabla\overline{v}^{\hat{h}}, \nabla\chi_{\hat{h}}) + (\overline{v}^{\hat{h}}, \chi_{\hat{h}}) - (\lambda_{\hat{h}}, \nabla\cdot\chi_{\hat{h}}) &= (v, \chi_{\hat{h}}) \ \forall\chi_{\hat{h}} \in X_{\hat{h}} \\ (\nabla\cdot\overline{v}^{\hat{h}}, q_{\hat{h}}) &= 0 \ \forall q_{\hat{h}} \in Q_{\hat{h}} \end{aligned} \tag{2.5}$$

LEMMA 2.3. *For incompressible $u \in H_0^k(\Omega)$, the discrete filter error satisfies*

$$\|u - \overline{u}^h\| \le C(\alpha^2 + h^k + \alpha^{-1}h^k|\lambda|_k) \tag{2.6}$$

REMARK 2.4. It can be seen from (2.3) that $\Delta\lambda = -\alpha^2\nabla\cdot\Delta\bar{v}$, so for smooth v in the periodic case, or if derivatives of v vanish at the boundary, we expect $|\lambda| = O(\alpha^2)$, and thus $\|u - \overline{u}^h\| \le C(\alpha^2 + h^k + \alpha h^k)$.

PROOF. Write $u - \overline{u}^h = (u - \overline{u}) + (\overline{u} - \overline{u}^h)$, and consider $(u - \overline{u})$ and $(\overline{u} - \overline{u}^h)$ separately. We then have

$$u - \overline{u} = \alpha^2\Delta\overline{u} + \nabla\lambda.$$

Multiply by $u - \overline{u}$ and integrate over Ω to get

$$\begin{aligned} \|u - \overline{u}\|^2 &= \alpha^2(\nabla\overline{u}, \nabla(u - \overline{u})) - (\lambda, \nabla\cdot(u - \overline{u})) \\ &= \alpha^2(\nabla u, \nabla\overline{u}) - \alpha^2\|\nabla\overline{u}\|^2 \\ &\le \alpha^2(\nabla u, \nabla\overline{u}) \\ &\le \frac{\alpha^4}{2}(\|\nabla u\|^2 + \|\nabla\overline{u}\|^2) \\ &\le \alpha^4\|\nabla u\|^2. \end{aligned} \tag{2.7}$$

Now multiply both the continuous and discrete filters by $v_h \in V_h$, and integrate to obtain

$$\alpha^2(\nabla\overline{u}, \nabla v_h) + (\overline{u}, v_h) = (u, v_h) - (\lambda - q_h, \nabla\cdot\phi_h),$$

$$\alpha^2(\nabla\overline{u}^h, \nabla v_h) + (\overline{u}^h, v_h) = (u, v_h).$$

Subtract to get

$$\alpha^2(\nabla(\overline{u} - \overline{u}^h), \nabla v_h) + (\overline{u} - \overline{u}^h, v_h) + (\lambda - q_h, \nabla\cdot\phi_h) = 0.$$

Decompose the error as, for some $w_h \in V_h$, $\overline{u} - \overline{u}^h = \overline{u} - w_h + w_h - \overline{u}^h = \eta + \phi_h$, and choose $v_h = \phi_h$ yielding

$$\alpha^2\|\nabla\phi_h\|^2 + \|\phi_h\|^2 = \alpha\|\nabla\eta\|^2 + \|\eta\|^2 + (\lambda - q_h, \nabla\cdot\phi_h) \le (\alpha^2+1)\|\nabla\eta\|^2 + C\alpha^{-2}\|\lambda - q_h\|^2.$$

Thus,

$$\begin{aligned} \|\overline{u} - \overline{u}^h\| &\le C(\alpha\|\nabla\eta\| + \alpha^{-1}\|\lambda - q_h\|) \\ &\le C(\alpha h^k|\bar{u}|_{k+1} + h^k|\bar{u}|_{k+1} + \alpha^{-1}h^k|\lambda|_k). \end{aligned} \tag{2.8}$$

□

3. Scheme and Stability

ALGORITHM 3.1. Given a kinematic viscosity $\nu > 0$, an end-time $T > 0$, a time step Δt chosen so that $\Delta t < T = M\Delta t$, $f \in L^\infty(0,T;(L^2(\Omega))^d)$, initial condition $u_0 \in X$, filtering radius $\alpha \leq O(\hat{h})$, find $(u_h^n, p_h^n) \in (X_h, Q_h)$ for $n = 1, 2, ..., M$ satisfying, $\forall(v_h, q_h) \in (X_h, Q_h)$,

$$\frac{1}{\Delta t}(u_h^{n+1} - u_h^n, v_h) + b^*(\overline{\frac{3}{2}u_h^n - \frac{1}{2}u_h^{n-1}}^{\hat{h}}, u_h^{n+1/2}, v_h)$$

$$-(p_h^{n+1}, \nabla \cdot v_h) + \nu(\nabla u_h^{n+1/2}, \nabla v_h) = (f^{n+1/2}, v_h), \tag{3.1}$$

$$(\nabla \cdot u_h^{n+1}, q_h) = 0, \tag{3.2}$$

REMARK 3.2. For the trilinear term, care must be taken to avoid integration errors. However, in our computations, inexact integration of this term did not cause observable errors.

LEMMA 3.3. *Solutions to Algorithm (3.1) exist, are unique, and satisfy*

$$\|u_h^M\|^2 + \Delta t \sum_{n=0}^{M-1} \nu \|\nabla u_h^{n+1/2}\|^2 \leq C(\nu, f, u_h^0, T). \tag{3.3}$$

PROOF. Begin by choosing $v_h = u_h^{n+1/2}$ in (3.1). The trilinear term vanishes, and then the rest follows identically to the NSE case [**8**]. □

4. Convergence

This section proves convergence of the scheme to the NSE solution. The result gives guidance in choice of parameter α to achieve optimal accuracy. For simplicity in stating the following convergence theorem, we summarize here the necessary regularity assumptions for the solution $(u(x,t), p(x,t))$ to the NSE

$$u \in L^\infty(0,T;H^{k+1}(\Omega) \cap H^3(\Omega)), u_{tt} \in L^4(0,T;H^1(\Omega)), \tag{4.1}$$

$$u_{ttt} \in L^2(0,T;L^2(\Omega)), p \in L^\infty(0,T;H^k(\Omega)). \tag{4.2}$$

THEOREM 4.1. *Let $(u(t), p(t))$ be a solution of the NSE satisfying no-slip boundary conditions and the first k derivatives of u vanish at the boundary (or periodic boundaries), and (4.1)-(4.2), with given $f \in L^\infty(0,T;H^{-1}(\Omega))$ and $u_0 \in H^1(\Omega)$. Let (u_h^n, p_h^n), $n = 0, 1, \ldots M$ be the solution of Algorithm (3.1), using (P_k, P_{k-1}) elements on a coarse mesh, and $(P_{\hat{k}}, P_{\hat{k}-1})$ on a fine mesh. Then for Δt small enough, the error in the discrete solution satisfies*

$$\|u(T) - u_h^M\|^2 + \Delta t \sum_{n=0}^{M-1} \nu \|\nabla(u^{n+1/2} - u_h^{n+1/2})\|^2 \leq C(\Delta t^4 + h^{2k} + \alpha^4 + \hat{h}^{2\hat{k}} + \alpha^2 \hat{h}^{2\hat{k}})$$

REMARK 4.2. With $\alpha = O(\hat{h})$, estimate of the theorem reduces to

$$\|u(T) - u_h^M\| + \Delta t \sum_{n=0}^{M-1} \nu \|\nabla(u^{n+1/2} - u_h^{n+1/2})\| \leq C(\Delta t^2 + h^k + \hat{h}^2)$$

Hence, using degree polynomial $k = 3$, $\hat{k} \geq 2$, $\hat{h} = O(h^{3/2})$ is the correct choice for optimal convergence.

PROOF. We begin by multiplying the NSE at $t = t^{n+\frac{1}{2}}$ by $v_h \in V_h$, and define $u^{n+1/2} := \frac{u(t^{n+1})+u(t^n)}{2}$. Then integrate to get, $\forall v_h \in V_h$,

$$(4.3)\quad \left(\frac{u(t^{n+1}) - u(t^n)}{\Delta t}, v_h\right) - (p(t^{n+1/2}), \nabla \cdot v_h) + \nu(\nabla u^{n+1/2}, \nabla v_h) + b^*(\overline{u^{n+1/2}}^{\hat{h}}, u^{n+1/2}, v_h) = (f(t^{n+1/2}), v_h) + G(u, n, v_h),$$

where

$$(4.4)\quad G(u, n, v_h) := \left(\frac{u(t^{n+1}) - u(t^n)}{\Delta t} - u_t(t^{n+1/2}), v_h\right) + \nu(\nabla(u^{n+1/2} - u(t^{n+2/2})), v_h) + b^*(\overline{u^{n+1/2}}^{\hat{h}}, u^{n+1/2}, v_h) - b^*(u(t^{n+1/2}), u(t^{n+1/2}), v_h).$$

Denote $e^n = u(t^n) - u_h^n$ and subtract (4.3) from (3.1) to get the error equation

$$(4.5)\quad \frac{1}{\Delta t}(e^{n+1} - e^n, v_h) + b^*(\overline{\frac{3}{2}u_h^n - \frac{1}{2}u_n^{n-1}}^{\hat{h}}, e^{n+1/2}, v_h) + b^*(\overline{e^{n+1/2}}^{\hat{h}}, u^{n+1/2}, v_h) + \nu(\nabla e^{n+1/2}, \nabla v_h) = (p(t^{n+1/2}), \nabla \cdot v_h) + G(u, n, v_h).$$

Decompose the velocity error as, $e^n = (u(t^n) - P_{V_h}^{L^2}(u(t^n))) + (P_{V_h}^{L^2}(u(t^n)) - u_h^n) =: \eta^n + \phi_h^n$. Expanding and choosing $v_h = \phi_h^{n+1/2}$, and reducing by Cauchy-Schwarz and Young's inequalities gives

$$(4.6)\quad \frac{1}{2\Delta t}(\|\phi_h^{n+1}\|^2 - \|\phi_h^n\|^2) + \frac{\nu}{2}\|\nabla\phi_h^{n+1/2}\|^2 \leq C\nu\|\nabla\eta^{n+1/2}\|^2 + C\nu^{-1}\inf_{q_h\in Q_h}\|p - q_h\|^2 + b^*\left(\overline{\eta^{n+1/2}}^{\hat{h}}, u^{n+1/2}, \phi_h^{n+1/2}\right) + b^*\left(\overline{\phi_h^{n+1/2}}^{\hat{h}}, u^{n+1/2}, \phi_h^{n+1/2}\right) + b^*\left(\overline{u_h^{n+1/2}}^{\hat{h}}, \eta^{n+1/2}, \phi_h^{n+1/2}\right) + G(u, n, \phi_h^{n+1/2}).$$

We now bound the three trilinear terms using Young's inequality:

$$(4.7)\quad b^*\left(\overline{\eta^{n+1/2}}^{\hat{h}}, u^{n+1/2}, \phi_h^{n+1/2}\right) \leq \frac{\nu}{16}\|\nabla\phi_h^{n+1/2}\|^2 + C\nu^{-1}\|\nabla\eta^{n+1/2}\|^2\|\nabla(u(t^{n+1}) + u(t^n))\|^2$$

$$(4.8)\quad b^*\left(\overline{u_h^{n+1/2}}^{\hat{h}}, \eta^{n+1/2}, \phi_h^{n+1/2}\right) \leq \frac{\nu}{16}\|\nabla\phi_h^{n+1/2}\|^2 + C\nu^{-1}\|\nabla\eta^{n+1/2}\|^2\|\nabla u_h^{n+1/2}\|^2$$

$$(4.9)\quad b^*\left(\overline{\phi_h^{n+1/2}}^{\hat{h}}, u^{n+1/2}, \phi_h^{n+1/2}\right)$$

$$= \frac{1}{2}\left(\overline{\phi_h^{n+1/2}}^{\hat{h}} \cdot \nabla u^{n+1/2}, \phi_h^{n+1/2}\right) - \frac{1}{2}\left(\overline{\phi_h^{n+1/2}}^{\hat{h}} \cdot \nabla \phi_h^{n+1/2}, u^{n+1/2}\right)$$

$$\leq C\|\overline{\phi_h^{n+1/2}}^{\hat{h}}\| \|\nabla u^{n+1/2}\|_{L^\infty} \|\phi_h^{n+1/2}\| + C\|\overline{\phi_h^{n+1/2}}^{\hat{h}}\| \|u^{n+1/2}\|_{L^\infty} \|\nabla \phi_h^{n+1/2}\|$$

$$\leq C\|\phi_h^{n+1/2}\|(\|\nabla u^{n+1/2}\|_{L^\infty} + \|u^{n+1/2}\|_{L^\infty})\|\nabla \phi_h^{n+1/2}\|$$

$$\leq \frac{\nu}{16}\|\nabla \phi_h^{n+1/2}\|^2 + C\|\phi_h^{n+1/2}\|^2 \|u^{n+1/2}\|^2_{H^3}$$

Using bounds on the three trilinear terms and the $G(u, n, \phi_h^{n+1/2})$ from **[9]** and **[1]**, and using the assumption of the true solution at the boundary and Remark 2.4

$$(4.10)\quad \frac{1}{2\Delta t}(\|\phi_h^{n+1}\|^2 - \|\phi_h^n\|^2) + \nu\|\nabla \phi_h^{n+1/2}\|^2 \leq C\nu\|\nabla \eta^{n+1/2}\|^2$$

$$+ C\nu^{-1} \inf_{q_h \in Q_h} \|p - q_h\|^2 + C\nu^{-1}\|\nabla u^{n+1/2}\|^2 \|\nabla \eta^{n+1/2}\|^2 + C\nu^{-1}\|\nabla u_h^{n+1/2}\|^2 \|\nabla \eta^{n+1/2}\|^2$$

$$+ C\|u^{n+1/2}\|^2_{H^3}\|\phi_h^{n+1/2}\|^2 + C(\Delta t)^3 \int_{t^n}^{t^{n+1}} \|u_{ttt}\|^2 dt$$

$$+ C\nu(\Delta t)^3 \int_{t^n}^{t^{n+1/2}} \|\nabla u_{tt}\|^2 dt + C\nu^{-1}(\Delta t)^4(\|\nabla u^{n+1/2}\|^4 + \|\nabla u(t^{n+1/2})\|^4)$$

$$+ C\nu^{-1}(\Delta t)^3 \int_{t^n}^{t^{n+1}} \|\nabla u_{tt}\|^4 dt + C\nu^{-1}\|u^{n+1/2}\|_{H^3}(\alpha^4 + \hat{h}^{2\hat{k}})$$

Standard analysis, Lemma 3.3, and Gronwall's inequality finish the proof. □

5. Numerical Experiments

In this section, we present numerical experiments that illustrate the effectiveness of our approach in giving good approximations of incompressible flows. In particular, we show that using fine mesh filtering gives significantly better results than filtering on the same coarse mesh.

5.1. Convergence Rate Verification. Our first numerical experiment is to verify our predicted convergence rates for Algorithm 3.1. The test problem we choose is to compute approximations to the chosen analytical solution

$$u_1(x, y, t) = \sin(2\pi x)(1 + 0.01t),$$
$$u_2(x, y, t) = \cos(2\pi y)(1 + 0.01t),$$
$$p(x, y, t) = x + y,$$

on $\Omega = (0, 1)^2$ and $t \in [0, 1]$. We calculate f from the NSE, the initial condition $u_0 = u(0)$, set $\nu = 1$, and use this data to compute on successfully refined meshes. We solve Algorithm 3.1 using (P_3, P_2) elements for the momentum-mass system, (P_3, P_2) elements for the filter on the same mesh, and (P_2, P_1) for the filter on a finer mesh. The increased order of convergence for the fine mesh filter solution can be seen in Table 1. On the left, where fine mesh filtering is used with appropriate filtering radius and timestep guided by Theorem 4.1, we see $O(h^3)$ convergence as expected, whereas when finemesh filtering is not used (right), we see only $O(h^2)$.

TABLE 1. $L^2(0,T;H^1(\Omega))$ errors and rates using (P_3,P_2) elements for the momentum-mass system, (P_3,P_2) elements for the filter on the same mesh, and (P_2,P_1) for the filter on a finer mesh for experiment 1, with $\hat{h}$ cut by 3 for each mesh refinement (LEFT) and with $\hat{h}=h$ (RIGHT).

h	Δt	$\hat{h}$	$\Vert u-u_h\Vert_{2,1}$	Rate		$h=\hat{h}$	Δt	$\Vert u-u_h\Vert_{2,1}$	Rate
$\frac{1}{2}$	T	$\frac{1}{2}$	4.88e-01	-		$\frac{1}{2}$	T	4.877e-01	-
$\frac{1}{4}$	$\frac{T}{3}$	$\frac{1}{6}$	7.63e-02	2.68		$\frac{1}{4}$	$\frac{T}{3}$	8.66e-02	2.49
$\frac{1}{8}$	$\frac{T}{9}$	$\frac{1}{18}$	1.15e-02	2.73		$\frac{1}{8}$	$\frac{T}{9}$	2.95e-02	1.55
$\frac{1}{16}$	$\frac{T}{27}$	$\frac{1}{54}$	1.46e-03	2.98		$\frac{1}{16}$	$\frac{T}{27}$	9.25e-03	1.67
$\frac{1}{24}$	$\frac{T}{61}$	$\frac{1}{122}$	4.25e-04	3.04		$\frac{1}{24}$	$\frac{T}{61}$	4.33e-03	1.87

FIGURE 1. NSE on a fine mesh. "Truth" solution.

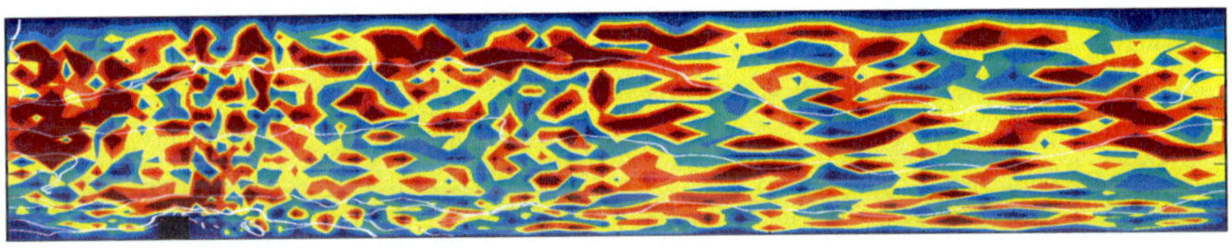

FIGURE 2. NSE on a coarse mesh. Solution incorrect.

FIGURE 3. Leray-α on a coarse mesh with (P_3,P_2) coarse mesh filter.

5.2. 2d Flow Over a Step. Our second experiment is for two-dimensional flow over a forward and backward facing step. This is a benchmark problem used, e.g., in **[9]** and **[1]**. The domain Ω is a 40 x 10 channel with a 1 x 1 step five units into the channel at the bottom. We assume no-slip boundary conditions on the top and bottom boundaries, and parabolic inflow and outflow profiles, given by $(y(10-y)/25,0)^T$. Figure 1 shows the correct $T=40$ direct solution, as streamlines over speed contours, of the NSE using (P_3,P_2) elements on a fine mesh of 21,953 degrees of freedom (dof).

Figure 2 shows the solution of the NSE on a coarse mesh of 8,023 dof. It is clearly not correct, as oscillations have completely destroyed it. Figure 3 shows

FIGURE 4. Finemesh filtering with (P_2, P_1) fine mesh filter.

the coarse mesh Leray-α solution, with filtering performed on the coarse mesh using (P_3, P_2) elements. The solution shows a smooth velocity field, but does not capture eddy detachment behind the step. Figure 4 shows the Leray-α solution, filtering with (P_2, P_1) elements on a uniform refinement of the coarse mesh (which provides 12,969 dof). Here we see a smooth velocity profile, with eddies forming and detaching behind the step, a clear improvement over the solution found with the "usual" Leray-α model.

6. Conclusion

We have developed, analyzed, and tested an efficient algorithm for approximating Navier-Stokes solutions on coarser meshes than a DNS can be performed. We have shown our algorithm to be unconditionally stable, and optimally convergent, using an element choice and filtering radius guided by our analysis. Numerical experiments were provided confirming our analysis and showing our schemes effectiveness.

References

[1] Abigail L. Bowers and Leo G. Rebholz, *Increasing accuracy and efficiency in FE computations of the Leray-deconvolution model*, Numer. Methods Partial Differential Equations **28** (2012), no. 2, 720–736, DOI 10.1002/num.20653. MR2879803

[2] A. Bowers, L. Rebholz, A. Takhirov, and C. Trenchea, *Improved accuracy in regularization models of incompressible flow via adaptive nonlinear filtering*, International Journal for Numerical Methods in Fluids **to appear** (2012).

[3] Alexey Cheskidov, Darryl D. Holm, Eric Olson, and Edriss S. Titi, *On a Leray-α model of turbulence*, Proc. R. Soc. Lond. Ser. A Math. Phys. Eng. Sci. **461** (2005), no. 2055, 629–649, DOI 10.1098/rspa.2004.1373. MR2121928 (2006b:76041)

[4] Argus Adrian Dunca, *A two-level multiscale deconvolution method for the large eddy simulation of turbulent flows*, Math. Models Methods Appl. Sci. **22** (2012), no. 6, 1250001, 30, DOI 10.1142/S0218202512500017. MR2914661

[5] V. J. Ervin and E. W. Jenkins, *Stabilized approximation to degenerate transport equations via filtering*, Appl. Math. Comput. **217** (2011), no. 17, 7282–7294, DOI 10.1016/j.amc.2011.02.020. MR2781120 (2012b:76110)

[6] Bernard J. Geurts and Darryl D. Holm, *Regularization modeling for large-eddy simulation*, Phys. Fluids **15** (2003), no. 1, L13–L16, DOI 10.1063/1.1529180. MR1971789 (2004b:76088)

[7] Bernard J. Geurts and Darryl D. Holm, *Leray and LANS-α modelling of turbulent mixing*, J. Turbul. **7** (2006), Paper 10, 33 pp. (electronic), DOI 10.1080/14685240500501601. MR2207369 (2006i:76047)

[8] William Layton, *Introduction to the numerical analysis of incompressible viscous flows*, Computational Science & Engineering, vol. 6, Society for Industrial and Applied Mathematics (SIAM), Philadelphia, PA, 2008. With a foreword by Max Gunzburger. MR2442411 (2009m:76056)

[9] William Layton, Carolina C. Manica, Monika Neda, and Leo G. Rebholz, *Numerical analysis and computational testing of a high accuracy Leray-deconvolution model of turbulence*, Numer. Methods Partial Differential Equations **24** (2008), no. 2, 555–582, DOI 10.1002/num.20281. MR2382797 (2009b:76064)
[10] Jean Leray, *Sur le mouvement d'un liquide visqueux emplissant l'espace*, Acta Math. **63** (1934), no. 1, 193–248, DOI 10.1007/BF02547354 (French). MR1555394

Department of Mathematical Sciences, Clemson University, Clemson, South Carolina 29634
E-mail address: `abowers@clemson.edu`

A High Order Schema for the Numerical Solution of Ordinary Fractional Differential Equations

Junying Cao and Chuanju Xu

ABSTRACT. We propose in this paper a high order schema to solve a family of ordinary fractional differential equations of order $\alpha, \alpha > 0$. The proposed schema is based on an improved block-by-block approach, which consists in finding 4 unknowns simultaneously at each step block through solving a 4×4 system. The truncation error analysis and numerical examples show that the numerical solution converges to the exact solution with order $5 + \alpha$ for $0 < \alpha \leq 1$, and order 6 for $\alpha > 1$.

1. Introduction

This paper addresses the numerical solution of the ordinary fractional differential equations (OFDEs) of the form $D_t^\alpha u(t) = f(t, u(t))$, where $\alpha > 0$ can be any real number, D_t^α is the fractional differential operator. This type of equations has been subject of many research, see e.g. [**1–8**], and etc. Unlike integer-order differential equations, the theoretical investigation of OFDEs is relatively sparse in the literature; see [**9**] for a review.

For a general right hand side function f, it is difficult to obtain the analytical solution of a fractional differential equation. Thus there is a need to develop numerical methods for OFDEs. Diethelm et al. proposed a predictor-corrector schema and a fractional Adams method [**10, 11**]. Other different approaches can be found in [**12–15**]. However the convergence order of the above mentioned schemes is no more than 3. One of main difficulties in solving fractional differential equations is due to the non local feature of the fractional operator. The fact that all previous solutions have to be saved to compute the current solution would make the storage very expensive if a low-order method is employed.

In this paper, we construct and analyze a fractional high order schema for nonlinear OFDEs. The method follows the idea of the block-by-block approach [**16**], which will be modified to yield a less expensive schema. The main advantage of this schema is that the unknowns are decoupled at each step block, while maintaining the high order convergence. The truncation error analysis shows that convergence order of the schema is $5 + \alpha$ for $0 < \alpha \leq 1$, and 6 for $\alpha > 1$, which is the highest in

---

2010 *Mathematics Subject Classification.* Primary 65L06, 65L12; Secondary 65L20.

Key words and phrases. Fractional differential equation, high order methods.

The research of this paper was partially supported by National NSF of China (Grants 11071203 and 91130002).

the literature to the authors' knowledge. Several numerical tests are conducted to support the theoretical results.

2. High order block-by-block schema

We consider the following ordinary fractional differential equation

$$D_t^\alpha u(t) = f(t, u(t)), \quad 0 \le t \le T, \quad \alpha > 0, \tag{2.1}$$

subject to the initial conditions:

$$u^{(k)}(0) = u_0^{(k)}, \quad k = 0, \dots, n-1, \tag{2.2}$$

where n is the integer such that $n-1 < \alpha \le n$, $u^{(k)}$ means the kth derivative of u, and the real numbers $u_0^{(k)}, k = 0, \dots, n-1$, are assumed to be given. The fractional derivative $D_t^\alpha u$ in (2.1) is defined, in the Caputo sense [**17**], by

$$D_t^\alpha u(t) = \frac{1}{\Gamma(n-\alpha)} \int_0^t (t-\tau)^{n-\alpha-1} u^{(n)}(\tau) d\tau,$$

where $\Gamma(\cdot)$ denotes Gamma function.

It has been proved [**18**] that the initial value problem (2.1)-(2.2) is equivalent to the following Volterra integral equation

$$u(t) = g(t) + \frac{1}{\Gamma(\alpha)} \int_0^t (t-\tau)^{\alpha-1} f(\tau, u(\tau)) d\tau, \quad \text{with } g(t) = \sum_{k=0}^{n-1} u_0^{(k)} \frac{t^k}{k!}. \tag{2.3}$$

We now divide the interval $[0, T]$ into $4N$ equal sub-intervals of size $\Delta t = \frac{T}{4N}$, and denote $t_j = j\Delta t, j = 0, \dots, 4N$. The numerical solution of (2.3) at t_j is denoted by u_j. Set $g_j = g(t_j), f_j = f(t_j, u_j)$.

Using quartic interpolation, $f(t, u(t))$ can be approximated in the interval $[t_0, t_1]$ as

$$f(t, u(t)) \approx \psi_0(t) f_0 + \psi_1(t) f_{1/4} + \psi_2(t) f_{2/4} + \psi_3(t) f_{3/4} + \psi_4(t) f_1, \; t \in [t_0, t_1], \tag{2.4}$$

where $f_{i/4} = f(t_{i/4}, u(t_{i/4})), t_{i/4} = t_0 + i\Delta t/4, i = 1, 2, 3$, and the quartic interpolating function $\psi_j(t)$ is defined by

$$\psi_j(t) = \prod_{i=0, i\ne j}^{4} \frac{t - \hat{t}_i}{\hat{t}_j - \hat{t}_i}, \quad \text{with } \hat{t}_j = t_0 + \frac{j\Delta t}{4}, j = 0, \dots, 4. \tag{2.5}$$

Substituting (2.4) into (2.3) yields

$$u(t_1) \approx g(t_1) + \frac{1}{\Gamma(\alpha)} \int_0^{t_1} (t_1 - \tau)^{\alpha-1} \Big[\sum_{i=0}^{4} \psi_i(\tau) f_{i/4}\Big] d\tau = g(t_1) + \sum_{i=0}^{4} \beta_1^i f_{i/4}, \tag{2.6}$$

where

$$\beta_1^i = \frac{1}{\Gamma(\alpha)} \int_{t_0}^{t_1} (t_1 - \tau)^{\alpha-1} \psi_i(\tau) d\tau, i = 0, \dots, 4,$$

which can be exactly computed. Furthermore, the values $f_{1/4}, f_{2/4}, f_{3/4}$ are approximated by using the interpolations

$$
\begin{aligned}
f_{1/4} &\approx \frac{1155}{2048}f_0 + \frac{385}{512}f_1 - \frac{495}{1024}f_2 + \frac{105}{512}f_3 - \frac{77}{2048}f_4 \doteq \sum_{i=0}^{4} a_i f_i,\\
f_{2/4} &\approx \frac{35}{128}f_0 + \frac{35}{32}f_1 - \frac{35}{64}f_2 + \frac{7}{32}f_3 - \frac{5}{128}f_4 \doteq \sum_{i=0}^{4} b_i f_i,\\
f_{3/4} &\approx \frac{195}{2048}f_0 + \frac{585}{512}f_1 - \frac{351}{1024}f_2 + \frac{65}{512}f_3 - \frac{45}{2048}f_4 \doteq \sum_{i=0}^{4} c_i f_i.
\end{aligned} \tag{2.7}
$$

Combining the approximations (2.6) and (2.7), we obtain the schema for the first step:

$$u_1 = g_1 + \gamma_1^{0,0} f_0 + \gamma_1^{1,0} f_1 + \gamma_1^{2,0} f_2 + \gamma_1^{3,0} f_3 + \gamma_1^{4,0} f_4, \tag{2.8}$$

where

$$
\begin{aligned}
\gamma_1^{0,0} &= \beta_1^0 + a_0\beta_1^1 + b_0\beta_1^2 + c_0\beta_1^3, \quad \gamma_1^{1,0} = a_1\beta_1^1 + b_1\beta_1^2 + c_1\beta_1^3 + \beta_1^4,\\
\gamma_1^{i,0} &= a_i\beta_1^1 + b_i\beta_1^2 + c_i\beta_1^3, i = 2,3,4.
\end{aligned}
$$

Note that computing u_1 through (2.8) requires the values of f (or indirectly, the values of u) at $t_i, i = 0, \ldots, 4$. Now we derive the schema for a few next steps. We use the approximation as follows:

$$f(t,u(t)) \approx \tilde{\psi}_0(t) f_0 + \tilde{\psi}_1(t) f_{1/2} + \tilde{\psi}_2(t) f_1 + \tilde{\psi}_3(t) f_{3/2} + \tilde{\psi}_4(t) f_2, \; t \in [t_0, t_2], \tag{2.9}$$

where

$$\tilde{\psi}_j(t) = \prod_{i=0, i\neq j}^{4} \frac{t - \tilde{t}_i}{\tilde{t}_j - \tilde{t}_i}, \quad \text{with } \tilde{t}_j = t_0 + \frac{j\Delta t}{2}, j = 0,1,2,3,4. \tag{2.10}$$

Substituting (2.9) into (2.3) leads to the approximation:

$$u(t_2) \approx g(t_2) + \frac{1}{\Gamma(\alpha)} \int_{t_0}^{t_2} (t_2 - \tau)^{\alpha-1} \Big[\sum_{i=0}^{4} \tilde{\psi}_i(\tau) f_{i/2} \Big] d\tau = g(t_2) + \sum_{i=0}^{4} \beta_2^i f_{i/2},$$

where

$$\beta_2^i = \frac{1}{\Gamma(\alpha)} \int_{t_0}^{t_2} (t_2 - \tau)^{\alpha-1} \tilde{\psi}_i(\tau) d\tau, i = 0,1,2,3,4.$$

Then we approximate $f_{1/2}$ by using the second formula of (2.7), and $f_{3/2}$ by

$$f_{3/2} \approx -\frac{5}{128}f_0 + \frac{15}{32}f_1 + \frac{45}{64}f_2 - \frac{5}{32}f_3 + \frac{3}{128}f_4 \doteq \sum_{i=0}^{4} d_i f_i.$$

Thus we obtain the schema for the second step as follows:

$$u_2 = g_2 + \gamma_2^{0,0} f_0 + \gamma_2^{1,0} f_1 + \gamma_2^{2,0} f_2 + \gamma_2^{3,0} f_3 + \gamma_2^{4,0} f_4, \tag{2.11}$$

where

$$
\begin{aligned}
\gamma_2^{0,0} &= \beta_2^0 + b_0\beta_2^1 + d_0\beta_2^3, \quad \gamma_2^{1,0} = b_1\beta_2^1 + \beta_2^2 + d_1\beta_2^3,\\
\gamma_2^{2,0} &= b_2\beta_2^1 + d_2\beta_2^3 + \beta_2^4, \quad \gamma_2^{3,0} = b_3\beta_2^1 + d_3\beta_2^3, \quad \gamma_2^{4,0} = b_4\beta_2^1 + d_4\beta_2^3.
\end{aligned}
$$

Similarly, by using the interpolation

$$f_{9/4} \approx \frac{35}{2048}f_0 - \frac{63}{512}f_1 + \frac{945}{1024}f_2 + \frac{105}{512}f_3 - \frac{45}{2048}f_4 \doteq \sum_{i=0}^{4} h_i f_i,$$

we arrive at the schema for the third and fourth steps as follows:

$$u_i = g_i + \gamma_i^{0,0} f_0 + \gamma_i^{1,0} f_1 + \gamma_i^{2,0} f_2 + \gamma_i^{3,0} f_3 + \gamma_i^{4,0} f_4,\ i = 3, 4, \tag{2.12}$$

where

$$\begin{aligned}
\gamma_3^{0,0} &= \beta_3^0 + c_0\beta_3^1 + d_0\beta_3^2 + h_0\beta_3^3,\ \gamma_3^{i,0} = c_i\beta_3^1 + d_i\beta_3^2 + h_i\beta_3^3, i = 1, 2,\\
\gamma_3^{3,0} &= c_3\beta_3^1 + d_3\beta_3^2 + h_3\beta_3^3 + \beta_3^4,\ \ \gamma_3^{4,0} = c_4\beta_3^1 + d_4\beta_3^2 + h_4\beta_3^3,\\
\beta_3^j &= \frac{1}{\Gamma(\alpha)}\int_{t_0}^{t_3}(t_3-\tau)^{\alpha-1}\prod_{i=0,i\neq j}^{4}\frac{\tau-\bar{t}_i}{\bar{t}_j-\bar{t}_i}d\tau,\ \bar{t}_j = t_0 + \frac{3j\Delta t}{4},\ j = 0, \ldots, 4,\\
\gamma_4^{j,0} &= \frac{1}{\Gamma(\alpha)}\int_{t_0}^{t_4}(t_4-\tau)^{\alpha-1}\prod_{i=0,i\neq j}^{4}\frac{\tau-t_i}{t_j-t_i}d\tau,\ j = 0, \ldots, 4.
\end{aligned}$$

We see that (2.8), (2.11), and (2.12) form the coupled system for the first four step solutions u_1, u_2, u_3, and u_4.

Now we construct the schema for general solutions $u(t_{4m+1}), u(t_{4m+2}), u(t_{4m+3})$, and $u(t_{4m+4}), m \geq 1$. By following the same lines as for the first four steps, we make use of the approximation:

$$\begin{aligned}
u(t_{4m+1}) \ &= \ g(t_{4m+1}) + \frac{1}{\Gamma(\alpha)}\Big[\int_0^{t_1}(t_{4m+1}-\tau)^{\alpha-1}f(\tau,u(\tau))d\tau\\
&\quad + \sum_{k=1}^{m}\int_{t_{4k-3}}^{t_{4k+1}}(t_{4m+1}-\tau)^{\alpha-1}f(\tau,u(\tau))d\tau\Big]\\
&\approx \ g(t_{4m+1}) + \frac{1}{\Gamma(\alpha)}\int_0^{t_1}(t_{4m+1}-\tau)^{\alpha-1}\Big[\sum_{i=0}^{4}\psi_i(\tau)f_{i/4}\Big]d\tau\\
&\quad + \frac{1}{\Gamma(\alpha)}\sum_{k=1}^{m}\int_{t_{4k-3}}^{t_{4k+1}}(t_{4m+1}-\tau)^{\alpha-1}\Big[\sum_{i=0}^{4}\varphi_i^{4k-3}(\tau)f_{4k-3+i}\Big]d\tau,
\end{aligned} \tag{2.13}$$

where $\psi_i(t)$ are defined in (2.5), and $\varphi_i^j(t), i = 0, \ldots, 4$, are quartic Lagrange polynomials associated with the points $t_j, t_{j+1}, t_{j+2}, t_{j+3}, t_{j+4}$. This leads to the following schema:

$$u_{4m+1} = g_{4m+1} + \sum_{i=0}^{4}\gamma_{4m+1}^{i,0}f_i + \sum_{k=1}^{m}\sum_{j=0}^{4}\gamma_{4m+1}^{j,k}f_{4k-3+j}, \tag{2.14}$$

where

$$\begin{aligned}
\gamma_{4m+1}^{0,0} &= \beta_{4m+1}^0 + a_0\beta_{4m+1}^1 + b_0\beta_{4m+1}^2 + c_0\beta_{4m+1}^3,\\
\gamma_{4m+1}^{1,0} &= a_1\beta_{4m+1}^1 + b_1\beta_{4m+1}^2 + c_1\beta_{4m+1}^3 + \beta_{4m+1}^4,\\
\gamma_{4m+1}^{i,0} &= a_i\beta_{4m+1}^1 + b_i\beta_{4m+1}^2 + c_i\beta_{4m+1}^3, i = 2, 3, 4,\\
\beta_{4m+1}^i &= \frac{1}{\Gamma(\alpha)}\int_{t_0}^{t_1}(t_{4m+1}-\tau)^{\alpha-1}\psi_i(\tau)d\tau, i = 0, \ldots, 4,\\
\gamma_{4m+1}^{i,k} &= \frac{1}{\Gamma(\alpha)}\int_{t_{4k-3}}^{t_{4k+1}}(t_{4m+1}-\tau)^{\alpha-1}\varphi_i^{4k-3}(\tau)d\tau, i = 0, \ldots, 4; k = 1, \ldots, m.
\end{aligned}$$

To compute u_{4m+2}, we use the approximation:

$$\begin{aligned} u(t_{4m+2}) &\approx g(t_{4m+2}) + \frac{1}{\Gamma(\alpha)}\int_0^{t_2}(t_{4m+2}-\tau)^{\alpha-1}\Big[\sum_{i=0}^{4}\tilde{\psi}_i(\tau)f_{i/2}\Big]d\tau \\ &+\frac{1}{\Gamma(\alpha)}\sum_{k=1}^{m}\int_{t_{4k-2}}^{t_{4k+2}}(t_{4m+2}-\tau)^{\alpha-1}\Big[\sum_{i=0}^{4}\varphi_i^{4k-2}(\tau)f_{4k-2+i}\Big]d\tau, \end{aligned}$$

which results in the schema:

$$u_{4m+2} = g_{4m+2} + \sum_{i=0}^{4}\gamma_{4m+2}^{i,0}f_i + \sum_{k=1}^{m}\sum_{j=0}^{4}\gamma_{4m+2}^{j,k}f_{4k-2+j}, \tag{2.15}$$

where

$$\begin{aligned} \gamma_{4m+2}^{0,0} &= \beta_{4m+2}^{0} + b_0\beta_{4m+2}^{1} + d_0\beta_{4m+2}^{3},\ \gamma_{4m+2}^{1,0} = b_1\beta_{4m+2}^{1} + \beta_{4m+2}^{2} + d_1\beta_{4m+2}^{3}, \\ \gamma_{4m+2}^{2,0} &= b_2\beta_{4m+2}^{1} + d_2\beta_{4m+2}^{3} + \beta_{4m+2}^{4},\ \gamma_{4m+2}^{i,0} = b_i\beta_{4m+2}^{1} + d_i\beta_{4m+2}^{3}, i = 3,4, \\ \gamma_{4m+2}^{i,k} &= \frac{1}{\Gamma(\alpha)}\int_{t_{4k-2}}^{t_{4k+2}}(t_{4m+2}-\tau)^{\alpha-1}\varphi_i^{4k-2}(\tau)d\tau, i=0,\dots,4; k=1,\dots,m. \\ \beta_{4m+2}^{i} &= \frac{1}{\Gamma(\alpha)}\int_{t_0}^{t_2}(t_{4m+2}-\tau)^{\alpha-1}\tilde{\psi}_i(\tau)d\tau, i=0,\dots,4. \end{aligned}$$

Similarly, we have the following two equations:

$$u_{4m+l} = g_{4m+l} + \sum_{i=0}^{4}\gamma_{4m+l}^{i,0}f_i + \sum_{k=1}^{m}\sum_{j=0}^{4}\gamma_{4m+l}^{j,k}f_{4k-4+l+j}, l = 3,4, \tag{2.16}$$

where

$$\begin{aligned} \gamma_{4m+3}^{0,0} &= \beta_{4m+3}^{0} + c_0\beta_{4m+3}^{1} + d_0\beta_{4m+3}^{2} + h_0\beta_{4m+3}^{3}, \\ \gamma_{4m+3}^{i,0} &= c_i\beta_{4m+3}^{1} + d_i\beta_{4m+3}^{2} + h_i\beta_{4m+3}^{3}, i=1,2, \\ \gamma_{4m+3}^{3,0} &= c_3\beta_{4m+3}^{1} + d_3\beta_{4m+3}^{2} + h_3\beta_{4m+3}^{3} + \beta_{4m+3}^{4}, \\ \gamma_{4m+3}^{4,0} &= c_4\beta_{4m+3}^{1} + d_4\beta_{4m+3}^{2} + h_4\beta_{4m+3}^{3}, \\ \beta_{4m+3}^{j} &= \frac{1}{\Gamma(\alpha)}\int_{t_0}^{t_3}(t_{4m+3}-\tau)^{\alpha-1}\prod_{i=0,i\neq j}^{4}\frac{\tau-\bar{t}_i}{\bar{t}_j-\bar{t}_i}d\tau,\ \bar{t}_j = t_0 + \frac{3j\Delta t}{4},\ j=0,\dots,4, \\ \gamma_{4m+3}^{i,k} &= \frac{1}{\Gamma(\alpha)}\int_{t_{4k-1}}^{t_{4k+3}}(t_{4m+3}-\tau)^{\alpha-1}\varphi_i^{4k-1}(\tau)d\tau, i=0,\dots,4; k=1,\dots,m, \\ \gamma_{4m+4}^{i,k} &= \frac{1}{\Gamma(\alpha)}\int_{t_{4k}}^{t_{4k+4}}(t_{4m+4}-\tau)^{\alpha-1}\varphi_i^{4k}(\tau)d\tau, i=0,\dots,4; k=0,\dots,m. \end{aligned}$$

The overall schema (2.8), (2.11), (2.12), (2.14), (2.15), and (2.16), can be rewritten under a unified form as follows:

$$\left\{ \begin{aligned} u_l &= g_l + \sum_{i=0}^{4}\gamma_l^{i,0}f_i,\ l=1,2,3,4, \\ u_{4m+l} &= g_{4m+l} + \sum_{i=0}^{4}\gamma_{4m+l}^{i,0}f_i + \sum_{k=1}^{m}\sum_{j=0}^{4}\gamma_{4m+l}^{j,k}f_{4(k-1)+l+j}, \\ &\quad l=1,2,3,4;\ m=1,\dots,N-1. \end{aligned} \right. \tag{2.17}$$

3. Estimates of the truncation errors

In this section, we give an analysis for the local errors. Without loss of generality, we will only derive an error estimate for the error

$$r_{4m+1}(\Delta t) := u(t_{4m+1}) - \bar{u}_{4m+1}, m \geq 1, \tag{3.1}$$

where $\bar{u}_{4m+1}$ is given by

$$\bar{u}_{4m+1} = g_{4m+1} + \sum_{i=0}^{4} \gamma_{4m+1}^{i,0} f(t_i, u(t_i)) + \sum_{k=1}^{m}\sum_{j=0}^{4} \gamma_{4m+1}^{j,k} f(t_{4k-3+j}, u(t_{4k-3+j})). \tag{3.2}$$

First we need a technical lemma.

LEMMA 3.1. *Let*

$$\begin{aligned} s_k \;=\; & 3[(k+1)^{\alpha+1} - k^{\alpha+1}] - \frac{50[(k+1)^{\alpha+2} + k^{\alpha+2}]}{\alpha+2} + \frac{420[(k+1)^{\alpha+3} - k^{\alpha+3}]}{(\alpha+2)(\alpha+3)} \\ & - \frac{1920[(k+1)^{\alpha+4} + k^{\alpha+4}]}{(\alpha+2)(\alpha+3)(\alpha+4)} + \frac{3840[(k+1)^{\alpha+5} - k^{\alpha+5}]}{(\alpha+2)(\alpha+3)(\alpha+4)(\alpha+5)}. \end{aligned}$$

Then for all positive integer m, there exists a constant c, which may depend on α, such that

$$\sum_{k=0}^{m} |s_k| \leq c \quad \text{if } 0 \leq \alpha \leq 1, \tag{3.3}$$

$$\sum_{k=0}^{m} |s_k| \leq c m^{\alpha-1} \quad \text{if } \alpha > 1. \tag{3.4}$$

PROOF. A direct calculation gives, for $k \geq 2$,

$$\begin{aligned} |s_k| = & k^{\alpha+1}\Big|3\Big[\Big(1+1/k\Big)^{\alpha+1} - 1\Big] - \frac{50k}{\alpha+2}\Big[\Big(1+1/k\Big)^{\alpha+2} + 1\Big] \\ & + \frac{420k^2[(1+1/k)^{\alpha+3} - 1]}{(\alpha+2)(\alpha+3)} - \frac{1920k^3[(1+1/k)^{\alpha+4} + 1]}{(\alpha+2)(\alpha+3)(\alpha+4)} \\ & + \frac{3840k^4[(1+1/k)^{\alpha+5} - 1]}{(\alpha+2)(\alpha+3)(\alpha+4)(\alpha+5)}\Big| \\ = & k^{\alpha+1}\Big|(1+\alpha)\alpha \sum_{i=0}^{\infty} \frac{\prod_{j=1}^{i+1}(\alpha-j)}{(i+7)!k^{i+3}}(i+2)(i+1)(3i^2+7i+30)\Big|. \end{aligned}$$

It is obvious that (3.3) holds if $\alpha = 0$ or 1 since $s_k = 0$ for all $k \geq 2$ in these cases. For $0 < \alpha < 1$, $k \geq 2$, we have

$$|s_k| \leq \Big|\frac{(1+\alpha)\alpha(\alpha-1)}{84}\Big| \frac{1}{k^{2-\alpha}}\Big(1 + \frac{1}{k} + \frac{1}{k^2} + \cdots\Big) \leq \frac{\sqrt{3}}{189}\frac{1}{k^{1+(1-\alpha)}}.$$

Thus the series $\sum_{k=2}^{\infty} |s_k|$ converges, which implies that (3.3) holds for any $m > 0$. Now we prove (3.4). For $\alpha > 1$, let

$$z_i := \frac{\prod_{j=1}^{i+1}(\alpha-j)}{(i+7)!}(i+2)(i+1)(3i^2+7i+30),$$

then for $k \geq 2$,

$$|s_k| = \frac{(1+\alpha)\alpha}{k^{2-\alpha}}\Big|\sum_{i=0}^{\infty} \frac{\prod_{j=1}^{i+1}(\alpha-j)}{(i+7)!k^i}(i+2)(i+1)(3i^2+7i+30)\Big| = \frac{(1+\alpha)\alpha}{k^{2-\alpha}}\Big|\sum_{i=0}^{\infty} \frac{z_i}{k^i}\Big|.$$

It is an easy matter to check that $|z_i|$ is uniformly bounded for any i. Therefore $|\sum_{i=0}^{\infty} \frac{z_i}{k^i}| \le c$ with c independent of k. Consequently, we have

$$\sum_{k=2}^{m} |s_k| \le c \sum_{k=2}^{m} \frac{1}{k^{2-\alpha}} \le c \int_1^m k^{\alpha-2} dk = \frac{c}{\alpha - 1}(m^{\alpha-1} - 1) \le cm^{\alpha-1}.$$

This completes the proof. □

THEOREM 3.2. *For all $m \ge 1$, it holds*

$$\begin{aligned} |r_{4m+1}(\Delta t)| &\le C\Delta t^{5+\alpha} \quad \text{if } 0 < \alpha \le 1, \\ |r_{4m+1}(\Delta t)| &\le C\Delta t^{6} \quad \text{if } \alpha > 1. \end{aligned}$$

PROOF. From (2.14) and (3.2), we have

$$\begin{aligned} r_{4m+1}(\Delta t) &= u(t_{4m+1}) - \Big\{ g_{4m+1} + f(t_0, u(t_0))\beta^0_{4m+1} \\ &+ \sum_{i=0}^{4} a_i f(t_i, u(t_i))\beta^1_{4m+1} + \sum_{i=0}^{4} b_i f(t_i, u(t_i))\beta^2_{4m+1} + \sum_{i=0}^{4} c_i f(t_i, u(t_i))\beta^3_{4m+1} \\ &+ f(t_1, u(t_1))\beta^4_{4m+1} + \sum_{k=1}^{m}\sum_{j=0}^{4} \gamma^{j,k}_{4m+1} f(t_{4k-3+j}, u(t_{4k-3+j}))\Big\} \\ &= \frac{1}{\Gamma(\alpha)} \int_0^{t_1} (t_{4m+1} - \tau)^{\alpha-1}[R_0(\tau) + R_1(\tau)\psi_1(\tau) + R_2(\tau)\psi_2(\tau) + R_3(\tau)\psi_3(\tau)]d\tau \\ &+ \frac{1}{\Gamma(\alpha)} \sum_{k=1}^{m} \int_{t_{4k-3}}^{t_{4k+1}} (t_{4m+1} - \tau)^{\alpha-1} R_{4k+1}(\tau)d\tau, \end{aligned}$$

where

$$R_0(\tau) = f(\tau, u(\tau)) - \sum_{i=0}^{4} f(t_{\frac{i}{4}}, u(t_{\frac{i}{4}}))\psi_i(\tau), R_1(\tau) = f(t_{\frac{1}{4}}, u(t_{\frac{1}{4}})) - \sum_{i=0}^{4} a_i f(t_i, u(t_i)),$$

$$R_2(\tau) = f(t_{\frac{1}{2}}, u(t_{\frac{1}{2}})) - \sum_{i=0}^{4} b_i f(t_i, u(t_i)), R_3(\tau) = f(t_{\frac{3}{4}}, u(t_{\frac{3}{4}})) - \sum_{i=0}^{4} c_i f(t_i, u(t_i)),$$

$$R_{4k+1}(\tau) = f(\tau, u(\tau)) - \sum_{i=0}^{4} \varphi_i^{4k-3}(\tau) f(t_{4k-3+i}, u(t_{4k-3+i})).$$

By using Taylor theorem, there exist $\xi_1(\tau), \xi_2(\tau), \xi_3(\tau), \xi_4(\tau) \in [t_0, t_1]$, and $\xi_k(\tau) \in [t_{4k-3}, t_{4k+1}]$, such that

$$\begin{aligned} r_{4m+1}(\Delta t) &= \frac{1}{\Gamma(\alpha)} \int_{t_0}^{t_1} (t_{4m+1} - \tau)^{\alpha-1} \frac{f^{(5)}(\xi_1(\tau), u(\xi_1(\tau)))}{5!} \prod_{i=0}^{4} (\tau - t_{i/4}) \\ &- \frac{1}{\Gamma(\alpha)} \int_{t_0}^{t_1} \frac{231}{8192} (t_{4m+1} - \tau)^{\alpha-1} \Delta t^5 f^{(5)}(\xi_2(\tau), u(\xi_2(\tau)))\psi_1(\tau)d\tau \\ &- \frac{1}{\Gamma(\alpha)} \int_{t_0}^{t_1} \frac{7}{256} (t_{4m+1} - \tau)^{\alpha-1} \Delta t^5 f^{(5)}(\xi_3(\tau), u(\xi_3(\tau)))\psi_2(\tau)d\tau \qquad (3.5) \\ &- \frac{1}{\Gamma(\alpha)} \int_{t_0}^{t_1} \frac{117}{8192} (t_{4m+1} - \tau)^{\alpha-1} \Delta t^5 f^{(5)}(\xi_4(\tau), u(\xi_4(\tau)))\psi_3(\tau)d\tau \\ &+ \frac{1}{\Gamma(\alpha)} \sum_{k=1}^{m} \int_{t_{4k-3}}^{t_{4k+1}} (t_{4m+1} - \tau)^{\alpha-1} \frac{f^{(5)}(\xi_k(\tau), u(\xi_k(\tau)))}{5!} \prod_{i=0}^{4} (\tau - t_{4k-3+i})d\tau. \end{aligned}$$

It remains to estimate the right hand side term by term. For the first term, denoted by $R1$, we have

$$
\begin{aligned}
|R1| &\leq \frac{1}{\Gamma(\alpha)} \int_{t_0}^{t_1} (t_{4m+1}-\tau)^{\alpha-1} \Big| \frac{f^{(5)}(\xi_1(\tau), u(\xi_1(\tau)))}{5!} \Big| \prod_{i=0}^{4} (\tau - t_{i/4}) \\
&\leq \frac{M_1 \Delta t^5}{\Gamma(\alpha)} \int_{t_0}^{t_1} (t_{4m+1}-\tau)^{\alpha-1} d\tau \leq \frac{M_1 \Delta t^6}{\Gamma(\alpha)} (t_{4m+1}-\tau^*)^{\alpha-1},
\end{aligned} \tag{3.6}
$$

where $M_1 = \sup_{\tau\in[0,T]} |f^{(5)}(\tau, u(\tau))|, \tau^* \in (t_0, t_1)$.

Similarly, the second, third, and fourth terms can be bounded by $\frac{M_1\Delta t^6}{\Gamma(\alpha)}(t_{4m+1} - \tau^*)^{\alpha-1}$. For the last term, denoted by $R5$, it holds

$$
\begin{aligned}
|R5| \leq \frac{1}{\Gamma(\alpha)} \sum_{k=1}^{m} \Big\{ \Big| \int_{t_{4k-3}}^{t_{4k+1}} (t_{4m+1}-\tau)^{\alpha-1} \frac{f^{(5)}(\widetilde{\xi}_k, u(\widetilde{\xi}_k))}{5!} \prod_{i=0}^{4} (\tau - t_{4k-3+i}) d\tau \Big| \\
+ \Big| \int_{t_{4k-3}}^{t_{4k+1}} (t_{4m+1}-\tau)^{\alpha-1} \frac{f^{(5)}(\xi_k(\tau), u(\xi_k(\tau))) - f^{(5)}(\widetilde{\xi}_k, u(\widetilde{\xi}_k))}{5!} \prod_{i=0}^{4} (\tau - t_{4k-3+i}) d\tau \Big| \Big\},
\end{aligned} \tag{3.7}
$$

where $\widetilde{\xi}_k = t_{4k-1}$. The first term in the right hand side of (3.7) satisfies

$$
\begin{aligned}
& \frac{1}{\Gamma(\alpha)} \sum_{k=1}^{m} \Big| \int_{t_{4k-3}}^{t_{4k+1}} (t_{4m+1}-\tau)^{\alpha-1} \frac{f^{(5)}(\widetilde{\xi}_k, u(\widetilde{\xi}_k))}{5!} \prod_{i=0}^{4} (\tau - t_{4k-3+i}) d\tau \Big| \\
\leq\ & \frac{M_1}{\Gamma(\alpha)} \sum_{k=1}^{m} \Big| \int_{t_{4k-3}}^{t_{4k+1}} (t_{4m+1}-\tau)^{\alpha-1} \prod_{i=0}^{4} (\tau - t_{4k-3+i}) d\tau \Big| \\
=\ & \frac{2^{2\alpha+5}\Delta t^{\alpha+5} M_1}{\Gamma(\alpha+2)} \sum_{k=0}^{m-1} \Big| 3[(k+1)^{\alpha+1} - k^{\alpha+1}] - \frac{50}{\alpha+2}[(k+1)^{\alpha+2} + k^{\alpha+2}] \\
& + \frac{420[(k+1)^{\alpha+3} - k^{\alpha+3}]}{(\alpha+2)(\alpha+3)} - \frac{1920[(k+1)^{\alpha+4} + k^{\alpha+4}]}{(\alpha+2)(\alpha+3)(\alpha+4)} \\
& + \frac{3840[(k+1)^{\alpha+5} - k^{\alpha+5}]}{(\alpha+2)(\alpha+3)(\alpha+4)(\alpha+5)} = \frac{2^{2\alpha+5}\Delta t^{\alpha+5} M_1}{\Gamma(\alpha+2)} \sum_{k=0}^{m-1} |s_k|,
\end{aligned}
$$

where s_k is defined in Lemma 3.1. By using the inequality:

$$
\left| \frac{f^{(5)}(\xi_k(\tau), u(\xi_k(\tau))) - f^{(5)}(\widetilde{\xi}_k, u(\widetilde{\xi}_k))}{5!} \right| \leq M_2 \Delta t, \forall \tau \in [t_{4k-3}, t_{4k+1}],
$$

where $M_2 = \sup_{\tau\in[0,T]} |f^{(6)}(\tau, u(\tau))|$, The second term in the right hand side of (3.7), denoted by $R6$, is bounded by

$$
\begin{aligned}
R6 &\leq \frac{M_2 \Delta t}{\Gamma(\alpha)} \sum_{k=1}^{m} \int_{t_{4k-3}}^{t_{4k+1}} (t_{4m+1}-\tau)^{\alpha-1} \Big| \prod_{i=0}^{4} (\tau - t_{4k-3+i}) \Big| d\tau \\
&\leq \frac{M_2 \Delta t^6}{\Gamma(\alpha)} \sum_{k=1}^{m} \int_{t_{4k-3}}^{t_{4k+1}} (t_{4m+1}-\tau)^{\alpha-1} d\tau \leq \frac{M_2 \Delta t^6}{\alpha\Gamma(\alpha)} (t_{4m+1} - t_1)^{\alpha} \leq \frac{M_2 \Delta t^6}{\alpha\Gamma(\alpha)} T^{\alpha}.
\end{aligned}
$$

Thus for $R5$, we have

$$|R5| \leq \frac{2^{2\alpha+5}\Delta t^{\alpha+5} M_1}{\Gamma(\alpha+2)} \sum_{k=0}^{m-1} |s_k| + \frac{M_2 \Delta t^6}{\alpha \Gamma(\alpha)} T^\alpha.$$

Combining all above estimates together yields

$$|r_{4m+1}(\Delta t)| \leq \frac{6M_1\Delta t^6}{\Gamma(\alpha)}(t_{4m+1} - \tau^*)^{\alpha-1} + \frac{2^{2\alpha+5}\Delta t^{\alpha+5} M_1}{\Gamma(\alpha+2)} \sum_{k=0}^{m-1} |s_k| + \frac{M_2 \Delta t^6}{\alpha \Gamma(\alpha)} T^\alpha.$$

Finally, applying Lemma 3.1 to the above inequality completes the proof. □

4. Numerical results

Although the stability of the schema is not rigorously established, we verify the convergence of the proposed method by means of numerical tests. To this end, we consider the initial value problem (2.1)-(2.2) with

$$f(t, u(t)) = \frac{\Gamma(7+\alpha)}{720} t^6 + t^{12+2\alpha} - u^2(t).$$

It can be verified that the exact solution is $u(t) = t^{6+\alpha}$.

All the results reported below correspond to the numerical solution captured at $T = 1$. In Table 1 and Table 2 we list the maximum errors $\max_i |u(t_i) - u_i|$ as a function of the time step size for several α ranging from 0.2 to 2. Also shown are the corresponding rates. From these two tables, it is observed that for all α smaller than 1, the convergence rate is close to $5 + \alpha$, while the error decay rates are all close to 6 for $\alpha \geq 1$. This is in a good agreement with the theoretical prediction demonstrated in Theorem 3.2.

TABLE 1. Maximum errors and decay rate with $\alpha = 0.2, 0.5$, and 0.8.

Δt	$\alpha = 0.2$	*Rate*	$\alpha = 0.5$	*Rate*	$\alpha = 0.8$	*Rate*
$\frac{1}{20}$	$2.0063E-006$	–	$1.1100E-006$	–	$6.0662E-007$	–
$\frac{1}{40}$	$2.7364E-008$	6.1960	$2.6788E-008$	5.3728	$1.2472E-008$	5.6040
$\frac{1}{80}$	$7.7650E-010$	5.1392	$6.2506E-010$	5.4214	$2.4161E-010$	5.6899
$\frac{1}{160}$	$2.1619E-011$	5.1666	$1.4304E-011$	5.4495	$4.6102E-012$	5.7117
$\frac{1}{320}$	$5.9566E-013$	5.1817	$3.2361E-013$	5.4661	$8.7024E-014$	5.7273
$\frac{1}{640}$	$1.6320E-014$	5.1898	$7.2626E-015$	5.4776	$1.6281E-015$	5.7402

TABLE 2. Maximum errors and decay rate with $\alpha = 1.0, 1.5$, and 2.0.

Δt	$\alpha = 1.0$	*Rate*	$\alpha = 1.5$	*Rate*	$\alpha = 2.0$	*Rate*
$\frac{1}{20}$	$1.5994E-007$	–	$1.7340E-006$	–	$4.0242E-006$	–
$\frac{1}{40}$	$2.2977E-009$	6.1212	$2.5785E-008$	6.0715	$5.7923E-008$	6.1184
$\frac{1}{80}$	$3.4497E-011$	6.0576	$3.9383E-010$	6.0328	$8.5655E-010$	6.0795
$\frac{1}{160}$	$5.3027E-013$	6.0236	$6.1134E-012$	6.0095	$1.2966E-011$	6.0457
$\frac{1}{320}$	$8.2225E-015$	6.0110	$9.5614E-014$	5.9986	$1.9919E-013$	6.0245
$\frac{1}{640}$	$1.2800E-016$	6.0054	$1.4995E-015$	5.9947	$3.0852E-015$	6.0126

To summarize, we have proposed a very high order schema for the fractional differential equations. The schema was constructed by using a block-by-block approach, which consists in solving a 4×4 system at each step block. We derived some estimates for the truncation errors, showing that the proposed schema is of order $5 + \alpha$ for $0 < \alpha \leq 1$, and order 6 for $\alpha > 1$. The carried out numerical example confirmed the theoretical prediction.

References

[1] B. Mandelhrot. Some noises with 1/f spectrum, a bridge between direct current and white noise. *Information Theory, IEEE Transactions on*, 13(2):289–298, 1967.

[2] A. Oustaloup. *La dérivation non entière: théorie, synthèse et applications.* Hermes, Paris, 1995.

[3] H. Sun, A. Abdelwahab, and B. Onaral. Linear approximation of transfer function with a pole of fractional power. *Automatic Control, IEEE Transactions on*, 29(5):441–444, 1984.

[4] R. C. Koeller, *Applications of fractional calculus to the theory of viscoelasticity*, Trans. ASME J. Appl. Mech. **51** (1984), no. 2, 299–307, DOI 10.1115/1.3167616. MR747787 (85k:73045)

[5] F. Mainardi. Fractional diffusive waves in viscoelastic solids. *Nonlinear Waves in Solids*, 93–97, 1995.

[6] F. Amblard, A.C. Maggs, B. Yurke, A.N. Pargellis, and S. Leibler. Subdiffusion and anomalous local viscoelasticity in actin networks. *Phys. Rev. Lett.*, 77(21):4470–4473, 1996.

[7] R.L. Bagley and R.A. Calico. Fractional order state equations for the control of viscoelastically damped structures. *Journal of Guidance, Control, and Dynamics*, 14(2):304–311, 1991.

[8] R. Klages, G. Radons, and I.M. Sokolov. *Anomalous transport: foundations and applications.* Wiley-VCH, Weinheim, 2008.

[9] Kai Diethelm, *The analysis of fractional differential equations*, Lecture Notes in Mathematics, vol. 2004, Springer-Verlag, Berlin, 2010. An application-oriented exposition using differential operators of Caputo type. MR2680847 (2011j:34005)

[10] Kai Diethelm, Neville J. Ford, and Alan D. Freed, *A predictor-corrector approach for the numerical solution of fractional differential equations*, Nonlinear Dynam. **29** (2002), no. 1-4, 3–22, DOI 10.1023/A:1016592219341. Fractional order calculus and its applications. MR1926466 (2003h:65085)

[11] Kai Diethelm, Neville J. Ford, and Alan D. Freed, *Detailed error analysis for a fractional Adams method*, Numer. Algorithms **36** (2004), no. 1, 31–52, DOI 10.1023/B:NUMA.0000027736.85078.be. MR2063572 (2005d:65110)

[12] T. A. M. Langlands and B. I. Henry, *The accuracy and stability of an implicit solution method for the fractional diffusion equation*, J. Comput. Phys. **205** (2005), no. 2, 719–736, DOI 10.1016/j.jcp.2004.11.025. MR2135000 (2005m:65174)

[13] Zhi-zhong Sun and Xiaonan Wu, *A fully discrete difference scheme for a diffusion-wave system*, Appl. Numer. Math. **56** (2006), no. 2, 193–209, DOI 10.1016/j.apnum.2005.03.003. MR2200938 (2006k:65227)

[14] Yumin Lin and Chuanju Xu, *Finite difference/spectral approximations for the time-fractional diffusion equation*, J. Comput. Phys. **225** (2007), no. 2, 1533–1552, DOI 10.1016/j.jcp.2007.02.001. MR2349193 (2008i:65200)
[15] R. Lin and F. Liu, *Fractional high order methods for the nonlinear fractional ordinary differential equation*, Nonlinear Anal. **66** (2007), no. 4, 856–869, DOI 10.1016/j.na.2005.12.027. MR2288436 (2008c:26007)
[16] P. Kumar and O.P. Agrawal. An approximate method for numerical solution of fractional differential equations. *Signal Process.*, 86(10):2602–2610, 2006.
[17] Igor Podlubny, *Fractional differential equations*, Mathematics in Science and Engineering, vol. 198, Academic Press Inc., San Diego, CA, 1999. An introduction to fractional derivatives, fractional differential equations, to methods of their solution and some of their applications. MR1658022 (99m:26009)
[18] K. Diethelm and N.J Ford, *Analysis of Fractional Differential Equations*, J. Math. Anal. Appl. **265** (2002), no. 2, 229–248.

School of Mathematical Sciences, Xiamen University, 361005 Xiamen; College of Science, Guizhou Minzu University, 550025 Guiyang, China

Corresponding author, School of Mathematical Sciences, Xiamen University, 361005 Xiamen, China

E-mail address: cjxu@xmu.edu.cn

3-D patterns in bacterial biofilms

Chen Chen and Qi Wang

ABSTRACT. We use the multiphase hydrodynamic model and the 3-D simulation tool developed for bacterial biofilms to study 3-D pattern formation in bacterial colonies. In this model, bacterial motility is effectively treated using an intermolecular mixing potential. The cell-substrate adhesion is modeled using an adhesion energy. Ring/shell like patterns are predicted in the bacterial dispersal problem with an initially concentrated and yet isolated bacterial colony. Patterns due to cell-adhesion to the specifically treated substrate is studied using the adhesion energy effectively. This study demonstrates the capability of the 3-D model and its accompanying simulation tool for pattern formation in live microorganisms.

1. Introduction

Bacteria are often found in biofilms or other bacterial colonies, which can grow into spectacular patterns visible under the microscope [12] [8]. In the laboratory, bacteria such as E. coli, can form regular geometric patterns like simple concentric rings and some other well-ordered forms [3] [2] [15]. The pattern formation results from bacterial collective behavior, such as chemotactic aggregation, anisotropic bacterial motion, competition for food or changes in phenotypes according to density etc. A diverse range of phenomenological models have been established [15] [1] [7], and many involve a large number of parameters due to the complexity of the detailed mechanism that the models are specifically designed for. Most of the pattern formation models are based on reaction-diffusion mechanisms [11], in which diffusion is often assumed as a proxy for the cell self-propelled or other subtle chemotaxical motion. However, bacterial system is such a complex biological system and often consists of their own system specificity determined by the microscopic details.

Recently, Cates et al. proposed a model aiming to subsume the complex, system-specific microscopic details into a small number of effective parameters that control the macroscopic behavior. Their model considered two mechanisms on scales intermediate between the microscopic dynamics of bacteria and the macroscopic scale of the patterns: bacterial mobility on density and logistic growth. Mathematically, their model can be recast into a Cahn-Hilliard equation with a logistic growth reactive term and an extended interaction potential.

1991 *Mathematics Subject Classification*. Primary 35B36, 92C15; Secondary 97M10, 00A71.

Key words and phrases. Biofilms, pattern formation, phase field modeling.

The work of Qi Wang's research is partially supported by AFOSR grant FA9550-12-1-0178, NSF grants DMS-0819051, DMS-0908330, and a SC EPSCOR award.

We have developed a suite of hydrodynamic models along with the accompanying 3-D simulation tools based on them for biofilms and interacting surrounding fluid matrix systems [**13, 14, 16, 17, 9, 10**]. The mixture system is derived using a kinetic theory for biomass-solvent mixtures, where the biomass consisting of the bacteria and EPS polymers is modeled effectively as a polymer solution and the solvent and nutrient substances are collectively treated as a viscous fluid. In some regimes, the kinetic theory can be coarse-grained to a self-consistent continuum theory. In the biofilm model, the biomass-solvent interaction is modeled as polymer-solvent molecular interaction via a Flory-Huggins mixing free energy. One feature of our biofilm models is that the interaction potential can be readily calibrated against biofilm growth experiments.

In our biofilm models, we "average out" the specific microscopic aspects and focus on the coarse-grained dynamics. Experimental evidence indicates that the pattern formation and the disruption of the pattern soon after it is formed may be regarded as a phase separation in the complex fluid mixture. We then treat this pattern formation as phase separation in polymer blends which our models are fully capable of describing. What we have to do next is to play around with the free energy density functional to best approximate the patterns. By using a double well free energy density function and adding a logistic growth model, we found our model captures the pattern formation experiments well qualitatively. The other model "parameter" we can adjust in our models is the nutrient boundary condition and the potentially important adhesion energy between the biomass and the solid substrate that on which the biofilm resides. In our models, we can also couple the phase-separation driven pattern formation with hydrodynamics of the complex fluid system, which is rarely done in pattern formation studies to our knowledge. In the following, we first give a brief overview of a hydrodynamic biofilm model and then discuss some interesting pattern formation numerical experiments obtained using a 3-D simulation tool based on the biofilm model.

2. Mathematical model

In [**13, 14, 16, 17, 9, 10, 6**], we developed a suite of hydrodynamic models for multiphase biological fluids, including biofilms residing in an ambient or buffer fluid. In the slow time scale, much longer than the EPS (extracellular polymeric substance) relaxation time scale in biofilms, the multiphase material mixture behaves like a viscous fluid. In this paper, we are interested in the slow growth time regime. We adopt the viscous constitutive law. In addition, we reinterpret the model by associating the loose (unattached from bacteria) EPS with the solvent and name it the effective solvent and the bacteria together with the attached EPS as the bacteria or biomass. In this model, we use an average velocity $\mathbf{v}$ to track the collective motion of the biofilm ensemble. The motion of each individual component relative to the average motion is described by the excessive velocity due to the intermixing of the two distinctive material components. The cell motility is then effectively modeled by the entropy and mixing of the two effective components in proxy.

We summarize the governing system of equations in dimensionless form below, which consists of the continuity equation, linear momentum balance equation, the phase transport equation, and the chemical (nutrient) transport equation. The phase transport equation is a singular Cahn-Hilliard equation for phase variable ϕ, which is defined as the volume fraction of the biomass in the biofilm,

$$\frac{\partial\phi}{\partial t} = \nabla\cdot(\lambda\phi\nabla(-\Gamma_1\nabla^2\phi + \Gamma_2\frac{\delta\hat{f}}{\delta\phi})) + \mu\phi(1-\frac{\phi}{\phi_{max}})\frac{c}{K_c+c}, \tag{2.1}$$

where $\Lambda = \frac{\lambda\rho_0}{t_0}$, λ is the mobility parameter, t_0 is a characteristic time, ρ_0 is a characteristic density, $\Gamma_1 = \frac{\gamma_1 k_B T t_0^2}{\rho_0 h^4}$, h is a characteristic length determined by the geometry of the fluid system considered, $\Gamma_2 = \frac{\gamma_2 k_B T t_0^2}{\rho_0 h^2}$, γ_1 and γ_2 measure the strength of the entropic and bulk intermolecular free energy density $\hat{f}$, respectively, k_B is the Boltzmann constant, T is the absolute temperature, $\mu := \mu t_0$ is the maximum growth rate in dimensionless form, ϕ_{max} is the maximum carrying capacity for the biomass volume fraction, $c := \frac{c}{c_0}$ is the nutrient concentration normalized to a characteristic concentration c_0, and $K_c := \frac{K_c}{c_0}$ is the half saturation constant in the monod model for the nutrient fueled biomass growth. This has a logistic growth term dictated by the existing nutrient density. The logistic growth is necessary to confine the volume fraction ϕ to within its physical range. The phase transport equation is then coupled with the velocity and the nutrient transport equation listed below.

$$\begin{aligned}
&\nabla\cdot\mathbf{v} = 0,\\
&\rho\frac{d\mathbf{v}}{dt} = \nabla\cdot(\phi\tau_n + (1-\phi)\tau_s) - [\nabla p + \Gamma_1\nabla\cdot(\nabla\phi\nabla\phi)],\\
&\tau_n = \frac{2}{Re_1}\mathbf{D}, \tau_s = \frac{2}{Re_2}\mathbf{D},\\
&\frac{\partial(1-\phi)c}{\partial t} + \nabla\cdot((1-\phi)c\mathbf{v}) = \nabla\cdot(D_s(1-\phi)\nabla c) - A\phi\frac{c}{K_s+c},
\end{aligned} \tag{2.2}$$

where $\rho = \frac{\rho_1\phi+\rho_2(1-\phi)}{\rho_0}$, $t := \frac{t}{t_0}, \mathbf{x} := \frac{\mathbf{x}}{h}, \mathbf{v} := \frac{\mathbf{v}t_0}{h}, p := \frac{pt_0^2}{\rho_0 h^2}$, ρ_1 is the mass of the bacteria and ρ_2 is the mass of the effective solvent consisting of the EPS and the solvent, $Re_1 = \frac{\rho_0 h^2}{\eta_1 t_0}, Re_2 = \frac{\rho_0 h^2}{\eta_2 t_0}$ are the Reynolds number for the biomass and the effective solvent, respectively, η_1 is the viscosity of the biomass and η_2 is the viscosity of the effective solvent, $D_s :- \frac{D_s t_0}{h^2}$ is the diffusion coefficient, $A := At_0$ is the maximum consumption rate of the nutrient, and $K_s := \frac{K_s}{c_0}$ is the half saturation constant in the monod model used for nutrient consumption.

We adopt the Flory-Huggins mixing free energy density for the binary mixture

$$\hat{f}(\phi) = \frac{\phi}{N_A}ln\phi + \frac{1-\phi}{N_B}ln(1-\phi) + \chi\phi(1-\phi), \tag{2.3}$$

where N_A and N_B are generalized polymer indices related to the molecular weight of the biomass and the effective solvent, respectively, and χ is a mixing parameter. By rewriting the diffusive part $\nabla\cdot(\phi\nabla\frac{\delta\hat{f}}{\delta\phi})$ as $\nabla\cdot\phi(\frac{\delta^2\hat{f}}{\delta\phi^2})\nabla\phi$, we arrive at the same format as in the dissipative model developed by Cates et al. [5].

In the paper by Cates et al., the governing equation for the measure of bacterial concentration can be recast into a Cahn-Hilliard equation with a bulk free energy density given by $\hat{f} = -\frac{R}{8\Phi^2}e^{-2\Phi u}2\Phi u + \tilde{C}_1 u + \tilde{C}_2$, where u is the measure for the concentration of bacteria in the model. For appropriate choices of C_1 and C_2, the energy function shows one local minimum and a decaying behavior for large u. In this model, the competition of the free energy density and the logistic growth leads to the formation of various bacterial patterns.

TABLE 1. Parameter values used in the simulation

Symbol	Parameter	value	Unit
T	Temperature	303	Kelvin
γ_1	Distortional energy	8×10^6	$kgm^{-1}s^{-2}$
γ_2	Mixing free energy	3×10^{17}	kgm^2s^{-2}
χ	Flory-Huggins parameter	0.60	
λ	Mobility parameter	1×10^{-9}	$kg^{-1}m^3s$
N_A	Molecular weight index for biomass	4	
N_B	Molecular weight index for solvent	4	
μ	Max. Production rate	6×10^{-4}	$kgm^{-3}s^{-1}$
K_c	Half saturation constant	1×10^{-4}	kgm^{-3}
K_d	Half saturation constant	1×10^{-4}	kgm^{-3}
A	Max. Consumption rate	0.1	$kgm^{-3}s^{-1}$
D_s	Substrate diffusion coefficient	2.3×10^{-11}	m^2s^{-1}
η_1	Dynamic viscosity of biomass	4.3×10^2	$kgm^{-1}s^{-1}$
η_2	Dynamic viscosity of solvent	1.002×10^{-3}	$kgm^{-1}s^{-1}$
ρ_1	biomass density	1×10^3	kgm^{-3}
ρ_2	Solvent density	1×10^3	kgm^{-3}
ϕ_{max}	Maximum carrying capacity	0.5	
c_0	Characteristic substrate concentration	1×10^{-3}	kgm^{-3}
h	Characteristic length scale	1×10^{-3}	m
t_0	Characteristic time scale	1×10^3	s
L	x-direction size of Ω	0.001	m

Qualitatively, the Cahn-Hilliard model can create the same mechanism like the Cates' model does by selecting appropriate model parameters. For example, if we choose $N_A = N_B = 4$ and $\chi = 0.6$, then, the free energy forms a double well potential. With this setting, we expect phase separation to occur should the carrying capacity for the volume fraction in the logistic growth is sufficiently high, since the system can lower its free energy by separating into two phases. In addition, the logistic growth for bacteria can control the growth of bacteria to a preset value for its volume fraction.

Table 1 lists the range of the dimensional parameter values used in our simulations presented next. Some specific parameter values differ from this table in certain simulations are listed in the corresponding figure captions.

3. Numerical Results and Discussion

We use the 3-D simulation tool developed based on the model to conduct a series numerical investigations for the pattern formation problem in bacterial colonies. The boundary conditions we adopt are the no-slip boundary condition for the velocity and no-flux for the volume fraction and the nutrient concentration. A benchmark numerical experiment of pattern formation is the formation of rings after a single small bacterial droplet is deposited on a solid substrate or in the middle of a solvent matrix. Cates et al. used their model to simulate the ring pattern formation in 2 space dimensions, in which the concentric rings break into dots due to the weak cell-cell interaction [**5**].

We first conduct a 3-D numerical simulation of a concentrated bacterial droplet deposited in the middle of the 3-D domain with model parameter $\chi = 0.6$ and the other parameters fixed at the values specified in this paper (see Fig. 1). We decouple the hydrodynamics together with the effect of the nutrient in this simulation and focus only on the phase equation in this simulation. In this simulation, we notice that the bacteria first spread rapidly, forming a round nearly uniformly distributed bacterial colony with the highest density at the center. This background volume fraction of the bacteria increases logistically until the onset of instability ensues via the generic phase separation mechanism in the Flory-Huggins energy. With the spherical symmetry, the instability causes concentric shells of high bacterial volume fraction in one shell and low in adjacent shells. This spatially heterogeneous structure is very stable in time due to the strong cell-cell interaction in the intermolecular potential. By reducing χ to $\chi = 0.55$, on the other hand, effectively corresponding to a weaker interaction, we observe that shells initially form but then get rapidly destabilized (see Fig. 1). The outer shells eventually break into spherical drops. These simulations show the sensitivity of predicted heterogeneous patterns to the interaction strength in the model. We remark that the excessive structures in the corners in the simulation are the artifact of the physical boundary conditions we imposed.

In the 3-D simulation of bacterial biomass dynamics, the break-up of shells into spherical drops shows a tendency to begin from the outside of the nested shell structure simultaneously. Of course, the setting for the simulation is distinct from the 2-D setting in Cates et al.'s work. With respect to this simulation, we do find some related experimental findings in the experiment on pattern formation with S. typhimurium by Budrene and Berg [**3**, **4**].

Next, we simulate the ring/shell formation with hydrodynamics fully coupled. I.e., we include the convection term into our phase transport equation model, and solve a coupled Navier-stokes equation for the average velocity. By plotting the 2-D slice at $z = 0.5$ and $t = 200$ with biomass velocity and effective solvent velocity respectively, we can explore the hydrodynamics of the biomass-effective solvent interaction and pattern formation. See Figure 2. The direction of v_n points from rings with lower concentration to rings with higher concentration, which means the biomass migrates to high volume fraction region. Corresponding to this biomass migration, the effective solvent is expelled outward from these high concentrated rings to reduce its concentration. We note that the cellular mobility mechanism that decreases the volume fraction is balanced by the logistic growth that regulates the bacteria population. We note that the coupled hydrodynamics does not change the volume fraction patterns much when compared to the decoupled case alluded to earlier.

4. Biomass-substrate interaction: adhesion

Biomass and the substrate on which they survive interact extensively from the cellular level all the way to the macroscopic level. To account for this important interaction, in addition to controls of the nutrient transport and boundary conditions, we also include an active biomass-substrate interaction via the adhesion

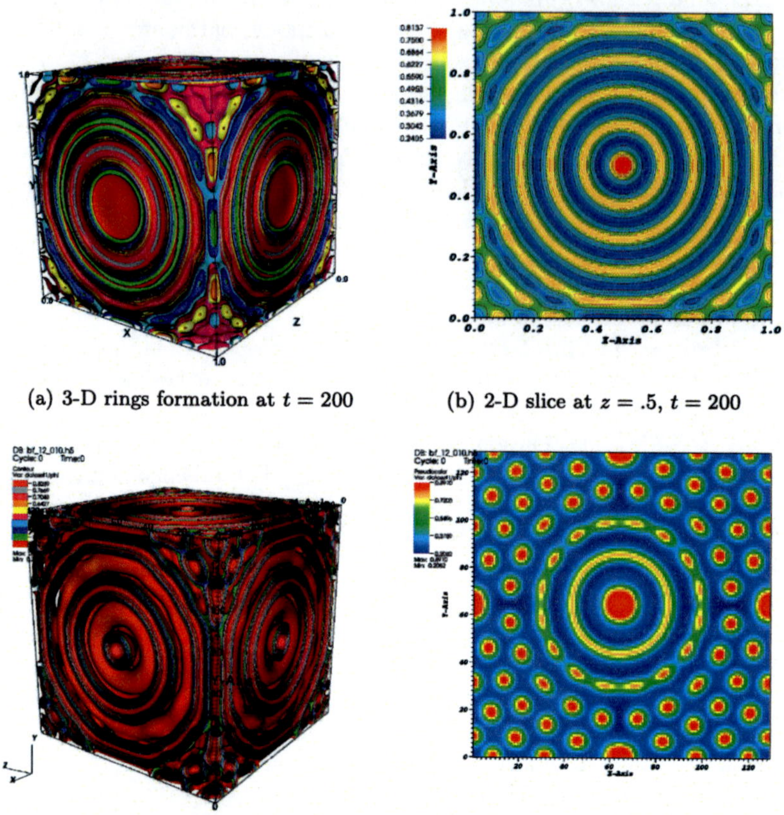

(a) 3-D rings formation at $t = 200$

(b) 2-D slice at $z = .5$, $t = 200$

(c) 3-D rings formation at $t = 200$, $\chi = 0.55$

(d) 2-D slice at $z = .5$, $t = 200$, $\chi = 0.55$

FIGURE 1. 3D-shell (2D ring) formation in our model with interaction strength $\chi = 0.6$ in (a-b) and $\chi = 0.55$ in (c-d), respectively. 3-D and 2-D slices are shown at $t = 200$. In (a-b), the shells are concentric and stable. In (c-d), the rings/shells break into dots/spheres at the outer range.

potential:

$$f_{a2} = -w_0 \phi(2 - \phi) e^{\frac{-y}{\delta}}, \tag{4.1}$$

where w_0 measures the strength of the adhesion potential and δ does the range away from the substrate. We assume the potential is nondimensionalized in this paper. We deposit the initial biomass drop at the bottom of our computational domain, trying to mimic the actual experimental situation. Also, we enable the nutrient transportation in the model.

In the adhesion energy functional, the strength w_0 can be made a function of (x, y, z) to reflect the potential hetergeneous property of the biomass-substrate interaction. In this way, the property of the substrate can be tailored to attract or expel bacterial to form the so-called designer's patterns of bacteria. We next present a numerical simulation where bacteria form an "NSF" pattern as we let the value

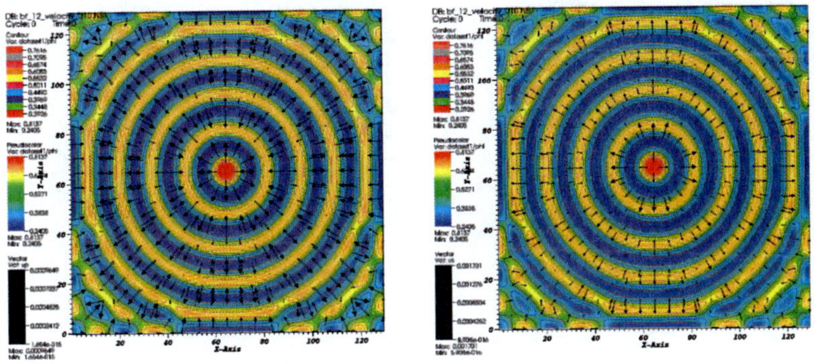

(a) 2-D slice with polymer network velocity at $z = .5$, $t = 200$

(b) 2-D slice with effective solvent velocity at $z = .5$, $t = 200$

FIGURE 2. 2-D slice of ring/shell formation with componentwise velocity v_n and v_s superimposed, respectively.

of w_0 be larger at the spatial region occupied by "NSF". As a result, the adhesion energy is smaller at the spot "NSF" occupied, leading to higher concentration of bacteria adhered to the "NSF" region. The visual effect is a pattern in the form of the word "NSF". See Figure 3. Another effect the adhesion interaction brings to the biomass is to collectively attract more bacteria to the substrate than to the open ambient fluid region, thereby reducing the bacterial concentration in the ambient.

We further investigate the benchmarking concentric ring pattern formation phenomena with this newly added adhesion energy. In this case, we are able to simulate the 3-D bacterial pattern formation in a more realistic situation. We initiate a bacteria drop at the center of the substratum, and investigate the concentric rings formed on the surface of the substrate. This is what ones observe in experiments as a bacteria drop is deposited in the middle of a plate with Succinic acid or other chemicals. With the adhesion energy added, the bacteria drop grows into concentric rings pattern on the substrate. The numerically simulated ring formation using our model is shown in Figures 4 and 5 with $\chi = 0.6$. When we set $\chi = 0.55$, which corresponds to the weaker phase separation regime, we can see that the concentric rings form and then break into droplets with the cascade starting from the middle. The following three figures depict the result of the simulation at different viewing angles (Figure 6,7,8).

5. Conclusion

In this paper, we use the mathematical model developed for bacteria and buffer/ambient fluid mixture to study bacterial pattern formation due to cell motility and cell-substrate interaction. A 3-D numerical simulation tool based on a 2nd order projection method for solving fluid flows is used to predict several ring-like patterns and designer's patterns due to substrate design. It provides a hydrodynamic simulation tool for ones to study complex bacterial systems in various conditions, where cell-substrate interaction is important.

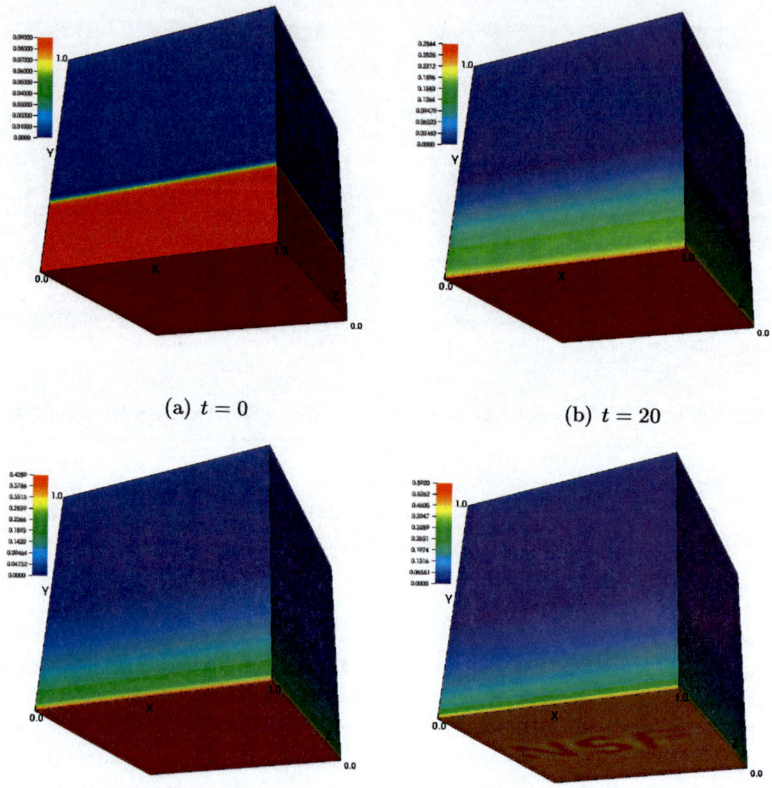

(a) $t = 0$ (b) $t = 20$

(c) $t = 40$ (d) $t = 60$

FIGURE 3. Formation of "NSF" pattern in 3-D at t=0,20,40,60, respectively, where $w_0 = 1500$ on the "NSF" pattern and $w_0 = 1000$ elsewhere, $\delta = 0.01$.

References

[1] Eshel Ben-Jacoba, Inon Cohena, and Herbert Levine. Cooperative self-organization of microorganisms. *Advances in Physics*, 49(4):395–554, 2000.

[2] Michael P. Brenner, Leonid S. Levitov, and Elena O. Budrene. Physical mechanisms for chemotactic pattern formation by bacteria. *Biophysical Journal*, 74(4):1677–1693, 1998.

[3] EO Budrene and HC Berg. Complex patterns formed by motile cells of Escherichia coli. *Nature*, 349:630–633, 1991.

[4] EO Budrene and HC Berg. Dynamics of formation of symmetrical patterns by chemotactic bacteria. *Nature*, 376:49–53, 1995.

[5] M. E. Cates, D. Marenduzzo, I. Pagonabarraga, and J. Tailleur. Arrested phase separation in reproducing bacteria creates a generic route to pattern formation. *PNAS*, 107(26):11715–11720, 2010.

[6] Chen Chen, Mingming Ren, Ashok Srinivansan, and Qi Wang. 3-D Numerical Simulations of Biofilm Flows. *East Asian Journal on Applied Mathematics*, 1(3):197–214, 2011.

[7] Sergei E. Esipov and J. A. Shapiro. Kinetic model of proteus mirabilis swarm colony development. *Advances in Physics*, 36(3):249–268, 1998. MR1608609

[8] Rasika M. Harshey. Bacterial motility on a surface: Many ways to a common goal. *Annual Review of Microbiology*, 57:249–273, 2003.

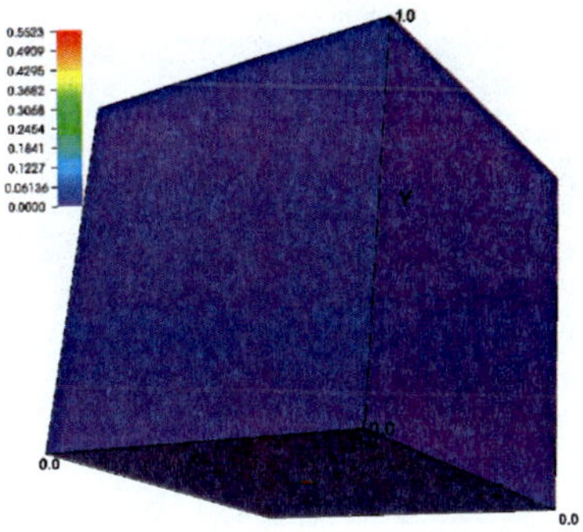

(a) $t = 0$

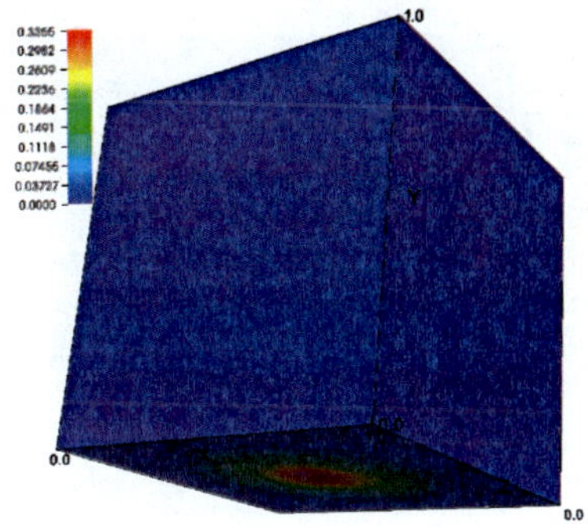

(b) $t = 10$

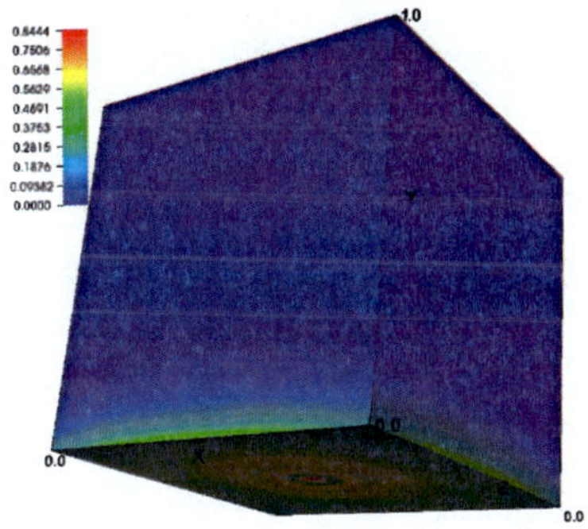

(c) $t = 200$

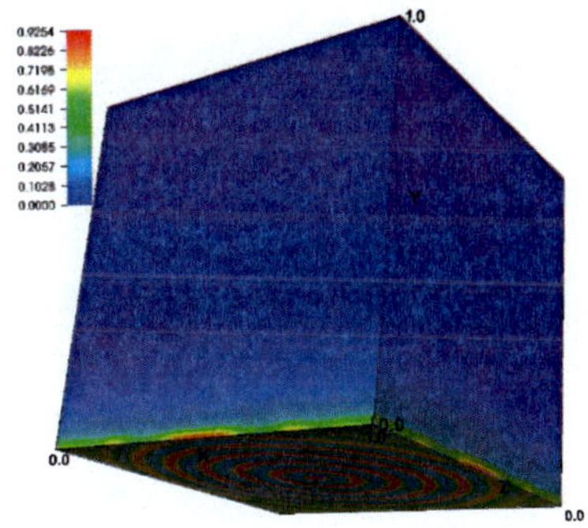

(d) $t = 300$

FIGURE 4. 3-D plot of the concentric rings/donuts. $\chi = 0.6$ and the parameters in the adhesion energy are identical to the ones used in Figure 3. Bacteria grow upward and expand horizontally due to the adhesion energy.

[9] Brandon Lindley, Qi Wang, and Tianyu Zhang. A multicomponent model for biofilmdrug interaction. *Discrete and Continuous Dynamical Systems - Series B*, 15(2):417–456, 2011. MR2754093 (2011k:92009)

[10] Brandon Lindley, Qi Wang, and Tianyu Zhang. Multicomponent hydrodynamic model for heterogeneous biofilms: Two-dimensional numerical simulations of growth and interaction with flows. *Physical Review E*, 85(3), 2012.

[11] J D Murray. *Mathematical biology. II Spatial models and biomedical applications.* New York : Springer, 2003. MR1952568 (2004b:92001)

[12] James A. Shapiro. The significance of bacterial colony patterns. *BioEssays*, 17(7):597–607, 1995.

[13] Qi Wang and Tianyu Zhang. Review of mathematical models for biofilms. *Solid State Communications*, 150(21):1009–1022, 2010.

[14] Qi Wang and Tianyu Zhang. Kinetic theories for biofilms. *DCDS-B*, 17(3):1027–1059, 2012. MR2873126 (2012k:92007)

[15] D.E. Woodward, R. Tyson, M.R. Myerscough, J.D. Murray, E.O. Budrene, and H.C. Berg. Spatio-temporal patterns generated by salmonella typhimurium. *Biophysical Journal*, 68(5):2181–2189, 2009.

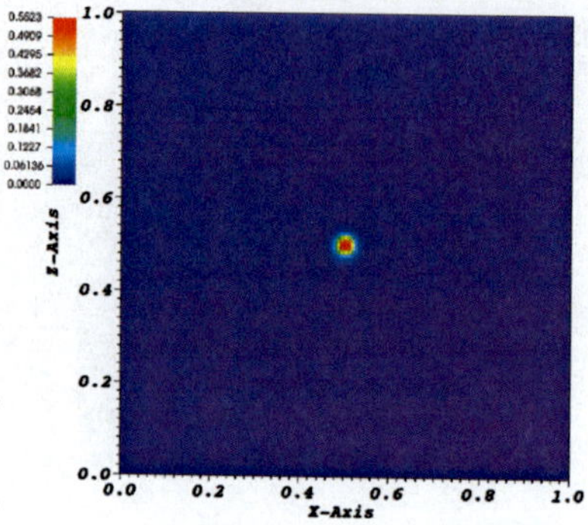

(a) $y = 0.0$, $t = 0$

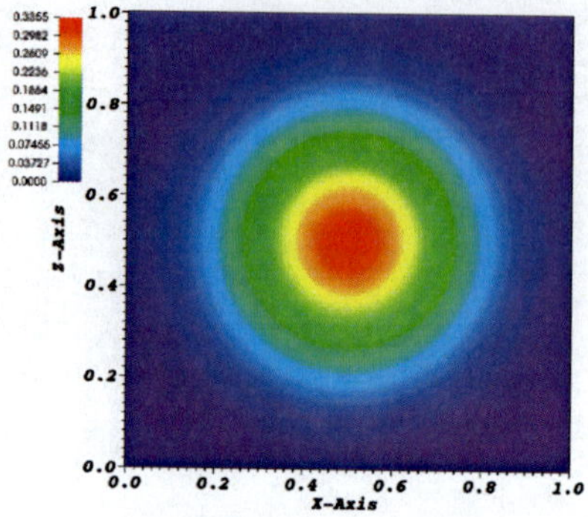

(b) $y = 0.0$, $t = 100$

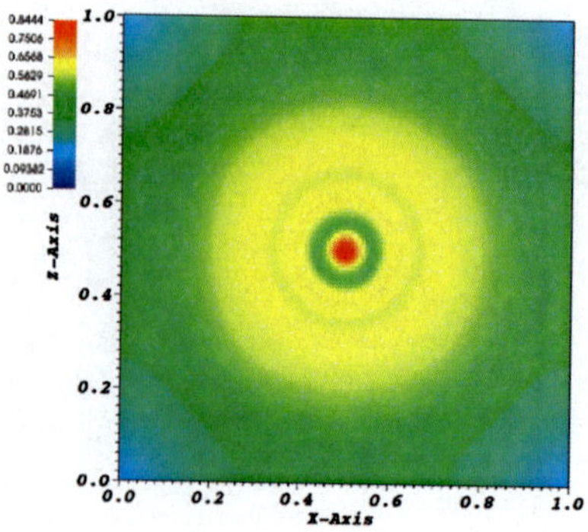

(c) $y = 0.0$, $t = 200$

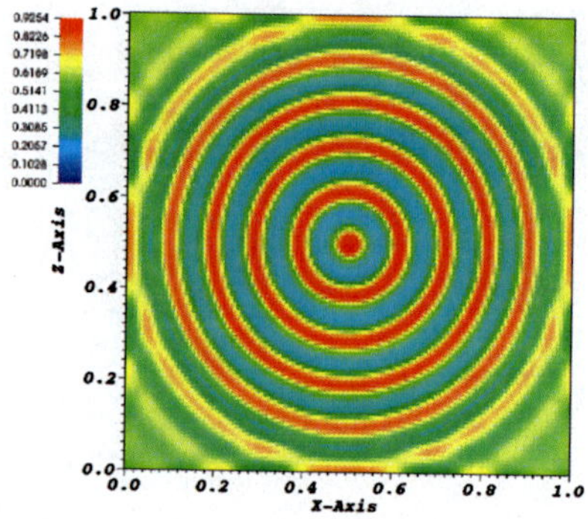

(d) $y = 0.0$, $t = 300$

FIGURE 5. 2-D cross sectional plots of the result in Figure 4 at $y = 0.0$.

[16] T. Y. Zhang, N. Cogan, and Q. Wang. Phase Field Models for Biofilms. I. Theory and 1-D simulations. *Siam Journal on Applied Math*, 69(3):641–669, 2008. MR2471198 (2010a:92072)

[17] T. Y. Zhang, N. Cogan, and Q. Wang. Phase Field Models for Biofilms. II. 2-D Numerical Simulations of Biofilm-Flow Interaction. *Communications in Computational Physics*, 4:72–101, 2008.

DEPARTMENT OF MATHEMATICS, UNIVERSITY OF SOUTH CAROLINA, COLUMBIA, SOUTH CAROLINA 29208

DEPARTMENT OF MATHEMATICS, INTERDISCIPLINARY MATHEMATICS INSTITUTE, AND NANOCENTER AT USC, UNIVERSITY OF SOUTH CAROLINA, COLUMBIA, SC 29208. SCHOOL OF MATHEMATICS, NANKAI UNIVERSITY, TIANJIN, CHINA. BEIJING COMPUTATIONAL SCIENCE RESEARCH CENTER, BEIJING, CHINA

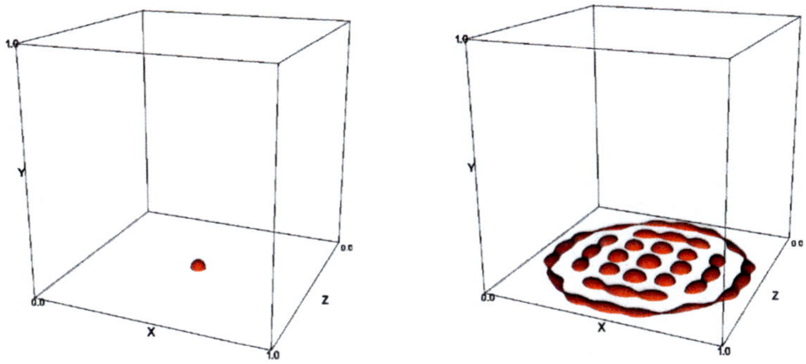

(a) $t = 0$ (b) $t = 400$

FIGURE 6. Ring/donuts formation in 3-D: the contour plot of the concentric rings/donuts on a substrate. $\chi = 0.55$ and the parameters in the adhesion energy are identical to the ones used in Figure 3. We can see that the concentric rings/donuts break into droplets from the middle.

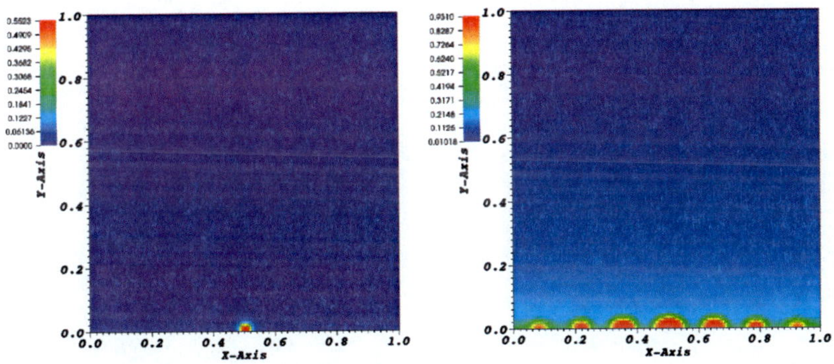

(a) $z = 0.5$, $t = 0$ (b) $z = 0.5$, $t = 400$

FIGURE 7. 2-D cross sectional plots of the result in Figure 6 at $z = 0.5$.

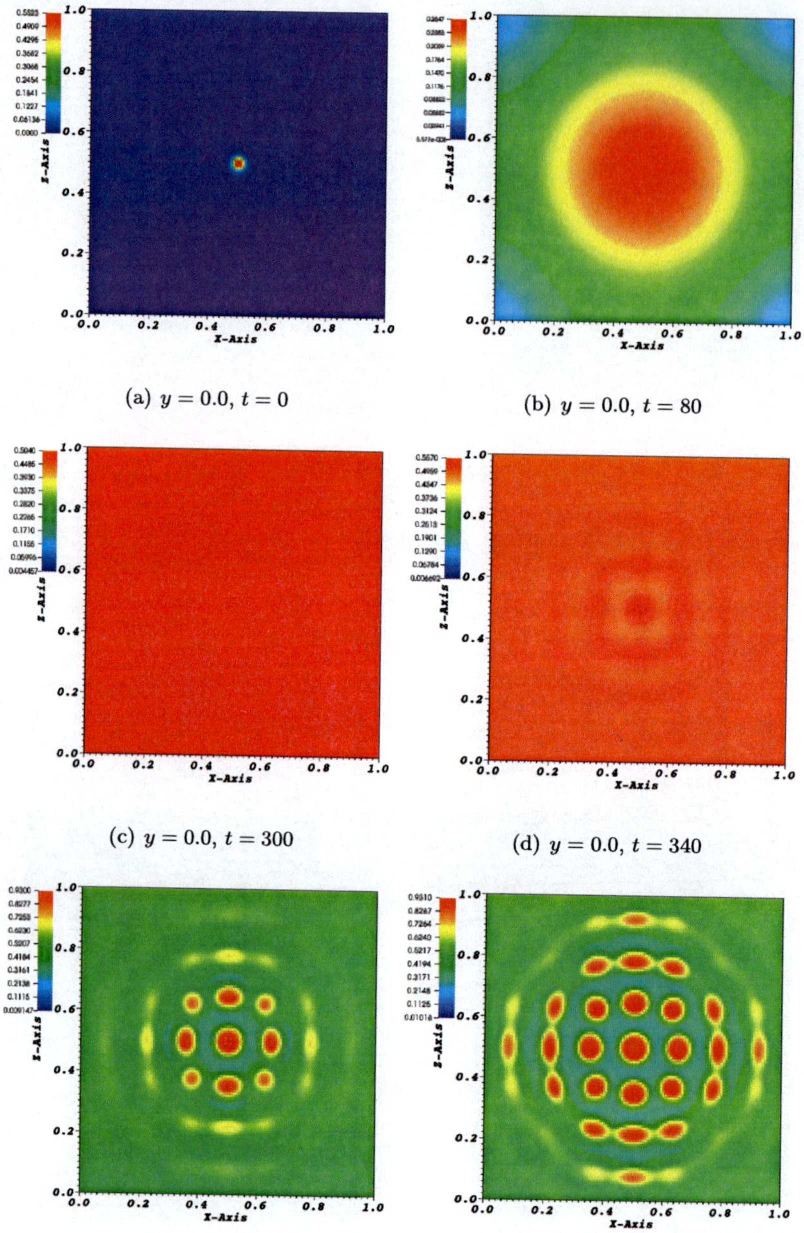

(a) $y = 0.0$, $t = 0$

(b) $y = 0.0$, $t = 80$

(c) $y = 0.0$, $t = 300$

(d) $y = 0.0$, $t = 340$

(e) $y = 0.0$, $t = 380$

(f) $y = 0.0$, $t = 400$

FIGURE 8. 2-D cross sectional plots of the result in Figure 6 at $y = 0.0$.

Critical path for an optimal hedging strategy

Jing Chen and Zhijian Wu

ABSTRACT. This paper presents the critical paths for the optimal hedging strategy for hedging a long term commitment with short-term futures contracts established by Larcher and Leobacher in (Larcher, Leobacher, 2003).

1. Introduction

If a firm commits to supplying at each time t in the interval $[0, T]$ a commodity at rate q with a deterministic price a_t, the firm is taking the risk of the underlying commodity's future price movement. To reduce the risk, the firm might enter into a sequence of short-dated futures contracts to protect itself from the effects of price slump. Assume that the market price of the underlying commodity is governed by a simple stochastic differential equation, i.e.,

$$dS_t = \mu dt + \sigma dW_t, \tag{1.1}$$

where W_t is the Wiener process on $[0, T]$, μ and $\sigma > 0$ are constants. Without loss of generality, we assume $T = 1$, $q = 1$ and $\sigma = 1$.

At time t, the cumulative cash balance D_t is the sum of the cumulative cash flow C_t from the commitment and H_t from the hedging strategy G, which can be formulated by, respectively, (see [**2**] by Glasserman, assuming basis is zero)

$$C_t = \int_0^t (a_s - S_s)ds \quad \text{and} \quad H_t = \int_0^t G(s)dS_s.$$

The deviation of D_t from its expected value is

$$D_t - E[D_t] = \int_0^t (G(s) + s - t)dW_s.$$

By Itô isometry, if G is a deterministic function over $[0, 1]$, then

$$\text{Var}[D_t] = \int_0^t (G(s) + s - t)^2 ds. \tag{1.2}$$

Suppose the goal of hedging is to prevent the actual cash balance from falling short of the expected cash balance by an amount $x > 0$. We define the shortfall risk over $[0, 1]$ as

$$P(\min_{0\le t\le 1}(D_t - E[D_t]) \le -x). \tag{1.3}$$

2010 *Mathematics Subject Classification.* Primary 91G80, 49K30, 60H05, 60H10.
Key words and phrases. Critical path, futures contract.

It is known by the theory of large deviations that the above shortfall risk can be written as (see also, [2] by Glasserman)

$$P(\min_{0\le t\le 1}(D_t - E[D_t]) \le -x) = e^{-\gamma x^2 + o(x^2)}, \tag{1.4}$$

where γ depends on the hedging strategy G. Clearly, a larger γ leads to a less risk. Moreover, by Dembo and Zeitouni [1] (page 152-164), γ can be calculated as

$$\gamma = \frac{1}{x^2} \inf_{\mathcal{A}_x} \frac{1}{2} \int_0^1 |\dot{\phi}(t)|^2 dt. \tag{1.5}$$

Here $\mathcal{A}_x$ (see precise definition later) is certain event of shortfalls of magnitude greater than or equal to $x > 0$. A path ϕ_* in $\mathcal{A}_x$ satisfying

$$\gamma = \frac{1}{2x^2} \int_0^1 |\dot{\phi}_*(t)|^2 dt$$

is called a critical path (of the corresponding hedging strategy), which gives information about how risky events occur at the lowest energy (or cost). It is important to identify the critical paths of a shortfall under different hedging strategies (see, for example, Glasserman [2]).

The object of this paper is to find all critical paths for the optimal hedging strategy provided by the following theorem of Larcher and Leobacher in [3] (see also [4] and [5]). Our main results are Theorems 2.1, 2.2 and 2.4 in next section.

THEOREM. *A unique continuous function $G_0 : [0,1] \to \mathbb{R}$ exists so that*

$$M := \max_{0\le t\le 1} \int_0^t (G_0(s) + s - t)^2 ds = \inf_{\substack{G:[0,1]\to\mathbb{R} \\ G \text{ integrable}}} \sup_{0\le t\le 1} \int_0^t (G(s) + s - t)^2 ds.$$

Moreover, $M = \frac{e^{-\frac{\pi}{2\sqrt{3}}}}{6\sqrt{3}} = 0.0388532\cdots$, *and G_0 is given by the fomulae*

$$G_0(t) = \begin{cases} 3t_0 - t, & \text{if } t \in [0, t_0); \\ e^{-\frac{\eta}{2}} cos(\frac{\sqrt{3}\eta}{2}) - t, & \text{if } t = \frac{1}{\sqrt{3}} e^{-\frac{\eta}{2}} \cos(\frac{\sqrt{3}}{2}\eta + \frac{\pi}{6}) \text{ with } \eta \in [0, \frac{\pi}{3\sqrt{3}}]; \\ 1 - t, & \text{if } t \in (\frac{1}{2}, 1]; \end{cases}$$

where $t_0 = \frac{e^{-\frac{\pi}{6\sqrt{3}}}}{2\sqrt{3}}$.

2. Critical path for the optimal hedging strategy

In the rest of this paper, we denote

$$g(t) = G_0(t) + t \quad \text{and} \quad t_0 = \frac{e^{-\frac{\pi}{6\sqrt{3}}}}{2\sqrt{3}},$$

and assume that the function $\eta : [0, t_0] \to [0, \frac{\pi}{3\sqrt{3}}]$ is defined implicitly by

$$t = \frac{1}{\sqrt{3}} e^{-\frac{\eta(t)}{2}} \cos(\frac{\sqrt{3}\eta(t)}{2} + \frac{\pi}{6}).$$

As discussed above, a critical path solves the following optimization problem, called also minimum energy problem,

$$\min_{\phi \in \mathcal{A}_x} \frac{1}{2} \int_0^1 |\dot{\phi}(t)|^2 dt \tag{2.1}$$

where $\mathcal{A}_x$, a subset of the shortfall $\min_{0\le t\le 1}(D_t - E[D_t]) \le -x$, is defined as

$$\left\{\phi : \phi(0) = 0, \dot{\phi} \in L^1([0,1]), \text{ and } \int_0^t (g(s) - t)d\phi(s) \le -x, \text{ for some } t \in [0,1]\right\}.$$

It is easy to see that for our optimization problem (2.1), the constraint set $\mathcal{A}_x$ can be refined as the following, denoted by $\mathcal{A}_x$ also,

$$\mathcal{A}_x = \left\{\phi : \phi(0) = 0, \dot{\phi} \in L^1([0,1]), \text{ and } \max_{0\le t\le 1}\left\{t\phi(t) - \int_0^t g(s)\dot{\phi}(s)ds\right\} = x\right\}.$$

In fact, if ϕ satisfies

$$x < -\int_0^t (g(s) - t)d\phi(s) = t\phi(t) - \int_0^t g(s)\dot{\phi}(s)ds,$$

for some $t \in [0,1]$, then there exists a constant $a > 1$ such that the above inequality holds with ϕ being replaced by $\psi = \phi/a$, and clearly ψ has smaller energy.

Without loss of generality, we assume $x = 1$, since x merely serves to scale the solution in our optimization problem.

Denote $\mathcal{A} = \mathcal{A}_1$ and

$$E = \min_{\phi\in\mathcal{A}} \frac{1}{2}\int_0^1 |\dot{\phi}(t)|^2 dt. \tag{2.2}$$

Let ϕ_* be a critical path of (2.2) and τ be the first time for the constraint function

$$F_*(t) = t\phi_*(t) - \int_0^t g(s)\dot{\phi}_*(s)ds \tag{2.3}$$

attains the maximum value 1 on $[0,1]$, i.e., $F_*(t) < 1$ if $t \in [0,\tau)$, $F_*(\tau) = 1$, and $F_*(t) \le 1$ if $t \in (\tau, 1]$. It is clear that $\tau > 0$.

THEOREM 2.1. *A critical path for the optimal hedging strategy G_0 has the form*

$$\phi_\tau(t) = \begin{cases} B\int_0^t g(s)ds + \int_0^t C(s)ds, & \text{if } t \in [0,\tau]; \\ B\int_0^\tau g(s)ds + \int_0^\tau C(s)ds, & \text{if } t \in (\tau, 1]. \end{cases}$$

where $\tau \in (0,1]$, B is a constant, and

$$C(t) = \begin{cases} C_1, & \text{if } t \in [0, t_0]; \\ C_2, & \text{if } t \in (t_0, \frac{1}{2}]; \\ C_3, & \text{if } t \in (\frac{1}{2}, 1]. \end{cases}$$

with $C_i, i = 1,2,3$ are constants.

PROOF. Let ϕ_* be a critical path and τ be the first time the corresponding constraint function F_* attains 1. Suppose $h \in C^1[0,1]$ with supp $h \subset (0,\tau]$ and satisfying $\int_0^\tau h(t)dg(t) = 0$.

Clearly, $\tau h(\tau) - \int_0^\tau g(s)\dot{h}(s)ds = 0$ and hence $\phi_* + \epsilon h \in \mathcal{A}$ holds for any $\epsilon \in \mathbb{R}$. Since ϕ_* is critical, we have

$$\frac{1}{2}\int_0^1 |\dot{\phi}_*(t) + \epsilon\dot{h}(t)|^2 dt \ge \frac{1}{2}\int_0^1 |\dot{\phi}_*(t)|^2 dt,$$

which implies (since ϵ is arbitrary)

$$\int_0^\tau \dot{\phi}_*(t)\dot{h}(t)dt = 0.$$

By integration by parts, we have

$$\int_0^\tau \ddot{\phi}_*(t)h(t)dt = 0. \tag{2.4}$$

We consider the following three cases for τ.

Case 1, $\tau \in (0, t_0]$. Since $dg(t) = 0$ on $[0, t_0]$, we have $\int_0^\tau h(t)dg(t) = 0$ holds for any $h \in C^1[0,1]$ with supp $h \subset (0, \tau]$. Therefore by (2.4), we have

$$\ddot{\phi}_*(t) = 0, \quad \text{if } t \in [0, \tau].$$

Since ϕ_* is critical, and $\frac{1}{2}\int_0^1 |\dot{\phi}_*(t)|^2 dt \geq \frac{1}{2}\int_0^\tau |\dot{\phi}_*(t)|^2 dt$, we have

$$\dot{\phi}_*(t) = 0, \text{ if } t \in (\tau, 1].$$

Case 2, $\tau \in (t_0, \frac{1}{2}]$. Since $\int_0^\tau h(t)dg(t) = 0$ holds for any $h \in C^1[0,1]$ with supp $h \subset (0, t_0]$, we have, as above

$$\ddot{\phi}_*(t) = 0, \quad \text{if } t \in [0, t_0].$$

Therefore

$$\int_{t_0}^\tau \ddot{\phi}_*(t)h(t)dt = \int_0^\tau \ddot{\phi}_*(t)h(t)dt = 0. \tag{2.5}$$

For any $H \in C^2[0,1]$ with supp $H \subset (t_0, \tau]$, let

$$h(t) = \frac{\dot{H}(t)}{\dot{g}(t)},$$

where $\dot{g}(t_0) = \dot{g}(t_0+)$ and $\dot{g}(\tau) = \dot{g}(\tau-)$.

It is easy to see that $h \in C^1[0,1]$, supp $h \subset (t_0, \tau]$, and

$$\int_0^\tau h(t)dg(t) = \int_{t_0}^\tau \frac{\dot{H}(t)}{\dot{g}(t)}\dot{g}(t)dt = H(\tau) - H(t_0) = 0.$$

Since

$$\int_{t_0}^\tau \frac{d}{dt}\left(\frac{\ddot{\phi}_*(t)}{\dot{g}(t)}\right) H(t)dt = \int_{t_0}^\tau \frac{\ddot{\phi}_*(t)}{\dot{g}(t)}\dot{H}(t)dt = \int_{t_0}^\tau \ddot{\phi}_*(t)h(t)dt = 0,$$

here the last equal sign is obtained by (2.5), we have

$$\frac{d}{dt}\left(\frac{\ddot{\phi}_*(t)}{\dot{g}(t)}\right) = 0, \quad \text{if } t \in (t_0, \tau].$$

Similar to Case 1 above, we have $\dot{\phi}_*(t) = 0$, if $t \in (\tau, 1]$.

Therefore, we have

$$\begin{aligned} &\ddot{\phi}_*(t) = 0, && \text{if } t \in [0, t_0]; \\ &\tfrac{d}{dt}\left(\tfrac{\ddot{\phi}_*(t)}{\dot{g}(t)}\right) = 0, && \text{if } t \in (t_0, \tau]; \\ &\dot{\phi}_*(t) = 0, && \text{if } t \in (\tau, 1]. \end{aligned}$$

Case 3, $\tau \in (\frac{1}{2}, 1]$. Since $dg(t) = 0$ on $(\frac{1}{2}, \tau]$, we have $\int_0^\tau h(t)dg(t) = 0$ holds for any $h \in C^1[0,1]$ with supp $h \subset (\frac{1}{2}, \tau]$. Therefore by (2.4), we obtain

$$\ddot{\phi}_*(t) = 0, \quad \text{if } t \in (\frac{1}{2}, \tau].$$

By the same argument as Case 2 above, we have

$$\begin{aligned}
&\ddot{\phi}_*(t) = 0, && \text{if } t \in [0, t_0];\\
&\tfrac{d}{dt}\left(\tfrac{\ddot{\phi}_*(t)}{\dot{g}(t)}\right) = 0, && \text{if } t \in (t_0, \tfrac{1}{2}];\\
&\ddot{\phi}_*(t) = 0, && \text{if } t \in (\tfrac{1}{2}, \tau];\\
&\dot{\phi}_*(t) = 0, && \text{if } t \in (\tau, 1].
\end{aligned}$$

Since $\frac{d}{dt}\left(\frac{\ddot{\phi}_*(t)}{\dot{g}(t)}\right) = 0$ is equivalent to $\ddot{\phi}_*(t) = B\dot{g}(t)$ with B being a constant, we conclude that for the three cases above

$$\dot{\phi}_*(t) = \begin{cases} Bg(t) + C(t), & \text{if } t \in [0, \tau];\\ 0, & \text{if } t \in (\tau, 1]. \end{cases} \quad \text{with } C(t) = \begin{cases} C_1, & \text{if } t \in [0, t_0];\\ C_2, & \text{if } t \in (t_0, \frac{1}{2}];\\ C_3, & \text{if } t \in (\frac{1}{2}, 1]. \end{cases}$$

Here $C_i, i = 1, 2, 3$ are constants. Since $\phi_*(0) = 0$ and ϕ_* is continuous, we have

$$\phi_*(t) = \begin{cases} B\int_0^t g(s)ds + \int_0^t C(s)ds, & \text{if } t \in [0, \tau];\\ B\int_0^\tau g(s)ds + \int_0^\tau C(s)ds, & \text{if } t \in (\tau, 1]. \end{cases}$$

The proof is complete. □

THEOREM 2.2. *For the optimal hedging strategy G_0, there are only two types of critical paths ϕ_τ and ϕ_1 corresponding to $\tau \in [t_0, \frac{1}{2}]$ or $\tau = 1$, respectively.*

(1) *If $\tau \in [t_0, \frac{1}{2}]$, then*

$$\phi_\tau(t) = \begin{cases} \frac{1}{4t_0^3}(\tau - 3t_0)t, & \text{if } t \in [0, t_0];\\ \frac{1}{4t_0^3}\left(\tau t - \frac{1}{2}\left(g(t) - 3t\right)\left(g(t) + t\right) - 3t^2\right), & \text{if } t \in (t_0, \tau];\\ \frac{1}{4t_0^3}\left(-\frac{1}{2}\left(g(\tau) - 3\tau\right)\left(g(\tau) + \tau\right) - 2\tau^2\right), & \text{if } t \in (\tau, 1]. \end{cases}$$

(2) *If $\tau = 1$, then*

$$\phi_1(t) = \begin{cases} \frac{1}{4t_0^3}(1 - 3t_0)t, & \text{if } t \in [0, t_0];\\ \frac{1}{4t_0^3}\left(t - \frac{1}{2}\left(g(t) - 3t\right)\left(g(t) + t\right) - 3t^2\right), & \text{if } t \in (t_0, \frac{1}{2}];\\ \frac{1}{32t_0^3}, & \text{if } t \in (\frac{1}{2}, 1]. \end{cases}$$

The minimum energy is $E = \frac{1}{8t_0^3}$.

The graphs of the critical paths are showed in figure 1.

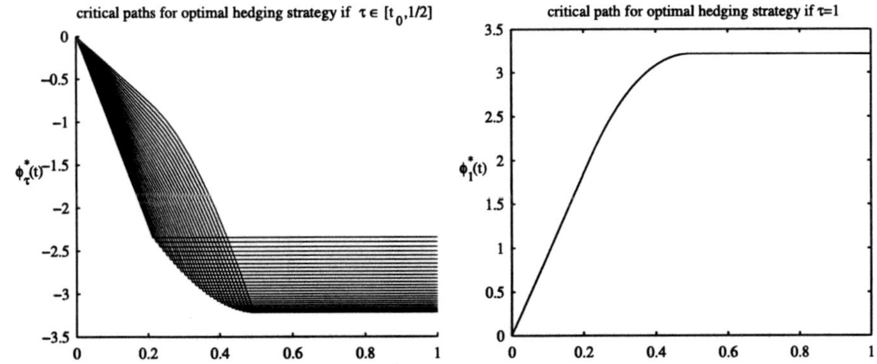

FIGURE 1. critical paths for optimal hedging strategy

The following lemma is needed and can be proved directly.

LEMMA 2.3. *For $t \in [t_0, \frac{1}{2}]$, we have*

$$\int_{t_0}^{t} g(s)ds = \frac{1}{2}\left(g(t) - 3t\right)\left(g(t) + t\right) + 3(t^2 - t_0^2),$$

$$\int_{t_0}^{t} g^2(s)ds = t\left(g(t) - 3t\right)\left(g(t) + t\right) + 5(t^3 - t_0^3).$$

PROOF OF THEOREM 2.2. By Theorem 2.1, it is easy to prove that the energy of ϕ_τ, for $\tau \in (0, t_0]$, attains its minimum at the endpoint. Therefore ϕ_{t_0} is the only possible critical path in this case, which will be included in the case of $\tau \in [t_0, \frac{1}{2}]$.

Case 1, $\tau \in [t_0, \frac{1}{2}]$. By Theorem 2.1, we have

$$\phi_\tau(t) = \begin{cases} (3Bt_0 + C_1)t, & \text{if } t \in [0, t_0]; \\ (3Bt_0 + C_1)t_0 + \int_{t_0}^{t}(Bg(s) + C_2)ds, & \text{if } t \in (t_0, \tau]; \\ (3Bt_0 + C_1)t_0 + \int_{t_0}^{\tau}(Bg(s) + C_2)ds, & \text{if } t \in (\tau, 1]. \end{cases}$$

Therefore the energy of ϕ_τ is

$$E_\tau = \frac{1}{2}\int_0^1 |\dot{\phi}_\tau(t)|^2 dt = \frac{1}{2}(3Bt_0 + C_1)^2 t_0 + \frac{1}{2}\int_{t_0}^{\tau}(Bg(t) + C_2)^2 dt.$$

By (2.3), the constraint function is, for $t \in [t_0, \tau]$,

$$F_\tau(t) = t_0(3Bt_0 + C_1)(t - 3t_0) + \int_{t_0}^{t}(Bg(s) + C_2)(t - g(s))ds.$$

Viewing E_τ and $F_\tau(\tau)$ as functions of B, C_1 and C_2, since E_τ is the minimum energy under the constraint $F_\tau(\tau) = 1$, by the method of Lagrange multipliers, we have

$$\begin{cases} 3t_0^2(3Bt_0 + C_1) + \int_{t_0}^{\tau}(Bg(t) + C_2)g(t)dt = \lambda\left(3t_0^2(\tau - 3t_0) + \int_{t_0}^{\tau}(\tau - g(t))g(t)dt\right), \\ t_0(3Bt_0 + C_1) = \lambda t_0(\tau - 3t_0), \\ \int_{t_0}^{\tau}(Bg(t) + C_2)dt = \lambda\int_{t_0}^{\tau}(\tau - g(t))dt, \\ t_0(3Bt_0 + C_1)(\tau - 3t_0) + \int_{t_0}^{\tau}(Bg(t) + C_2)(\tau - g(t))dt = 1. \end{cases}$$

Solving this system by using Lemma 2.3, we get

$$B = -\lambda, \quad C_1 = C_2 = \lambda\tau, \quad \text{and} \quad \lambda = \frac{1}{4t_0^3}.$$

Case 2, $\tau \in (\frac{1}{2}, 1]$. By Theorem 2.1, we have

$$\phi_\tau(t) = \begin{cases} (3Bt_0 + C_1)t, & \text{if } t \in [0, t_0]; \\ (3Bt_0 + C_1)t_0 + \int_{t_0}^{t}(Bg(s) + C_2)ds, & \text{if } t \in (t_0, \frac{1}{2}]; \\ (3Bt_0 + C_1)t_0 + \int_{t_0}^{\frac{1}{2}}(Bg(s) + C_2)ds + (B + C_3)(t - \frac{1}{2}), & \text{if } t \in (\frac{1}{2}, \tau]; \\ (3Bt_0 + C_1)t_0 + \int_{t_0}^{\frac{1}{2}}(Bg(s) + C_2)ds + (B + C_3)(\tau - \frac{1}{2}), & \text{if } t \in (\tau, 1]. \end{cases}$$

Therefore the energy of ϕ_τ is

$$E_\tau = \frac{1}{2}\left((3Bt_0 + C_1)^2 t_0 + \int_{t_0}^{\frac{1}{2}}(Bg(t) + C_2)^2 dt + (B + C_3)^2(\tau - \frac{1}{2})\right).$$

By (2.3), the constraint function is, for $t \in (\frac{1}{2}, \tau]$,

$$F_\tau(t) = t_0(3Bt_0+C_1)(t-3t_0)+\int_{t_0}^{\frac{1}{2}}(Bg(s)+C_2)(t-g(s))ds+(B+C_3)(t-1)(t-\frac{1}{2}).$$

Viewing E_τ and $F_\tau(\tau)$ as functions of B, C_1, C_2 and C_3, since E_τ is the minimum energy under the constraint $F_\tau(\tau) = 1$, by the method of Lagrange multipliers, we have

$$\begin{cases} 3t_0^2(3Bt_0 + C_1) + \int_{t_0}^{\frac{1}{2}}(Bg(t) + C_2)g(t)dt + (B + C_3)(\tau - \frac{1}{2}) \\ \quad = \lambda\left(3t_0^2(\tau - 3t_0) + \int_{t_0}^{\frac{1}{2}}(\tau - g(t))g(t)dt + (\tau - 1)(\tau - \frac{1}{2})\right), \\ t_0(3Bt_0 + C_1) = \lambda t_0(\tau - 3t_0), \\ \int_{t_0}^{\frac{1}{2}}(Bg(t) + C_2)dt = \lambda\int_{t_0}^{\frac{1}{2}}(\tau - g(t))dt, \\ (B + C_3)(\tau - \frac{1}{2}) = \lambda(\tau - 1)(\tau - \frac{1}{2}), \\ t_0(3Bt_0 + C_1)(\tau - 3t_0) + \int_{t_0}^{\frac{1}{2}}(Bg(t) + C_2)(\tau - g(t))dt \\ \quad +(B + C_3)(\tau - 1)(\tau - \frac{1}{2}) = 1. \end{cases}$$

Solving this system by using Lemma 2.3, we get

$$B = -\lambda, \quad C_1 = C_2 = C_3 = \lambda\tau, \quad \text{and} \quad \lambda = \frac{1}{4t_0^3}.$$

It is easy to check that E_τ attains its minimum only at $\tau = 1$ on $(\frac{1}{2}, 1]$.

It can be calculated directly that $E_\tau = \frac{1}{8t_0^3}$ for any $\tau \in [t_0, \frac{1}{2}]$ or $\tau = 1$. □

THEOREM 2.4. *The constraint functions for the critical paths for the optimal hedging strategy G_0 have the following forms for $\tau \in [t_0, \frac{1}{2}]$ or $\tau = 1$, respectively.*

$$F_\tau(t) = \begin{cases} \frac{1}{4t_0^3}(\tau - 3t_0)(t^2 - 3t_0t), & \text{if } t \in [0, t_0], \\ \frac{1}{4t_0^3}\left(\frac{1}{2}(g(t) - 3t)(g(t) + t)(t - \tau) + 2(t^3 - \tau t^2 + 2t_0^3)\right), & \text{if } t \in [t_0, \tau], \\ \frac{1}{4t_0^3}\left(\frac{1}{2}(g(\tau) - 3\tau)(g(\tau) + \tau)(\tau - t) + 2(\tau^3 - \tau^2 t + 2t_0^3)\right), & \text{if } t \in [\tau, 1]; \end{cases}$$

$$F_1(t) = \begin{cases} \frac{1}{4t_0^3}(1 - 3t_0)(t^2 - 3t_0t), & \text{if } t \in [0, t_0]; \\ \frac{1}{4t_0^3}\left(\frac{1}{2}(g(t) - 3t)(g(t) + t)(t - 1) + 2(t^3 - t^2 + 2t_0^3)\right), & \text{if } t \in (t_0, \frac{1}{2}]; \\ \frac{1}{4t_0^3}\left(\frac{1}{8}t + 4t_0^3 - \frac{1}{8}\right), & \text{if } t \in (\frac{1}{2}, 1]; \end{cases}$$

PROOF. Direct computation using (2.3) and Theorem 2.2 is enough. □

3. Final remark

Under the optimal hedging strategy G_0, the spot variance (1.2) is

$$\mathrm{Var}[D_t] = \begin{cases} (t - 3t_0)^2 t, & \text{if } t \in [0, t_0]; \\ 4t_0^3, & \text{if } t \in (t_0, \frac{1}{2}]; \\ t^3 - 2t^2 + \frac{5}{4}t + 4t_0^3 - \frac{1}{4}, & \text{if } t \in (\frac{1}{2}, 1]. \end{cases}$$

Comparing the constraint functions for critical paths and the spot variance for the optimal hedging strategy, we see that the constraint functions and spot variance attain their maximum at the same time, as shown in figure 2 below. This phenomena can be further studied.

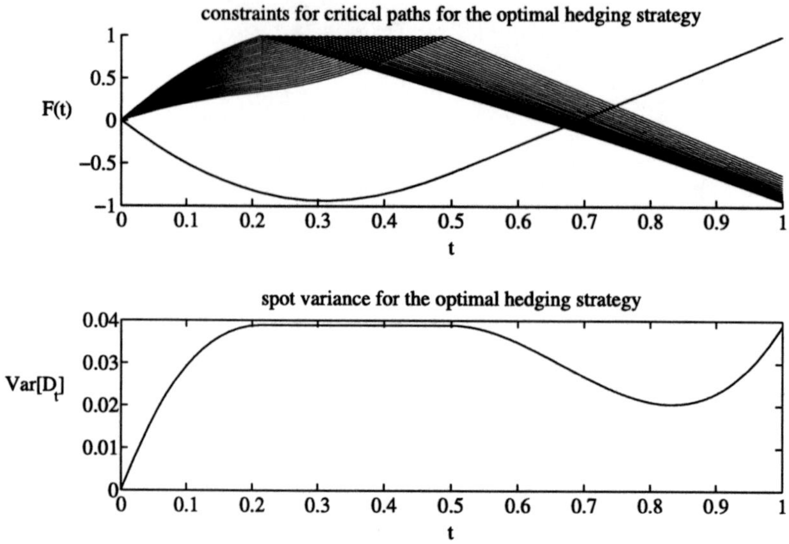

FIGURE 2. comparison of constraint functions for critical paths and spot variance for optimal hedging strategy

In addition, it would be interesting to find the critical paths for the optimal hedging strategy with the constraint of terminal risk by Wu, Yu and Zheng in [**5**], which will show dynamics for the critical paths.

Furthermore, it would be desirable to consider that the commodity price follows geometric Brownian motion $dS_t = \mu S_t dt + \sigma S_t dW_t$ and a general model with mean reversion $dS_t = -\alpha(S_t - c_t)dt + \sigma dW_t$, where $0 \le \alpha < 1$ measures the speed of mean reversion, and c_t is the level toward which the price reverts at time t.

References

1. A. Dembo, and O. Zeitouni, *Large Deviations Techniques and Applications*, Jones and Bartlett Publishers, 1992 MR1202429 (95a:60034)
2. P. Glasserman (2001), *Shortfall risk in long-term hedging with short-term futures contracts*, Option Pricing, Interest Rates and Risk Management, eds. J. Cvitanic, E. Jouini, and M. Musiela. Cambridge: Cambridge University Press, 477-508 MR1848560
3. G. Larcher, and G. Leobacher (2003), *An optimal strategy for hedging with short-term futures contracts*, Math Finance, 13(2): 311-344 MR1967779
4. G. Leobacher (2008), *On a class of optimization problems emerging when hedging with short term futures contracts*, Mathematical Methods of Operations Research, 67: 65-90 MR2372998 (2008k:90147)
5. Z. Wu, C. Yu, and X. Zheng (2011), *Managing risk with short-term futures contracts*, SIAM J. Financial Math., Vol. 2, pp. 715-726 MR2836498 (2012k:91366)

DEPARTMENT OF MATHEMATICS, THE UNIVERSITY OF WEST ALABAMA, LIVINGSTON, ALABAMA 35470

E-mail address: jchen@uwa.edu

DEPARTMENT OF MATHEMATICS, THE UNIVERSITY OF ALABAMA, TUSCALOOSA, ALABAMA 35487

E-mail address: zwu@as.ua.edu

Reservoir Simulation on NVIDIA Tesla GPUs

Zhangxin Chen, Hui Liu, Song Yu, Ben Hsieh, and Lei Shao

ABSTRACT. In this paper, we introduce our work on accelerating a black oil simulator using GPU-based parallel iterative linear solvers. We develop iterative linear solvers and several commonly used preconditioners on NVIDIA Tesla GPUs. These solvers and preconditioners are coupled with our in-house reservoir simulator. Numerical experiments show that our GPU-based black oil simulator is sped up around six times faster than a pure CPU-based simulator.

1. Introduction

For large scale reservoir simulation, especially when the number of grid blocks is over millions, the running time of reservoir simulators can be very long. To our experience, solving a linear system arising from reservoir simulation is the most time-consuming part. For SPE 10 [**CMFPM**], for example, over 95% running time is spent on the solution of linear systems. It is clear that if the linear systems are solved efficiently, the whole simulation can be sped up.

GPUs are now much more powerful in float point calculation than conventional CPUs [**NVCUDAPG, CUDABPG**]. They have become very popular nowadays and have been used in many scientific applications [**ESMV, ISMV, ECMCP, GPILS**]. In this paper, we introduce our work on accelerating a black oil simulator on GPUs. We develop a new matrix vector multiplication kernel [**ESMV, ISMV**] for NVIDIA GPUs and other related BLAS 1/2 subroutines. Based on these subroutines, seven Krylov subspace solvers [**TLS, IMSLS, GPILS, ECMCP**] are developed. Several commonly used preconditioners, such as polynomial, block ILU(k), ILU(k), block ILUT, ILUT [**TLS, IMSLS**] and domain decomposition preconditioners [**RAS**], are also developed. Our solvers and preconditioners are applied to our in-house black oil simulator. The SPE 10 problem is chosen as a benchmark. The number of grid blocks in SPE 10 is over 1.1 million and the number of all unknowns is over 2.2 millions. Numerical experiments show that we can speed the whole simulation up around six times faster than our pure CPU-based simulator.

The layout is as follows. In §2, our new matrix format and sparse matrix-vector multiplication kernel is proposed first, then linear solvers and preconditioners are

---

1991 *Mathematics Subject Classification.* Primary 568W10, 68N30; Secondary 65F10, 65F50.

Key words and phrases. Reservoir simulation, linear solver, GPU.

The support of Department of Chemical and Petroleum Engineering, University of Calgary and Reservoir Simulation Group is gratefully acknowledged. The research is partly supported by NSERC/AIEE/Foundation CMG and AITF Chairs.

introduced. In §3, numerical experiments are employed to test the efficiency of the GPU-based linear solvers and preconditioners.

2. Iterative Linear Solvers and Preconditioners

In this section, we introduce our sparse matrix-vector multiplication kernel first, and then the GPU-based linear solvers and preconditioners.

2.1. Sparse Matrix-Vector Multiplication Kernel. The matrix format used for GPUs is HEC (hybrid ELL and CSR), which is developed in [**PPRS**] and is demonstrated by Figure 1. From this figure, we can see that an HEC matrix contains two submatrices, an ELL matrix and a CSR matrix. The ELL matrix has two matrices, one for the column indices and the other one for non-zeros. The length of each row in these two matrices is the same. A CSR matrix contains three arrays, the first one for the offset of each row, the second one for the column indices and the last one for non-zeros.

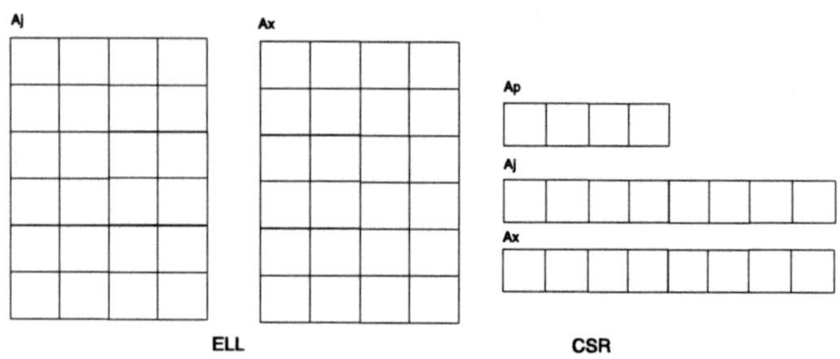

FIGURE 1. HEC matrix format.

The ELL matrix is in column-major order and is aligned in GPUs, which ensure that the data access pattern of global memory for NVIDIA Tesla GPUs is well coalesced [**ESMV, ISMV, NVCUDAPG, CUDABPG**]. In this case the data access speed for the ELL matrix is high. A disadvantage of the ELL format is that even if only one row has too many elements, the length of all rows must be the same. Hence it is a waste of memory. Then a CSR matrix is applied to overcome this problem.

For a HEC format matrix, the corresponding sparse matrix-vector multiplication kernel is described in Algorithm 1. This is a two-step algorithm. In the first step, the ELL part is calculated, where each CUDA thread [**NVCUDAPG, CUDABPG, ESMV, ISMV**] is responsible for one row. Then the CSR part is calculated. Other BLAS 2 subroutines are developed similarly. One BLAS 1 subroutine is described in Algorithm 2. For this algorithm, each CUDA thread calculates only one element. Other BLAS 1 subroutines are similar.

2.2. Iterative Linear Solvers. We consider the following linear system:

$$Ax = b, \tag{2.1}$$

where A is a nonsingular $n \times n$ matrix, b is the right-hand side and x is the solution to be solved for. Several Krylov subspace linear solvers are listed in [**TLS, IMSLS**].

Algorithm 1 Sparse Matrix-Vector Multiplication Kernel, $y = Ax$

```
for i = 1: n do                    ▷ ELL, Use one GPU kernel to deal with this loop
    the ith thread calculates the ith row of ELL matrix;        ▷ Use one thread
end for

for i = 1: n do                    ▷ CSR, Use one GPU kernel to deal with this loop
    the ith thread calculates the ith row of CSR matrix;        ▷ Use one thread
end for
```

Algorithm 2 BLAS 1 subroutine, $y = \alpha x + \beta y$

```
for i = 1: n do                    ▷ Use one GPU kernel to deal with this loop
    y[i] = αx[i] + βy[i];                                        ▷ Use one thread
end for
```

From the descriptions of these solvers, we can see that these solvers share the following common operations:

$$y = \alpha Ax + \beta y, \quad \alpha, \beta \in R, \tag{2.2}$$

$$z = \alpha Ax + \beta y, \quad \alpha, \beta \in R, \tag{2.3}$$

$$y = \alpha x + \beta y, \quad \alpha, \beta \in R, \tag{2.4}$$

$$z = \alpha x + \beta y, \quad \alpha, \beta \in R, \tag{2.5}$$

$$\alpha = \langle x, y \rangle, \tag{2.6}$$

where A is a matrix, x, y and z are vectors, α and β are real numbers, and $\langle \cdot, \cdot \rangle$ is the scalar product.

These subroutines are simple variants of Algorithm 1 and Algorithm 2. With these BLAS 1/2 operations, the linear solvers can be developed in a straightforward manner. Seven GPU-based Krylov subspace solvers are developed, including GMRES, CG, BICGSTAB, GCR, CGS, ORTHOMIN and ORTHODIR [**IMSLS, TLS**]. The CPU-based versions are also developed.

2.3. Preconditioners. In practice, an equivalent linear system of equations (2.1) is solved:

$$M^{-1}Ax = M^{-1}b, \tag{2.7}$$

where M is called a preconditioner or left-preconditioner. When choosing preconditioner M, a general principle is that M is an approximation of A and in this case, it means that the product of M^{-1} and A approximates the unit matrix I. The condition number of $M^{-1}A$ is smaller than that of A and the linear system (2.7) is much easier to solve compared to the original equation (2.1). Meanwhile, M should be easy to construct and be easy to solve.

When the spectrum of $N = I - A$ is less than 1, we have the Neumann expansion [**TLS**]

$$A^{-1} = I + N + N^2 + N^3 + N^4 + \cdots. \tag{2.8}$$

For any positive integer s, a Neumann polynomial preconditioner is defined as follows:

$$M^{-1} = I + N + N^2 + N^3 + \cdots + N^s. \tag{2.9}$$

When we solve the preconditioned system, only the matrix-vector multiplication is involved.

A simple idea of constructing a preconditioner is to apply LU factorization. However, for a given sparse matrix A, the accurate L and U are usually much denser than the lower and upper parts of A, respectively. Alternatively, incomplete-LU is applied. The ILU factorization computes a sparse lower triangular matrix L and a sparse upper triangular matrix U for a given matrix A. If the non-zero pattern of L and U is the same as that of the lower and upper parts of A, respectively, we obtain the so-called ILU(0) preconditioner and higher order ILU(k) is obtained similarly [**IMSLS**]. Another method is ILUT, which drops entries based on the numerical values of the fill-in elements [**IMSLS, GPILS**], where L and U are controlled by the drop tolerance and the maximal number of fill-ins in each row.

The solution procedure for L and U is sequential. In this paper, block ILU(0) and block ILUT are implemented. If the number of blocks increases, both preconditioners have better parallel performance. The matrix A is partitioned by METIS [**METIS**]. The lower and upper triangular problems are solved by a modified level schedule method [**PPRS**].

Cai et al. developed a restricted additive Schwarz preconditioner (RAS) for solving general sparse matrices [**RAS**]. The basic idea is to partition the original problem to some smaller problems and then to solve these smaller problems simultaneously. In this paper, the matrix is also partitioned by METIS [**METIS**]. The submatrices are extended according to the topology of the original matrix. Each smaller problem is solved by ILU(0) or ILUT.

2.4. Package Structure. Figure 2 is the basic structure of our linear solver package. This package has a multi-level structure. The bottom is the infrastructure, where memory management, communication, input, output, and preprocessing modules are developed. These modules serve the whole package. The middle level includes the matrix and vector operations. The top level includes our solvers and preconditioners. These solvers and preconditioners are designed in such a way that each solver or preconditioner is independent of each other. In this case, this package is friendly to the user, who can choose the proper solver and preconditioner depending on the individual application, and if one solver or preconditioner has bugs, these bugs do not affect other solvers or preconditioners.

3. Numerical Results

In this section, numerical experiments are performed on our workstation with Intel Xeon X5570 CPUs and NVIDIA Tesla C2050/C2070 GPUs. The operating system is Fedora 13 X86_64 with CUDA Toolkit 4.0 and GCC 4.4. All CPU codes are compiled with *-O3* option. The type of float point number is double.

EXAMPLE 3.1. In this example, the matrix is from SPE 10 [**CMFPM**]. The dimension of this matrix is 2,188,851 and the number of non-zeros is 29,915,573. Three solvers are tested without using any preconditioner, and the number of iteration is fixed at 20. Performance data is collected in Table 1.

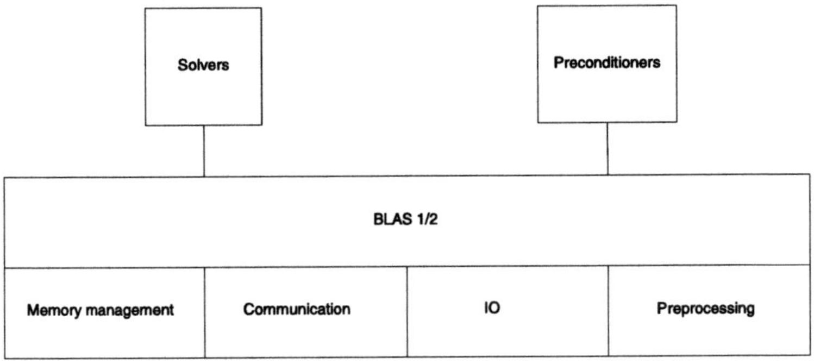

FIGURE 2. Structure of our package.

TABLE 1. Performance of solvers without preconditioner

Solver	CPU (s)	GPU (s)	Speedup
BICGSTAB	3.27	0.31	9.95
ORTHOMIN(20)	5.95	0.52	10.61
ORTHOMIN(40)	5.71	0.53	9.92
GMRES(20)	60.39	5.61	10.72
GMRES(40)	178.08	17.01	10.45
GMRES(60)	361.34	34.32	10.52

This example is designed to test the framework of our package. From Table 1, we can see that when no preconditioner is applied, the average speedup for each solver is around 10.4. We have a maximal speedup of 10.72 when GMRES(40) solver is employed. The table also indicates that the BLAS 1/2 subroutines are efficient, and the whole framework of our package works well.

EXAMPLE 3.2. The matrix used in this example is the same as that in Example 3.1. Here the Neumann polynomial preconditioner is applied, and the order, s, of the polynomial preconditioner is 8. The number of iterations is also 20. Performance data is collected in Table 2.

TABLE 2. Performance of solvers with Neumann polynomial preconditioner

Solver	CPU (s)	GPU (s)	Speedup
BICGSTAB	20.64	2.06	9.90
ORTHOMIN(40)	22.83	2.33	9.62
GMRES(20)	251.97	24.06	10.46
GMRES(40)	619.94	53.07	11.67

This example is employed to test the performance of the developed sparse matrix-vector multiplication kernel, which is fundamental to a linear solver package. From Table 2, we can conclude that the performance of our sparse matrix-vector multiplication kernel is high, and for this example, a maximal speedup of 11.67 is achieved. The average speedup is around 10.5.

EXAMPLE 3.3. Here only the solver GMRES(20) is employed. The preconditioner is block ILU(0) with a different number of blocks. The matrix used here is the same as that in Example 3.1. The terminating criteria is $2e-2$. Performance data is collected in Table 3.

TABLE 3. Performance of GMRES(20) with block ILU(0)

Blks	CPU (s)	GPU (s)	Speedup	IT
1	122.33	14.99	8.14	21
4	124.33	15.00	8.27	21
8	126.40	15.31	8.23	23
16	180.06	19.03	9.44	29

The combination of GMRES and ILU(0) is the most commonly used method for sequential reservoir simulation. Since the solution of ILU(0) is sequential in nature, it is hard to parallelize. This example is to test the parallel performance of our GPU-based block ILU(0) preconditioner. When the number of blocks is one, then the block ILU(0) is the so-called ILU(0). Though ILU(0) is sequential, we can still speed up this preconditioner around 8.14 times faster than the CPU-based ILU(0). When we increase the number of blocks, the speedup increases. It means that the block ILU(0) has better parallel performance. However, the number of iteration increases, too. For this matrix, we have an average speedup of 8.3. When the number of blocks is 16, a maximal speedup of 9.44 is achieved.

EXAMPLE 3.4. Here the block ILUT is applied. All other settings are the same as those in Example 3.3. Performance data is collected in Table 4.

TABLE 4. Performance of GMRES(20) with block ILUT

Blks	CPU (s)	GPU (s)	Speedup	IT
1	34.19	11.70	2.92	5
4	45.52	10.34	4.39	7
8	45.78	9.57	4.76	7
16	63.12	12.42	5.07	10

The ILUT preconditioner is computed by dropping small elements of lower and upper triangular matrices. The non-zero pattern of L and U is less regular than that of ILU(0), which means that their data dependency is more complicated than that in ILU(0). This is also reflected from Table 4. The speedup of block ILUT is lower compared to that of block ILU(0). An average speedup of 4.2 is achieved. However, comparing the data in Table 3 and Table 4, we find that block ILUT is better than block ILU(0) in terms of total running time and the number of iterations. The block ILUT is also sensitive to the number of blocks. The number of iterations increases when the number of blocks increases.

EXAMPLE 3.5. The RAS preconditioner is tested. The solver is GMRES(20) and the matrix is also the same as above. For the RAS preconditioner, the smaller problems are solved by ILU(0) and ILUT here. Data is collected in Tables 5 and 6.

TABLE 5. Performance of RAS using ILU(0)

Blks	overlap	CPU (s)	GPU (s)	Speedup	IT
4	1	101.64	15.25	6.65	21
8	1	134.96	15.18	8.87	21
16	1	142.18	16.14	8.78	22

TABLE 6. Performance of RAS using ILUT

Blks	overlap	CPU (s)	GPU (s)	Speedup	IT
4	1	36.28	8.56	4.22	5
8	1	36.88	8.11	4.53	5
16	1	45.21	9.13	4.93	5

From Tables 5 and 6, we find that ILU(0) has better speedup than ILUT. But in terms of the solution time and the number of iterations, ILUT is still better. Since the subdomain is enlarged, the data from Tables 5 and 6 shows that the number of iterations does not change largely when we increase the number of blocks. It means that the RAS preconditioner is not as sensitive as block ILU(0) and block ILUT. In addition, we can increase the number of blocks to have better performance. For ILU(0), we have an average speedup of 8, and meanwhile, we have an average speedup of 4.5 for ILUT.

EXAMPLE 3.6. The SPE 10 problem is tested. SPE 10 is a standard benchmark for the black oil simulator [**CMFPM**]. The problem is highly heterogenous and it has been designed to be difficult to solve. The grid size for SPE 10 is 60x220x85. The number of unknowns is 2,188,851 and the number of non-zeros is 29,915,573. The time period is 100 days. The solver is GMRES(20). Performance data is collected in Table 7.

TABLE 7. Performance of the SPE10

Preconditioner	Blks	CPU (s)	GPU (s)	Speedup
BILU(0)	1	49610.28	7721.09	6.43
BILU(0)	4	53350.63	8524.31	6.26
BILU(0)	8	54286.07	8720.25	6.23
BILUT	1	19533.45	9008.22	2.17
BILUT	4	23187.85	8670.53	2.67
BILUT	8	21718.45	7908.42	2.75
RAS + ILU(0)	8	47855.24	8451.55	5.66
RAS + ILU(0)	16	49315.97	8812.98	6.00
RAS + ILUT	8	18553.33	7730.54	2.40
RAS + ILUT	16	19541.72	7419.27	2.63

From Table 7, we can see that when the block ILU(0) and RAS with ILU(0) are applied, the average speedup is around 6. This means that we can speed up the black oil simulator 6 times faster. When the block ILUT and RAS with ILUT are applied, the average speedup is lower, which is only about 2.5. We can still speed up the simulator 2.5 times faster than the pure CPU simulator. The parallel performance of these preconditioners is similar, but for sequential performance, the ILUT-related preconditioners are much better than ILU(0)-related preconditioners.

4. Conclusion

We have presented our work on accelerating a black oil simulator using GPU-based linear solvers and preconditioners. The numerical experiments show that these solvers and preconditioners are efficient. The simulator can be sped up around 6 times faster with these solvers and preconditioners.

References

[CMFPM] Zhangxin Chen, Guanren Huan, and Yuanle Ma, *Computational methods for multiphase flows in porous media*, Computational Science & Engineering, Society for Industrial and Applied Mathematics (SIAM), Philadelphia, PA, 2006. MR2217767 (2007c:76070)

[IUG] R. Grimes, D. Kincaid, and D. Young, ITPACK 2.0 User's Guide, Technical Report CNA-150, Center for Numerical Analysis, University of Texas, August 1979.

[TLS] Richard Barrett, Michael Berry, Tony F. Chan, and et al., *Templates for the solution of linear systems: building blocks for iterative methods*, Society for Industrial and Applied Mathematics (SIAM), Philadelphia, PA, 1994. MR1247007 (94m:65002)

[IMSLS] Yousef Saad, *Iterative methods for sparse linear systems*, 2nd ed., Society for Industrial and Applied Mathematics, Philadelphia, PA, 2003. MR1990645 (2004h:65002)

[PPRS] H. Liu, S. Yu, Z. Chen, B. Hsieh and L. Shao, Parallel Preconditioners for Reservoir Simulation on GPU, SPE Latin American and Caribbean Petroleum Engineering Conference held in Mexico City, Mexico, 16-18 April 2012, SPE 152811-PP.

[ESMV] N. Bell and M. Garland, Efficient sparse matrix-vector multiplication on CUDA, NVIDIA Technical Report, NVR-2008-004, NVIDIA Corporation, 2008.

[ISMV] N. and M. Garland, Implementing sparse matrix-vector multiplication on throughput-oriented processors, Proc. Supercomputing, 2009, 1-11.

[ECMCP] H. Klie, H. Sudan, R. Li, and Y. Saad, Exploiting capabilities of many core platforms in reservoir simulation, SPE RSS Reservoir Simulation Symposium, 21-23 February 2011

[GPILS] R. Li and Y. Saad, GPU-accelerated preconditioned iterative linear solvers, Technical Report umsi-2010-112, Minnesota Supercomputer Institute, University of Minnesota, Minneapolis, MN, 2010.

[NVCUDAPG] NVIDIA Corporation, Nvidia CUDA Programming Guide (version 3.2), 2010.

[CUDABPG] NVIDIA Corporation, CUDA C Best Practices Guide (version 3.2), 2010.

[RAS] Xiao-Chuan Cai and Marcus Sarkis, *A restricted additive Schwarz preconditioner for general sparse linear systems*, SIAM J. Sci. Comput. **21** (1999), no. 2, 792–797 (electronic), DOI 10.1137/S106482759732678X. MR1718707 (2000f:65133)

[METIS] George Karypis and Vipin Kumar, *A fast and high quality multilevel scheme for partitioning irregular graphs*, SIAM J. Sci. Comput. **20** (1998), no. 1, 359–392 (electronic), DOI 10.1137/S1064827595287997. MR1639073 (99f:68158)

Department of Chemical and Petroleum Engineering, University of Calgary, Alberta, Canada
E-mail address: zhachen@ucalgary.ca

Department of Chemical and Petroleum Engineering, University of Calgary, Alberta, Canada
E-mail address: hui.j.liu@ucalgary.ca

Department of Chemical and Petroleum Engineering, University of Calgary, Alberta, Canada
E-mail address: yusong0926@gmail.com

Department of Chemical and Petroleum Engineering, University of Calgary, Alberta, Canada
E-mail address: bhsieh@ucalgary.ca

Department of Chemical and Petroleum Engineering, University of Calgary, Alberta, Canada
E-mail address: jedyfun@gmail.com

Two–grid hp–DGFEM for second order quasilinear elliptic PDEs based on an incomplete Newton iteration

Scott Congreve and Paul Houston

ABSTRACT. In this paper we propose a class of so-called two-grid hp-version discontinuous Galerkin finite element methods for the numerical solution of a second-order quasilinear elliptic boundary value problem based on the application of a single step of a nonlinear Newton solver. We present both the *a priori* and *a posteriori* error analysis of this two-grid hp–version DGFEM as well as performing numerical experiments to validate the bounds.

1. Introduction

In our recent articles [**4, 5**] we have considered a class of two-grid finite element methods for strongly monotone partial differential equations. Here, the underlying problem is first approximated on a coarse finite element space; the resulting coarse solution is then used to linearise the underlying problem on a finer finite element space, so that only a linear system of equations is solved on this richer space. In this paper we consider an alternative two-grid interior penalty (IP) discontinuous Galerkin finite element method (DGFEM), based on employing a single step of a Newton solver on the finer space, cf. [**1**], [**9**, Section 5.2], for the numerical solution of the following quasilinear elliptic boundary value problem:

$$-\nabla \cdot (\mu(\boldsymbol{x}, |\nabla u|)\nabla u) = f \quad \text{in } \Omega, \qquad u = 0 \quad \text{on } \Gamma, \tag{1.1}$$

where Ω is a bounded polygonal domain in $\mathbb{R}^2$, with boundary Γ and $f \in L^2(\Omega)$.

We assume that $\mu \in C^2(\bar{\Omega} \times [0, \infty))$ satisfies the condition: there exists positive constants m_μ and M_μ such that the following monotonicity property is satisfied:

$$m_\mu(t - s) \le \mu(\boldsymbol{x}, t)t - \mu(\boldsymbol{x}, s)s \le M_\mu(t - s), \qquad t \ge s \ge 0, \quad \boldsymbol{x} \in \bar{\Omega}. \tag{1.2}$$

For ease of notation we write $\mu(t)$ instead of $\mu(\boldsymbol{x}, t)$. The outline of this article is as follows. In Section 2 we state the proposed two-grid IP DGFEM. In Sections 3 and 4 we consider the *a priori* and *a posteriori* error analysis, respectively, of the two-grid IP DGFEM. Finally, in Section 5 we present some numerical results to validate the theoretical error bounds.

2010 *Mathematics Subject Classification*. Primary 65N12, 65N30, 65N50; Secondary 65N35.

The second author acknowledges the financial support of the EPSRC under the grant EP/H005498.

2. Two-Grid *hp*–Version IP DGFEM

We consider shape-regular meshes $\mathcal{T}_h$ that partition $\Omega \subset \mathbb{R}^2$ into open disjoint elements κ such that $\overline{\Omega} = \bigcup_{\kappa \in \mathcal{T}_h} \overline{\kappa}$. By h_κ we denote the element diameter of $\kappa \in \mathcal{T}_h$, $h = \max_{\kappa \in \mathcal{T}_h} h_\kappa$, and $\boldsymbol{n}_\kappa$ signifies the unit outward normal vector to κ. We allow the meshes $\mathcal{T}_h$ to be *1-irregular*; further, we suppose that $\mathcal{T}_h$ is of *bounded local variation*, i.e., there exists a constant $\rho_1 \geq 1$, independent of the element sizes, such that $\rho_1^{-1} \leq h_\kappa/h_{\kappa'} \leq \rho_1$, for any pair of elements $\kappa, \kappa' \in \mathcal{T}_h$ which share a common edge $e = \partial\kappa \cap \partial\kappa'$. To each $\kappa \in \mathcal{T}_h$ we assign a polynomial degree $p_\kappa \geq 1$ and define the degree vector $\boldsymbol{p} = \{p_\kappa : \kappa \in \mathcal{T}_h\}$. We suppose that $\boldsymbol{p}$ is also of bounded local variation, i.e., there exists a constant $\rho_2 \geq 1$, independent of the element sizes and $\boldsymbol{p}$, such that, for any pair of neighbouring elements $\kappa, \kappa' \in \mathcal{T}_h$, $\rho_2^{-1} \leq p_\kappa/p_{\kappa'} \leq \rho_2$.

With this notation, we introduce the finite element space

$$V(\mathcal{T}_h, \boldsymbol{p}) = \{v \in L^2(\Omega) : v|_\kappa \in \mathcal{S}_{p_\kappa}(\kappa) \quad \forall \kappa \in \mathcal{T}_h\},$$

where $\mathcal{S}_{p_\kappa}(\kappa) = \mathcal{P}_{p_\kappa}(\kappa)$ if κ is a triangle and $\mathcal{S}_{p_\kappa}(\kappa) = \mathcal{Q}_{p_\kappa}(\kappa)$ if κ is a parallelogram. Here, for $p \geq 0$, $\mathcal{P}_p(\kappa)$ denotes the space of polynomials of degree at most p on κ, while $\mathcal{Q}_p(\kappa)$ is the space of polynomials of degree at most p in each variable on κ.

For the mesh $\mathcal{T}_h$, we write $\mathcal{E}_h^{\mathcal{I}}$ to denote the set of all interior edges of the partition $\mathcal{T}_h$ of Ω, $\mathcal{E}_h^{\mathcal{B}}$ the set of all boundary edges of $\mathcal{T}_h$, and set $\mathcal{E}_h = \mathcal{E}_h^{\mathcal{B}} \cup \mathcal{E}_h^{\mathcal{I}}$. Let v and $\boldsymbol{q}$ be scalar- and vector-valued functions, respectively, which are sufficiently smooth inside each element $\kappa \in \mathcal{T}_h$. Given two adjacent elements, $\kappa^+, \kappa^- \in \mathcal{T}_h$ which share a common edge $e \in \mathcal{E}_h^{\mathcal{I}}$, i.e., $e = \partial\kappa^+ \cap \partial\kappa^-$, we write $v^\pm$ and $\boldsymbol{q}^\pm$ to denote the traces of the functions v and $\boldsymbol{q}$, respectively, on the edge e, taken from the interior of $\kappa^\pm$, respectively. With this notation, the averages of v and $\boldsymbol{q}$ at $\boldsymbol{x} \in e$ are given by $\{\!\{v\}\!\} = 1/2(v^+ + v^-)$ and $\{\!\{\boldsymbol{q}\}\!\} = 1/2(\boldsymbol{q}^+ + \boldsymbol{q}^-)$, respectively. Similarly, the jumps of v and $\boldsymbol{q}$ at $\boldsymbol{x} \in e$ are given by $[\![v]\!] = v^+\boldsymbol{n}_{\kappa^+} + v^-_{\kappa^-}\boldsymbol{n}_{\kappa^-}$ and $[\![\boldsymbol{q}]\!] = \boldsymbol{q}^+ \cdot \boldsymbol{n}_{\kappa^+} + \boldsymbol{q}^- \cdot \boldsymbol{n}_{\kappa^-}$, respectively, where $\boldsymbol{n}_{\kappa^\pm}$ denotes the unit outward normal vector on $\partial\kappa^\pm$, respectively. On a boundary edge $e \in \mathcal{E}_h^{\mathcal{B}}$, we set $\{\!\{v\}\!\} = v$, $\{\!\{\boldsymbol{q}\}\!\} = \boldsymbol{q}$, $[\![v]\!] = v\boldsymbol{n}$ and $[\![\boldsymbol{q}]\!] = \boldsymbol{q} \cdot \boldsymbol{n}$, with $\boldsymbol{n}$ denoting the unit outward normal vector on the boundary Γ. For $e \in \mathcal{E}_h$, we define h_e to be the length of the edge; moreover, we set $p_e = \max(p_\kappa, p_{\kappa'})$, if $e = \partial\kappa \cap \partial\kappa' \in \mathcal{E}_h^{\mathcal{I}}$, and $p_e = p_\kappa$, if $e = \partial\kappa \cap \Gamma \in \mathcal{E}_h^{\mathcal{B}}$.

2.1. Standard IP DGFEM discretisation.

Given a fine mesh partition $\mathcal{T}_h$ of Ω, with the corresponding polynomial degree vector $\boldsymbol{p}$, the standard IP DGFEM is defined as follows: find $u_{h,p} \in V(\mathcal{T}_h, \boldsymbol{p})$ such that

$$A_{h,p}(u_{h,p}, v_{h,p}) = F_{h,p}(v_{h,p}) \tag{2.1}$$

for all $v_{h,p} \in V(\mathcal{T}_h, \boldsymbol{p})$, where $F_{h,p}(v) = \int_\Omega fv \,\mathrm{d}\boldsymbol{x}$ and

$$A_{h,p}(u,v) = \int_\Omega \mu(|\nabla_h u|)\nabla_h u \cdot \nabla_h v \,\mathrm{d}\boldsymbol{x} - \sum_{e \in \mathcal{E}_h} \int_e \{\!\{\mu(|\nabla_h u|)\nabla_h u\}\!\} \cdot [\![v]\!] \,\mathrm{d}s$$
$$+ \theta \sum_{e \in \mathcal{E}_h} \int_e \{\!\{\mu(h_e^{-1}|[\![u]\!]|)\nabla_h v\}\!\} \cdot [\![u]\!] \,\mathrm{d}s + \sum_{e \in \mathcal{E}_h} \int_e \sigma_{h,p}[\![u]\!] \cdot [\![v]\!] \,\mathrm{d}s.$$

Here, $\theta \in [-1, 1]$, ∇_h is the element-wise gradient operator and $\sigma_{h,p} = \gamma p_e^2/h_e$, where $\gamma > 0$ is a sufficiently large constant. We define the *energy norm* on $V(\mathcal{T}_h, \boldsymbol{p})$:

$$\|v\|_{\mathrm{DG}}^2 = \|\nabla_h v\|_{L^2(\Omega)}^2 + \sum_{e \in \mathcal{E}_h} \int_e \sigma_{h,p}|[\![v]\!]|^2 \,\mathrm{d}s.$$

LEMMA 2.1 (See [6]). *The semilinear form $A_{h,\boldsymbol{p}}(\cdot,\cdot)$ is strongly monotone in the sense that, there exists $\gamma_{\min} > 0$, such that for any $\gamma \geq \gamma_{\min}$*

$$A_{h,\boldsymbol{p}}(w_1, w_1 - w_2) - A_{h,\boldsymbol{p}}(w_2, w_1 - w_2) \geq C_m \|w_1 - w_2\|_{\mathrm{DG}}^2 \;\forall w_1, w_2 \in V(\mathcal{T}_h, \boldsymbol{p}), \tag{2.2}$$

where C_m is a positive constant, independent of the discretisation parameters.

2.2. Two-grid IP DGFEM discretisation. We now introduce a two-grid IP DGFEM based on employing a single step of the Newton iteration on the fine mesh. To this end, we consider two partitions $\mathcal{T}_h$ and $\mathcal{T}_H$ of Ω, with granularity h and H, respectively. We assume that $\mathcal{T}_h$ and $\mathcal{T}_H$ are nested in that sense that for any element $\kappa_h \in \mathcal{T}_h$ there exists an element $\kappa_H \in \mathcal{T}_H$ such that $\overline{\kappa}_h \subseteq \overline{\kappa}_H$. Moreover for each mesh, $\mathcal{T}_h$ and $\mathcal{T}_H$, we have a corresponding polynomial degree vector $\boldsymbol{p} = \{p_\kappa : \kappa \in \mathcal{T}_h\}$ and $\boldsymbol{P} = \{p_\kappa : \kappa \in \mathcal{T}_H\}$, respectively, where given an element $\kappa_h \in \mathcal{T}_h$ and an element $\kappa_H \in \mathcal{T}_H$, such that $\overline{\kappa}_h \subseteq \overline{\kappa}_H$, the polynomial degree vectors satisfy the condition that $p_{\kappa_h} \geq p_{\kappa_H}$. Thereby, the finite element spaces $V(\mathcal{T}_h, \boldsymbol{p})$ and $V(\mathcal{T}_H, \boldsymbol{P})$ satisfy the following the condition: $V(\mathcal{T}_H, \boldsymbol{P}) \subseteq V(\mathcal{T}_h, \boldsymbol{p})$.

Using this notation we introduce the hp-version two-grid IP DGFEM discretisation of (1.1) based on a single Newton iteration step, cf. [1], [9, Section 5.2]:

(1) Compute the coarse grid approximation $u_{H,P} \in V(\mathcal{T}_H, \boldsymbol{P})$ such that

$$A_{H,P}(u_{H,P}, v_{H,P}) = F_{H,P}(v_{H,P}) \qquad \text{for all } v_{H,P} \in V(\mathcal{T}_H, \boldsymbol{P}). \tag{2.3}$$

(2) Determine the fine grid solution $u_{2G} \in V(\mathcal{T}_h, \boldsymbol{p})$ such that

$$A'_{h,p}[u_{H,P}](u_{2G}, v_{h,p}) = A'_{h,p}[u_{H,P}](u_{H,P}, v_{h,p}) - A_{h,p}(u_{H,P}, v_{h,p}) + F_{h,p}(v_{h,p}) \tag{2.4}$$
$$\text{for all } v_{h,p} \in V(\mathcal{T}_h, \boldsymbol{p}).$$

Here, $A'_{h,p}[u](\phi, v)$ denotes the Fréchet derivative of $u \to A_{h,p}(u, v)$, for fixed v, evaluated at u; thereby, given ϕ we have $A'_{h,p}[u](\phi, v) = \lim_{t\to 0} \frac{A_{h,p}(u+t\phi,v) - A_{h,p}(u,v)}{t}$.

REMARK 2.2. For simplicity of presentation, throughout the rest of this article we shall only consider the incomplete IP variation of the DGFEM, i.e., when $\theta = 0$.

LEMMA 2.3. *Under the assumptions on μ, the following inequality holds:*

$$A'_{h,p}[u](v, v) \geq C_m \|v\|_{\mathrm{DG}}^2 \qquad \forall u, v \in V(\mathcal{T}_h, \boldsymbol{p}).$$

PROOF. Setting $w_1 = u + tv$ and $w_2 = u$ in Lemma 2.1, $u, v \in V(\mathcal{T}_h, \boldsymbol{p})$, $t > 0$:

$$\frac{A_{h,p}(u + tv, v) - A_{h,p}(u, v)}{t} \geq C_m \|v\|_{\mathrm{DG}}^2.$$

Taking the limit as $t \to 0$, we deduce the statement of the Lemma. □

3. *A Priori* Error Analysis

For simplicity of presentation, in this section we assume that the mesh is quasiuniform with mesh size h and that $\boldsymbol{p}$ is uniform over the mesh, i.e., $\boldsymbol{p} \equiv p$.

THEOREM 3.1. *Assuming that $u \in C^1(\Omega)$ and $u \in H^k(\Omega)$, $k \geq 2$, the solution of $u_{2G} \in V(\mathcal{T}_h, \boldsymbol{p})$ of the two-grid IP DGFEM satisfies*

$$\|u_{h,p} - u_{2G}\|_{\mathrm{DG}} \leq C \frac{p^{7/2}}{h} \frac{H^{2S-2}}{P^{2k-3}} \|u\|_{H^k(\Omega)}^2, \tag{3.1}$$

$$\|u - u_{2G}\|_{\mathrm{DG}} \leq C \frac{h^{s-1}}{p^{k-3/2}} \|u\|_{H^k(\Omega)} + C \frac{p^{7/2}}{h} \frac{H^{2S-2}}{P^{2k-3}} \|u\|_{H^k(\Omega)}^2, \tag{3.2}$$

with $1 \leq s \leq \min\{p+1,k\}$, $p \geq 1$ *and* $1 \leq S \leq \min\{P+1,k\}$, $P \geq 1$, *where* $C > 0$ *is independent of the discretisation parameters.*

3.1. Auxiliary Results. We first state the following auxiliary results.

LEMMA 3.2. *For a function* $v \in V(\mathcal{T}_h, \boldsymbol{p})$ *we have the inverse inequality*

$$\|v\|_{L^4(\Omega)} \leq Cph^{-1/2}\|v\|_{L^2(\Omega)},$$

where C *is a positive constant, independent of the discretisation parameters.*

PROOF. Given $\kappa \in \mathcal{T}_h$, employing standard inverse inequalities, see [**8**], gives

$$\int_\kappa |v|^4 \,\mathrm{d}\boldsymbol{x} \leq \|v\|^2_{L^\infty(\kappa)}\|v\|^2_{L^2(\kappa)} \leq Cp^4h^{-2}\|v\|^2_{L^2(\kappa)}\|v\|^2_{L^2(\kappa)} = Cp^4h^{-2}\|v\|^4_{L^2(\kappa)}.$$

Summing over $\kappa \in \mathcal{T}_h$, employing the inequality $\sum_{i=1}^n a_i \leq \left(\sum_{i=1}^n \sqrt{a_i}\right)^2$, $a_i \geq 0$, $i = 1, \ldots, n$, and taking the fourth root of both sides, completes the proof. □

LEMMA 3.3. *For any* $v, w, \phi \in V(\mathcal{T}_h, \boldsymbol{p})$,

$$A_{h,\boldsymbol{p}}(w,\phi) = A_{h,\boldsymbol{p}}(v,\phi) + A'_{h,\boldsymbol{p}}[v](w-v,\phi) + \mathcal{Q}(v,w,\phi), \tag{3.3}$$

where the remainder $\mathcal{Q}$ *satisfies*

$$|\mathcal{Q}(v,w,\phi)| \leq Cp^2h^{-1}\left(1 + \|\nabla w\|_{L^\infty(\Omega)} + \|\nabla v\|_{L^\infty(\Omega)}\right)\|\nabla(w-v)\|^2_{\mathrm{DG}}\|\nabla\phi\|_{\mathrm{DG}},$$

and C *is a positive constant, independent of the discretisation parameters.*

PROOF. We follow the proof outlined by [**9**, Lemma 3.1]; to this end, setting $\xi(t) = v + t(w-v)$ and $\eta(t) = A_{h,\boldsymbol{p}}(\xi(t),\phi)$, we note that the first equation follows from the identity

$$\eta(1) = \eta(0) + \eta'(0) + \int_0^1 \eta''(t)(1-t)\,\mathrm{d}t,$$

where $\mathcal{Q}(v,w,\phi) = \int_0^1 \eta''(t)(1-t)\,\mathrm{d}t$ and $\eta''(t) = A''_{h,\boldsymbol{p}}[\xi(t)](w-v,w-v,\phi)$. Thereby,

$$\begin{aligned}
\mathcal{Q}(v,w,\phi) =& 2\int_0^1\int_\Omega \mu'_{\nabla u}(|\nabla\xi(t)|)\cdot\nabla(w-v)\nabla(w-v)\cdot\nabla\phi\,\mathrm{d}\boldsymbol{x}(1-t)\,\mathrm{d}t \\
&+ \int_0^1\int_\Omega \mu''_{\nabla u}(|\nabla\xi(t)|)|\nabla(w-v)|^2\nabla\xi(t)\cdot\nabla\phi\,\mathrm{d}\boldsymbol{x}(1-t)\,\mathrm{d}t \\
&- 2\int_0^1\sum_{e\in\mathcal{E}_h}\int_e \{\!\!\{\mu'_{\nabla u}(|\nabla\xi(t)|)\cdot\nabla(w-v)\nabla(w-v)\}\!\!\}\cdot[\![\phi]\!]\,\mathrm{d}s(1-t)\,\mathrm{d}t \\
&- \int_0^1\sum_{e\in\mathcal{E}_h}\int_e \{\!\!\{\mu''_{\nabla u}(|\nabla\xi(t)|)|\nabla(w-v)|^2\nabla\xi(t)\}\!\!\}\cdot[\![\phi]\!]\,\mathrm{d}s(1-t)\,\mathrm{d}t \\
\equiv& T_1 + T_2 + T_3 + T_4.
\end{aligned}$$

Here, $\mu'_{\nabla u}(|\cdot|)$ and $\mu''_{\nabla u}(|\cdot|)$ denote the first and second derivatives of $\mu(|\cdot|)$, respectively. First consider T_1: given that $\mu \in \mathrm{C}^2(\bar{\Omega}\times[0,\infty))$, Lemma 3.2 gives

$$T_1 \leq C\|\nabla(w-v)\|^2_{L^4(\Omega)}\|\nabla\phi\|_{L^2(\Omega)} \leq Cp^2h^{-1}\|\nabla(w-v)\|^2_{L^2(\Omega)}\|\nabla\phi\|_{L^2(\Omega)}.$$

Secondly, term T_2 is bounded in an analogous fashion as follows:

$$\begin{aligned}
T_2 &\leq C\left(\|\nabla w\|_{L^\infty(\Omega)} + \|\nabla v\|_{L^\infty(\Omega)}\right)\|\nabla(w-v)\|^2_{L^4(\Omega)}\|\nabla\phi\|_{L^2(\Omega)} \\
&\leq C\left(\|\nabla w\|_{L^\infty(\Omega)} + \|\nabla v\|_{L^\infty(\Omega)}\right)p^2h^{-1}\|\nabla(w-v)\|^2_{L^2(\Omega)}\|\nabla\phi\|_{L^2(\Omega)}.
\end{aligned}$$

Term T_3 is bounded via the inverse trace inequality, see [**8**], and Lemma 3.2:

$$\begin{aligned}T_3 &\le C\left\{\sum_{e\in\mathcal{E}_h} h_e p_e^{-2}\|\{\!\{|\nabla(w-v)|^2\}\!\}\|^2_{L^2(e)}\right\}^{\frac{1}{2}}\left\{\sum_{e\in\mathcal{E}_h}\int_F p_e^2 h_e^{-1}|[\![\phi]\!]|^2\,\mathrm{d}s\right\}^{\frac{1}{2}}\\ &\le C\|\nabla(w-v)\|^2_{L^4(\Omega)}\|\phi\|_{\mathrm{DG}} \le Cp^2h^{-1}\|\nabla(w-v)\|^2_{L^2(\Omega)}\|\phi\|_{\mathrm{DG}}.\end{aligned}$$

We can bound T_4 in an analogous manner as follows:

$$\begin{aligned}T_4 \le\, & C\left\{\sum_{e\in\mathcal{E}_h} h_e p_e^{-2}\|\{\!\{|\nabla(w-v)|^2|\nabla w|\}\!\}\|^2_{L^2(F)}\right\}^{\frac{1}{2}}\left\{\sum_{e\in\mathcal{E}_h}\int_F p^2h^{-1}|[\![\phi]\!]|^2\,\mathrm{d}s\right\}^{\frac{1}{2}}\\ &+C\left\{\sum_{e\in\mathcal{E}_h} h_e p_e^{-2}\|\{\!\{|\nabla(w-v)|^2|\nabla v|\}\!\}\|^2_{L^2(F)}\right\}^{\frac{1}{2}}\left\{\sum_{e\in\mathcal{E}_h}\int_F p^2h^{-1}|[\![\phi]\!]|^2\,\mathrm{d}s\right\}^{\frac{1}{2}}\\ \le\, & C\left\{\||\nabla(w-v)|^2|\nabla w|\|_{L^2(\Omega)}+\||\nabla(w-v)|^2|\nabla v|\|_{L^2(\Omega)}\right\}\|\phi\|_{\mathrm{DG}}\\ \le\, & Cp^2h^{-1}\left\{\|\nabla w\|_{L^\infty(\Omega)}+\|\nabla v\|_{L^\infty(\Omega)}\right\}\|\nabla(w-v)\|^2_{L^2(\Omega)}\|\phi\|_{\mathrm{DG}}.\end{aligned}$$

Combining these bounds for terms T_1, T_2, T_3 and T_4 completes the proof. □

LEMMA 3.4. *Let $u\in H^2(\Omega)$ be the analytical solution of* (1.1)*, such that $\nabla u\in[L^\infty(\Omega)]^2$, and $u_{h,p}\in V(\mathcal{T}_h,\mathbf{p})$ be the IP DGFEM defined by* (2.1)*, we have that*

$$\|\nabla u_{h,p}\|_{L^\infty(\Omega)}\le Cp^{3/2},$$

where C is a positive constant, independent of the discretisation parameters.

PROOF. Writing $\mathcal{P}_u$ to denote the projection of u onto the finite element space $V(\mathcal{T}_h,\mathbf{p})$ defined in [**2**], we have that $\|u-\mathcal{P}_u\|_{H^q(\Omega)}\le C\frac{h^{2-q}}{p^{2-q}}\|u\|_{H^2(\Omega)}$ and $\|\nabla(u-\mathcal{P}_u)\|_{L^\infty(\Omega)}\le C\|u\|_{H^2(\Omega)}$ for all $q\le 2$. Exploiting these bounds, standard inverse inequalities, [**8**], and the *a priori* bound for the IP DGFEM, [**6**], gives

$$\begin{aligned}\|\nabla u_{h,p}\|_{L^\infty(\Omega)} &\le \|\nabla(u_{h,p}-\mathcal{P}_u)\|_{L^\infty(\Omega)}+\|\nabla\mathcal{P}_u\|_{L^\infty(\Omega)}\\ &\le Cp^2h^{-1}\|\nabla(u_{h,p}-\mathcal{P}_u)\|_{L^2(\Omega)}+\|\nabla(u-\mathcal{P}_u)\|_{L^\infty(\Omega)}+\|\nabla u\|_{L^\infty(\Omega)}\\ &\le Cp^{3/2}\left\{\|u\|_{H^2(\Omega)}+\|\nabla u\|_{L^\infty(\Omega)}\right\}.\end{aligned}$$

Since $u\in H^2(\Omega)$ and $\nabla u\in[L^\infty(\Omega)]^2$, the quantities $\|u\|_{H^2(\Omega)}$ and $\|\nabla u\|_{L^\infty(\Omega)}$ are both bounded uniformly by a constant; this then completes the proof. □

3.2. Proof of Theorem 3.1. We now exploit the above results to prove Theorem 3.1. For the first bound (3.1), we employ Lemma 2.3, (2.1), (2.4) and (3.3); thereby, with $\phi=u_{h,p}-u_{2G}$, we deduce that

$$\begin{aligned}C_m\|u_{h,p}-u_{2G}\|^2_{\mathrm{DG}} &\le A'_{h,p}[u_{H,P}](u_{h,p}-u_{2G},\phi)\\ &= A'_{h,p}[u_{H,P}](u_{h,p}-u_{H,P},\phi)+A'_{h,p}[u_{H,P}](u_{H,P}-u_{2G},\phi)\\ &= A'_{h,p}[u_{H,P}](u_{h,p}-u_{H,P},\phi)+A_{h,p}(u_{H,P},\phi)-F_{h,p}(\phi)\\ &= A'_{h,p}[u_{H,P}](u_{h,p}-u_{H,P},\phi)+A_{h,p}(u_{H,P},\phi)-A_{h,p}(u_{h,p},\phi)\\ &= -\mathcal{Q}(u_{H,P},u_{h,p},\phi).\end{aligned}$$

Hence, from Lemma 3.3 we get that

$$\|u_{h,p} - u_{2G}\|_{\mathrm{DG}} \leq Cp^2h^{-1}\Big(1+\|\nabla u_{h,p}\|_{L^\infty(\Omega)}+\|\nabla u_{H,P}\|_{L^\infty(\Omega)}\Big)\|u_{h,p} - u_{H,P}\|^2_{\mathrm{DG}}.$$

Applying Lemma 3.4, noting that $p^{3/2} \geq P^{3/2} \geq 1$, and the *a priori* bound for the standard IP DGFEM, cf. [**6**, Theorem 3.3], gives

$$\begin{aligned}\|u_{h,p} - u_{2G}\|_{\mathrm{DG}} &\leq Cp^2h^{-1}\left(1+p^{3/2}+P^{3/2}\right)\left\{\|u-u_{h,p}\|^2_{\mathrm{DG}} + \|u-u_{H,P}\|^2_{\mathrm{DG}}\right\} \\ &\leq Cp^{7/2}h^{-1}\left\{\frac{h^{2s-2}}{p^{2k-3}}\|u\|^2_{H^k(\Omega)} + \frac{H^{2S-2}}{P^{2k-3}}\|u\|^2_{H^k(\Omega)}\right\}.\end{aligned}$$

Noting that $h \leq H$ and $p \geq P$ completes the proof of the first bound (3.1). To prove the second inequality (3.2), we first employ the triangle inequality

$$\|u - u_{2G}\|_{\mathrm{DG}} \leq \|u - u_{h,p}\|_{\mathrm{DG}} + \|u_{h,p} - u_{2G}\|_{\mathrm{DG}}.$$

Thereby, applying the *a priori* error bound for the standard IP DGFEM, together with the bound (3.1), completes the proof of Theorem 3.1.

4. *A Posteriori* Error Analysis

Here, we state an *a posteriori* error bound for the two-grid IP DGFEM.

THEOREM 4.1. *Let $u \in \mathrm{H}^1_0(\Omega)$ be the analytical solution of* (1.1), *$u_{H,P} \in V(\mathcal{T}_H, \boldsymbol{P})$ and $u_{2G} \in V(\mathcal{T}_h, \boldsymbol{p})$ the numerical approximations obtained from* (2.3) *and* (2.4), *respectively; then the following hp–a* posteriori *error bound holds*

$$\|u - u_{2G}\|^2_{\mathrm{DG}} \leq C\sum_{\kappa\in\mathcal{T}_h}\left(\eta^2_\kappa + \xi^2_\kappa\right), \tag{4.1}$$

with a constant $C > 0$, which is independent of h, H, $\boldsymbol{p}$ and $\boldsymbol{P}$. Here, for $\kappa \in \mathcal{T}_h$,

$$\begin{aligned}\eta^2_\kappa &= h^2_\kappa p_\kappa^{-2}\|\Pi_{\kappa,p_\kappa} f + \nabla\cdot\{\mu(|\nabla u_{H,P}|)\nabla u_{2G}\}\|^2_{L^2(\kappa)} \\ &\quad + h_e p_e^{-1}\|[\![\mu(|\nabla u_{H,P}|)\nabla u_{2G}]\!]\|^2_{L^2(\partial\kappa\setminus\Gamma)} + \gamma^2 h_e^{-1}p_e^3\|[\![u_{2G}]\!]\|^2_{L^2(\partial\kappa)}, \\ \xi^2_\kappa &= \|(\mu(|\nabla u_{H,P}|) - \mu(|\nabla u_{2G}|))\nabla u_{2G}\|^2_{L^2(\kappa)} \\ &\quad + \|(\mu'_{\nabla u}(|\nabla u_{H,P}|)\cdot(\nabla u_{2G} - \nabla u_{H,P}))\,\nabla u_{H,P}\|^2_{L^2(\kappa)} \\ &\quad + h_e p_e^{-1}\|(\mu'_{\nabla u}(|\nabla u_{H,P}|)\cdot(\nabla u_{2G} - \nabla u_{H,P}))\,\nabla u_{H,P}\|^2_{L^2(\partial\kappa)},\end{aligned}$$

and Π_{κ,p_κ} denotes the (elementwise) L^2-projection onto $V(\mathcal{T}_h, \boldsymbol{p})$.

PROOF. The proof of this error bound follows in an analogous manner to the *a posteriori* proof presented in [**5**], cf. also [**7**]. For details, we refer to [**3**]. □

5. Numerical Experiments

In this section we perform numerical experiments to validate the *a priori* error bound, Theorem 3.1 and demonstrate the performance of the *a posteriori* error bound, Theorem 4.1; here, we set $\gamma = 10$ and $\theta = 0$. Throughout this section, we let Ω be the unit square $(0,1)^2 \subset \mathbb{R}^2$ and define the nonlinear coefficient as $\mu(\boldsymbol{x}, |\nabla u|) = 2+(1+|\nabla u|)^{-1}$. We select the right-hand forcing function f so that the analytical solution to (1.1) is given by $u(x,y) = x(1-x)y(1-y)(1-2y)\mathrm{e}^{-20(2x-1)^2}$.

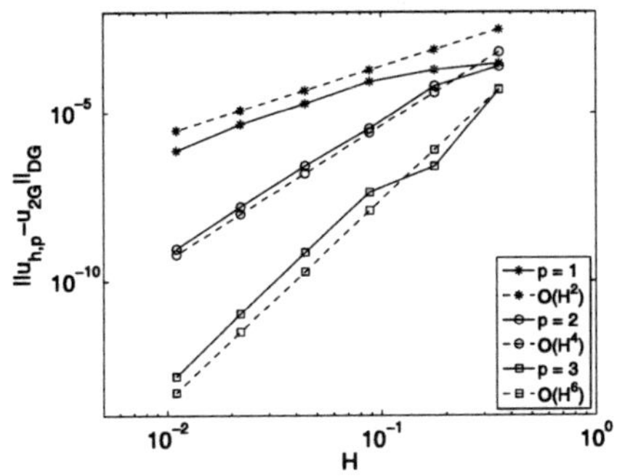

FIGURE 1. Convergence of error between u_{2G} and $u_{h,p}$.

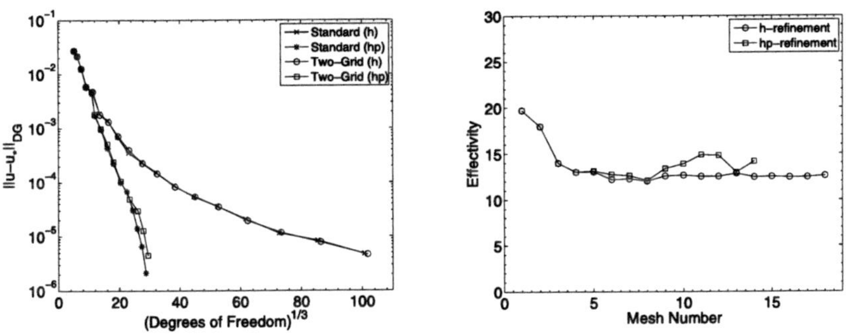

(a) (b)

FIGURE 2. (a) Comparison of the error in the DGFEM norm, using both the IP DGFEM ($u_* = u_{h,p}$) and the two-grid IP DGFEM ($u_* = u_{2G}$); (b) Effectivity indices of the two-grid IP DGFEM.

5.1. Validation of Theorem 3.1. We first validate the bound given in Theorem 3.1; to this end we first solve the standard IP DGFEM on a 256 × 256 uniform mesh of quadrilaterals to compute $u_{h,p}$ for a fixed constant polynomial degree $p = 1, 2, 3$. We then compute the solution u_{2G} to (2.3)–(2.4), for $p = 1, 2, 3$, on a fixed fine 256 × 256 mesh, while performing uniform h-refinement of the coarse mesh, starting from a 4 × 4 mesh with polynomial degree $P = p$. Figure 1 shows the convergence rate of the error between $u_{h,p}$ and u_{2G}, measured in the DG norm, compared to the size of the coarse mesh. Here, we observe that $\|u_{h,p} - u_{2G}\|_{\mathrm{DG}}$ tends to zero at the optimal rate $\mathcal{O}(H^{2P})$, for each fixed P, cf. Theorem 3.1.

5.2. Adaptive Refinement using Theorem 4.1. For this experiment we use the two-grid mesh adaptation algorithm from [**5**], with the *local error indicators* η_κ and *local two-grid error indicators* ξ_κ from Theorem 4.1, to automatically refine the coarse and fine meshes employing both h– and hp–adaptive mesh refinement. Figure 2 shows $\|u - u_{2G}\|_{\mathrm{DG}}$ compared to the third root of the degrees of freedom, as well as the effectivity indices of the error estimator. As can be seen for both h– and hp–adaptive refinement, the effectivity indices are roughly constant, indicating that the error bound overestimates the error by a roughly constant factor. For reference purposes, we also calculate the standard IP DGFEM solution $u_{h,p}$, using

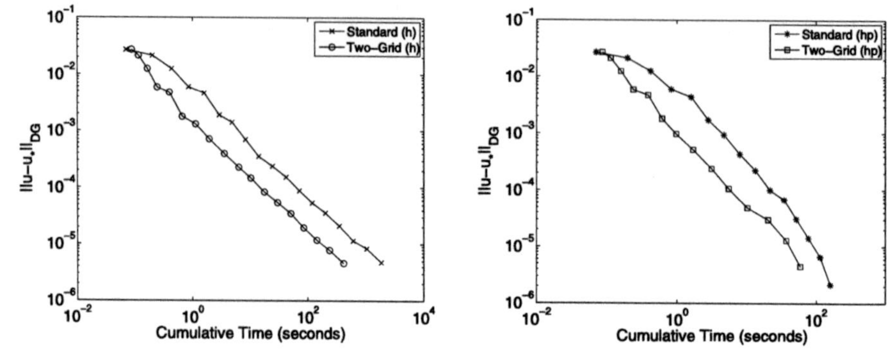

FIGURE 3. CPU timing of the IP DGFEM ($u_* = u_{h,p}$) and the two-grid IP DGFEM ($u_* = u_{2G}$) employing h– and hp–refinement.

both h– and hp–adaptive refinement; cf. Figure 2(a). Finally, in Figure 3 we compare the error in the standard and two-grid IP DGFEMs against the cumulative CPU time when both h– and hp–adaptive refinement are employed; here, we observe that the two-grid IP DGFEM is more efficient than the standard IP DGFEM.

References

1. O. Axelsson and W. Layton, *A two-level method for the discretization of nonlinear boundary value problems*, SIAM J. Numer. Anal. **33** (1996), no. 6, 2359–2374. MR1427468 (98c:65181)
2. I. Babuška and M. Suri, *The $h-p$ version of the finite element method with quasiuniform meshes*, RAIRO Modél. Math. Anal. Numér. **21** (1987), no. 2, 199–238. MR896241 (88d:65154)
3. S. Congreve, *Discontinuous Galerkin finite element methods for quasilinear partial differential equations*, Ph.D. thesis, University of Nottingham, in preparation.
4. S. Congreve, P. Houston, and T. P. Wihler, *Two-grid hp-version DGFEMs for strongly monotone second-order quasilinear elliptic PDEs*, Proc. Appl. Math. Mech. **11** (2011), 3–6.
5. ———, *Two-grid hp-version discontinuous Galerkin finite element methods for second-order quasilinear elliptic PDEs*, J. Sci. Comput. (in press).
6. P. Houston, J. Robson, and E. Süli, *Discontinuous Galerkin finite element approximation of quasilinear elliptic boundary value problems I: the scalar case*, IMA J. Numer. Anal. **25** (2005), 726–749. MR2170521 (2006k:65322)
7. P. Houston, E. Süli, and T. P. Wihler, *A posteriori error analysis of hp-version discontinuous Galerkin finite-element methods for second-order quasi-linear PDEs*, IMA J. Numer. Anal. **28** (2008), no. 2, 245–273. MR2401198 (2009c:65274)
8. C. Schwab, *p- and hp-FEM — theory and applications in solid and fluid mechanics*, Oxford University Press, Oxford, 1998. MR1695813 (2000d:65003)
9. J. Xu, *Two-grid discretization techniques for linear and nonlinear PDEs*, SIAM J. Numer. Anal. **33** (1996), no. 5, 1759–1777. MR1411848 (97i:65169)

SCHOOL OF MATHEMATICAL SCIENCES, UNIVERSITY OF NOTTINGHAM, UNIVERSITY PARK, NOTTINGHAM, NG7 2RD, UNITED KINGDOM
E-mail address: pmxsc@nottingham.ac.uk

SCHOOL OF MATHEMATICAL SCIENCES, UNIVERSITY OF NOTTINGHAM, UNIVERSITY PARK, NOTTINGHAM, NG7 2RD, UNITED KINGDOM
E-mail address: Paul.Houston@nottingham.ac.uk

On the enforcement of discrete mass conservation in incompressible flow simulations with continuous velocity approximation

Erica M. D'Agnillo and Leo G. Rebholz

ABSTRACT. We consider the problem of numerical instability and inaccuracy caused by poor discrete mass conservation in finite element simulations of the incompressible Navier-Stokes and related equations. We analytically and numerically compare three different finite element implementations with continuous velocity field approximations, that give different levels of mass conservation: Taylor-Hood elements, grad-div stabilized Taylor-Hood elements, and Scott-Vogelius elements.

1. Introduction

An important subtask in many industrial problems is the numerical simulation of fluid flow, and herein we consider incompressible fluids modeled with the Navier-Stokes equations (NSE). There are many numerical methods that can be used to accomplish this, but one common choice is the finite element method (FEM) implementation, which is attractive because of its rigorous mathematical foundation and flexibility for complicated geometries and boundary conditions. However, one potential drawback of FEM implementations of incompressible NSE and related equations is that mass conservation is only enforced weakly, which has recently been found to play a major role in a computed solution's accuracy for certain flow problems [**Lin09, CELR11**]. Indeed, this weak enforcement of mass conservation is not always sufficient for many common element choices such as Taylor-Hood (TH, defined in Section 2), where mass conservation is only enforced globally and controlled only by the H^1 velocity error. One potential 'fix' for this problem is to use grad-div stabilization, which directly controls the size of the L^2 norm of the divergence error, but at the cost of increasing the condition number of the resulting linear systems and potentially making their sparsity structure worse.

Scott-Vogelius elements (SV, defined in Section 2) have recently been found to be of practical use on meshes with mild restrictions on their structure [**Qin94, Zha05**]. These elements differ in construction from TH only in that their pressure space is discontinuous, but this difference allows SV elements to provide pointwise mass conservation in their solutions, even though it is only weakly enforced by the FEM. However, SV elements have some drawbacks. First, they are likely not

2000 *Mathematics Subject Classification.* Primary 35Q30, 65M12, 65N30, 76D05. 76F65.

This work was partially supported by NSF grant DMS1112593.

as efficient as TH elements in developed codes, since TH elements are much more common and therefore preconditioners and linear solvers are already tailored to their use; converting a code to be as efficient with SV could be costly. Second, to use SV with lower order elements, a barycenter (or possibly more complex) mesh refinement of a regular mesh must be used to guarantee stability. This can be quite impractical when one has to create a mesh in 3D just from boundary points of a domain (e.g. when meshing an aorta or carotid artery obtained from an image of a patient), and reasonable quality Delaunay meshes already give millions of degrees of freedom. Finally, because it is comprised of discontinuous piecewise polynomials instead of continuous piecewise polynomials, the pressure space of SV is significantly larger than for TH. However, if one is using SV with lower order elements that require a macrostructure on the mesh, static condensation can be used to alleviate this potential problem by significantly reducing the system size [**CBL+12**].

In this work, we will give both analysis and numerical examples to compare the accuracy provided by TH, TH with grad-div stabilization (THGD), and SV elements in FEM simulations of Navier-Stokes and related flows. We first give an analysis that shows what causes the differences in accuracy for TH, THGD, and SV solutions, and then give numerical tests for an analytical test problem with known solution, channel flow around a cylinder and heated cavity.

2. Analytical comparison

We begin our analytical comparison by defining notation, and properly defining the spaces and elements. For simplicity, we consider the domain to be a convex polygon/polyhedra, and we will enforce homogeneous Dirichlet velocity boundary conditions.

Recall the natural Hilbert spaces natural for velocity and pressure in this setting are $X := H_0^1$ and $Q := L_0^2$ We denote the L^2 inner product and norm by $(\cdot,\cdot)$ and $\|\cdot\|$, respectively. Due to the Poincare inequality, we have equivalence of the H^1 norm on $H_0^1(\Omega)$ is equivalent to $\|\nabla\cdot\|$. $T_h(\Omega)$ will denote a regular, conforming triangulation/tetrahedralization of the domain Ω, with maximum element diameter h. The finite element spaces $(X_h, Q_h^{TH}) \subset (X,Q)$ and $(X_h, Q_H^{SV}) \subset (X,Q)$ correspond to TH and SV elements respectively, and are defined by

$$
\begin{aligned}
X_h &= X_h(T_h(\Omega)) = \{v_h \in C^0(\Omega) \cap P_k(e)\, \forall e \in T_h(\Omega)\},\\
Q_h^{TH} &= Q_h^{TH}(T_h(\Omega)) = \{q_h \in C^0(\Omega) \cap P_{k-1}(e)\, \forall e \in T_h(\Omega)\},\\
Q_h^{SV} &= Q_h^{SV}(T_h(\Omega)) = \{q_h \in P_{k-1}(e)\, \forall e \in T_h(\Omega)\},
\end{aligned}
$$

where $P_k(e)$ denotes degree k polynomials on an element e. We note TH and SV use the same velocity space, andX_h is the discrete velocity space for both.

2.1. Stokes equations. Consider now the Stokes equations with homogeneous Dirichlet boundary conditions for velocity:

$$-\nu\Delta u - \nabla p = f. \;\; \nabla\cdot u = 0, \;\; u|_{\partial\Omega} = 0.$$

This leads to the following finite element formulation

$$
\begin{aligned}
&(2.1) \qquad \nu(\nabla u_h, \nabla v_h) + (p_h, \nabla\cdot u_h) = (f, v_h) \quad \forall v_h \in X_h,\\
&(2.2) \qquad (\nabla\cdot u_h, q_h) = 0 \quad \forall q_h \in Q_h.
\end{aligned}
$$

For more complicated equations such as Navier-Stokes, Boussinesq, magnetohydrodynamics, etc., equation (2.1) will become more complicated, but equation (2.2) will stay the same. Hence this is a good starting place to consider mass conservation in difference element choices.

TH elements, given by $(X_h, Q_h) = ((P_k)^d, P_{k-1})$, can only be shown to conserve mass weakly. Since $\nabla \cdot X_h \not\subseteq Q_h$, we cannot choose $q_h = \nabla \cdot u_h$ in (2.2). However, since $u_h = 0$ on $\partial\Omega$, $\int_{\partial\Omega} u_h \cdot n\, ds = 0$, so by the divergence theorem we get $\int_\Omega \nabla \cdot u_h = 0$, and thus we see these elements do enforce mass conservation globally. Also, finite element error analysis (see Theorem 2.1 below) provides us with the error estimate

$$\|\nabla(u - u_h)\| \leq Ch^k$$

where C depends on data and seminorms of the true solution pressure and velocity, but is independent of h and k. Since $\nabla \cdot u = 0$ for the true Stokes solution, this implies that

$$\|\nabla \cdot u_h\| = \|\nabla \cdot (u - u_h)\| \leq \|\nabla(u - u_h)\| \leq Ch^k$$

Thus the mass conservation in TH solution is controlled by the (optimal) H^1 velocity error. However, in practice, there is a minimum h which can be used, and as we will see below this apparently optimal estimate may not be good enough.

To better control the divergence error when using TH elements, grad-div stabilization can be used to improve discrete mass conservation and overall accuracy [**LMN+09, OLL09, OR04**]. This is a consistent stabilization in the sense that it can be derived at the continuous level by adding $-\gamma\nabla(\nabla \cdot u) = 0$ to the momentum equation. In the FEM, this term becomes $\gamma(\nabla \cdot u_h, \nabla \cdot v_h)$. This "extra" term penalizes the divergence and thus forces $\|\nabla \cdot u_h\|$ to be "small." To see this idea, consider the (2.1)-(2.2) with the grad-div term added to (2.1). Choosing the test function $v_h = u_h$ and using standard inequalities gives

$$\text{(2.3)}\quad \gamma\|\nabla \cdot u_h\|^2 + \nu\|\nabla u_h\|^2 = (f, u_h) \leq \|f_{-1}\|\|\nabla u_h\| \leq \frac{\nu^{-1}}{2}\|f\|_{-1}^2 + \frac{\nu}{2}\|\nabla u_h\|^2,$$

which implies

$$\gamma\|\nabla \cdot u_h\|^2 \leq C(\nu, f).$$

Hence with THGD, we can control discrete mass conservation with the parameter γ. A potential downside to this is that the condition number scales linearly with γ, so choosing γ too large will cause inaccuracy in the solution of the linear solve.

SV elements, denoted by $(X_h, Q_h^{SV}) = ((P_k)^d, P_{k-1}^{disc})$, provide pointwise mass conservation when used with (2.1)-(2.2). With SV, $\nabla \cdot X_h \subseteq Q_h$, so we can choose $q_h = \nabla \cdot u_h$ in (2.2), which gives us

$$\|\nabla \cdot u_h\|^2 = 0$$

and therefore $\nabla \cdot u_h = 0$.

SV elements, which were developed in the 1980's [**Vog83b, Vog83a, SV85**] were originally only known to be inf-sup stable for high degree polynomials. In 1994, Qin and Arnold showed that using (P_2, P_1^{disc}) on a barycenter triangular mesh in two dimensions was LBB-stable [**AQ92**]. Then in 2005, Zhang showed that in three dimensions, (P_3, P_2^{disc}) on a barycenter tetrahedra mesh was LBB-stable [**Zha05**]. It was later shown by Zhang in 2008 that using a Powell-Sabin mesh in three dimensions, which is a special refinement of one tetrahedral into twelve, (P_2, P_1^{disc}) is inf-sup stable. Due to these results, in recent years have people began using these elements. All references to SV elements will assume that the mesh and polynomial degree are sufficient to allow inf-sup stability.

2.2. Error analysis. We now compare the a priori error analysis of (2.1)-(2.2) for the three methods TH, TH with grad-div, and SV. This will show us how there can be differences between the methods, and in which cases to use one method instead of another.

THEOREM 2.1. *Let $(u,p) \in (X,Q)$ solve the Stokes equation, and $(u_h,p_h) \in (X_h,Q_h)$ solve (2.1)-(2.2), with $Q_h = Q_h^{SV}$ if SV elements are used, and Q_h^{TH} otherwise. Then we have that if*

- *using Taylor-Hood elements and $\gamma = 0$*

$$\nu\|\nabla(u-u_h)\|^2 \leq C\left(\nu \inf_{v_h\in X_h}\|\nabla(u-v_h)\|^2 + \nu^{-1}\inf_{q_h\in Q_h^{TH}}\|p-q_h\|^2\right),$$

- *using Taylor-Hood elements and $\gamma > 0$*

$$\nu\|\nabla(u-u_h)\|^2 \leq C\left(\inf_{v_h\in X_h}(\gamma\|\nabla\cdot(u-v_h)\|^2 + \nu\|\nabla(u-v_h)\|^2) + \gamma^{-1}\inf_{q_h\in Q_h^{TH}}\|p-q_h\|^2\right),$$

- *and using Scott-Vogelius elements*

$$\|\nabla(u-u_h)\| \leq C\inf_{v_h\in X_h}\|\nabla(u-v_h)\|$$

where the C in the above equations is a generic constant independent of ν.

REMARK 2.2. For example, if $k=2$, $\inf_{q_h\in Q_h}\|p-q_h\|^2 \leq Ch^2|p|_2$. So if $|p|_2$ is very large, then we can deduce from the Theorem 3.1 that Scott-Vogelius elements might be worth the extra computational cost. If it is very small, then TH might be the best option.

PROOF. The proof for the TH case is well known [**Lay08**]. For TH with grad-div stabilization, the key difference is how to handle the pressure term that arises, $(p-q_h, \nabla\cdot\phi_h)$, where $\phi_h = u_h - w_h$ for some arbitrary w_h in the discretely divergence free subspace of X_h. When grad-div stabilization is used, standard analysis reveals there will be a left hand side term $\gamma\|\nabla\cdot\phi_h\|^2$, and so from here Cauchy-Schwarz and Young can be used in the straightforward way to get the stated result. For the SV case, the main difference is that the discretely divergence free subspace of X_h is pointwise divergence free, and so the pressure term simply vanishes. □

3. Numerical comparison

We now compare the methods on three test problems. In all of our tests, we use $\gamma = 1$ with THGD. Also, we use (P_2, P_1) elements for TH and THGD, and (P_2, P_1^{disc}) for SV.

3.1. Test problem 1: Stokes problem with known analytical solution. Consider the Stokes problem with true solution:

$$u = \begin{pmatrix}\cos(\pi y)\\ \sin(\pi x)\end{pmatrix} \quad p = \sin(n\pi(x+y)),\ \ n\in\mathbb{Z}$$

with domain taken to be the unit square. We compute (2.1)-(2.2) using barycenter refined triangular meshes that has 24,843 degrees of freedom (dof) for TH and THGD, and 26,452 dof for SV. We compute the above problem with $n = 1$ and

TABLE 1. Velocity and divergence errors for test problem 1

$\nu = 1$ and $n = 1$:

Method	$\|\nabla(u - u_h)\|$	$\|\nabla \cdot u_h\|$
TH	1.01538e-3	6.77393e-4
THGD	9.22284e-4	4.10547e-4
SV	1.84529e-3	7.07111e-9

$\nu = 0.001$ and $n = 5$:

Method	$\|\nabla(u - u_h)\|$	$\|\nabla \cdot u_h\|$
TH	13.749	13.2173
THGD	2.02068e-2	1.50882e-2
SV	4.4977e-3	7.06944e-9

$n = 5$ for large and small ν. The results are shown in Table 1, and agree with the above analysis that as the pressure gets more complex and ν decreases, mass conservation becomes more important in the H^1 velocity error.

3.2. Test problem 2: 2D channel flow around a cylinder. We now consider 2D NSE channel flow around a cylinder as in [**Joh04**]. We compute the NSE with the FEM

$$((u_h)_t, v_h) + (u_h \cdot \nabla u_h, v_h) + \frac{1}{2}((\nabla \cdot u_h)u_h, v_h)$$

$$+\nu(\nabla u_h, \nabla v_h) - (p_h, \nabla \cdot u_h) = (f, v_h) \; \forall v_h \in X_h, \tag{3.1}$$

$$(\nabla \cdot u_h, q_h) = 0 \; \forall q_h \in Q_h, \tag{3.2}$$

with the grad-div stabilization term for that case, and discretize temporally with linearly extrapolated Crank-Nicolson.

The domain is a 2.2×0.41 rectangular channel with a cylinder of radius 0.05 centered at $(0.2, 0.2)$. The inflow and outflow are given by

$$\begin{aligned} u_1(0, y, t) &= u_1(2.2, y, t) = \frac{6}{0.41^2} \sin\left(\frac{\pi t}{8}\right) y(0.41 - y) \\ u_2(0, y, t) &= u_2(2.2, y, t) = 0. \end{aligned}$$

Using $\nu = 10^{-3}$, $\Delta t = 0.001$, $f = 0$, and $T = 8$ we will again study the solutions using TH, TH with grad-div stabilization (THGD) with $\gamma = 1$, and SV on a 4 different barycenter refined triangular meshes having approximately 10k, 20k, 30k, and 40k total dof with SV elements (somewhat less for TH cases), labeled 1-4 respectively. The reference values for max drag and lift coefficients of a resolved solution (using upwards of one million dof) are 2.93-2.97 for drag and 0.47-0.49 for lift [**Joh04**]. Table 2 shows the drag and lift values calculated for each case, and we observe that THGD and SV give much better predictions on these relatively coarse meshes.

Figures 1-3 shows the velocity contours on Mesh 2 at time T = 6. What is interesting is that we only observe a slight difference in the TH solution around the cylinder compared to the THGD and SV solution, even though the fundamentally important quantities for maximum lift and drag coefficients differ greatly.

TABLE 2. Maximum values of lift and drag coefficients for the flow around a cylinder test problem

Mesh	TH (drag,lift)	THGD (drag,lift)	SV (drag,lift)
1	2.90649 , 0.042845	2.78777 , 0.462946	2.789 , 0.463285
2	2.92038 , 0.218153	2.85115 , 0.461426	2.85753 , 0.456316
3	2.93199 , 0.354614	2.89991 , 0.447685	2.90547 , 0.460096
4	2.93957 , 0.420598	2.9261 , 0.471722	2.92613 , 0.471739

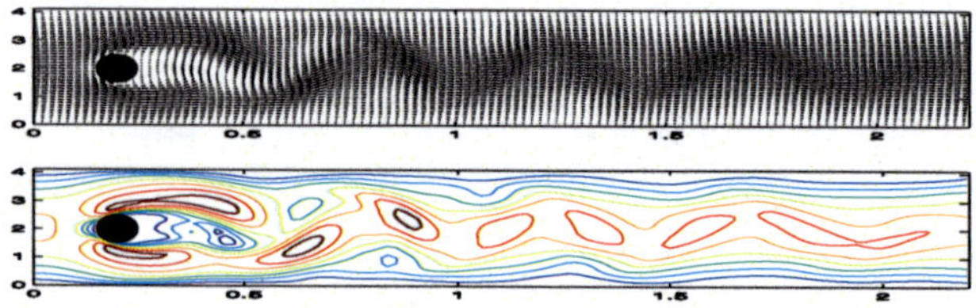

FIGURE 1. Velocity field and speed contours using TH t=6.

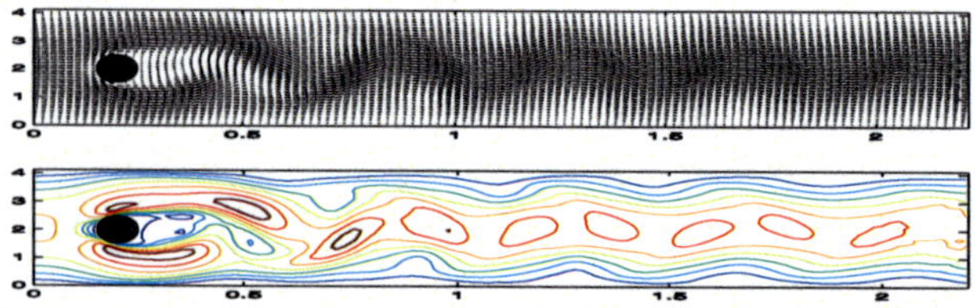

FIGURE 2. Velocity field and speed contours using THGD at t=6.

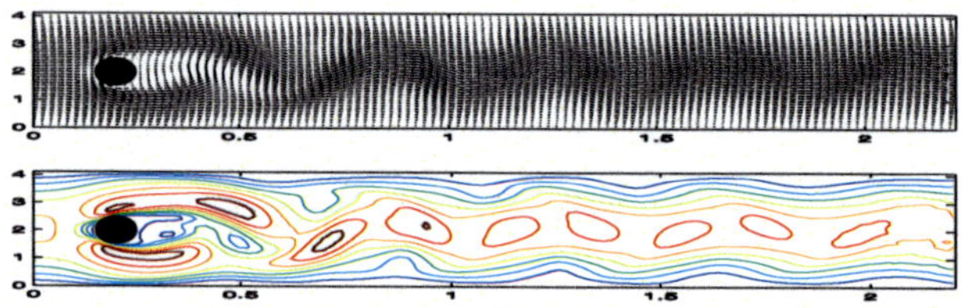

FIGURE 3. Velocity field and speed contours using SV at t=6.

3.3. Heated Cavity Problem. The Navier-Stokes-Boissinesq equations, given by

$$\begin{aligned} \frac{1}{Pr}(u_t + (u \cdot \nabla)u) - \Delta u + \nabla p &= \mathrm{Ra} T e_y \\ \nabla \cdot u &= 0 \\ T_t - \Delta T + u \cdot \nabla T &= 0, \end{aligned}$$

can be used to model natural convection in a differentially heated cavity. We implement this system in the usual way with the FEM (see, e.g. [**GLRWar**]), and using P_2 to approximate temperature. Consider the domain $\Omega = (0,1)^2$ where the left side is heated, the right cooled, and the top and bottom are insulated. No slip boundary conditions are enforced on all wall for the velocity. We will choose Pr $= \infty$ and Ra $= 10^6$, which represents silicon oil, and study the results using TH, THGD, and SV all on a barycenter refined mesh with 2,778 velocity dof. For this problem, the size of the pressure is of the order of Ra, and so this problem has the potential to see better solutions with SV elements than TH or even grad-div stabilized TH. Figures 4-6 shows the velocity and temperature plots for each of the three cases, and we observe a much better solution for SV compared to the other cases. We note for the case of air (Pr=0.71, Ra=10^6) in [**GLRWar**] saw similar behavior among these methods, but in this case it happens without convection in the momentum equation.

FIGURE 4. Velocity Streamlines and Temperature Contours for TH, $\|\nabla \cdot u_h^{TH}\| = 332.845$

FIGURE 5. Velocity Streamlines and Temperature Contours for THGD, $\|\nabla \cdot u_h^{THGD}\| = 199.753$

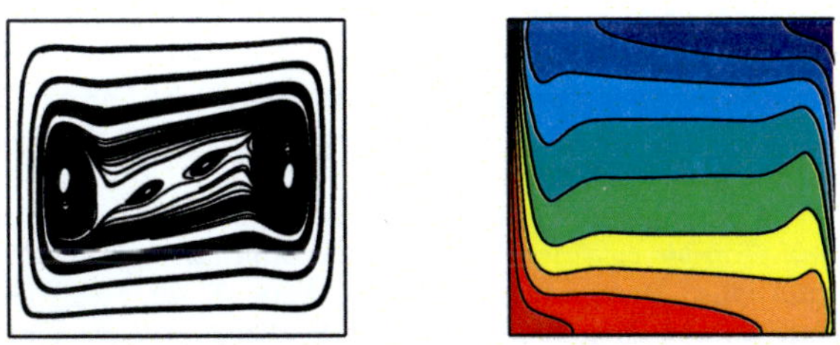

FIGURE 6. Velocity Streamlines and Temperature Contours for SV, $\|\nabla \cdot u_h^{SV}\| = 1.46457e-7$

4. Conclusion

We have observed that even though SV elements could possibly be more expensive than TH, in the problems where the pressure is large, SV can give a much better solution that TH or THGD. In benchmark flow around a cylinder, we observed TH give a poor solution, but the addition of grad-div stabilization was enough to 'fix' the solution, and the SV solution looked identical to the grad-div stabilized TH solution.

References

[AQ92] D.N. Arnold and J. Qin, *Quadratic velocity/linear pressure Stokes elements*, Advances in Computer Methods for Partial Differential Equations VII (R. Vichnevetsky, D. Knight, and G. Richter, eds.), IMACS, 1992, pp. 28–34.

[CBL+12] B. Cousins, S. Le Borne, A. Linke, L. Rebholz, and Z. Wang, *On incompressible flow simulations using Scott-Vogelius finite elements*, Submitted (2012).

[CELR11] Michael A. Case, Vincent J. Ervin, Alexander Linke, and Leo G. Rebholz, *A connection between Scott-Vogelius and grad-div stabilized Taylor-Hood FE approximations of the Navier-Stokes equations*, SIAM J. Numer. Anal. **49** (2011), no. 4, 1461–1481, DOI 10.1137/100794250. MR2831056

[GLRWar] K. Galvin, A. Linke, L. Rebholz, and N. Wilson, *Stabilizing poor mass conservation in incompressible flow problems with large irrotational forcing and application to thermal convection*, Computer Methods in Applied Mechanics and Engineering (to appear.).

[Joh04] V. John, *Reference values for drag and lift of a two dimensional time-dependent flow around a cylinder*, International Journal for Numerical Methods in Fluids **44** (2004), 777–788.

[Lay08] William Layton, *Introduction to the numerical analysis of incompressible viscous flows*, Computational Science & Engineering, vol. 6, Society for Industrial and Applied Mathematics (SIAM), Philadelphia, PA, 2008. With a foreword by Max Gunzburger. MR2442411 (2009m:76056)

[Lin09] Alexander Linke, *Collision in a cross-shaped domain—a steady 2d Navier-Stokes example demonstrating the importance of mass conservation in CFD*, Comput. Methods Appl. Mech. Engrg. **198** (2009), no. 41-44, 3278–3286, DOI 10.1016/j.cma.2009.06.016. MR2571343

[LMN+09] William Layton, Carolina C. Manica, Monika Neda, Maxim Olshanskii, and Leo G. Rebholz, *On the accuracy of the rotation form in simulations of the Navier-Stokes equations*, J. Comput. Phys. **228** (2009), no. 9, 3433–3447, DOI 10.1016/j.jcp.2009.01.027. MR2513841 (2010h:76117)

[OLL09] Maxim Olshanskii, Gert Lube, Timo Heister, and Johannes Löwe, *Grad-div stabilization and subgrid pressure models for the incompressible Navier-Stokes equations*, Comput. Methods Appl. Mech. Engrg. **198** (2009), no. 49-52, 3975–3988, DOI 10.1016/j.cma.2009.09.005. MR2557485 (2010k:76070)

[OR04] Maxim A. Olshanskii and Arnold Reusken, *Grad-div stabilization for Stokes equations*, Math. Comp. **73** (2004), no. 248, 1699–1718, DOI 10.1090/S0025-5718-03-01629-6. MR2059732 (2005a:65137)

[Qin94] Jinshui Qin, *On the convergence of some low order mixed finite elements for incompressible fluids*, ProQuest LLC, Ann Arbor, MI, 1994. Thesis (Ph.D.)–The Pennsylvania State University. MR2691498

[SV85] L.R. Scott and M. Vogelius, *Conforming finite element methods for incompressible and nearly incompressible continua*, Large-scale computations in fluid mechanics, Part 2, Lectures in Applied Mathematics, vol. 22-2, Amer. Math. Soc., 1985, pp. 221–244.

[Vog83a] Michael Vogelius, *An analysis of the p-version of the finite element method for nearly incompressible materials. Uniformly valid, optimal error estimates*, Numer. Math. **41** (1983), no. 1, 39–53, DOI 10.1007/BF01396304. MR696549 (85f:65113b)

[Vog83b] Michael Vogelius, *A right-inverse for the divergence operator in spaces of piecewise polynomials. Application to the p-version of the finite element method*, Numer. Math. **41** (1983), no. 1, 19–37, DOI 10.1007/BF01396303. MR696548 (85f:65113a)

[Zha05] Shangyou Zhang, *A new family of stable mixed finite elements for the 3D Stokes equations*, Math. Comp. **74** (2005), no. 250, 543–554, DOI 10.1090/S0025-5718-04-01711-9. MR2114637 (2005j:65151)

DEPARTMENT OF MATHEMATICAL SCIENCES, CLEMSON UNIVERSITY, CLEMSON, SOUTH CAROLINA 29634

E-mail address: `edagnil@clemson.edu`

DEPARTMENT OF MATHEMATICAL SCIENCES, CLEMSON UNIVERSITY, CLEMSON, SOUTH CAROLINA29634

E-mail address: `rebholz@clemson.edu`

An adaptive treecode algorithm for computing the evolution of microstructures in an elastic media

Hualong Feng, Amlan Barua, Xiaofan Li, and Shuwang Li

Abstract. This paper is concerned with computing the evolution of microstructures in an elastic media. The problem consists of a system of boundary integral equations describing the diffusion and elastic fields. We develop an adaptive treecode algorithm to speed up the evaluations of these integrals, and demonstrate its efficiency through long time simulations of complicated microstrutures.

1. Introduction

This paper is motivated by a crystal growth problem arising from the production of binary alloys via solid-solid phase transformations. Mullins and Sekerka proposed a linear analytic approach to this problem [**MS63**]. Their linear stability analysis suggests that the diffusion process is unstable in the sense that the evolving microstructures are susceptible to short wavelength perturbations. At long times, the nonlinear diffusional instability leads the evolution to a variety of complicated morphologies, which may be addressed from a computational perspective. The problem has been studied using the boundary integral method (BIM), the phase field method, and many others. In this paper, we use a boundary integral method to compute the evolution of precipitates in an elastic media. The main advantage of the BIM is its high accuracy, dimension reduction, and exact treatment of the boundary conditions.

It would be difficult to conduct a comprehensive literature review of the boundary integral method applied to microstructure computations, so we only mention works that came to our attention (for details, see a review article [**HLS01**]). The most related work is presented in [**JLL97**], where the authors studied a precipitate/matrix phase transition problem with elasticity. Note that the direct summation method in their work has a time complexity $O(N^2)$. To speed up computations (especially the matrix-vector multiplications in an iterative solver like GMRES [**SS86**]), fast summation methods have been developed over the past several decades. Examples include the fast multipole method (FMM) [**GR87, CGR99**] and the treecode [**BH86, LK01**]. For the specific kernel functions considered

---

2010 *Mathematics Subject Classification.* Primary 65N35, 74B05; Secondary 65D30.

Key words and phrases. Treecode, elasticity, boundary integral method.

The third author was supported in part by NSF Grant DMS#0923111.

The fourth author (for correspondence) is supported in part by NSF Grant DMS#0914923.

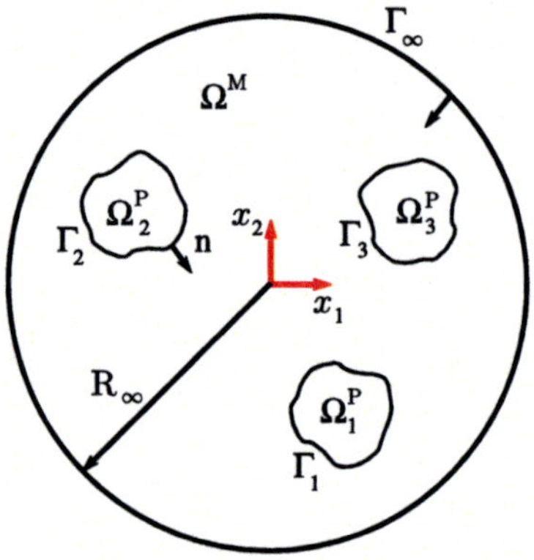

FIGURE 1. A two phase domain showing 3 precipitates surrounded by matrix

in our problem, we develop a fast adaptive treecode algorithm. Adaptivity is of importance as the distribution of computational points is not uniform in our problem. The treecode has a time complexity $O(N \log N)$. The fundamental idea in a treecode is that particle-particle interactions are approximated with particle-cluster or cluster-cluster interactions. We only consider particle-cluster interactions in the current work. Cluster-cluster interactions can also be used in a treecode. Due to their complexity, they will be incorporated in future work.

This paper is organized as follows. In section 2, we present the mathematical formulation of the problem. In section 3, we develop a treecode algorithm. In section 4, we present the results and give concluding remarks.

2. Mathematical formulation

2.1. Governing equations. We study microstructural evolutions in a two-dimensional space. The matrix phase Ω^M extends to infinity. The precipitate phase Ω^P consists of p precipitates occupying separate bounded regions Ω_i^P. The boundary of the ith precipitate is denoted by $\Gamma_i, i = 1, \cdots, p$. The union of all individual boundaries forms the interface between the matrix and the precipitates, and is denoted by Γ. A schematic description of the problem is given by Figure 1.

Let U be the concentration of the diffusing species in the matrix phase. Under a quasi-static assumption, U satisfies the Laplace's equation $\nabla^2 U = 0$ in $\Omega^M \subset R^2$ subject to $U = \tau(\theta)\kappa + ZG_{\text{el}}$ on $\Gamma = \cup_{i=1}^n \Gamma_i$, where τ is the anisotropic surface tension, κ is the mean curvature, G_{el} is the elastic energy density **[JLL97]**, and Z characterizes the relative contribution of elasticity. Mathematically, $\tau(\theta) = \gamma(\theta) + \gamma''(\theta)$, where the m-fold anisotropic surfae energy $\gamma(\theta) = 1 + \gamma_m \cos m\theta$. Hence, $\tau(\theta) = 1 - (m^2 - 1)\gamma_m \cos m\theta$, where γ_m is the strength. When $\gamma_m = 0$, the surface tension is isotropic. The far-field condition is taken to be $\lim_{R_\infty \to \infty} \int_{\Gamma_\infty} (-\nabla U \cdot n) ds = 2\pi J$, where s is arclength and $2\pi J$ is the total mass flux. Once the concentration U is obtained by solving the Laplace equation, the velocity of the interface can be computed by taking normal derivative of the concentration, $V_i = \frac{\partial U}{\partial n}$ on Γ_i.

The G_{el} term in the boundary condition requires a solution to the two phase elasticity problem for domain Ω^M and Ω^P. Assuming there is no body force, the elasticity problem is posed as $\sigma^{\chi}_{ij,j} = 0$ in $\Omega^{\chi}, u^P_i = u^M_i$ on $\Gamma, \sigma^P_{ij}\widehat{n}_j = \sigma^M_{ij}\widehat{n}_j$ on Γ, subject to the boundary condition at infinity $\lim_{r\to\infty} \epsilon^M_{ij} = \epsilon^0_{ij}$, where χ can be either matrix or precipitate. Once the elasticity equations are solved, we can compute the elastic energy density G_{el} in the boundary condition for the diffusion equation [**JLL97**].

2.2. Boundary integral formulation of the elasticity problem. Using complex variable notations $z = x_1 + ix_2$ to denote computational points on the interface, we write the solutions to the elasticity problems in terms of boundary integral equations [**JLL97**] ,

$$\frac{1}{2}u_j + \int_{\Gamma_i} u_k T^P_{jk} ds' - \int_{\Gamma_i} t_k U^P_{jk} ds' = \int_{\Gamma_i} t^T_k U_{jk} ds', \tag{2.1}$$

$$\frac{1}{2}u_j - \int_{\Gamma} u_k T^M_{jk} ds' + \int_{\Gamma} t_k U^M_{jk} ds' = \frac{1}{2}u^0_j - \int_{\Gamma} u^0_k T^M_{jk} ds' + \int_{\Gamma} t^0_k U^M_{jk} ds', \tag{2.2}$$

where u_j and t_j are the unknown displacement vector and traction vector respectively. Functions $U_{jk}(z, z') = \frac{1}{8\pi\mu(1-\nu)}\left[(3-4\nu)\ln\left(\frac{1}{r}\right)\delta_{jk} + r_{,j'}r_{,k'}\right]$ and $T_{jk}(z, z') = \frac{-1}{4\pi(1-\nu)r}\left[\frac{\partial r}{\partial n'}\left((1-2\nu)\delta_{jk} + 2r_{,j'}r_{,k'}\right) + (1-2\nu)(n'_j r_{,k'} - n'_k r_{,j'})\right]$ are the fundamental Kelvin solutions, where $r = |z' - z|, r_{,k'} = \frac{\partial r}{\partial x'_k}$ and n'_k is the kth component of the normal at point z'. The quantity $t^T_j = \sigma^T_{jk}n_k$ on the right hand side is a given traction due to the misfit strain. In this paper, we study the applied strain and neglect misfit strain, since these two effects can be made equivalent under certain conditions [**Mur87**].

Note that in Eq. (2.2), the integration is over the entire interface, while the integration in Eq. (2.1) is only over the interface Γ_i. These $2p+2$ integrals equations must be solved for u_1, u_2, t_1, and t_2 to compute the stress and strains. To solve these equations numerically, we discretize the integrals using a spectrally accurate alternating point quadrature [**SI88, JLL97**], and solve the discrete system iteratively using a preconditioned GMRES method [**SS86, JLL97**].

2.3. Boundary integral formulation of the diffusion problem. In terms of an unknown dipole density potential function ϕ on Γ and unknown p source terms $A_1, A_2, ..., A_p$, we write the solution to the diffusion equation

$$\left(-\frac{1}{2}\mathcal{I} + K\right)[\phi] + \sum_{k=1}^{p} A_k \log|z(s) - S_k| = \kappa + ZG_{\text{el}}, \tag{2.3}$$

$$\sum_{k=1}^{p} A_k = J, \tag{2.4}$$

$$\int_{\Gamma_j} \phi(s')ds' = 0, \quad j = 1, \cdots, p-1, \tag{2.5}$$

where $S_k = x_{1k} + ix_{2k}$ is a point inside the closed interface Γ_k, and the integral kernel K is such that

$$K[\phi](s) = \frac{1}{2\pi}\int_{\Gamma} \phi(s')\left[\frac{\partial}{\partial n(s')}\log\left|z(s') - z(s)\right| + 1\right] ds'. \tag{2.6}$$

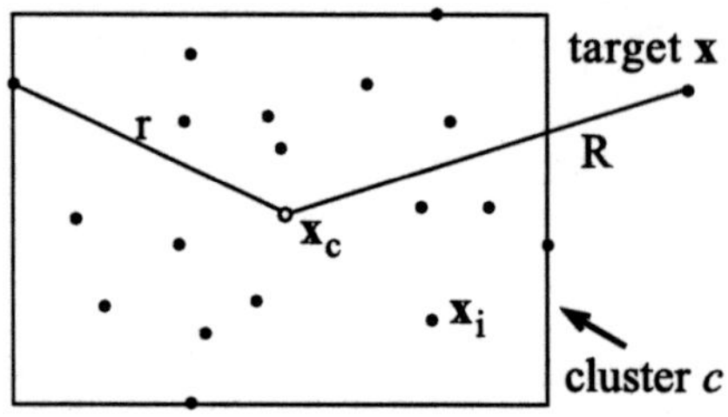

FIGURE 2. A source cluster and a target, a q-th order Taylor approximation incurs error $\sim (r/R)^q$ [**FBLL**]

Similar to the elasticity problem, we discretize the Eqs. (2.3) and (2.5), and solve the discretized system for ϕ and A_k using a preconditioned GMRES method [**JLL97**]. Note that the integral kernel in Eq. (2.6) is smooth even if both $z(s^{'})$ and $z(s)$ lie on the same Γ_i. Once the diffusion problem is solved, we can compute the normal velocity V_i of the interface Γ_i using the Dirichlet-Neumann map and update the interface.

3. The treecode

In this section, we present an adaptive treecode for evaluating the boundary integrals in (2.1), (2.2), and (2.6) following [**BH86, LK01**]. When developing a treecode, two issues need to be addressed. First, one has to derive recurrence relations between the Taylor coefficients for the kernel functions. Thus all Taylor coefficients can be evaluated. Second, one needs to perform error analysis of the Taylor approximations. In the treecode, we call points making contributions sources, and points receiving contributions targets, though they may coincide. The treecode is adaptive in the sense that it accommodates non-uniform point distribution well.

When a target point is distant from a cluster of source points (see Figure 2), the interaction between the target and the cluster can be computed with a Taylor approximation. We note that we write $\mathbf{x}$ to mean (x_1, x_2), and we also use the complex variable notation $z = x_1 + ix_2$, so $\mathbf{x}$ and z are used interchangeably. Let $c = \{\mathbf{x}_i, i = 1, \ldots, N_c\}$ be a cluster of N_c computational points with center $\mathbf{x}_c$ and w_i be their corresponding weights. If $\Phi(\mathbf{x})$ is the kernel function, then the interaction between a target point $\mathbf{x}$ and a distant cluster c of source points is

$$\sum_{i=1}^{N_c} \Phi(\mathbf{x} - \mathbf{x}_i) w_i. \tag{3.1}$$

Next, we expand the kernel in a Taylor series about $\mathbf{x} - \mathbf{x}_c$.

$$\begin{aligned} \sum_{i=1}^{N_c} \Phi(\mathbf{x} - \mathbf{x}_i) w_i &= \sum_{i=1}^{N_c} \Phi(\mathbf{x} - \mathbf{x}_c + \mathbf{x}_c - \mathbf{x}_i) w_i \\ &= \sum_{i=1}^{N_c} \sum_{\mathbf{k}} \frac{1}{\mathbf{k}!} D^{\mathbf{k}} \Phi(\mathbf{x} - \mathbf{x}_c)(\mathbf{x}_c - \mathbf{x}_i)^{\mathbf{k}} w_i = \sum_{\mathbf{k}} b_{\mathbf{k}}(\mathbf{x} - \mathbf{x}_c) m_{\mathbf{k}}(c), \end{aligned} \tag{3.2}$$

where $\mathbf{k} = (k_1, k_2)$ is the 2D multi-index, $D^{\mathbf{k}}$ is the $\mathbf{k}$-th differentiation operator,

$$b_{\mathbf{k}}(\mathbf{x} - \mathbf{x}_c) = \frac{1}{\mathbf{k}!} D^{\mathbf{k}} \Phi(\mathbf{x} - \mathbf{x}_c) \tag{3.3}$$

is the $\mathbf{k}$-th Taylor coefficient, and

$$m_{\mathbf{k}}(c) = \sum_{i=1}^{N_c} (\mathbf{x}_c - \mathbf{x}_i)^{\mathbf{k}} w_i \tag{3.4}$$

is the $\mathbf{k}$th moment of the cluster c about the center $\mathbf{x}_c$. We retain the terms of order $|\mathbf{k}| < q$ in the infinite series in (3.2) and get the q-th order point-cluster approximation:

$$\sum_{i=1}^{N_c} \Phi(\mathbf{x} - \mathbf{x}_i) w_i \approx \sum_{|\mathbf{k}|<q} b_{\mathbf{k}}(\mathbf{x} - \mathbf{x}_c) m_{\mathbf{k}}(c). \tag{3.5}$$

When evaluating a point-cluster approximation in (3.5), one needs the Taylor coefficients of the kernel functions. We derive a recurrence relation to compute higher order Taylor coefficients based on the lower order ones. As an example, we derive a recurrence relation for the kernel function $\phi = \log r$. Note that ϕ satisfies the equation

$$r^2 D_{x_1} \phi - x_1 = 0. \tag{3.6}$$

We apply $D_{x_1}^{k_1 - 1}$ to (3.6) and use product rule of differentiation to obtain

$$r^2 D_{x_1}^{k_1} \phi + 2(k_1 - 1) x_1 D_{x_1}^{k_1 - 1} \phi + (k_1 - 1)(k_1 - 2) D_{x_1}^{k_1 - 2} \phi = 0. \tag{3.7}$$

Applying $D_{x_2}^{k_2}$ to the above equation and using the definition of $b_{\mathbf{k}}$ we get

$$k_1 r^2 b_{\mathbf{k}} + 2k_1 \sum_{i=1}^{2} x_i b_{\mathbf{k}-\mathbf{e}_i} - 2x_1 b_{\mathbf{k}-\mathbf{e}_1} + k_1 \sum_{i=1}^{2} b_{\mathbf{k}-2\mathbf{e}_i} - 2b_{\mathbf{k}-2\mathbf{e}_1} = 0. \tag{3.8}$$

This recurrence relation is used to compute the Taylor coefficients of the kernel $\phi = \log r$. The relation involves three consecutive orders of derivatives. If $k_i < 0$ for either i in $\mathbf{k} = (k_1, k_2)$, let $b_{\mathbf{k}} = 0$. Recurrence relations between Taylor coefficients for all other kernel functions involved can be derived similarly [**FBLL**].

To summarize, the accuracy and efficiency of the treecode are controlled by several parameters. First, when the root cluster is divided into hierarchical levels of clusters, the treecode sets a parameter N_0 so that any cluster containing less than N_0 points is not divided further. This often outperforms a 'full' space division. Second, the treecode uses varying order Taylor expansions. A maximum allowable order of expansion $q_{\max}$ is set in the treecode, and the code chooses adaptively whether to perform an expansion of order $q \leq q_{\max}$ or a direct summation. Third, there is error control available for the treecode. The code takes a tolerance ϵ as a parameter and the absolute error in the evaluated quantity is bounded by ϵ.

4. Results

4.1. A comparison between the treecode and the direct summation. All computations and tests are performed on a cluster in which each node has eight Intel E5540 2.53 GHz processors. We make a comparison between the treecode and the direct summation method. We use the treecode for the elasticity equation (2.1) and (2.2). The elasticity part is more time consuming and takes over 80% CPU

Algorithm 1 The treecode

```
program main
construct tree
for i = 1 : N do
  compute_interaction(x_i, root, ε)
end for

subroutine compute_interaction(x, c, ε)
if particle-cluster interaction b/w x and cluster c incurs error (r/R)^q < ε then
  use q-th order Taylor approximation
else
  if c is a leaf then
    use direct summation
  else
    for each child of c (# children of c: chd_cnt) do
      call compute_interaction(x, child, ε/chd_cnt)
    end for
  end if
end if
```

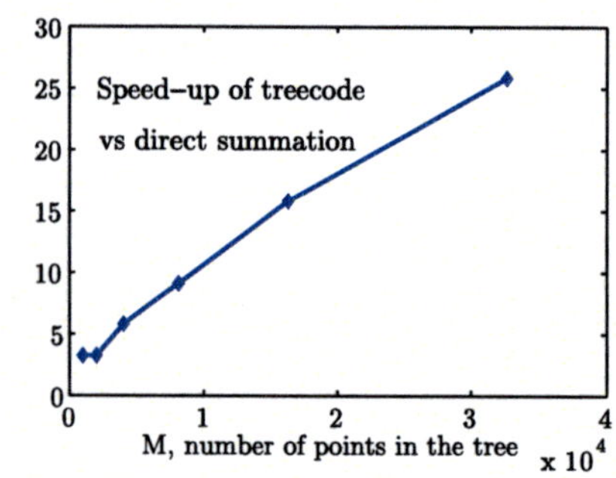

FIGURE 3. Speedup of the treecode vs the direct summation method

time, hence performance of the elasticity treecode will mirror its counterpart for the diffusion equation (2.6).

To test the performance of the treecode, we simulated the evolution of a single precipitate. The treecode is used to compute matrix-vector product in the GMRES solver. We use an alternating point quadrature when discretizing the integrals in Eqs. (2.1), (2.2), and (2.6), hence the treecode constructs two trees in each GMRES iteration. For example, if $N = 256$ is the total number of computational points on the precipitate interface, there are $M = 128$ points in each tree. In the test, we used N ranging from 2^{11} to 2^{16}. Equivalently, M ranges from 2^{10} to 2^{15}. For M between 2^{10} and 2^{12}, the test data come from interface in Figure 4(b)-(d). While for M between 2^{13} to 2^{15}, test data is obtained by taking the interface in Figure 4(d) and doubling the number of points successively through interpolations. This is because when the simulation ends, $M = 2^{12}$, which is relatively small to study the asymptotic time complexity of the treecode.

In Figure 3, we plot the speed-up of treecode against direct summation. The error tolerance of the treecode is 10^{-8}. The break-even point between the treecode and direct summation is around $M = 1024$. We note that when $M = 32768$, the treecode is about 25 times faster than the direct summation method.

4.2. Evolution of a single precipitate with elasticity. In this subsection, we present a numerical simulation of a single precipitate with elasticity. Our intention is to compare with the work in [**JLL97**]. The initial interface is a perturbed circle with radius $r = 1 + 0.05(\cos 6\theta + \sin 2\theta), \theta \in (0, 2\pi)$. The flux is $J = 10$, and the elasticity coefficient is $Z = 500$. The GMRES tolerance $\epsilon = 10^{-8}$, treecode error tolerance $\epsilon = 10^{-8}$. The computation starts with $N = 256$ points on the interface. The initial time step is $\Delta t = 5 \times 10^{-4}$. The small scale decomposition method (SSD) described in [**HLS94**] is applied here to remove the stiffness. Note that the SSD allows much larger time steps. For the purpose of accuracy, we use small time step in our simulations.

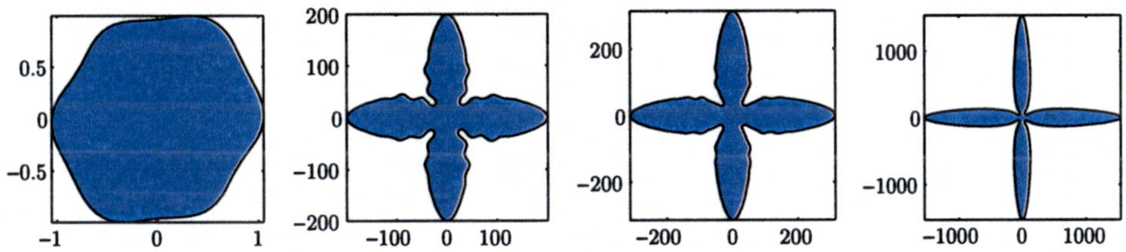

(a) M=256 (b) M=1024 (c) M=2048 (d) M=4096

FIGURE 4. A time sequence of interface between the precipitate and the matrix

In Figure 4(a)-(d), we plot a time evolution sequence of interface with the initial shape shown in (a). In simulations with different elasticities, we found that a strong elasticity tends to damp out small features on the precipitate interface quickly. We found that for the case of $Z = 500$, which is smaller than $Z = 3000$ used in [**JLL97**], there are relatively richer morphologies developed at early times. In the end, the interface appears similar to the one in [**JLL97**].

4.3. Multiple preci-pitates with anisotropic surface tension. As an example of a long time simulation with anisotropic surface tension using the treecode, we performed computations of three precipitates interacting with each other (see Figure 5 for a time sequence). We note that the simulation is purely diffusion driven. At time $t = 0$, we center these precipitates respectively at $(-14, 14)$, $(-7, -7)$, and $(7, 7)$, with shapes $3(1 + 0.05(\cos 3\theta + \sin 2\theta))$, $2(1 + 0.05(\cos 3\theta + \sin 2\theta))$, and $1 + 0.05(\cos 3\theta + \sin 2\theta)$. We use flux $J = 10$, GMRES tolerance $\epsilon = 10^{-8}$, treecode error tolerance $\epsilon = 10^{-8}$, initial time step $\Delta t = 5 \times 10^{-5}$, and surface tension $\gamma = 1 + 0.005 \cos 4\theta$. We started with 512 computational points on each precipitate, and went up to 8192 points in the end.

If the growth happens under isotropic surface tension [**LLL07**], tip splitting is the dominant feature of precipitate morphology. Since in our case the surface tension is anisotropic, we find that side branching is the most prominent feature, although some tip splitting occurs in middle branches of the precipitate originally centered at $(-14, 14)$. Overall, the precipitates try to align themselves in the x-

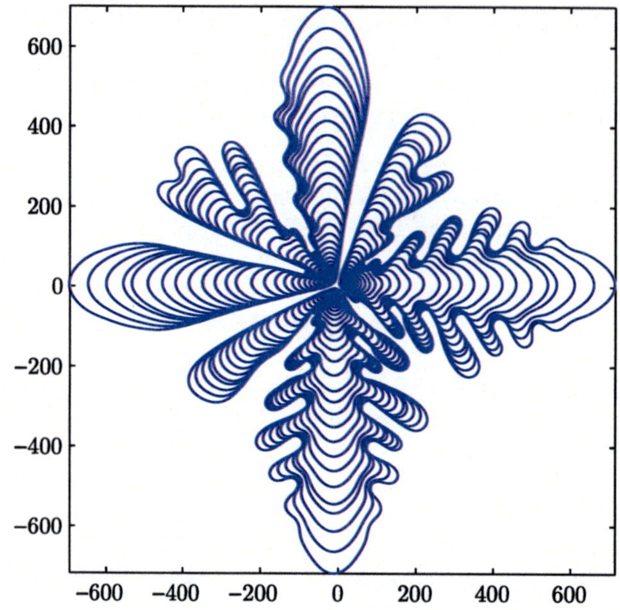

FIGURE 5. Three precipitates of different initial sizes with anisotropic surface tension

and y- axes. Interestingly, the branch aligned along the negative x- axis does not undergo tip splitting or side branching, because it is in close proximity to two other well developed branches, and the tendency to develop side branches is suppressed.

In summary, we developed an adaptive treecode for computing a solid/solid phase transformation problem by a boundary integral method. We tested the efficiency by simulating long time evolution of microstructures in an elastic media, which is not practical with a direct summation method due to a prohibitive time cost.

References

[BH86] J. Barnes and P. Hut, *A hierarchical $O(N \log N)$ force-calculation algorithm*, Nature **324** (1986), no. 6096, 446–449.

[CGR99] H. Cheng, L. Greengard, and V. Rokhlin, *A fast adaptive multipole algorithm in three dimensions*, J. Comput. Phys. **155** (1999), no. 2, 468–498. MR1723309 (2000h:65178)

[FBLL] H. Feng, A. Barua, S. Li, and X. Li, *A parallel adaptive treecode algorithm for elastic kernel functions*, in review.

[GR87] L. Greengard and V. Rokhlin, *A fast algorithm for particle simulations*, J. Comput. Phys. **73** (1987), no. 2, 325–348. MR918448 (88k:82007)

[HLS94] T. Hou, J.S. Lowengrub, and M. Shelley, *Removing the stiffness from interfacial flows with surface tension*, J. Comput. Phys. **114** (1994), 312. MR1294935 (95e:76069)

[HLS01] ———, *Boundary integral methods for multicomponent fluids and multiphase materials*, J. Comput. Phys. **169** (2001), 302. MR1836520 (2002c:76069)

[JLL97] H.-J. Jou, P.H. Leo, and J.S. Lowengrub, *Microsructural evolution in inhomogenous elastic media*, J. Comput. Phys. **131** (1997), 109.

[LK01] K. Lindsay and R. Krasny, *A particle method and adaptive treecode for vortex sheet motion in three-dimensional flow*, J. Comput. Phys. **172** (2001), no. 2, 879–907. MR1857621 (2002h:76112)

[LLL07] S. Li, J. Lowengrub, and P.H. Leo, *A rescaling scheme with application to the long-time simulation of viscous fingering in a hele-shaw cell*, J. Comput. Phys. **225** (2007), no. 1, 554–567. MR2346691 (2008f:76073)

[MS63] W. Mullins and R.F. Sekerka, *Morphological stability of a particle growing by diffusion or heat flow*, J. Appl. Phys. **34** (1963), 323.

[Mur87] T. Mura, *Micromechanics of defects in solids*, Nijhoff, Boston, 1987.

[SI88] A. Sidi and M. Israel, *Quadrature methods for periodic singular and weakly singular fredholm integral equations*, J. Sci. Comput. **3** (1988), 201. MR981760 (90e:65194)

[SS86] Y. Saad and M. Schultz, *A generalized minimum residual method for solving non symmetric linear systems*, SIAM J. Sci. Stat. Comput. **7** (1986), 856. MR848568 (87g:65064)

School of Science, Nanjing University of Science and Technology, Nanjing, Jiangsu, China 210094

Current address: Department of Applied Mathematics, Illinois Institute of Technology, Chicago, Illinois 60616

E-mail address: hfeng8@iit.edu

Department of Applied Mathematics, Illinois Institute of Technology, Chicago, Illinois 60616

E-mail address: abarua@hawk.iit.edu

Department of Applied Mathematics, Illinois Institute of Technology, Chicago, Illinois 60616

E-mail address: lix@iit.edu

Department of Applied Mathematics, Illinois Institute of Technology, Chicago, Illinois 60616

E-mail address: sli@math.iit.edu

Chebyshev spectral-collocation method for Volterra integral equations

Zhendong Gu and Yanping Chen

ABSTRACT. A Chebysheve spectral-collocation method is proposed to solve the Volterra integral equations of the second kind. The biggest advantage for the proposed method is that only the Chebyshev Gauss points and Chebyshev weights, which can be obtained easily and accurately, are used in this method. This enables us to avoid the complex computation of the Legendre Gauss points and the Legendre weights. Numerical examples are presented to show that the numerical errors decay in exponential rate. Our next work is the convergence analysis for the proposed method, which is in progress.

1. Introduction

The Volterra integral equations (VIEs) arise in many problems of physics and in other areas, such as in the particle transport problems of astrophysics, potential theory and Dirichlet problems, electrostatics and radiative heat transfer problems. There are many existing numerical methods for solving the VIEs, such as collocation methods, product integration methods, see, e.g., Brunner [4] and references therein.

In order to obtain spectral accuracy and to provide theoretical analysis, Tang, et al. in [**13**] proposed a Legendre spectral-collocation method to solve VIEs of the second kind whose kernel and solutions are sufficiently smooth. Subsequently, in [**6**, **7**], Chen, et al. proposed and analyzed a Jacobi spectral-collocation approximation for the linear VIEs of the second kind with weakly singular kernels provided that the underlying solutions of the VIEs are sufficiently smooth. Then, in [**11**], the above Jacobi spectral-collocation method is extended to solve the VIEs with the Abel type kernel. Recently, another spectral method, i.e., the Legendre spectral Galerkin method, is investigated in [**15**] for VIEs. The spectral-collocation methods also attract the interest of those people, who study the Volterra type integral and related functional differential equations (see, e.g., [**1, 2, 3, 10, 12, 16, 17**]).

In this paper, we propose Chebyshev spectral-collocation method for the VIEs of the second kind. The biggest advantage for the proposed method is that only the Chebyshev Gauss points and Chebyshev weights, which can be obtained easily and

---

2010 *Mathematics Subject Classification.* Primary 35Q99, 35R35, 65M12, 65M70.

Key words and phrases. Chebyshev spectral-collocation method, Volterra integral equation.

The second author (corresponding author) is supported by National Science Foundation of China, Foundation for Talent Introduction of Guangdong Provincial University, Guangdong Province Universities and Colleges Pearl River Scholar Funded Scheme (2008), and Specialized Research Fund for the Doctoral Program of Higher Education (20114407110009).

accurately, are used in this method. This enables us to avoid the complex computation of the Legendre Gauss points and the Legendre weights. Numerical results are also presented to show that the proposed method obtains spectral accuracy. The numerical method we proposed in this paper is some different from the ones in [**8, 9, 14**].

The VIEs considered in this paper are as follows:

$$y(\tau) = f(\tau) + \int_0^T R(\tau, \xi) y(\xi) d\xi, \ \tau \in [0, T], \tag{1.1}$$

where the unknown function $y(\tau)$ is defined on $[0, T]$, $T < +\infty$, and the given functions satisfy

$$f(\tau) \in C^m([0, T]), \ R(\tau, \xi) \in C^m(\Omega), \ m \geq 1, \tag{1.2}$$

where

$$\Omega := \Big\{ (\tau, \xi) : 0 \leq \xi \leq \tau \leq T \Big\}.$$

This paper is organized as follows. In the following section, we introduce the Chebyshev spectral-collocation method for the Volterra integral equations, and the numerical examples are shown in section 3. Finally, in Section 4, we end with the conclusion and the future work.

2. Chebyshev spectral-collocation method

Consider (1.1). For ease of analysis, we will transform the problem (1.1) to an equivalent problem defined in [-1,1]. More precisely, we use the changes of variables

$$\tau = \frac{T}{2}(1 + x), \ x := \frac{2}{T}\tau - 1,$$
$$\xi = \frac{T}{2}(1 + s), \ s := \frac{2}{T}\xi - 1,$$

to rewrite (1.1) as follows

$$u(x) = g(x) + \int_{-1}^{x} \widetilde{K}(x, s) u(s) ds, \tag{2.1}$$

where $x \in [-1, 1]$, and

$$u(x) := y\Big(\frac{T}{2}(1 + x)\Big), \ \ g(x) := f\Big(\frac{T}{2}(1 + x)\Big),$$
$$\widetilde{K}(x, s) := \frac{T}{2} R\Big(\frac{T}{2}(1 + x), \frac{T}{2}(1 + s)\Big).$$

Set the collocation points as the set of $N+1$-point Chebyshev Gauss, or Chebyshev Gauss-Radau, or Chebyshev Gauss-Lobatto points, $\{x_i\}_{i=0}^N$ (see, e.g., [**5**]). Then (2.1) holds at x_i:

$$u(x_i) = g(x_i) + \int_{-1}^{x_i} \widetilde{K}(x_i, s) u(s) ds, \tag{2.2}$$

for $0 \leq i \leq N$. We will transfer the integral interval $[-1, x_i]$ to a fixed interval $[-1, 1]$ using the following simple linear transformation:

$$s_x(z) = \frac{x + 1}{2} z + \frac{x - 1}{2}, \ z \in [-1, 1].$$

Then (2.2) becomes

$$u(x_i) = g(x_i) + \int_{-1}^{1} K(x_i, z)u(s_{x_i}(z))dz, \tag{2.3}$$

for $i = 0, 1, \cdots, N$, where

$$K(x, z) := \frac{x+1}{2}\widetilde{K}(x, s_x(z)).$$

We use u_i to approximate the function value $u(x_i)$, and use

$$u^N(x) := \sum_{j=0}^{N} u_j F_j(x)$$

to approximate the function $u(x)$, where $F_j(x)$ is the j-th Lagrange basic function associated with $\{x_i\}_{i=0}^N$. We approximate the integration by replacing the integral function by its interpolation polynomial. Then our goal is to find $u_i, i = 0, 1, \cdots, N$ such that

$$u_i = g(x_i) + \int_{-1}^{1} I_N(K(x_i, z)u^N(s_{x_i}(z)))dz, i = 0, 1, \cdots, N, \tag{2.4}$$

where I_N is the interpolation operater associated with the $N + 1$-point Chebyshev Gauss, or Chebyshev Gauss-Radau, or Chebyshev Gauss-Lobatto points $\{x_i\}_{i=0}^N$, namely,

$$I_N v := \sum_{i=0}^{N} v(x_i)F_i(x), v \in C([-1, 1]).$$

Then (2.4) can be rewrote as follows

$$u_i = g(x_i) + \sum_{j=0}^{N} u_j \sum_{k=0}^{N} K(x_i, z_k)F_j(s_{x_i}(z_k)) \int_{-1}^{1} F_k(z)dz, i = 0, 1, \cdots, N, \tag{2.5}$$

where $z_k := x_k, k = 0, 1, 2, \cdots, N$. Write the above equations in matrix form:

$$U = G + AU,$$

where

$$\begin{aligned} U &:= [u_0, u_1, \cdots, u_N]', \\ G &:= [g(x_0), g(x_1), \cdots, g(x_N)]', \\ A &:= (a_{ij})_{(N+1)\times(N+1)}, \\ a_{ij} &:= \sum_{k=0}^{N} K(x_i, z_k)F_j(s_{x_i}(z_k)) \int_{-1}^{1} F_k(z)dz. \end{aligned}$$

We now discuss an efficient computation of $F_j(s)$ and $\int_{-1}^{1} F_k(z)dz$. The idea is to express $F_j(s)$ in terms of the Chebyshev functions:

$$F_j(s) = \sum_{p=0}^{N} a_{p,j}T_p(s), \tag{2.6}$$

where $a_{p,j}$ is called the discrete polynomial coefficients of F_j. The inverse relation is (see, e.g., [**5**]):

$$a_{p,j} = \frac{1}{\gamma_p}\sum_{i=0}^{N} F_j(x_i)T_p(x_i)\omega_i^c = T_p(x_j)\omega_j^c/\gamma_p, \tag{2.7}$$

where ω_i^c is the weight corresponding to $x_i, i = 0, 1, \cdots, N$, and

$$\gamma_p = \sum_{i=0}^{N} T_p^2(x_i)\omega_i^c = \begin{cases} \pi, & p = 0, \\ \frac{\pi}{2}, & 1 \le p < N, \end{cases}$$

and $\gamma_N = \frac{\pi}{2}$ if $\{x_j\}_{j=0}^N$ are the $N+1$-point Chebyshev Gauss or the Chebyshev Gauss-Radau points, $\gamma_N = \pi$ if $\{x_j\}_{j=0}^N$ are the $N+1$-point Chebyshev Gauss-Lobatto points. It follows from (2.6) and (2.7) that

$$F_j(s) = \omega_j^c \sum_{p=0}^{N} \Big(T_p(x_j)/\gamma_p\Big) T_p(s),$$

which, together with the known recurrence formulas for $T_p(s)$, can be used to evaluate $F_j(s)$ in an efficient way. The above equation also deduces that

$$\int_{-1}^{1} F_j(z)dz = \omega_j^c \sum_{p=0}^{N} \Big(T_p(x_j)/\gamma_p\Big) \int_{-1}^{1} T_p(z)dz.$$

To compute $\int_{-1}^{1} T_p(z)dz$, we use the following formula

$$2T_p(z) = \frac{1}{p+1}T'_{p+1}(z) - \frac{1}{p-1}T'_{p-1}(z), \ p \ge 1,$$
$$T_0(z) = 1, T_1(z) = z.$$

Then

$$\int_{-1}^{1} T_p(z)dz = \begin{cases} \frac{1}{p+1} - \frac{1}{p-1}, & \text{p is nonnegative even number,} \\ 0, & \text{p is nonnegative odd number.} \end{cases}$$

3. Numerical examples

Without lose of generality, we will only use the Chebyshev Gauss-Lobatto points as the collocation points. Our numerical evidence show that the other two kinds of Chebyshev Gauss points produce results with similar accuracy. More details about three kinds of Chebyshev Gauss points and the corresponding weights can be found in [**5**].

EXAMPLE 3.1. Consider (2.1) with

$$\widetilde{K}(x,s) = -e^{xs}, \quad g(x) = e^{4x} + \frac{1}{x+4}\Big(e^{x(x+4)} - e^{-(x+4)}\Big).$$

The corresponding exact solution is given by $u(x) = e^{4x}$, $x \in [-1, 1]$.

This example has appeared in [**13**]. We use the numerical scheme (2.5). Numerical errors versus several values of N are displayed in Table 1 and Fig.1. These results indicate that the desired spectral accuracy is obtained.

For the nonlinear Volterra integral equation of the following form

$$u(x) = g(x) + \int_{-1}^{x} \widetilde{K}(x, s, u(s))ds, \tag{3.1}$$

TABLE 1. Example 3.1: The errors $u - u^N$ versus the number of collocation points in L^∞ norms.

N	6	8	10	12	14
L^∞-error	3.36e-02	1.24e-03	3.31e-05	6.53e-07	1.01e-08
N	16	18	20	22	24
L^∞-error	1.25e-10	1.26e-12	4.26e-14	2.84e-14	4.97e-14

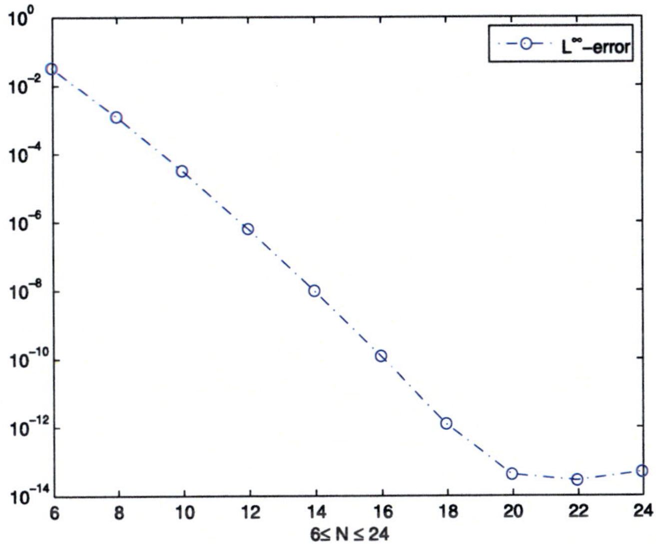

FIGURE 1. Example 3.1: The errors $u - u^N$ versus the number of collocation points in L^∞ norms.

we can design a spectral collocation method similar to the linear case. More precisely, we assume that (3.1) holds at the Chebyshev collocation points and transform the interval $[-1, x]$ to $[-1, 1]$. This gives

$$u(x_i) = g(x_i) + \int_{-1}^{1} K(x_i, z, u(s_{x_i}(z)))dz, \tag{3.2}$$

for $i = 0, 1, \cdots, N$, where

$$K(x, z, u) := \frac{x+1}{2}\widetilde{K}(x, s_x(z), u).$$

Similarly to (2.4), (2.5) we have

$$u_i = g(x_i) + \int_{-1}^{1} I_N(K(x_i, z, u^N(s_{x_i}(z))))dz, \tag{3.3}$$

and

$$u_i = g(x_i) + \sum_{k=0}^{N} K(x_i, z_k, \sum_{j=0}^{N} u_j F_j(s_{x_i}(z_k))) \int_{-1}^{1} F_k(z)dz, \tag{3.4}$$

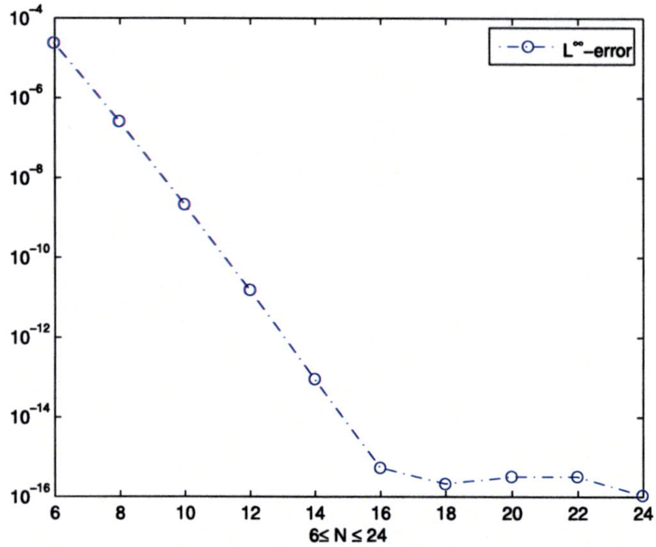

FIGURE 2. Example 3.2: The errors versus the number of collocation points in L^∞ norms.

where $z_k := x_k, k, i = 0, 1, 2, \cdots, N$.

EXAMPLE 3.2. Consider (3.1) with

$$\widetilde{K}(x, s, u(s)) = x(\cos s)u^2(s), \; g(x) = \sin x - \frac{x}{3}(\sin^3 x + \sin^3 1).$$

The corresponding exact solution is $u(x) = \sin x$.

The numerical scheme is

$$u_i = g(x_i) + \sum_{k=0}^{N} K(x_i, z_k)(\sum_{j=0}^{N} u_j F_j(s_{x_i}(z_k)))^2 \int_{-1}^{1} F_k(z)dz, \tag{3.5}$$

for $i = 0, 1, \cdots, N$. In our computations, we just use Newton iteration method to solve the above nonlinear system. Fig.2 plots the errors for $6 \leq N \leq 24$ in L^∞ norms. The corresponding errors versus several values of N are displayed in Table 2. As expected that the errors decay exponentially.

TABLE 2. Example 3.2 The errors versus the number of collocation points in L^∞ norms.

N	6	8	10	12	14
L^∞-error	2.44e-05	2.60e-07	2.21e-09	1.56e-11	9.11e-14
N	16	18	20	22	24
L^∞-error	5.55e-16	2.22e-16	3.33e-16	3.33e-16	1.11e-16

4. Conclusion and future work

From the numerical scheme (2.5), we know that only the Chebyshev Gauss points and Chebyshev weights, both of which can be easily and accurately obtained, are used in this scheme. Replacing the integral function by its interpolation polynomial to approximate the integration in (2.3) enables us to avoid the complex computation of the Legendre Gauss points and Legendre weights. This is the biggest advantage for the proposed method.

The convergence analysis for the proposed method is our next work in progress. In our future work, we will study the Chebyshev spectral-collocation method for VIEs with delay, and for the Volterra type integral-differential equations.

References

1. I. Ali, *Convergence analysis of spectral methods for integro-differential equations with vanishing proportional delays*, J. Comput. Math. , 28(2010), pp. 962–973. MR2723986 (2012b:65214)
2. I. Ali, H. Brunner, and T. Tang, *A spectral method for pantograph-type delay differential equations and its convergence analysis*, J. Comput. Math., 27(2009), pp. 254–265. MR2495059 (2010f:65197)
3. I. Ali, H. Brunner, and T. Tang, *Spectral methods for pantograph-type differential and integral equations with multiple delays*, Front. Math. China, 4(2009), pp. 49–61. MR2481596 (2009m:65099)
4. H. Brunner, *Collocation methods for Volterra integral and related functional differential equations*, Cambridge University Press, 2004. MR2128285 (2005k:65002)
5. C. Canuto, M. Y. Hussaini, A. Quarteroni, and T. A. Zang, *Spectral method fundamentals in single domains*, Spring-Verlag, 2006. MR2223552 (2007c:65001)
6. Y. Chen, and T. Tang, *Spectral methods for weakly singular Volterra integral equations with smooth solutions*, J. Comput. Appl. Math., 233(2009), pp. 938–950. MR2557286 (2010k:65304)
7. Y. Chen, and T. Tang, *Convergence analysis of the Jacobi spectral-collocation methods for Volterra integral equation with a weakly singular kernel*, Math. Comput. 792010, pp. 147–167. MR2552221 (2011c:65298)
8. G. N. Elnagar and M. Kazemi, *Chebyshev spectral solution of nonlinear Volterra-Hammerstein integral equations*, J. Comput. Appl. Math., 76 (1996), pp. 147–158. MR1423514 (97i:45008)
9. H. Fujiwara, *High-accurate numerical method for integral equations of the first kind under multiple-precision arithmetic*, Preprint, RIMS, Kyoto University, 2006.
10. Y. Jiang, *On spectral methods for Volterra-type integro-differential equations*, J. Comput. Appl. Math., 230 (2009), pp. 333–340. MR2532327 (2010d:45011)
11. X. Li, and T. Tang, *Convergence analysis of the Javobi collocation methods foe Abel-Volterra integral equations of the second kind*, Front. Math. China, 7(2012), pp. 69–84. MR2876899
12. T. Tang, and X. Xu, *Accuracy Enhancement Using Spectral Postprocessing for Differential Equations and Integral Equations*, Commun. Comput. Phys., 5(2009), pp. 779–792. MR2513715 (2010d:65178)
13. T. Tang, X. Xu, and J. Cheng, *On Spectral methods for Volterra integral equation and the convergence analysis*, J. Comput. Math., 26(2008), pp. 825–837. MR2464738 (2010c:65256)
14. H. C. Tian, *Spectral Method for Volterra Integral Equation*, MSc Thesis, Simon Fraser University, 1995.
15. Z. Wan, Y. Chen, and Y. Huang, *Legendre spectral Galerkin method for second-kind Volterra integral equations*, Front. Math. China, 4(2009), pp. 181–193. MR2481603 (2009k:65285)
16. Y. Wei, and Y. Chen, *Convergence analysis of the Legendre spectral collocation methods for second order Volterra integro-differential equations*, Numer. Math. Theory Methods Appl., 4(2011), pp. 419–438. MR2863103 (2012k:65177)
17. Y. Wei, and Y. Chen, *Convergence analysis of the spectral methods for weakly singular Volterra integro-differential equations with smooth solutions*, Adv. Appl. Math. Mech., 4(2012), pp. 1–20. MR2876648

School of Mathematics and Computational Science, Xiangtan University, Xiangtan 411105, Hunan, China

E-mail address: guzhd@qq.com

School of mathematics science, South China Normal University, Guangzhou 510631, China

E-mail address: yanpingchen@scnu.edu.cn

A numerical method for generalized Fokker-Planck equations

Weimin Han, Yi Li, Qiwei Sheng, and Jinping Tang

ABSTRACT. Generalized Fokker-Planck (GFP) equations have been employed to approximate the radiative transfer equation in applications of highly forward peaked biological media. In this paper, we discuss a numerical method for solving GFP equations. The numerical method is based on a variational formulation involving even and odd components of the solution. We show the well-posedness of the variational formulation and develop a Galerkin method where spherical harmonics are used for the angular approximation and finite elements are used for spatial discretization. An iteration procedure is introduced to solve the problem and its convergence is shown.

1. Introduction

The radiative transfer equation (RTE) arises in a wide range of applications, such as neutron transport, heat transfer, radiation in atmosphere and ocean, optics. The RTE also describes the light propagation within biological media ([**NW01**]), and for this reason, in recent years, there has been active research in the community of medical optics on direct and inverse problems of the RTE (cf. e.g. [**A99, B09, HEHL11**]). The steady state monochromatic form of the RTE is ([**LM84, A98**])

$$\boldsymbol{\omega}\cdot\nabla u + \mu_t u = \mu_s\, Su + f \quad \text{in } Q := X \times \Omega. \tag{1.1}$$

Here X is a bounded domain in $\mathbb{R}^3$, Ω is the unit sphere in $\mathbb{R}^3$, $\mu_t = \mu_a + \mu_s$, μ_a is the absorption coefficient, μ_s is the scattering coefficient, f is a source function, and S is an integral operator of the form

$$Su(\mathbf{x},\boldsymbol{\omega}) = \int_\Omega \eta(\mathbf{x},\boldsymbol{\omega}\cdot\boldsymbol{\omega}')\, u(\mathbf{x},\boldsymbol{\omega}')\, d\sigma(\boldsymbol{\omega}')$$

with the phase function $\eta(\mathbf{x},\boldsymbol{\omega}\cdot\boldsymbol{\omega}') \geq 0$ satisfying the normalization condition

$$\int_\Omega \eta(\mathbf{x},\boldsymbol{\omega}\cdot\boldsymbol{\omega}')\, d\sigma(\boldsymbol{\omega}') = 1 \quad \forall\boldsymbol{\omega}\in\Omega,\ \forall\mathbf{x}\in X.$$

A well-known example is the Henyey-Greenstein phase function (cf. [**HG41**])

$$\eta(t) = \frac{1-g^2}{4\pi(1+g^2-2gt)^{3/2}}, \quad t\in[-1,1]. \tag{1.2}$$

1991 *Mathematics Subject Classification.* Primary 65N12, 65N30, 35R09.

Key words and phrases. Radiative transfer equation, generalized Fokker-Planck equation, variational formulation, well-posedness, iteration, Galerkin approximation.

The work was supported in part by Simons Foundation Grants.

The parameter $g \in (-1,1)$ is the anisotropy factor of the scattering medium: $g > 0$ for forward scattering, and the bigger the value of g, the stronger the degree of forward scattering. In biomedical optics, g is close to 1.

We assume the boundary ∂X is C^1 and use $\boldsymbol{\nu}(\mathbf{x})$ to denote the unit outward normal vector at $\mathbf{x} \in \partial X$. Let Γ be the boundary of the set Q and define the inflow boundary $\Gamma_- = \{(\mathbf{x},\boldsymbol{\omega}) \in \Gamma : \boldsymbol{\nu}(\mathbf{x})\cdot\boldsymbol{\omega} < 0\}$ and outgoing boundary $\Gamma_+ = \{(\mathbf{x},\boldsymbol{\omega}) \in \Gamma : \boldsymbol{\nu}(\mathbf{x})\cdot\boldsymbol{\omega} > 0\}$. We assume

$$\mu_t, \mu_s \in L^\infty(X), \quad \mu_s \geq 0 \text{ a.e. in } X, \quad \mu_a \geq c_0 > 0 \text{ a.e. in } X, \tag{1.3}$$

$$f \in L^2(Q). \tag{1.4}$$

These assumptions are naturally valid in applications.

Due to the high dimensionality and integro-differential form of the equation, it is very challenging to numerically solve the RTE accurately. Moreover, in applications in biomedical optics, the light propagation within the biological media is highly forward peaked, leading to additional numerical difficulties for solving the RTE. It is shown in the literature (e.g. [**KimK03**]) that for applications in highly forward peaked biological media, the following generalized Fokker-Planck equation (GFPE) is a good approximation to RTE:

$$\boldsymbol{\omega}\cdot\boldsymbol{\nabla} u + \mu_t u = \mu_s (I - \alpha\,\Delta^*)^{-1} u + f \quad \text{in } Q. \tag{1.5}$$

Here, $\alpha(\mathbf{x}) \geq 0$ and Δ^* is the Laplace–Beltrami operator, i.e., the restriction of the Laplace operator on the unit sphere ([**AH12**]). For the Henyey-Greenstein phase function (1.2), $\alpha = (1-g)/(2g)$ is a constant. In this paper, we consider any GFPE of the form (1.5) with α bounded and positively-valued. The equation (1.5) is supplemented by the following boundary condition:

$$u = u_{\text{in}} \quad \text{on } \Gamma_-. \tag{1.6}$$

Introduce a function space

$$H^{1,2}(Q) := \{v \in L^2(Q) \mid \boldsymbol{\omega}\cdot\boldsymbol{\nabla} v \in L^2(Q)\},$$

$\boldsymbol{\omega}\cdot\boldsymbol{\nabla} v$ being the generalized directional derivative of v in the direction $\boldsymbol{\omega}$. We denote by $(u,v)_Q$ the integral of uv on Q, and similarly define $(\cdot,\cdot)_\Gamma$, $(\cdot,\cdot)_{\Gamma_-}$, $(\cdot,\cdot)_\Omega$. We assume

$$u_{\text{in}} \in L^2_*(\Gamma_-), \tag{1.7}$$

where $L^2_*(\Gamma_-)$ denotes the space of measurable functions v on Γ_- such that the norm $\|v\|_{L^2_*(\Gamma_-)} := (|\boldsymbol{\nu}\cdot\boldsymbol{\omega}|\, v, v)^{1/2}_{\Gamma_-} < \infty$. It is shown in [**HECW11**] that under the assumptions (1.3)–(1.4) and (1.7), the problem (1.5)–(1.6) has a unique solution $u \in H^{1,2}(Q)$ and this solution is Lipschitz continuous with respect to the source function f and the boundary condition u_{in}. Moreover, if $f \geq 0$ and $u_{\text{in}} \geq 0$, then $u \geq 0$. This property is desirable for the GFPE to be a physically meaningful model.

In [**ES12**], a mixed weak framework is studied for RTE. In this paper, we introduce a similar weak formulation for the problem (1.5)–(1.6) and study its numerical approximation. Although it is possible to give the presentation for any spatial dimension, for definiteness and due to the importance and relevance in applications, we focus on the case of three spatial dimension.

2. Spaces and operators

We introduce additional function spaces and operators that will be needed later. Let $V_0 := L^2(Q)$ and let V_1 be the subspace of $H^{1,2}(Q)$ with traces on Γ in $L^2_*(\Gamma)$.

Define the transport operator from V_1 to V_0 by $v(\mathbf{x},\boldsymbol{\omega}) \mapsto \boldsymbol{\omega}\cdot\nabla v(\mathbf{x},\boldsymbol{\omega})$, and the removal operator $R : V_0 \to V_0$ by

$$(Rv)(\mathbf{x},\boldsymbol{\omega}) := \mu_t(\mathbf{x})\, v(\mathbf{x},\boldsymbol{\omega}) - \mu_s(\mathbf{x})\left(I - \alpha\,\Delta^*\right)^{-1} v(\mathbf{x},\boldsymbol{\omega}).$$

We have the following properties for the removal operator R.

PROPOSITION 2.1. The operator $R : V_0 \to V_0$ is linear, and is
(a) self-adjoint: $(Ru, v)_Q = (u, Rv)_Q\ \forall\, u, v \in V_0$,
(b) bounded: $\|Rv\|_{V_0} \le c_1\|v\|_{V_0}\ \forall\, v \in V_0$,
(c) elliptic: $(Rv, v)_Q \ge c_0\|v\|^2_{V_0}$, $c_0 > 0$, $\forall\, v \in V_0$.

PROOF. The linearity and self-adjointness of R are easily seen.

Let $w = \left(I - \alpha\,\Delta^*\right)^{-1} v$. Then $w - \alpha\,\Delta^* w = v$, and

$$(w, z)_\Omega + \alpha\,(\nabla^* w, \nabla^* z)_\Omega = (v, z)_\Omega \quad \forall\, z \in H^1(\Omega),$$

where $H^1(\Omega) := \left\{z \in L^2(\Omega) : |\nabla^* z| \in L^2(\Omega)\right\}$. Take $z = w$ to obtain

$$\|w\|^2_{L^2(\Omega)} + \alpha\, \||\nabla^* w|\|^2_{L^2(\Omega)} \le \|v\|_{L^2(\Omega)}\|w\|_{L^2(\Omega)} \le \frac{1}{2}\|w\|^2_{L^2(\Omega)} + \frac{1}{2}\|v\|^2_{L^2(\Omega)}.$$

Thus,

$$\|w\|^2_{L^2(\Omega)} \le \|v\|^2_{L^2(\Omega)}. \tag{2.1}$$

Therefore, $\|w\|_{V_0} \le \|v\|_{V_0}$, and the boundedness (b) is valid:

$$\|Rv\|_{V_0} \le \|\mu_t v\|_{V_0} + \|\mu_s w\|_{V_0} \le c_1\|v\|_{V_0}, \quad c_1 = \|\mu_t\|_{L^\infty(X)} + \|\mu_s\|_{L^\infty(X)}.$$

By the Cauchy-Schwarz inequality,

$$(\mu_s w, v)_Q \le (\mu_s w, w)_Q^{1/2}\,(\mu_s v, v)_Q^{1/2} \le (\mu_s v, v)_Q\,.$$

Hence,

$$\begin{aligned}(Rv, v)_Q &= (\mu_t v, v)_Q - (\mu_s w, v)_Q \ge (\mu_t v, v)_Q - (\mu_s v, v)_Q\\ &= (\mu_a v, v)_Q \ge c_0\|v\|^2_{V_0},\end{aligned}$$

i.e., the ellipticity (c) holds. □

Given the properties stated in Proposition 2.1, we can apply the Lax-Milgram Lemma to conclude the following result.

COROLLARY 2.2. The operator R has an inverse $R^{-1} : V_0 \to V_0$ that is linear, self-adjoint, bounded and elliptic. Moreover, operators $R^{\pm 1/2}$ are well-defined. The expressions $\|v\|_R := (Rv, v)_Q^{1/2}$ and $\|v\|_{R^{-1}} := \left(R^{-1}v, v\right)_Q^{1/2}$ define norms on V_0 that are equivalent to the standard norm $\|v\|_{V_0}$.

The weak formulations studied in this paper involve the splitting of a function v into even part v^+ and odd part v^-, defined by

$$v^+(\mathbf{x},\boldsymbol{\omega}) := \frac{1}{2}\left(v(\mathbf{x},\boldsymbol{\omega}) + v(\mathbf{x},-\boldsymbol{\omega})\right), \qquad v^-(\mathbf{x},\boldsymbol{\omega}) := \frac{1}{2}\left(v(\mathbf{x},\boldsymbol{\omega}) - v(\mathbf{x},-\boldsymbol{\omega})\right).$$

Then given a function space V, we define $V^{\pm}$ to be the subspaces of V consisting of even and odd functions in V. In particular, we will use the space $W := V_1^+ \oplus V_0^-$. The norm in the space W is

$$\|v\|_W := \left(\|\boldsymbol{\omega}\cdot\nabla v^+\|^2_{R^{-1}} + \|v\|^2_R + \|v^+\|^2_{L^2_*(\Gamma)}\right)^{1/2},$$

where $\|v\|_R^2 = (Rv, v)_Q$, $\|v\|^2_{R^{-1}} = (R^{-1}v, v)_Q$.

It is easy to see that $\boldsymbol{\omega}\cdot\nabla^* v \in V_0^{\mp}$ for $v \in V_1^{\pm}$. Moreover, the removal operator R is parity preserving, i.e., $R : V_1^+ \to V_1^+$, and $V_0^- \to V_0^-$.

3. A weak formulation

To derive the weak formulation, rewrite the equation (1.5) as

$$\boldsymbol{\omega}\cdot\nabla u + Ru = f.$$

Multiply the equation by a smooth function v and integrate,

$$(\boldsymbol{\omega}\cdot\nabla u, v)_Q + (Ru, v)_Q = (f, v)_Q.$$

Use the decomposition $u = u^+ + u^-$ and $v = v^+ + v^-$ in terms of the even and odd components to obtain

$$(\boldsymbol{\omega}\cdot\nabla u, v)_Q = \left(\boldsymbol{\omega}\cdot\nabla u^+, v^-\right)_Q + \left(\boldsymbol{\omega}\cdot\nabla u^-, v^+\right)_Q.$$

Then perform an integration by parts,

$$\left(\boldsymbol{\omega}\cdot\nabla u^-, v^+\right)_Q = \left(\boldsymbol{\nu}\cdot\boldsymbol{\omega}\, u^-, v^+\right)_\Gamma - \left(u^-, \boldsymbol{\omega}\cdot\nabla v^+\right)_Q.$$

Note that

$$\begin{aligned}\left(\boldsymbol{\nu}\cdot\boldsymbol{\omega}\, u^-, v^+\right)_\Gamma &= 2\left(\boldsymbol{\nu}\cdot\boldsymbol{\omega}\, u^-, v^+\right)_{\Gamma_-} = 2\left(\boldsymbol{\nu}\cdot\boldsymbol{\omega}\,(u_{\rm in} - u^+), v^+\right)_{\Gamma_-}\\ &= 2\left(\boldsymbol{\nu}\cdot\boldsymbol{\omega}\, u_{\rm in}, v^+\right)_{\Gamma_-} + \left(|\boldsymbol{\nu}\cdot\boldsymbol{\omega}|\, u^+, v^+\right)_\Gamma,\end{aligned}$$

where the boundary condition (1.6) is applied. Then for a solution of the problem (1.5)–(1.6),

$$\begin{aligned}&(Ru, v)_Q - \left(u^-, \boldsymbol{\omega}\cdot\nabla v^+\right)_Q + \left(\boldsymbol{\omega}\cdot\nabla u^+, v^-\right)_Q + \left(|\boldsymbol{\nu}\cdot\boldsymbol{\omega}|\, u^+, v^+\right)_\Gamma\\ &\quad = (f, v)_Q - 2\left(\boldsymbol{\nu}\cdot\boldsymbol{\omega}\, u_{\rm in}, v^+\right)_{\Gamma_-}\end{aligned}$$

for any smooth function v. Define a bilinear form and a linear form over W as follows:

$$b(u, v) := (Ru, v)_Q - \left(u^-, \boldsymbol{\omega}\cdot\nabla v^+\right)_Q + \left(\boldsymbol{\omega}\cdot\nabla u^+, v^-\right)_Q + \left(|\boldsymbol{\nu}\cdot\boldsymbol{\omega}|\, u^+, v^+\right)_\Gamma, \tag{3.1}$$

$$\ell(v) := (f, v)_Q - 2\left(\boldsymbol{\nu}\cdot\boldsymbol{\omega}\, u_{\rm in}, v^+\right)_{\Gamma_-}. \tag{3.2}$$

Then the weak formulation is

$$u \in W, \qquad b(u, v) = \ell(v) \quad \forall\, v \in W. \tag{3.3}$$

THEOREM 3.1. *Under the assumptions* (1.3)–(1.4) *and* (1.7), *the problem* (3.3) *has a unique solution, and the solution depends continuously on the data.*

As in [**ES12**], Theorem 3.1 is proved by employing the following result adapted from [**B71**].

THEOREM 3.2. *Assume $b(\cdot,\cdot)$ is a bounded bilinear form on W and there exists a constant $b_0 > 0$ such that*

$$\inf_{u\in W}\sup_{v\in W}\frac{b(u,v)}{\|u\|_W\|v\|_W}\geq b_0,\qquad \inf_{v\in W}\sup_{u\in W}\frac{b(u,v)}{\|u\|_W\|v\|_W}\geq b_0. \tag{3.4}$$

Then for any $\ell\in W'$, the problem (3.3) *has a unique solution $u\in W$ and for some constant c, $\|u\|_W\leq c\,\|\ell\|_{W'}$.*

In applying Theorem 3.2 to prove Theorem 3.1, the crucial part is to show (3.4). For the bilinear form defined by (3.1), let us prove below the first inequality of (3.4); the second inequality can be proved similarly.

For $u\neq 0$, let $\overline{u}=u+R^{-1}(\boldsymbol{\omega}\cdot\nabla u^+)$. Then $\|\overline{u}\|_W\leq c\,\|u\|_W$ and

$$b(u,\overline{u})=(Ru,u)_Q+\left(u,\boldsymbol{\omega}\cdot\nabla u^+\right)_Q+\left(\boldsymbol{\omega}\cdot\nabla u^+,R^{-1}(\boldsymbol{\omega}\cdot\nabla u^+)\right)_Q+\left(|\boldsymbol{\nu}\cdot\boldsymbol{\omega}|\,u^+,u^+\right)_\Gamma.$$

Now

$$\begin{aligned}\left(u,\boldsymbol{\omega}\cdot\nabla u^+\right)_Q&=\left(u^-,\boldsymbol{\omega}\cdot\nabla u^+\right)_Q=\left(R^{1/2}u^-,R^{-1/2}(\boldsymbol{\omega}\cdot\nabla u^+)\right)_Q\\&\geq-\frac{1}{2}\,\|\boldsymbol{\omega}\cdot\nabla u^+\|^2_{R^{-1}}-\frac{1}{2}\,\|u^-\|^2_{R^{-1}}.\end{aligned}$$

Hence,

$$b(u,\overline{u})\geq\frac{1}{2}\,\|\boldsymbol{\omega}\cdot\nabla u^+\|^2_{R^{-1}}+\frac{1}{2}\,\|u\|^2_R+\|u^+\|^2_{L^2_*(\Gamma)}\geq\frac{1}{2}\,\|u\|^2_W.$$

This inequality, combined with $\|\overline{u}\|_W\leq c\,\|u\|_W$, implies the first inequality of (3.4) for some constant $b_0>0$.

The rest of the assumptions of Theorem 3.2 can be verified easily. Thus, Theorem 3.1 holds. As in **[ES12]**, it can be further proved that the solution u of the problem (3.3) satisfies the equation (1.5) a.e. in Q and the boundary condition (1.6) a.e. on Γ_-.

Using $w:=(I-\alpha\,\Delta^*)^{-1}u$ as an unknown, we can rewrite the problem (3.3) as: Find $u^+\in V_1^+$, $u^-\in V_0^-$ and $w(\mathbf{x},\cdot)\in H^1(\Omega)$ for a.e. $\mathbf{x}\in X$ such that

$$\begin{aligned}&\left(\mu_t u^+-\mu_s w^+,v^+\right)_Q-\left(u^-,\boldsymbol{\omega}\cdot\nabla v^+\right)_Q+\left(|\boldsymbol{\nu}\cdot\boldsymbol{\omega}|\,u^+,v^+\right)_\Gamma\\&\qquad=\left(f^+,v^+\right)_Q-2\left(\boldsymbol{\nu}\cdot\boldsymbol{\omega}\,u_{\rm in},v^+\right)_Q\quad\forall\,v^+\in V_1^+,\end{aligned}\tag{3.5}$$

$$\left(\mu_t u^--\mu_s w^-,v^-\right)_Q+\left(\boldsymbol{\omega}\cdot\nabla u^+,v^-\right)_Q=\left(f^-,v^-\right)_Q\quad\forall\,v^-\in V_0^-,\tag{3.6}$$

$$(w,v)_\Omega+\alpha\,(\nabla^*w,\nabla^*v)_\Omega=(u,v)_\Omega\quad\forall\,v\in H^1(\Omega).\tag{3.7}$$

4. Galerkin approximation

For a discretization of the problem (3.3), we use spherical harmonics of an arbitrary order for the angular approximation and finite elements for spatial discretization. For this purpose, denote by h the meshsize of a finite element partition of the domain X. Let $V_{1,h}$ be the linear element space of continuous piecewise linear functions and $V_{0,h}$ be the space of piecewise constant functions. For a positive integer n, let $\{Y_{n,m}\}_{m=-n}^{n}$ be a basis of spherical harmonics of order n, e.g., in spherical coordinates,

$$(-1)^{(m+|m|)/2}\left[\frac{(2n+1)\,(n-|m|)!}{4\pi\,(n+|m|)!}\right]^{\frac{1}{2}}(\sin\theta)^m P_n^{(|m|)}(\cos\theta)\,e^{im\phi},\quad -n\leq m\leq n,$$

where $P_n^{(|m|)}(t)$ is the $|m|^{th}$ derivative of $P_n(t)$, the Legendre polynomial of degree n. In real valued form, the following basis functions are used:

$$\left[\frac{(2n+1)\,(n-m)!}{2\pi\,(n+m)!}\right]^{\frac{1}{2}} (\sin\theta)^m P_n^{(m)}(\cos\theta)\,\cos(m\phi), \quad 0 \le m \le n,$$

$$\left[\frac{(2n+1)\,(n-m)!}{2\pi\,(n+m)!}\right]^{\frac{1}{2}} (\sin\theta)^m P_n^{(m)}(\cos\theta)\,\sin(m\phi), \quad 1 \le m \le n.$$

See **[AH12]** for an introduction of spherical harmonics. We then define the following finite dimensional spaces of combined finite elements and spherical harmonics:

$$V_{1,h,n}^{+} = \left\{ v_{h,n}^{+}(\mathbf{x},\boldsymbol{\omega}) = \sum_{j=-n}^{n} \sum_{i=-2j}^{2j} v_{h,2j,i}(\mathbf{x})\, Y_{2j,i}(\boldsymbol{\omega}) : v_{h,2j,i} \in V_{1,h} \right\},$$

$$V_{0,h,n}^{-} = \left\{ v_{h,n}^{-}(\mathbf{x},\boldsymbol{\omega}) = \sum_{j=-n-1}^{n} \sum_{i=-(2j+1)}^{2j+1} v_{h,2j+1,i}(\mathbf{x})\, Y_{2j+1,i}(\boldsymbol{\omega}) : v_{h,2j+1,i} \in V_{0,h} \right\},$$

and $W_{h,n} = V_{1,h,n}^{+} \oplus V_{0,h,n}^{-}$. Then any function $v_{h,n} \in W_{h,n}$ can be expressed as $v_{h,n} = v_{h,n}^{+} + v_{h,n}^{-}$ with $v_{h,n}^{+} \in V_{1,h,n}^{+}$ and $v_{h,n}^{-} \in V_{0,h,n}^{-}$. Note that $\boldsymbol{\omega}\cdot\nabla v_{h,n}^{+} \in V_{0,h,n}^{-}$ for any $v_{h,n}^{+} \in V_{1,h,n}^{+}$. This property ensures the discrete version of (3.4): For the same constant $b_0 > 0$,

$$\inf_{u_{h,n}\in W_{h,n}} \sup_{v_{h,n}\in W_{h,n}} \frac{b(u_{h,n},v_{h,n})}{\|u_{h,n}\|_W \|v_{h,n}\|_W} \ge b_0,$$

$$\inf_{v_{h,n}\in W_{h,n}} \sup_{u_{h,n}\in W_{h,n}} \frac{b(u_{h,n},v_{h,n})}{\|u_{h,n}\|_W \|v_{h,n}\|_W} \ge b_0.$$

Then the Galerkin approximation of the weak formulation (3.3)

$$u_{h,n} \in W_{h,n}, \qquad b(u_{h,n},v_{h,n}) = \ell(v_{h,n}) \quad \forall\, v_{h,n} \in W_{h,n} \tag{4.1}$$

has a unique solution and for the error,

$$\|u - u_{h,n}\|_W \le 2\,b_0 \inf_{v_{h,n}\in W_{h,n}} \|u - v_{h,n}\|_W. \tag{4.2}$$

It can be verified that similar to (3.5)–(3.7), the discrete problem (4.1) can be rewritten as: Find $u_{h,n}^{+} \in V_{1,h,n}^{+}$, $u_{h,n}^{-} \in V_{0,h,n}^{-}$ and $w_{h,n} \in W_{h,n}$ such that

(4.3)
$$\left(\mu_t u_{h,n}^{+} - \mu_s w_{h,n}^{+}, v_{h,n}^{+}\right)_Q - \left(u_{h,n}^{-}, \boldsymbol{\omega}\cdot\nabla v_{h,n}^{+}\right)_Q + \left(|\boldsymbol{\nu}\cdot\boldsymbol{\omega}|\, u_{h,n}^{+}, v_{h,n}^{+}\right)_\Gamma = \left(f^{+}, v_{h,n}^{+}\right)_Q - 2\left(\boldsymbol{\nu}\cdot\boldsymbol{\omega}\, u_{\rm in}, v_{h,n}^{+}\right)_Q \quad \forall\, v_{h,n}^{+} \in V_{1,h,n}^{+},$$

(4.4)
$$\left(\mu_t u_{h,n}^{-} - \mu_s w_{h,n}^{-}, v_{h,n}^{-}\right)_Q + \left(\boldsymbol{\omega}\cdot\nabla u_{h,n}^{+}, v_{h,n}^{-}\right)_Q = \left(f^{-}, v_{h,n}^{-}\right)_Q \quad \forall\, v_{h,n}^{-} \in V_{0,h,n}^{-},$$

(4.5)
$$(w_{h,n}, v_n)_\Omega + \alpha\,(\nabla^* w_{h,n}, \nabla^* v_n)_\Omega = (u_{h,n}, v_n)_\Omega \quad \forall\, v_n \in V_n.$$

Here,

$$V_n = \operatorname{span}\{Y_{j,i}(\boldsymbol{\omega}) : -j \le i \le j,\ -2n-1 \le j \le 2n+1\}.$$

5. An iteration procedure

The form (3.5)–(3.7) or the discrete version (4.3)–(4.5) naturally suggests an iteration method for solving the problems. Here, to simplify the notation, we discuss the iteration method for solving (3.5)–(3.7) as an example. With an initial guess w_0, say $w_0 = 0$, we define a sequence $\{(u_k, w_k)\}_{k\geq 1}$ by the following: $u_k^+ \in V_1^+$, $u_k^- \in V_0^-$ and $w_k(\mathbf{x}, \cdot) \in H^1(\Omega)$ for a.e. $\mathbf{x} \in X$ such that

$$(5.1)\qquad \left(\mu_t u_k^+ - \mu_s w_{k-1}^+, v^+\right)_Q - \left(u_k^-, \boldsymbol{\omega}\cdot\nabla v^+\right)_Q + \left(|\boldsymbol{\nu}\cdot\boldsymbol{\omega}|\, u_k^+, v^+\right)_\Gamma$$
$$= \left(f^+, v^+\right)_Q - 2\left(\boldsymbol{\nu}\cdot\boldsymbol{\omega}\, u_{\text{in}}, v^+\right)_Q \quad \forall v^+ \in V_1^+,$$

$$(5.2)\qquad \left(\mu_t u_k^- - \mu_s w_{k-1}^-, v^-\right)_Q + \left(\boldsymbol{\omega}\cdot\nabla u_k^+, v^-\right)_Q = \left(f^-, v^-\right)_Q \quad \forall v^- \in V_0^-,$$

$$(5.3)\qquad (w_k, v)_\Omega + \alpha\,(\nabla^* w_k, \nabla^* v)_\Omega = (u_k, v)_\Omega \quad \forall v \in H^1(\Omega).$$

The sequence is well defined. Here we focus on convergence of the iteration method.

THEOREM 5.1. *Under the assumptions* (1.3)–(1.4) *and* (1.7), *the iteration method converges:*

$$\|u - u_k\|_{L^2(Q)} + \|w - w_k\|_{L^2(Q)} + \|u^+ - u_k^+\|_{L^2_*(\Gamma)} \to 0 \quad \text{as } k \to \infty.$$

PROOF. Define the errors $e_{u,k} := u - u_k$, $e_{w,k} := w - w_k$, and their even and odd components $e_{u^+,k} = e_{u,k}^+ = u^+ - u_k^+$, etc. By subtracting the equations (5.1)–(5.3) from the corresponding equations (3.5)–(3.7), we obtain the error relations

$$(5.4)\qquad \left(\mu_t e_{u^+,k}, v^+\right)_Q - \left(e_{u^-,k}, \boldsymbol{\omega}\cdot\nabla v^+\right)_Q + \left(|\boldsymbol{\nu}\cdot\boldsymbol{\omega}|\, e_{u^+,k}, v^+\right)_\Gamma$$
$$= \left(\mu_s e_{w^+,k-1}, v^+\right)_Q \quad \forall v^+ \in V_1^+,$$

$$(5.5)\qquad \left(\mu_t e_{u^-,k}, v^-\right)_Q + \left(\boldsymbol{\omega}\cdot\nabla e_{u^+,k}, v^-\right)_Q = \left(\mu_s e_{w^-,k-1}, v^-\right)_Q \quad \forall v^- \in V_0^-,$$

$$(5.6)\qquad (e_{w,k}, v)_\Omega + \alpha\,(\nabla^* e_{w,k}, \nabla^* v)_\Omega = (e_{u,k}, v)_\Omega \quad \forall v \in H^1(\Omega).$$

Take $v^+ = e_{u^+,k}$ in (5.4), $v^- = e_{u^-,k}$ in (5.5), and add the two resulting inequalities,

$$(5.7)\qquad (\mu_t e_{u,k}, e_{u,k})_Q + \left(|\boldsymbol{\nu}\cdot\boldsymbol{\omega}|\, e_{u^+,k}, e_{u^+,k}\right)_\Gamma = (\mu_s e_{w,k-1}, e_{u,k})_Q\,.$$

For a.e. $\mathbf{x} \in X$, take $v = e_{w,k}$ in (5.6) to obtain

$$(5.8)\qquad (e_{w,k}, e_{w,k})_\Omega + (\alpha\,\nabla^* e_{w,k}, \nabla^* e_{w,k})_\Omega = (e_{u,k}, e_{w,k})_\Omega\,.$$

From the assumption (1.3), we know that

$$\kappa := \sup_{\mathbf{x}\in X} \frac{\mu_s(\mathbf{x})}{\mu_t(\mathbf{x})} < 1.$$

By (5.7), we then have

$$(5.9)\qquad (\mu_t e_{u,k}, e_{u,k})_Q + \left(|\boldsymbol{\nu}\cdot\boldsymbol{\omega}|\, e_{u^+,k}, e_{u^+,k}\right)_\Gamma \le \kappa\,(\mu_t e_{w,k-1}, e_{w,k-1})_Q^{1/2}\,(\mu_t e_{u,k}, e_{u,k})_Q^{1/2}.$$

Then,

$$(\mu_t e_{u,k}, e_{u,k})_Q \le \kappa^2\,(\mu_t e_{w,k-1}, e_{w,k-1})_Q\,.$$

Using (5.9) again,

$$(5.10)\qquad (\mu_t e_{u,k}, e_{u,k})_Q + \left(|\boldsymbol{\nu}\cdot\boldsymbol{\omega}|\, e_{u^+,k}, e_{u^+,k}\right)_\Gamma \le \kappa^2\,(\mu_t e_{w,k-1}, e_{w,k-1})_Q\,.$$

From (5.8),

$$(5.11)\qquad (\mu_t e_{w,k}, e_{w,k})_Q \le (\mu_t e_{u,k}, e_{w,k})_Q\,.$$

Combining (5.10) and (5.11),

$$(5.12)\qquad (\mu_t e_{u,k}, e_{u,k})_Q + \left(|\boldsymbol{\nu}\cdot\boldsymbol{\omega}|\, e_{u^+,k}, e_{u^+,k}\right)_\Gamma \le \kappa^2 (\mu_t e_{u,k-1}, e_{u,k-1})_Q .$$

An induction on (5.12) shows that

$$(\mu_t e_{u,k}, e_{u,k})_Q + \left(|\boldsymbol{\nu}\cdot\boldsymbol{\omega}|\, e_{u^+,k}, e_{u^+,k}\right)_\Gamma \le \kappa^{2k} (\mu_t e_{u,0}, e_{u,0})_Q \to 0 \text{ as } k \to \infty.$$

Moreover, by (5.11), we also have

$$(\mu_t e_{w,k}, e_{w,k})_Q \to 0 \text{ as } k \to \infty.$$

Thus, the stated convergence result holds. □

Implementation and simulation of the Galerkin method (4.1), as well as studies of related inverse problems, are future research topics.

References

[A98] Valeri Agoshkov, *Boundary value problems for transport equations*, Modeling and Simulation in Science, Engineering and Technology, Birkhäuser Boston Inc., Boston, MA, 1998. MR1638817 (99j:45007)

[A99] S. R. Arridge, *Optical tomography in medical imaging*, Inverse Problems **15** (1999), no. 2, R41–R93, DOI 10.1088/0266-5611/15/2/022. MR1684463 (2000b:78023)

[AH12] Kendall Atkinson and Weimin Han, *Spherical harmonics and approximations on the unit sphere: an introduction*, Lecture Notes in Mathematics, vol. 2044, Springer, Heidelberg, 2012. MR2934227

[B71] Ivo Babuška, *Error-bounds for finite element method*, Numer. Math. **16** (1970/1971), 322–333. MR0288971 (44 #6166)

[B09] Guillaume Bal, *Inverse transport theory and applications*, Inverse Problems **25** (2009), no. 5, 053001, 48, DOI 10.1088/0266-5611/25/5/053001. MR2501018 (2010d:78021)

[ES12] Herbert Egger and Matthias Schlottbom, *A mixed variational framework for the radiative transfer equation*, Math. Models Methods Appl. Sci. **22** (2012), no. 3, 1150014, 30, DOI 10.1142/S021820251150014X. MR2890452

[HECW11] Weimin Han, Joseph A. Eichholz, Xiaoliang Cheng, and Ge Wang, *A theoretical framework of x-ray dark-field tomography*, SIAM J. Appl. Math. **71** (2011), no. 5, 1557–1577, DOI 10.1137/100809039. MR2835363 (2012j:35448)

[HEHL11] Weimin Han, Joseph A. Eichholz, Jianguo Huang, and Jia Lu, *RTE-based bioluminescence tomography: a theoretical study*, Inverse Probl. Sci. Eng. **19** (2011), no. 4, 435–459, DOI 10.1080/17415977.2010.500383. MR2803147 (2012e:65243)

[HG41] L. Henyey and J. Greenstein, *Diffuse radiation in the galaxy*, Astrophysical J. **93** (1941), 70–83.

[KimK03] A. D. Kim and J. B. Keller, *Light propagation in biological tissue*, J. Opt. Soc. Amer. A **20** (2003), 92–98.

[LM84] E.E. Lewis and W.F. Miller, *Computational Methods of Neutron Transport*, John Wiley & Sons, New York, 1984.

[NW01] Frank Natterer and Frank Wübbeling, *Mathematical methods in image reconstruction*, SIAM Monographs on Mathematical Modeling and Computation, Society for Industrial and Applied Mathematics (SIAM), Philadelphia, PA, 2001. MR1828933 (2002c:94006)

Department of Mathematics and Program in Applied Mathematical and Computational Sciences, University of Iowa, Iowa City, Iowa 52242
E-mail address: weimin-han@uiowa.edu

Department of Mathematics, Wright State University, Daton, Ohio
E-mail address: yi.li@wright.edu

Department of Mathematics and Program in Applied Mathematical and Computational Sciences, University of Iowa, Iowa City, Iowa 52242
E-mail address: qiwei-sheng@uiowa.edu

Department of Mathematics, Harbin Institute of Technology (HIT), China
E-mail address: tangshuang107@126.com

Numerical study of a viscoelastic flow in a moving domain

Jason Howell, Hyesuk Lee, and Shuhan Xu

Abstract. A numerical method for viscoelastic fluid flow interacting with an elastic structure is considered. The fluid-structure system is decoupled using interface conditions and a staggered time-stepping scheme. The fluid subproblem is approximated by the finite element method based on the Arbitrary Lagrangian-Eulerian (ALE) formulation. A time discretization of each subproblem is discussed and a decoupling numerical algorithm for the system is presented.

1. Model equations

Consider the viscoelastic fluid governed by the *Johnson-Segalman model*

$$\boldsymbol{\sigma} + \lambda \left(\frac{\partial \boldsymbol{\sigma}}{\partial t} + \mathbf{u} \cdot \nabla \boldsymbol{\sigma} + g_a(\boldsymbol{\sigma}, \nabla \mathbf{u}) \right) - 2\,\alpha\, D(\mathbf{u}) = \mathbf{0} \quad \text{in } \Omega_t, \tag{1.1}$$

$$\rho \left(\frac{\partial \mathbf{u}}{\partial t} + \mathbf{u} \cdot \nabla \mathbf{u} \right) - \nabla \cdot \boldsymbol{\sigma} - 2(1-\alpha)\nabla \cdot D(\mathbf{u}) + \nabla p = \mathbf{f} \quad \text{in } \Omega_t, \tag{1.2}$$

$$div\, \mathbf{u} = 0 \quad \text{in } \Omega_t, \tag{1.3}$$

where $\boldsymbol{\sigma}$ denotes the extra stress tensor, $\mathbf{u}$ the velocity vector, p the pressure of fluid, ρ the density of the fluid, and λ is the Weissenberg number defined as the product of the relaxation time and a characteristic strain rate. Assume that p has zero mean value over Ω_t. In (1.1) and (1.2), $D(\mathbf{u}) := (\nabla \mathbf{u} + \nabla \mathbf{u}^T)/2$ is the rate of the strain tensor, α a number such that $0 < \alpha < 1$ which may be considered as the fraction of viscoelastic viscosity, and $\mathbf{f}$ the body force. In (1.1), $g_a(\boldsymbol{\sigma}, \nabla \mathbf{u})$ is defined by

$$g_a(\boldsymbol{\sigma}, \nabla \mathbf{u}) := \frac{1-a}{2}(\boldsymbol{\sigma}\nabla \mathbf{u} + \nabla \mathbf{u}^T \boldsymbol{\sigma}) - \frac{1+a}{2}(\nabla \mathbf{u}\, \boldsymbol{\sigma} + \boldsymbol{\sigma} \nabla \mathbf{u}^T) \tag{1.4}$$

for $a \in [-1, 1]$.

Initial and boundary conditions for $\mathbf{u}$ and $\boldsymbol{\sigma}$ are given as follows:

$$\mathbf{u}(\mathbf{x}, 0) = \mathbf{u}^0, \quad \boldsymbol{\sigma}(\mathbf{x}, 0) = \boldsymbol{\sigma}^0 \quad \text{in } \Omega_0 \tag{1.5}$$

$$\mathbf{u} = \mathbf{u}_{BC} \quad \text{on } \Gamma_t, \qquad \boldsymbol{\sigma} = \boldsymbol{\sigma}_{BC} \text{ on } \Gamma_{tin}, \tag{1.6}$$

where $\int_{\Gamma_t} \mathbf{u}_{BC} \cdot \mathbf{n}\, d\Gamma_t = 0$ and Γ_{tin} denotes the inflow boundary.

2010 *Mathematics Subject Classification.* Primary 65M60; Secondary 65M12.
Key words and phrases. Fluid-structure interaction, viscoelastic fluid flow, moving boundary.
Partially supported by the NSF under grant no. DMS-1016182.

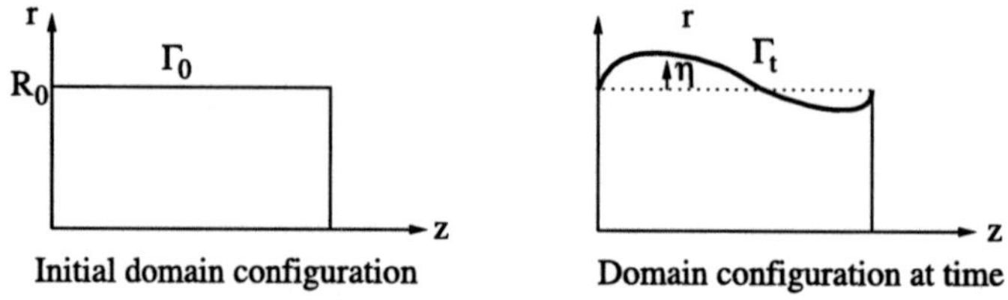

FIGURE 1.1. Domain configuration

For 2D-1D coupling of the fluid and elastic equations, we consider the one-dimensional *generalized rod model* [8]

$$\rho\frac{\partial^2\eta}{\partial t^2} - a\frac{\partial^2\eta}{\partial z^2} + b\eta - c\frac{\partial^3\eta}{\partial z^2\partial t} = \Phi\,, \tag{1.7}$$

where η represents the radial displacement of the structure with respect to the rest configuration

$$\Gamma_0 := \{(z,r)\in\mathbb{R}^2 : z\in(0,L), r\in(-R,R)\}\,.$$

In (1.7) Φ is the external force in the radial direction, and a, b, c are positive constants related to the physical properties of the elastic medium.

As the structure equation (1.7) is of a propagative type, we consider the first order absorbing boundary conditions [7]

$$\begin{cases} \frac{\partial\eta}{\partial t} - \sqrt{\frac{a}{\rho}}\frac{\partial\eta}{\partial z} = 0 & \text{at } z=0 \\ \frac{\partial\eta}{\partial t} + \sqrt{\frac{a}{\rho}}\frac{\partial\eta}{\partial z} = 0 & \text{at } z=L. \end{cases}$$

The interface conditions on the moving boundary Γ_t are given as:

$$\mathbf{u} = \frac{\partial\eta}{\partial t}\mathbf{n}, \tag{1.8}$$

$$-(\boldsymbol{\sigma} + 2(1-\alpha)D(\mathbf{u}) - p\mathbf{I})\cdot\mathbf{n} = \Phi\mathbf{n}. \tag{1.9}$$

where $\mathbf{n}$ is the outward unit normal vector to Γ_t. The first condition (1.8) guarantees the adherence of the fluid to the structure by making the velocity of the displacement exactly the vertical velocity of the fluid for every point on the interface. The second condition (1.9), according to a reaction principle, satisfies the continuity of the stresses on the interface.

2. ALE formulation and time discretization

We consider the Arbitrary Lagrangian Eulerian (ALE) [3] method for a numerical solution of the fluid equations in a moving domain. In the ALE formulation, a one-to-one coordinate transformation is introduced for the fluid domain, and the fluid equations can be rewritten with respect to a fixed reference domain. Specifically, we define the time-dependent bijective mapping Ψ_t which maps the reference domain Ω_0 to the physical domain Ω_t:

$$\Psi_t : \Omega_0 \to \Omega_t, \qquad \Psi_t(\mathbf{y}) = \mathbf{x}(\mathbf{y},t)\,, \tag{2.1}$$

where $\mathbf{y}$ and $\mathbf{x}$ are the spatial coordinates in Ω_0 and Ω_t, respectively. The coordinate $\mathbf{y}$ is often called the *ALE coordinate*. Using Ψ_t, the weak formulation of the flow equations in Ω_t can be recast into a weak formulation defined in the reference domain Ω_0. Thus the model equations in the reference domain can be considered for numerical simulation, and the transformation function Ψ_t needs to be determined at each time step as a part of computation.

For the variational formulation of the flow equations in the ALE framework, define function spaces for the reference domain:

$$\begin{aligned} \mathbf{U}_0 &:= \mathbf{H}^1(\Omega_0)\,, \\ Q_0 &:= L_0^2(\Omega_0) = \{\overline{q} \in L^2(\Omega_0) : \textstyle\int_{\Omega_0} \overline{q}\, d\Omega = 0\}\,, \\ \mathbf{\Sigma}_0 &:= \{\boldsymbol{\tau} \in \mathbf{L}^2(\Omega_0) : \tau_{ij} = \tau_{ji}\}\,. \end{aligned}$$

The function spaces for Ω_t are then defined as

$$\begin{aligned} \mathbf{U}_t &:= \{\mathbf{v} : \Omega_t \times [0,T] \to \mathbb{R}^2,\ \mathbf{v} = \overline{\mathbf{v}} \circ \Psi_t^{-1} \text{ for } \overline{\mathbf{v}} \in \mathbf{U}_0\}\,, \\ Q_t &:= \{q : \Omega_t \times [0,T] \to \mathbb{R},\ q = \overline{q} \circ \Psi_t^{-1} \text{ for } \overline{p} \in Q_0\}\,, \\ \mathbf{\Sigma}_t &:= \{\boldsymbol{\tau} : \Omega_t \times [0,T] \to \mathbb{R}^{2\times 2},\ \boldsymbol{\tau} = \overline{\boldsymbol{\tau}} \circ \Psi_t^{-1} \text{ for } \overline{\boldsymbol{\tau}} \in \mathbf{\Sigma}_0\}\,. \end{aligned}$$

The variational formulation in ALE framework is then given by: find $(\mathbf{u}, p, \boldsymbol{\sigma}) \in \mathbf{U}_t \times Q_t \times \mathbf{\Sigma}_t$ satisfying (1.6) such that

$$\begin{aligned} &(\boldsymbol{\sigma}, \boldsymbol{\tau})_{\Omega_t} + \lambda \frac{d}{dt}(\boldsymbol{\sigma}, \boldsymbol{\tau})_{\Omega_t} + \lambda\,(((\mathbf{u} - \mathbf{z}) \cdot \nabla)\boldsymbol{\sigma} - \boldsymbol{\sigma}(\nabla \cdot \mathbf{z}) + g_a(\boldsymbol{\sigma}, \nabla \mathbf{b}), \boldsymbol{\tau})_{\Omega_t} \\ &\qquad -2\alpha\,(D(\mathbf{u}), \boldsymbol{\tau})_{\Omega_t} = 0 \quad \forall \boldsymbol{\tau} \in \tilde{\mathbf{\Sigma}}_t\,, \end{aligned} \tag{2.2}$$

$$\begin{aligned} &\rho \frac{d}{dt}(\mathbf{u}, \mathbf{v})_{\Omega_t} + \rho\,((\mathbf{u} - \mathbf{z}) \cdot \nabla \mathbf{u} - \mathbf{u}(\nabla \cdot \mathbf{z}), \mathbf{v})_{\Omega_t} + (\boldsymbol{\sigma}, D(\mathbf{v}))_{\Omega_t} \\ &\qquad +2(1-\alpha)(D(\mathbf{u}), D(\mathbf{v}))_{\Omega_t} - (p, \nabla \cdot \mathbf{v})_{\Omega_t} = (\mathbf{f}, \mathbf{v})_{\Omega_t}\ \forall \mathbf{v} \in \tilde{\mathbf{U}}_t\,, \end{aligned} \tag{2.3}$$

$$(q, \nabla \cdot \mathbf{u})_{\Omega_t} = 0 \quad \forall q \in Q_t\,, \tag{2.4}$$

where $\mathbf{z} := \frac{\partial \mathbf{x}}{\partial t}|_{\mathbf{y}}$ represents the domain velocity and $\tilde{\mathbf{\Sigma}}_t$, $\tilde{\mathbf{U}}_t$ are subspaces of $\mathbf{\Sigma}_t$, and $\mathbf{U}_t$ satisfies the homogeneous boundary conditions on $\Gamma_{t in}$ and Γ_t, respectively.

In order to define the ALE mapping Ψ_t, we consider the boundary position function $h : \Gamma_0 \times [0,T] \to \Gamma_t$. The ALE mapping may be then determined by solving the Laplace equation

$$\Delta_{\mathbf{y}} \mathbf{x}(\mathbf{y}) = 0 \quad \text{in } \Omega_0\,, \qquad \mathbf{x}(\mathbf{y}) = h(\mathbf{y}) \quad \text{on } \Gamma_0\,. \tag{2.5}$$

This method is called the *harmonic extension technique*, where the boundary position function h is extended onto the entire domain [4]. For the fluid-structure interaction problem the boundary position function is determined by the elastic displacement η.

In implementing the ALE method we consider the time integration scheme, referred to as the *geometric conservation law* (GCL), which is known to be related to the stability of numerical solutions [**2**, **4**, **5**, **7**]. The GCL requires a numerical time discretization scheme to simulate a uniform flow exactly on a moving domain. The GCL in the finite element ALE framework suggests that a quadrature rule should be chosen so that the time integration is performed exactly for a chosen finite element polynomial [**8**]. For example, a quadrature formula with the degree of precision 1 or higher satisfies the GCL if piecewise linear elements are used for the ALE mapping Ψ_t in $\mathbb{R}^2$

Following the GCL, we consider the mid-point rule

$$\text{(2.6)}\quad \int_{\Omega_{t^{n+1}}} \mathbf{v}_h \, d\Omega - \int_{\Omega_{t^n}} \mathbf{v}_h \, d\Omega = \int_{t^n}^{t^{n+1}} \int_{\Omega_t} \mathbf{v}_h \nabla \cdot \mathbf{z}_h \, d\Omega \, dt \approx \Delta t \int_{\Omega_{t^{n+\frac{1}{2}}}} \mathbf{v}_h \nabla \cdot \mathbf{z}_h \, ds$$

for time integration. The implicit Euler scheme reads as:

$$\begin{aligned}\text{(2.7)}\quad & \lambda\left[(\boldsymbol{\sigma}^{n+1}, \boldsymbol{\tau})_{\Omega_{t_{n+1}}} - (\boldsymbol{\sigma}^{n}, \boldsymbol{\tau})_{\Omega_{t_n}}\right] \\ & + \Delta t\left[(\boldsymbol{\sigma}^{n+1}, \boldsymbol{\tau})_{\Omega_{t^{n+\frac{1}{2}}}} - 2\alpha(D(\mathbf{u}^{n+1}), \boldsymbol{\tau})_{\Omega_{t^{n+\frac{1}{2}}}}\right] \\ & + \lambda \Delta t\left[((\mathbf{u}^{n+1} - \mathbf{z}^{n+\frac{1}{2}}) \cdot \nabla \boldsymbol{\sigma}^{n+1}, \boldsymbol{\tau})_{\Omega_{t^{n+\frac{1}{2}}}} - (\boldsymbol{\sigma}^{n+1}(\nabla \cdot \mathbf{z}^{n+\frac{1}{2}}), \boldsymbol{\tau})_{\Omega_{t^{n+\frac{1}{2}}}} \right. \\ & \left. + (g_a(\boldsymbol{\sigma}^{n+1}, \nabla \mathbf{u}^{n+1}), \boldsymbol{\tau})_{\Omega_{t^{n+\frac{1}{2}}}}\right] = 0,\end{aligned}$$

$$\begin{aligned}\text{(2.8)}\quad & \rho\left[(\mathbf{u}^{n+1}, \mathbf{v})_{\Omega_{t_{n+1}}} - (\mathbf{u}^{n}, \mathbf{v})_{\Omega_{t_n}}\right] \\ & + \Delta t \, \rho\left[((\mathbf{u}^{n+1} - \mathbf{z}^{n+\frac{1}{2}}) \cdot \nabla \mathbf{u}^{n+1}, \mathbf{v})_{\Omega_{t^{n+\frac{1}{2}}}} - (\mathbf{u}^{n+1}(\nabla \cdot \mathbf{z}^{n+\frac{1}{2}}), \mathbf{v})_{\Omega_{t^{n+\frac{1}{2}}}}\right] \\ & + \Delta t\left[2(1-\alpha)(D(\mathbf{u}^{n+1}), D(\mathbf{v}))_{\Omega_{t^{n+\frac{1}{2}}}} + (\boldsymbol{\sigma}^{n+1}, D(\mathbf{v}))_{\Omega_{t^{n+\frac{1}{2}}}} \right. \\ & \left. + (p^{n+1}, \nabla \cdot \mathbf{v})_{\Omega_{t^{n+\frac{1}{2}}}}\right] = \Delta t \, (\mathbf{f}^{n+\frac{1}{2}}, \mathbf{v})_{\Omega_{t^{n+\frac{1}{2}}}},\end{aligned}$$

$$\text{(2.9)}\quad (q, \nabla \cdot \mathbf{u}^{n+1})_{\Omega_{t^{n+\frac{1}{2}}}} = 0,$$

for all $(\boldsymbol{\tau}, \mathbf{v}, q) \in \tilde{\boldsymbol{\Sigma}}_t \times \tilde{\mathbf{U}}_t \times Q_t$.

For the variational formulation of the structure model, define $\mathbf{W} := H^1(0, L)$. Then the weak problem for the structure reads as: find $\eta \in \mathbf{W}$ such that for all $\xi \in \mathbf{W}$

$$\begin{aligned}\text{(2.10)}\quad & \rho\left(\frac{\partial^2 \eta}{\partial t^2}, \xi\right) + b(\eta, \xi) + \left(\left(a\frac{\partial \eta}{\partial z} + c\frac{\partial^2 \eta}{\partial z \partial t}\right), \frac{\partial \xi}{\partial z}\right) \\ & + \sqrt{\frac{\rho}{a}}\left(a\frac{\partial \eta}{\partial t} + c\frac{\partial^2 \eta}{\partial t^2}\right)\xi|_{z=0,z=L} = (\hat{\Phi}, \xi),\end{aligned}$$

where

$$\text{(2.11)}\quad \hat{\Phi} = \Phi\sqrt{1 + \left(\frac{\partial \eta}{\partial z}\right)^2}$$

and Φ is the right hand side of (1.7).

Recall that the matching condition (1.9) is defined on Γ_t, hence the term under the square root is introduced due to the change in the surface measure passing from Γ_t to Γ_0.

For the time discretization, we consider the first-order Implicit Euler scheme:

$$(2.12)\quad \rho\left(\frac{(\eta^{n+1},\xi)-2(\eta^n,\xi)+(\eta^{n-1},\xi)}{\Delta t^2}\right)+b(\eta^{n+1},\xi)$$
$$+a\left(\frac{\partial\eta^{n+1}}{\partial z},\frac{\partial\xi}{\partial z}\right)+c\left(\frac{\frac{\partial\eta^{n+1}}{\partial z}-\frac{\partial\eta^{n}}{\partial z}}{\Delta t},\frac{\partial\xi}{\partial z}\right)$$
$$+\sqrt{\frac{\rho}{a}}\left(a\frac{\eta^{n+1}-\eta^n}{\Delta t}+c\frac{\eta^{n+1}-2\eta^n+\eta^{n-1}}{\Delta t^2}\right)\xi_h|_{z=0,z=L}=(\hat{\Phi},\xi).$$

3. Decoupling algorithm

We considered the *first order explicit algorithm* (also called the *loosely coupled scheme*). In this algorithm the structure can be solved in advanced time, providing a new boundary position and boundary velocity for the fluid, since its forcing term Φ is determined by $(\mathbf{u}^n, p^n)$ from the previous time step. Two subproblems communicate with each other by two matching conditions, one is the forcing term on the wall due to the fluid and the other is the boundary value of the fluid given by the velocity of the displacement of the structure. This simple splitting algorithm is often known as the *conventional serial staggered scheme (CSS)*. We summarize the algorithm as follows:

Algorithm: In each time interval $[t_n, t_{n+1}]$,

(Step 1) Solve the structure equation using the forcing term obtained from the previous time level t_n.

(Step 2) Update the domain, the boundary velocity of the fluid, and the ALE mapping. For each point (x, y) on the boundary, a new position of the boundary at time step $n+1$ can be computed as

$$x_{(n+1)} = x_0, \qquad y_{(n+1)} = y_0 + \eta_{(n+1)};$$

the ALE mapping is computed by the harmonic extension and the boundary velocity for the fluid is computed by a finite difference scheme.

(Step 3) Solve the fluid equations in the updated domain Ω_{n+1}. By the ALE method discussed in previous chapter, we compute $(\mathbf{u}_{n+1}, p_{n+1}, \boldsymbol{\sigma}_{n+1})$.

(Step 4) Compute the forcing term Φ of the structure equation by (1.9) using $(\mathbf{u}_{n+1}, p_{n+1}, \boldsymbol{\sigma}_{n+1})$.

The explicit algorithm has an advantage in efficiency, since for each time step it only requires one solution of each subproblem.

4. Numerical test

We present numerical results obtained by the algorithm discussed in the previous section.

Figure 4.2 shows the domain considered for numerical tests. Suppose both the fluid and structure are started from rest. A rectangle of height $D = 1$ and length $L = 6$ is defined as the initial domain for the fluid flow. Without loss of generality, suppose that the bottom boundary of the domain is a fixed wall, the top edge is deformable in the vertical direction, and the left and right edges are inflow and outflow boundaries, respectively.

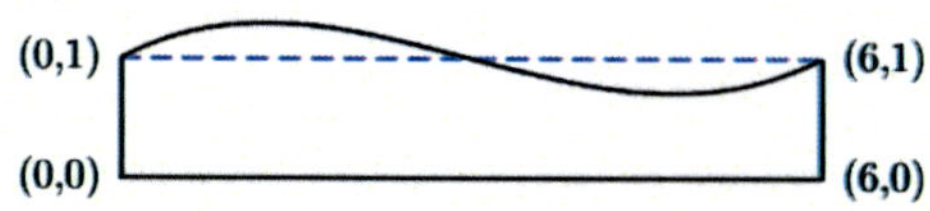

FIGURE 4.2. Model for numerical experiment

On the inflow and outflow boundaries, we consider a Neumann boundary condition for the fluid; the boundary condition imposed on the inflow section is

$$(\boldsymbol{\sigma} + 2(1-\alpha)D(\mathbf{u}) - p\mathbf{I}) \cdot \mathbf{n} = \begin{cases} 10^4[\cos(\frac{\pi t}{2.5ms}) - 1]\mathbf{n} & 0 \le t \le 5 \\ \mathbf{0} & 5 < t < T \end{cases},$$

while on the outflow section

$$(\boldsymbol{\sigma} + 2(1-\alpha)D(\mathbf{u}) - p\mathbf{I}) \cdot \mathbf{n} = \mathbf{0}.$$

For the fluid equations, the parameters are set as $\alpha = 0.5$, $\lambda = 0.5$, and $\rho = 50$. The parameters for the structure equation are $a = 25000$, $c = 0.01$, and $b = 400000$. For the spatial finite element discretization, we consider the (P_2, P_1, P_{1DG}) elements as described in [1] for the fluid equation and P_2 finite elements for the structure.

The numerical experiment was based on a 5×25 uniform grid mesh initially and re-meshed according to the ALE mapping. A time step $\Delta t = 0.0046$ was used. Figures 4.3-4.6 illustrate the displacement of the wall at the values $t = 0.09, 0.32, 0.46, 0.69$, respectively. For comparison, we performed similar numerical tests for a Newtonian fluid-structure problem based on the same mesh size and time step. These two problems share the same parameters with the exception of the value of λ - we set $\lambda = 0.5$ for the viscoelastic fluid and $\lambda = 0$ for the Newtonian case. In each figure, the blue curve represents the configurations of the moving boundary for the non-Newtonian case, while the red curve represents the Newtonian case.

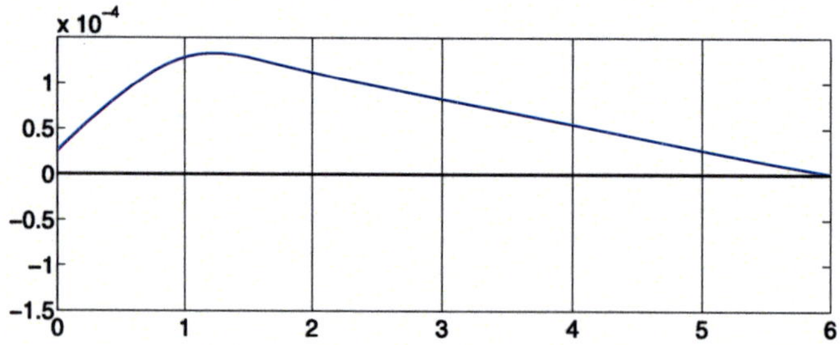

FIGURE 4.3. Displacement of the wall at t=0.09

From the above figures, we may observe the presence of a smooth wave inside the tube, which matches our expectation on the motion of the fluid. More precisely, we observe a bump moving horizontally from the inflow section to the outflow section. The magnitude of the summit value does not blow up in the procedure. The configuration went back to zero after a certain time since the force imposed on

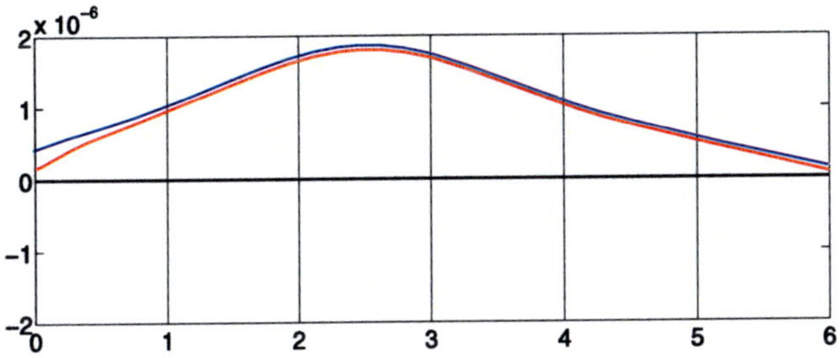

FIGURE 4.4. Displacement of the wall at t=0.32

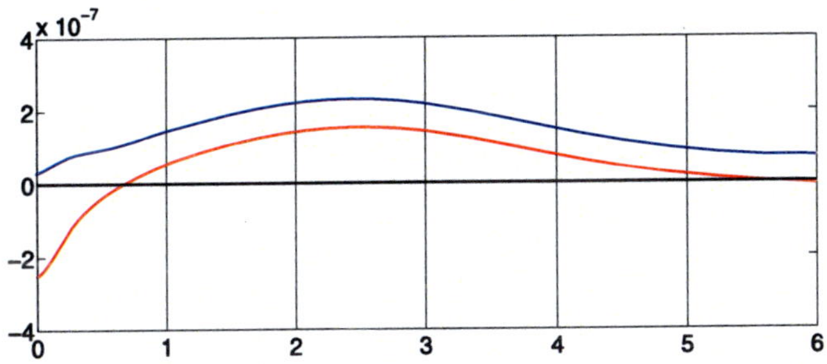

FIGURE 4.5. Displacement of the wall at t=0.46

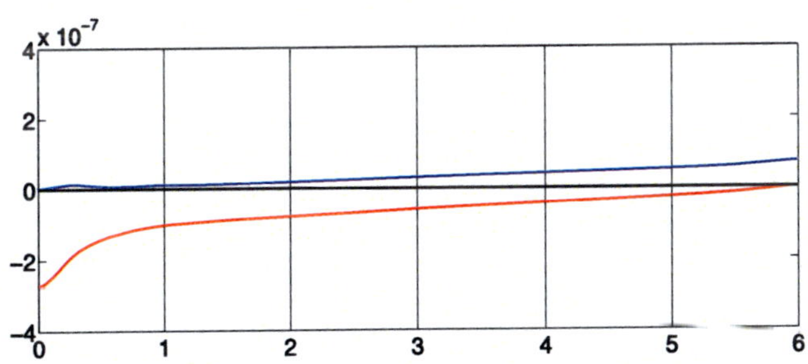

FIGURE 4.6. Displacement of the wall at t=0.69

the inflow section is set to be zero after 5ms in (4). Hence, the fluid depicted by the sequence of figures match the physical meaning of our numerical experiment.

At the same time, we may observe that the gap between the Newtonian and the non-Newtonian curves increases as time goes (it can be negligible in first two figures). This difference suggests that using a Newtonian model in blood flow simulation may lead to significant modeling and numerical errors.

However, we found that the explicit coupling algorithm was not stable enough over a long period. The numerical tests show significant instability and dramatic oscillation if the mesh size and time step are not carefully picked. In fact, the explicit coupling algorithm seems relatively more stable for a Newtonian fluid than non-Newtonian cases. An implicit coupling algorithm of second oder in time is currently under investigation.

References

[1] J. Baranger and D. Sandri, *Finite element approximation of viscoelastic fluid flow: existence of approximate solutions and error bounds. I. Discontinuous constraints*, Numer. Math. **63** (1992), no. 1, 13–27, DOI 10.1007/BF01385845. MR1182509 (93k:65090)

[2] Daniele Boffi and Lucia Gastaldi, *Stability and geometric conservation laws for ALE formulations*, Comput. Methods Appl. Mech. Engrg. **193** (2004), no. 42-44, 4717–4739, DOI 10.1016/j.cma.2004.02.020. MR2091121 (2005d:65164)

[3] J. Donea, S. Giuliani, J.P. Halleux, An arbitrary Lagrangian-Eulerian finite element method for transient dynamic fluid-structure interactions, *Comput. Methods Appl. Mech. Engrg.* 33, 689-723, 1982.

[4] L. Formaggia and F. Nobile, *A stability analysis for the arbitrary Lagrangian Eulerian formulation with finite elements*, East-West J. Numer. Math. **7** (1999), no. 2, 105–131. MR1699243 (2000d:65150)

[5] Luca Formaggia and Fabio Nobile, *Stability analysis of second-order time accurate schemes for ALE-FEM*, Comput. Methods Appl. Mech. Engrg. **193** (2004), no. 39-41, 4097–4116, DOI 10.1016/j.cma.2003.09.028. MR2087006 (2005d:65135)

[6] L. Gastaldi, *A priori error estimates for the arbitrary Lagrangian Eulerian formulation with finite elements*, East-West J. Numer. Math. **9** (2001), no. 2, 123–156. MR1836870 (2002m:65094)

[7] F. Nobile, *Numerical approximation of fluid-structure interaction problems with application to haemodynamics*, Ph.D. thesis, 2001.

[8] A. Quarteroni, M. Tuveri and A. Veneziani, Computational vascular fluid dynamics: problems, models, and methods, *Comput. Visual Sci.*, 2, 163-197, 2000.

Department of Mathematics, College of Charleston, Charleston, South Carolina 29424

E-mail address: howelljs@cofc.edu

Department of Mathematical Sciences, Clemson University, Clemson, South Carolina 29634-0975

E-mail address: hklee@clemson.edu

Department of Mathematical Sciences, Clemson University, Clemson, South Carolina 29634-0975

E-mail address: shuhanx@clemson.edu

A superconvergence of the Morley element via postprocessing

Jianguo Huang, Xuehai Huang, and Shangyou Zhang

ABSTRACT. Via an equivalence between the Morley element for the biharmonic equation and the Crouzeix-Raviart element for the Stokes equations, we establish a postprocessing algorithm which produces an half order superconvergence for the Morley element. Numerical tests verify the theory.

1. Introduction

The Morley finite element is a nonconforming element for fourth order elliptic partial differential equations ([**2, 10, 15**]), in particular, for the biharmonic equation:

$$(1.1)\qquad \begin{cases} \Delta^2\varphi = f & \text{in } \Omega, \\ \varphi = \dfrac{\partial}{\partial n}\varphi = 0 & \text{on } \partial\Omega. \end{cases}$$

Here $f \in L^2(\Omega)$, Ω is a polygonal domain and $\boldsymbol{n}$ is the unit outward normal vector at a boundary point. Due to simplicity, the Morley element is widely used and analyzed. It is interesting to mention that the extended Morley elements for higher order equations and in higher space dimensions were constructed and discussed in [**11, 12, 17, 18**].

In this paper, we study a superconvergence property of the Morley element. Superconvergence is discovered for many continuous finite elements [**1, 5, 6, 8, 13, 16, 23, 25, 24, 26**]. But for nonconforming elements, such as the Morley element, it is difficult to show the superconvergence. So far, there is only one superconvergence result for the Morley element. In [**9**], an half order superconvergence is proved for the Morley element solution on a type of uniform grids, cf. Figure 1. If the solution of (1.1) $\varphi \in H^4(\Omega) \cap H_0^2(\Omega)$,

$$|\varphi - I_{3h}^3\varphi_h|_{2,h} \le Ch^{3/2}(|\varphi|_{4,\Omega} + \|f\|_{0,\Omega}),$$

2010 *Mathematics Subject Classification.* Primary 65M60, 65N30.

Key words and phrases. Superconvergence, biharmonic equation, Stokes equations, nonconforming element, Morley element, Crouzeix-Raviart element.

The first author was supported in part by NSFC (Grant nos. 11171219, 11161130004) and E–Institutes of Shanghai Municipal Education Commission (E03004).

The second author was supported in part by NSFC (Grant nos. 11126226, 11171257) and Zhejiang Provincial Natural Science Foundation of China (Y6110240).

where I_{3h}^3 is the nodal, bicubic interpolation operator, defined on squares formed by 18 triangles (cf. Figure 1).

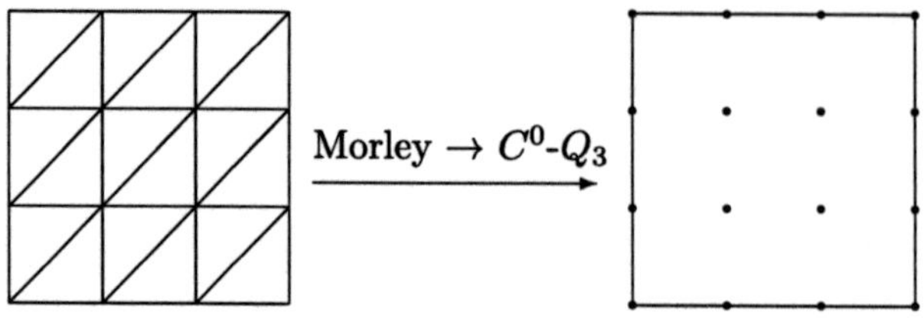

FIGURE 1. A nodal interpolation operator I_{3h}^3 used by [9].

Instead of analyzing the Morley solution, we derive a superconvergence from its equivalent solution to the Stokes equations. We note that the solution φ of (1.1) is related by $\boldsymbol{u}$ $(= \mathbf{curl}\,\varphi)$ to the unique solution of the Stokes equation:

$$\begin{cases} -\Delta \boldsymbol{u} + \nabla\lambda = \boldsymbol{g} & \text{in } \Omega, \\ \operatorname{div} \boldsymbol{u} = 0 & \text{in } \Omega, \\ \boldsymbol{u} = \boldsymbol{0} & \text{on } \partial\Omega, \end{cases} \tag{1.2}$$

where $\boldsymbol{g}$ satisfies $\operatorname{curl} \boldsymbol{g} = f$, for example, $\boldsymbol{g} = \langle 0, \int f dx\rangle$. Not only the continuous solutions are related, but also discrete solutions. Falk and Morley [7] showed that the curl of the Morley solution to (1.1) is also the Crouzeix-Raviart solution to (1.2). Based on this equivalence and a superconvergence result of Ye [20] on the Crouzeix-Raviart element, we derive an half order superconvergence ($k = 3$ in Theorem 4.2) for the Morley element. Here, the curl of a Morley solution is locally projected (cf. Figure 2) to provide a superconvergence approximation of φ in (semi, discrete) H^1 and H^2 norms. Neither [9] nor our result is optimal yet. Numerical results in [9] and here in last section indicate that a full order superconvergence is possible. A further study should be done.

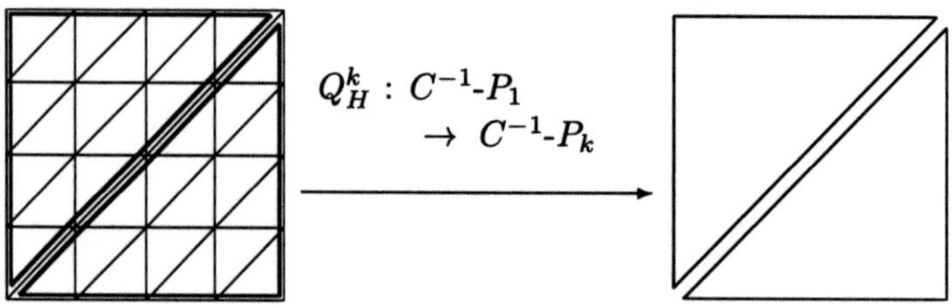

FIGURE 2. A local L^2-projection operator Q_H^2, from C^{-1}-P_1 to C^{-1}-P_2.

2. The continuous equivalent formulation

Consider the biharmonic problem (1.1). Suppose that there exists a vector field $\boldsymbol{g} \in (H^1(\Omega))^2$ satisfying $f = \operatorname{curl} \boldsymbol{g} := \partial_1 g_2 - \partial_2 g_1$. Then, by integration by parts, the variational formulation for (1.1) is available: Finding $\varphi \in H_0^2(\Omega)$ such that

$$\int_\Omega \boldsymbol{\nabla}^2\varphi : \boldsymbol{\nabla}^2\vartheta \, dx = \int_\Omega f\vartheta \, dx = \int_\Omega \boldsymbol{g} \cdot \mathbf{curl}\,\vartheta \, dx \quad \forall \vartheta \in H_0^2(\Omega), \tag{2.1}$$

where $\mathbf{curl}\,\vartheta := (\partial_2\vartheta, -\partial_1\vartheta)^{\mathrm{T}}$.

THEOREM 2.1. *Let $\varphi \in H_0^2(\Omega)$ be the solution of* (2.1), *and* $(\boldsymbol{u}, \lambda) \in (H_0^1(\Omega))^2 \times L_0^2(\Omega)$ *be the solution of the Stokes equation*

$$\tag{2.2} \begin{cases} \displaystyle\int_\Omega \boldsymbol{\nabla u} : \boldsymbol{\nabla v}\, dx + \int_\Omega \lambda \boldsymbol{\nabla} \cdot \boldsymbol{v}\, dx = \int_\Omega \boldsymbol{g} \cdot \boldsymbol{v}\, dx & \forall \boldsymbol{v} \in (H_0^1(\Omega))^2, \\ \displaystyle\int_\Omega \mu \boldsymbol{\nabla} \cdot \boldsymbol{u}\, dx = 0 & \forall \mu \in L_0^2(\Omega). \end{cases}$$

Then

$$\boldsymbol{u} = \mathbf{curl}\, \varphi.$$

PROOF. From the second equation of (2.2), we find that $\operatorname{div} \boldsymbol{u} = 0$. Thus, there exists a certain $\tilde{\varphi} \in H_0^2(\Omega)$ such that $\boldsymbol{u} = \mathbf{curl}\, \tilde{\varphi}$. Substituting the expression into the first equation of (2.2) and taking $\boldsymbol{v} = \mathbf{curl}\, \vartheta$ for any $\vartheta \in H_0^2(\Omega)$, we have

$$\int_\Omega \boldsymbol{\nabla}^2 \tilde{\varphi} : \boldsymbol{\nabla}^2 \vartheta\, dx = \int_\Omega \boldsymbol{g} \cdot \mathbf{curl}\, \vartheta\, dx \quad \forall \vartheta \in H_0^2(\Omega).$$

Therefore, $\tilde{\varphi}$ is identically equal to φ, in terms of the unique solvability of (2.1), leading to the desired result. □

3. The discrete equivalent formulation

Theorem 2.1 establishes an equivalence between the biharmonic equation and the Stokes equation. We are going to obtain a discrete analogue of this equivalence.

Let $\{\mathcal{T}_h(\Omega)\}_{h>0}$ be a non-degenerate family of triangulations of Ω [**3, 4**]; $h := \max_{\tau \in \mathcal{T}_h} h_\tau$ and $h_\tau := \operatorname{diam}(\tau)$. Let $\mathcal{E}_h$ be the union of all edges of the triangulation $\mathcal{T}_h(\Omega)$ and $\mathcal{E}_h^\partial$ the union of all edges of the triangulation $\mathcal{T}_h(\Omega)$ on $\partial\Omega$. Set $(\tau, P_\tau, \Sigma_\tau)$ be the Morley element with τ the geometric shape, P_τ the shape function space $P_2(\tau)$ and Σ_τ the set of nodal parameters $\{\vartheta(p_i), \partial_n \vartheta(m_i), 1 \le i \le 3\}$, where $P_k(\tau)$ is the space of all k-th order polynomials on τ for any nonnegative integer k, p_i is the vertex of τ and m_i the edge midpoint of τ. m_i is opposed to p_i.

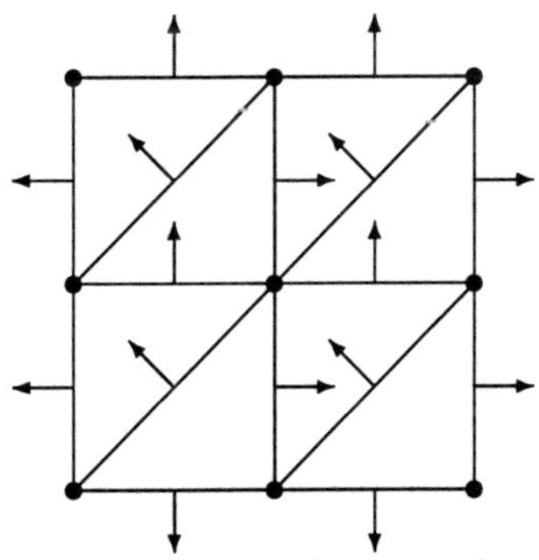

FIGURE 3. The nodal definition of the Morley element, cf. (3.1).

The Morley finite element space is defined, on the grid $\mathcal{T}_h(\Omega)$, as follows, cf. Figure 3.

$$\tag{3.1} V_h^M = \{\vartheta; \vartheta|_\tau \in P_2(\tau), \forall \tau \in \mathcal{T}_h(\Omega), \vartheta(\text{resp. } \partial_{\boldsymbol{n}}\vartheta) \text{ is continuous at each vertex } p \text{ of } \tau (\text{resp. each edge midpoint } m \text{ of } \tau)\},$$

$$\tag{3.2} V_{h,0}^M = \{\vartheta \in V_h^M(\Omega); \vartheta(p) = \partial_{\boldsymbol{n}}\vartheta(m) = 0, \forall p, m \in \partial\Omega\}.$$

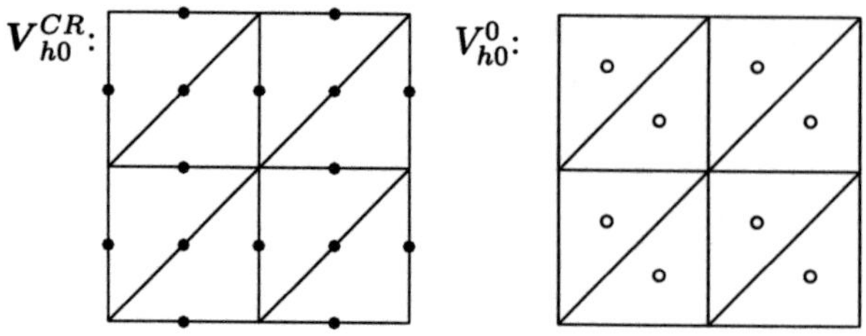

FIGURE 4. The nodes of Crouzeix-Raviart and P_0 elements, cf. (3.3).

Let the Crouzeix-Raviart element (P_1-nonconforming) spaces be defined as follows.

$$V_h^{CR} = \{v \mid v|_\tau \in P_1, \text{continuous at each edge midpoint } m \text{ of } \tau \in \mathcal{T}_h(\Omega)\}, \tag{3.3}$$

$$V_{h,0}^{CR} = \{v \in V_h^{CR}(\Omega); v(m) = 0, \forall m \in \partial\Omega\}. \tag{3.4}$$

For convenience, we let the recovery space for superconvergence be

$$V_h^k = \{v \in L^2(\Omega); v|_\tau \in P_k(\tau), \forall \tau \in \mathcal{T}_h(\Omega)\}, \qquad \text{for any } k \geq 0.$$

And let $V_{h,0}^0 = V_h^0 \cap L_0^2(\Omega)$ be the finite element space for the pressure, i.e., piecewise constant space on the grid $\mathcal{T}_h(\Omega)$. For any function space B, let $\boldsymbol{B} = B \times B$. For example, $\boldsymbol{V}_{h,0}^{CR} = V_{h,0}^{CR} \times V_{h,0}^{CR}$. Then, the mixed finite element space for the Stokes equations (2.2) is $\boldsymbol{V}_{h,0}^{CR} \times V_{h,0}^0$, cf. Figure 4.

With these notations, the Morley element method for problem (2.1) is defined as follows. Find $\varphi_h \in V_{h,0}^M$ such that

$$\int_\Omega \nabla_h^2 \varphi_h : \nabla_h^2 \vartheta \, dx = \int_\Omega \boldsymbol{g} \cdot \mathbf{curl}_h \vartheta \, dx \quad \forall \vartheta \in V_{h,0}^M. \tag{3.5}$$

Here and later, with discrete ∇_h operator, an integral is understood as its piecewise sum, i.e., $\int_\Omega \nabla_h^2 \varphi_h : \nabla_h^2 \vartheta \, dx = \sum_{\tau \in \mathcal{T}_h} \int_\tau \nabla^2 \varphi_h : \nabla^2 \vartheta \, dx$. The discrete variational formulation for (2.2) is to find $\boldsymbol{u}_h \in \boldsymbol{V}_{h,0}^{CR}$, $\lambda_h \in V_{h,0}^0$ such that

$$\begin{cases} \int_\Omega \nabla_h \boldsymbol{u}_h : \nabla_h \boldsymbol{v} \, dx - \int_\Omega \lambda_h \nabla_h \cdot \boldsymbol{v} \, dx = \int_\Omega \boldsymbol{g} \cdot \boldsymbol{v} \, dx & \forall \boldsymbol{v} \in \boldsymbol{V}_{h,0}^{CR}, \\ \int_\Omega \mu \nabla_h \cdot \boldsymbol{u}_h \, dx = 0 & \forall \mu \in V_{h,0}^0. \end{cases} \tag{3.6}$$

Before proving the equivalence between (3.5) and (3.6), we recall an important result, given in **[7]**. Define

$$\boldsymbol{Z}_h = \left\{ \boldsymbol{v}_h \in \boldsymbol{V}_{h,0}^{CR}; \int_\Omega \mu \nabla_h \cdot \boldsymbol{v}_h \, dx = 0, \forall \, \mu \in V_h^0 \right\}.$$

Then we have

$$\boldsymbol{Z}_h = \mathbf{curl}_h V_{h,0}^M. \tag{3.7}$$

THEOREM 3.1. *Let $\varphi_h \in V_{h,0}^M$ be the solution of (3.5) and $(\boldsymbol{u}_h, \lambda_h) \in \boldsymbol{V}_{h,0}^{CR} \times V_h^0$ the solution of (3.6). The discrete variational formulation (3.5) is equivalent to (3.6):*

$$\boldsymbol{u}_h = \mathbf{curl}_h \varphi_h.$$

PROOF. The proof is similar to the one for Theorem 2.1. It is obvious from the second equation of (3.6) that $\boldsymbol{u}_h \in \boldsymbol{Z}_h$. Then by (3.7), there exists $\tilde{u}_h \in V_{h,0}^M$ such that $\boldsymbol{u}_h = \mathbf{curl}_h \tilde{u}_h$. Substituting it into the first equation of (3.6) and taking $\boldsymbol{v} = \mathbf{curl}_h \vartheta$ for any $\vartheta \in V_{h,0}^M$, we have

$$\int_\Omega \nabla_h^2 \tilde{u}_h : \nabla_h^2 \vartheta \, dx = \int_\Omega \boldsymbol{g} \cdot \mathbf{curl}_h \vartheta \, dx \quad \forall \vartheta \in V_{h,0}^M.$$

Using the unique solvability of (3.5), we get $\tilde{\varphi}_h = \varphi_h$. Therefore $\boldsymbol{u}_h = \mathbf{curl}_h \varphi_h$. □

4. A postprocessing algorithm

In what follows, we always assume that $\varphi \in H_0^2(\Omega)$ is the solution of (2.1), and $(\boldsymbol{u}, \lambda) \in (H_0^1(\Omega))^2 \times L_0^2(\Omega)$ is the solution of the Stokes equation (2.2), $\varphi_h \in V_{h,0}^M$ is the solution of (3.5), and $(\boldsymbol{u}_h, \lambda_h) \in \boldsymbol{V}_{h,0}^{CR} \times V_h^0$ is the solution of (3.6). Assume the triangulation $\mathcal{T}_h(\Omega)$ is quasi-uniform in this section. Let $\mathcal{T}_H(\Omega)$ be a coarser quasi-uniform triangulation with mesh size H satisfying $h \ll H$ and

$$H = h^\alpha,$$

where $\alpha \in (0,1)$. Denote $\boldsymbol{Q}_H^k$ to be the L^2-orthogonal projector from $\boldsymbol{L}^2(\Omega)$ onto the space $\boldsymbol{V}_H^k$.

Hereafter we assume $k > 1$ and let $\alpha = \frac{2}{k+1}$. From [**20**], we have the following superconvergence result for problem (3.6).

LEMMA 4.1. *Assume Ω is so regular that for any $\boldsymbol{g} \in \boldsymbol{H}^{s-2}(\Omega)$, the problem (2.2) has a unique solution $\boldsymbol{u} \in \boldsymbol{H}_0^1(\Omega) \cap \boldsymbol{H}^s(\Omega)$ and $\lambda \in H^{s-1}(\Omega) \cap L_0^2(\Omega)$ satisfying a priori estimate*

$$\|\boldsymbol{u}\|_s + \|\lambda\|_{s-1} \leq C\|\boldsymbol{g}\|_{s-2},$$

where $1 \leq s \leq 2$ and C is a constant independent of $\boldsymbol{g}$. If $(\boldsymbol{u}, \lambda) \in \boldsymbol{H}^2(\Omega) \cap \boldsymbol{H}^{k+1}(\Omega) \cap \boldsymbol{H}_0^1(\Omega) \times H^1(\Omega) \cap L_0^2(\Omega)$, then

$$|\boldsymbol{u} - \boldsymbol{Q}_H^k \boldsymbol{u}_h|_{1,H} \leq Ch^{2k/(k+1)}(\|\boldsymbol{u}\|_2 + \|\boldsymbol{u}\|_{k+1} + \|\lambda\|_1).$$

Moreover, if Ω is convex, then

$$|\boldsymbol{u} - \boldsymbol{Q}_H^k \boldsymbol{u}_h|_{1,H} \leq Ch^{2k/(k+1)}(\|\boldsymbol{g}\|_0 + \|\boldsymbol{u}\|_{k+1}).$$

THEOREM 4.2. *Assume Ω is convex. If $\varphi \in H_0^2(\Omega) \cap H^{k+2}(\Omega)$, then*

$$|\mathbf{curl}\varphi - \boldsymbol{Q}_H^k \mathbf{curl}_h \varphi_h|_{1,H} \leq Ch^{2k/(k+1)}(\|\boldsymbol{g}\|_0 + \|\varphi\|_{k+2}). \tag{4.1}$$

PROOF. It follows from Theorems 2.1-3.1 and Lemma 4.1 that

$$|\,\mathbf{curl}\,\varphi - \boldsymbol{Q}_H^k \mathbf{curl}_h \varphi_h|_{1,H} = |\boldsymbol{u} - \boldsymbol{Q}_H^k \boldsymbol{u}_h|_{1,H} \leq Ch^{2k/(k+1)}(\|\boldsymbol{g}\|_0 + \|\varphi\|_{k+2}).$$

□

5. Numerical tests

We solve the biharmonic equation (1.1) on a unit square, $\Omega = [0,1]^2$, by the Morley element. The exact solutions of the biharmonic equation is

$$\varphi = 2^8(x - x^2)^2(y - y^2)^2. \tag{5.1}$$

We also solve the Stokes equations (1.2) on the unit square by the Crouzeix-Raviart mixed element, where the exact solution is, for φ being the solution above,

$$\boldsymbol{u} = \mathbf{curl}\,\varphi, \quad \lambda = 0. \tag{5.2}$$

The domain is subdivided by a multigrid refinement that the level 3 grid and level 1 grid are plotted in Figure 2. First, we compute the Crouzeix-Raviart solution for (5.2). As the mixed-element is locally divergence-free, cf. [**14, 21, 22**], the resulting linear systems of equations can be solved efficiently by the iterated penalty method (IPM). In Table 1, we list the number of iterated penalty iterations and the number of conjugate gradient iteration within each IPM iteration. By the second column, we can see there is a natural superconvergence for the CR element in H^1-norm for solving the Stokes equations. Here, the optimal order is supposed to be 1 for the H^1-norm convergence. But we get 2. In fact, the discrete H^1-convergence is even better than that of L^2.

TABLE 1. The error ($\boldsymbol{e_u} = \boldsymbol{I}_h\boldsymbol{u} - \boldsymbol{u}_h$) of CR element for (5.2).

grid	$\|\boldsymbol{e_u}\|_{L^2}$	h^k	$\|\boldsymbol{e_u}\|_{H^1}$	h^k	$\|e_\lambda\|_{L^2}$	h^k	#cg	#ipm
3	0.5374	1.6	0.3199	1.7	3.72	0.7	41	4
4	0.1921	1.5	0.0945	1.8	1.94	0.9	187	4
5	0.0596	1.7	0.0255	1.9	0.98	1.0	1048	3
6	0.0167	1.8	0.0066	2.0	0.49	1.0	3128	3
7	0.0044	1.9	0.0017	2.0	0.25	1.0	8266	3
8	0.0011	2.0	0.0004	2.0	0.12	1.0	18626	3

Next, we compute the Morley element solution and its equivalence to the Crouzeix-Raviart solution. When solving the biharmonic equation, there are two ways to compute the right hand side vector:

$$\int_\Omega \nabla_h^2 \varphi_h : \nabla_h^2 \phi_i \, dx = \int_\Omega f \phi_i \, dx, \tag{5.3}$$

$$\int_\Omega \nabla_h^2 \varphi_h : \nabla_h^2 \phi_i \, dx = \int_\Omega \boldsymbol{g} \cdot \mathbf{curl}\, \phi_i \, dx, \tag{5.4}$$

where $f = \mathbf{curl}\, \boldsymbol{g}$, cf. (2.1), and ϕ_i denotes a shape basis function of the Morley element space. If we use the first method (5.3), then the curl of Morley solution is not the same as the Crouzeix-Raviart solution. This is shown in the data on the left of Table 2. That is, the quadrature formula for (5.3) and for the Stokes equations (3.6) cannot produce equivalent integral values. But if we use method (5.4), the difference between the two solutions is 10^{-7} times that by (5.3). This can be seen from the last two columns in Table 2. The difference is caused by the iterative error (used the Iterated Penalty Method) in the Stokes equations, not by the two finite elements. Table 2 verifies Theorem 3.1.

TABLE 2. The difference between two solutions, $\boldsymbol{e}_h = \mathbf{curl}_h\, \varphi_h - \boldsymbol{u}_h$.

grid	$\|\boldsymbol{e}_h\|_{l^\infty}$	h^k	$\|\boldsymbol{e}_h\|_{L^2}$	h^k	$\|\boldsymbol{e}_h\|_{l^\infty}$	$\|\boldsymbol{e}_h\|_{L^2}$
	By (5.3)				By (5.4)	
3	1.25780		0.7225		0.1161E-12	0.1887E-13
4	0.41507	1.6	0.2276	1.7	0.1821E-13	0.1069E-13
5	0.11893	1.8	0.0611	1.9	0.8661E-11	0.5307E-11
6	0.03142	1.9	0.0156	2.0	0.2633E-11	0.1589E-11
7	0.00806	2.0	0.0039	2.0	0.1886E-11	0.1029E-11

Next, we compute the Morley element solution to the biharmonic equation. In Table 3, we list the error and the order of convergence for the Morley element. It converges at a linear order in H^2-seminorm. This is the optimal order. Finally, we locally L^2-project $\mathbf{curl}\,\varphi_h$ to P_2 polynomials (on the grids two levels lower.) Then we obtain one order higher convergence than the optimal order, both in H^1 and in H^2 semi-norms. The error and the order of superconvergence are listed in Table 4. This partially confirms our main theorem, Theorem 4.2, where the order can be only 1/2 higher. We note that people usually express superconvergence as $|I_h\varphi-\varphi_h|_{H^2}$ before post-processing. But we have post-processed φ_h in our method. Thus, from Table 4, $|I_h\varphi - Q_H^2\varphi_h|_{H^2} \le |I_h\varphi - \varphi|_{H^2} + |\varphi - Q_H^2\varphi_h|_{H^2} = O(h^2)$.

TABLE 3. The error ($e_h = \varphi - \varphi_h$) of the Morley element for (5.1).

grid	$\|e_h\|_{L^2}$	h^k	$\|e_h\|_{H^1}$	h^k	$\|e_h\|_{H^2}$	h^k	#cg	#dof
4	0.04158155	1.8	0.140097	1.8	2.68762	0.9	61	225
5	0.01083490	1.9	0.037031	1.9	1.36707	1.0	140	961
6	0.00274282	2.0	0.009430	2.0	0.68720	1.0	407	3969
7	0.00068801	2.0	0.002369	2.0	0.34409	1.0	1402	16129
8	0.00017215	2.0	0.000593	2.0	0.17211	1.0	5773	65025

TABLE 4. The superconvergence ($\tilde{e}_h = \varphi - Q_H^2\varphi_h$) of Morley element.

grid	$\|\tilde{e}_h\|_{H^1}$	h^k	$\|\tilde{e}_h\|_{H^2}$	h^k
4	0.39262	2.0	3.17743	1.2
5	0.06400	2.6	1.13909	1.5
6	0.00899	2.8	0.30133	1.9
7	0.00131	2.8	0.07651	2.0
8	0.00023	2.5	0.01945	2.0

References

[1] I. Babuska, T. Strouboulis, C.S. Upadhyay and S.K. Gangaraj, *Computer-based proof of the existence of superconvergence points in the finite element method; superconvergence of the derivatives in finite element solutions of Laplaces, Poissons, and the elasticity equations,* Numer. Meth. PDEs. **12** (1996), 347–392. MR1388445 (97c:65160)

[2] G.P. Bazeley, Y.K. Cheung, B.M., Irons and O.C. Zienkiewicz, *Triangular elements in plate bending — conforming and nonconforming solutions,* in Proceedings of the Conference on Matrix Methods in Structural Mechanics, Wright Patterson A. F. Base, Ohio, 1965, 547–576.

[3] S. C. Brenner and L. R. Scott, The Mathematical Theory of Finite Element Methods (3rd Edition). Springer, New York, 2008. MR2373954 (2008m:65001)

[4] P. G. Ciarlet, The Finite Element Method for Elliptic Problems. North-Holland, Amsterdam, 1978. MR0520174 (58:25001)

[5] J. Douglas, Jr. and T. Dupont, *Superconvergence for Galerkin methods for the two point boundary problem via local projections,* Numer. Math. **21** (1973), 270–278. MR0331798 (48:10130)

[6] R.E. Ewing, R.D. Lazarov and J. Wang, *Superconvergence of the velocity along the Gauss lines in mixed finite element methods,* SIAM J. Numer. Anal. **28** (1991), 1015–1029. MR1111451 (92e:65149)

[7] R. S. Falk and M. E. Morley, *Equivalence of finite element methods for problems in elasticity,* SIAM J. Numer. Anal. **27** (1990) No. 6, 1486–1505. MR1080333 (91i:65177)

[8] M. Křížek and P. Neittaanmäki, *On superconvergence techniques,* Acta Appl. Math. **9** (1987), 175–198. MR900263 (88h:65208)

[9] S. Mao and Z.-C. Shi, *High accuracy analysis of two nonconforming plate elements,* Numer. Math. **111** (2009), 407–443. MR2470146 (2009k:65254)

[10] L.S.D. Morley, *The triangular equilibrium element in the solution of plate bending problems,* Aero. Quart. **19** (1968), 149–169.

[11] V. Ruas, *A quadratic finite element method for solving biharmonic problems in R^n*, Numer. Math. **52** (1988), 33–43. MR918315 (89a:65173)

[12] V. Ruas, F. Ghadi and M. Wakrim, *External approximation of $H^3(\Omega)$ in a bounded domain of R^3 with piecewise cubics of weak C^2-class,* C. R. Acad. Sci. Paris, Ser. I **338** (2004), 969–974. MR2066360

[13] A.H. Schatz, I.H. Sloan and L.B. Wahlbin, *Superconvergence in finite element methods and meshes that are symmetric with respect to a point,* SIAM. J. Numer. Anal. **33** (1996), 505–521. MR1388486 (98f:65112)

[14] R. Scott and S. Zhang, *Multilevel ierated penalty method for mixed elements,* In: Domain Decomposition Methods in Sciences and Engineering, P.E. Bjorstad, M.S. Espedal and D.E. Keyes eds., Domain Decomposition Press, Bergen 1998, 133-139.

[15] Z.-C. Shi, *On the error estimates of Morley element,* Numerica Mathematica Sinica **12** (1990), No. 2, 113–118. MR1070298 (91i:65182)

[16] L.B. Wahlbin, Superconvergence in Galerkin Finite Element Methods. Lecture Notes in Mathematics, vol. 1605. Springer, Berlin, 1995. MR1439050 (98j:65083)

[17] M. Wang and J. Xu, *The Morley element for fourth order elliptic equations in any dimensions,* Numer. Math. **103** (2006) no. 1, 155–169. MR2207619 (2006i:65205)

[18] M. Wang and J. Xu, *Minimal finite element spaces for $2m$-th order partial differential equations in R^n,* preprint.

[19] R. K. Mohanty, *Single-cell compact finite-difference discretization of order two and four for multidimensional triharmonic problems,* Numer. Meth. PDE., **26** (2010) No. 6, 1420–1426. MR2732386 (2011i:65184)

[20] X. Ye, *Superconvergence of nonconforming finite element method for the Stokes equations,* Numer. Methods Partial Differential Equations, **18** (2002) No.2, 143-154. MR1902289 (2003c:65109)

[21] S. Zhang, *A family of $Q_{k+1,k} \times Q_{k,k+1}$ divergence-free finite elements on rectangular grids,* SIAM J. Num. Anal., **47** (2009), 2090–2107. MR2519595 (2010d:65338)

[22] S. Zhang, *Divergence-free finite elements on tetrahedral grids for $k \geq 6$,* Math. Comp. **80** (2011), 669-695. MR2772092 (2012b:65191)

[23] Z. Zhang, *Derivative superconvergence points in finite element solutions of Poissons equation for the serendipity and intermediate families: A theoretical justification,* Math. Comp. **67** (1998), 541–552. MR1459393 (98i:65104)

[24] O.C. Zienkiewicz and J.Z. Zhu, *The superconvergence patch recovery and a posteriori error estimates Part 1: The recovery technique,* Int. J. Numer. Methods Eng. **33** (1992), 1331–1364. MR1161557 (93c:73098)

[25] O.C. Zienkiewicz and J.Z. Zhu, *The superconvergence patch recovery and a posteriori error estimates Part 2: Error estimates and adaptivity,* Int. J. Numer. Methods Eng. **33** (1992), 1364–1382.

[26] M. Zlamal, *Superconvergence and reduced integration in the finite element method,* Math Comp. **32** (1978), 663–685. MR0495027 (58:13794)

Department of Mathematics, and MOE-LSC, Shanghai Jiao Tong University, Shanghai 200240, China; Division of Computational Science, E-Institute of Shanghai Universities, Shanghai Normal University, China

E-mail address: jghuang@sjtu.edu.cn

College of Mathematics and Information Science, Wenzhou University, Wenzhou 325035, China

E-mail address: xuehaihuang@wzu.edu.cn

Department of Mathematical Sciences, University of Delaware, Newark, Delaware 19716

E-mail address: szhang@udel.edu

A nonsmooth Newton multigrid method for a hybrid, shallow model of marine ice sheets

Guillaume Jouvet, Ed Bueler, Carsten Gräser, and Ralf Kornhuber

ABSTRACT. The time evolution of ice sheets and ice shelves is modelled by combining a shallow lubrication approximation for shear deformation with the shallow shelf approximation for basal sliding, along with the mass conservation principle. At each time step two p-Laplace problems and one transport problem are solved. Both p-Laplace problems are formulated as minimisation problems. They are approximated by a finite element truncated nonsmooth Newton multigrid method. As an illustration, we compute the steady state shape of an idealized ice sheet/shelf system.

1. Introduction

Because of questions related to sea level rise [**VA07**], the literature on modeling of ice sheets and ice shelves has expanded in recent years [**DPF11, DGDF+09, GHS09, PHDBDS06, VP05, WMH+11**]. Special attention has been given to the grounding lines between ice sheets and ice shelves because they mostly control the stability of such marine ice sheets [**Sch07**]. Indeed, small perturbations of thermal state or climatic mass balance might lead the grounding line to move substantially, causing considerable changes to the geometry of the entire ice mass. The design of reliable models and fast numerical schemes is, therefore, important for understanding the evolution of marine ice sheets.

Ice is usually modelled as an incompressible non-Newtonian fluid, in a nonlinear Stokes problem [**GB09**], but the small aspect ratio of ice sheets can yield substantial simplifications of the stress balance. The vertical shear component is dominant in the majority of shallow ice sheets, giving a lubrication-type flow, while longitudinal components dominate in shallow ice shelves yielding a "plug" or membrane-type flow [**WMH+11**]. However, at the grounding line both regimes are significant. The change of regime generally occurs within a few-kilometer zone of nested boundary layers [**Sch11**].

One can distinguish two kinds of marine ice sheet models. The first uses separate equations for the ice sheet and ice shelf, with an explicit evolution of the grounding line [**DPF11, Sch07**]. In two dimensions, such models are usually implemented with a moving grid such that the grounding line is one point of the mesh.

The first author was supported by the Swiss National Science Foundation (project 133349) and the Deutsche Forschungsgemeinschaft (project KL 1806 5-1).

The second author is supported by the NASA (grant NNX09AJ38G).

Unfortunately, in three dimensions, moving the mesh to follow the grounding line, a curve along the bedrock, is a harder task. Such two-mode models need to be connected by a continuity boundary condition at the grounding line.

The current paper adopts a second strategy. We use a unified model for both the ice sheet and shelf. The grounding line has an implicit description, namely the flotation criterion. We use an adaptive grid to deal with the sharp changes in flow regime close to the grounding line. Our model also derives the shape and evolution of grounded ice sheet margins from the same combination of equations, but in a parameter range where basal resistance dominates. Our results are for two-dimensional (plane) flows only; the three-dimensional case is in preparation.

2. Model

Let $\Omega \subset \mathbb{R}$ be the maximum horizontal extent of a two-dimensional ice sheet and $[0, T]$ be a time interval, with $T > 0$. We assume a fixed bedrock elevation function $z = b(x)$ in Ω. We denote by $l(x,t)$ and $s(x,t)$ the elevation of the lower and upper ice surfaces, respectively; $l = b$ where ice is grounded and $l > b$ where ice is floating (Fig. 1). The ice thickness is $h = s - l$, and $h \geq 0$ on Ω. Where the bedrock elevation b is above sea level $z = 0$, $h = 0$ is allowed. The functions s, l and h are continuous on Ω, and cliffs are allowed only at the boundary of Ω.

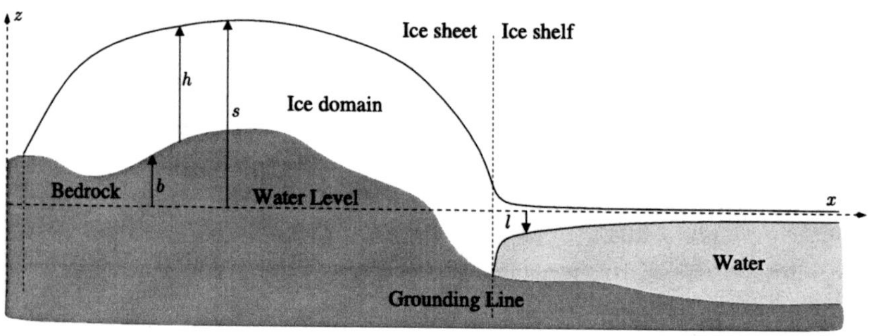

FIGURE 1. Notation for a two-dimensional ice sheet/shelf system.

Let ρ_i and ρ_w denote the densities of ice and water, respectively. Archimedes's principle for flotation implicitly determines the grounding line, so that the lower surface elevation is determined from other fields [**PSP**$^+$**12, WMH**$^+$**11**]:

$$l = \max\left\{b, -\frac{\rho_i}{\rho_w}h\right\}. \tag{2.1}$$

The grounding line is the abscissa where the maximum in (2.1) switches.

Following [**BB09, WMH**$^+$**11**], the ice flow is described by superposing velocity from the isothermal shallow ice approximation (SIA) model [**GB09**] that accounts for the vertical shear, and from the shallow shelf approximation (SSA) model [**Sch06a**] that accounts for the longitudinal stresses and basal friction (where grounded).

2.1. The shallow ice approximation. The isothermal SIA stress balance combines with mass conservation to give a single partial differential equation (PDE) for the ice thickness h [**GB09**]. Let $\Gamma = 2A(\rho_i g)^{p-1}/(p+1) > 0$, where $A > 0$

is the constant ice softness, $p = n_{\text{Glen}} + 1$ where the Glen flow law exponent has standard value $n_{\text{Glen}} = 3$ [**GB09**], and g the acceleration of gravity. The PDE is

$$\frac{\partial h}{\partial t} - \Gamma \frac{\partial}{\partial x}\left(h^{p+1}\left|\frac{\partial h}{\partial x} + \frac{\partial l}{\partial x}\right|^{p-2}\left(\frac{\partial h}{\partial x} + \frac{\partial l}{\partial x}\right)\right) + \frac{\partial}{\partial x}(hu) = a, \qquad \text{if } h > 0, \tag{2.2}$$

$$h = 0, \quad \text{otherwise.} \tag{2.3}$$

Here u is the basal sliding (defined later by the SSA model) and $a(x)$ is the climatic mass balance (accumulation and ablation). Equations (2.2), (2.3) should be interpreted as an obstacle problem that incorporates the free-boundary constraint $h \geq 0$ [**CDD$^+$02, JB**], however. The variational inequality form is

$$\int_\Omega \frac{\partial h}{\partial t}(v-h) + \Gamma \int_\Omega h^{p+1}\left|\frac{\partial h}{\partial x} + \frac{\partial l}{\partial x}\right|^{p-2}\left(\frac{\partial h}{\partial x} + \frac{\partial l}{\partial x}\right)\frac{\partial}{\partial x}(v-h) + \int_\Omega \frac{\partial}{\partial x}(h\,u)\,(v-h) \geq \int_\Omega a(v-h), \tag{2.4}$$

where $v \geq 0$ are test functions. The grounded ice sheet margin, the time-dependent free boundary of the ice domain $\{h > 0\}$, comes from solving (2.4).

2.2. The shallow shelf approximation. In this section, we describe the model which determines u in (2.4). The velocity u solves on the restricted ice domain $\{x \in \Omega,\ h(x) > 0\}$ the SSA equation [**Sch06a**]:

$$-2A^{1-q}\frac{\partial}{\partial x}\left(h\left|\frac{\partial u}{\partial x}\right|^{q-2}\frac{\partial u}{\partial x}\right) + C|u|^{m-1}u \times \mathbf{1}_{G(h)} = -\rho_i g h \frac{\partial s}{\partial x}, \quad \text{if } |u| > 0, \tag{2.5}$$

$$u = 0, \qquad \text{else,} \tag{2.6}$$

where $q = p/(p-1) \in (1,2)$ is the conjugate exponent to p and $m \geq 0$, $C \geq 0$ are given parameters. We denote the grounded set

$$G(h) := \{x \in \Omega,\ b(x) + (\rho_i/\rho_w)h(x) > 0\} \tag{2.7}$$

so that $\mathbf{1}_{G(h)}$ in (2.5) is equal to one in the grounded part and zero otherwise. Physically, the first term in (2.5) represents longitudinal stress gradients while the second term represents basal friction. The right-hand-side represents the gravitational forces in the form called the "driving stress" [**GB09**]. The $m = 0$ case is a Coulomb-type or "plastic till" friction law [**Sch06b, Sch06a, Sch09**]. On the boundary of $\{h > 0\}$, we have either a stress-free condition if the margin is grounded or a water-ice balance stress condition if the margin is floating, i.e. at the calving front. In both cases, the condition is:

$$2A^{1-q}h\left|\frac{\partial u}{\partial x}\right|^{q-2}\frac{\partial u}{\partial x} = \frac{1}{2}\left(1 - \frac{\rho_i}{\rho_w}\right)\rho_i g h^2 := F. \tag{2.8}$$

By extending u by zero where $h = 0$, equation (2.5) with boundary condition (2.8) is reformulated as the variational inequality [**Sch06a**]:

$$2A^{1-q}\int_\Omega h\left|\frac{\partial u}{\partial x}\right|^{q-2}\frac{\partial u}{\partial x}\frac{\partial}{\partial x}(v-u) + \frac{C}{m+1}\int_{G(h)}\left(|v|^{m+1} - |u|^{m+1}\right) + \rho_i g \int_\Omega h\frac{\partial s}{\partial x}(v-u) - \int_{\partial\Omega} F(v-u) \geq 0, \tag{2.9}$$

where v is a test function. When $m > 0$, inequality (2.9) can be actually rewritten as a variational equality. Like (2.4), variational inequality (2.9) is attractive since it does not involve the boundary of the ice domain $\{h > 0\}$.

3. Approximations

3.1. Discretization in time. Variational inequality (2.4) derives from a non-linear diffusion-advection equation (2.2). It is expected to be advection-dominated where ice is floating and diffusion-dominated where ice is grounded. Operator splitting techniques [**Glo03**, chapter 2] for solving (2.4) are used to decouple the advection and diffusion operators. This first-order splitting corresponds to first solving without the diffusion and source terms, and then solving without advection.

Let $N > 0$ and suppose $[t_0, ..., t_{N+1}]$ is a time discretization of $[0, T]$ with time steps $\tau_n = t_{n+1} - t_n, n = 0, 1, ..., N$. Denote by h_n an approximation of $h(t_n)$. Assuming h_n is known for some n, the following three-step scheme describes how to compute h_{n+1}.

I. Find u_n that minimizes:

$$\mathcal{J}_{SSA}(v) := \frac{2A^{1-q}}{q} \int_\Omega h_n \left| \frac{\partial v}{\partial x} \right|^q + \frac{C}{m+1} \int_{G(h_n)} |v|^{m+1} + \rho_i g \int_\Omega h_n \frac{\partial s_n}{\partial x} v - \int_{\partial\Omega} F v,$$

where $s_n = l_n + h_n$ and l_n is computed from (2.1) using h_n.

II. Find $h_{n+\frac{1}{2}}$, the solution at time t_{n+1} of the advection problem:

$$\begin{cases} \frac{\partial h}{\partial t} + \frac{\partial}{\partial x}(h u_n) & = 0, \qquad \text{on} \quad (t_n, t_{n+1}). \\ h(t_n) & = h_n, \end{cases}$$

III. Find $h_{n+1} \geq 0$ that minimizes over all $v \geq 0$:

$$\mathcal{J}_{SIA}(v) := \frac{1}{2\tau_n} \int_\Omega v^2 + \frac{\Gamma}{p} \int_\Omega (h_{n+\frac{1}{2}})^{p+2} \left| \frac{\partial v}{\partial x} + \frac{\partial l_{n+\frac{1}{2}}}{\partial x} \right|^p - \int_\Omega \left(\frac{h_{n+\frac{1}{2}}}{\tau_n} + a \right) v,$$

where $l_{n+\frac{1}{2}}$ is computed from (2.1) using $h_{n+\frac{1}{2}}$.

If $G(h_n)$ has a positive measure then one can show that the functional $\mathcal{J}_{SSA}$ is strictly convex and strongly-continuous in $W^{1,q}(\Omega)$ and therefore lower-semi-continuous [**Sch06b, Sch09**]. However, coerciveness in the $m > 0$ case would require h_n to be uniformly lower-bounded by a positive constant. The case $m = 0$, namely Coulomb friction, requires more hypotheses [**Sch09**]. As a consequence the well-posedness of the minimisation problem related to $\mathcal{J}_{SSA}$ is not guaranteed since h_n might tend to zero. One can show that $\mathcal{J}_{SIA}$ is strictly convex, strongly continuous in $\{v \in W^{1,p}(\Omega), v \geq 0\}$ and then lower-semi-continuous [**JB**]. However, since $h_{n+\frac{1}{2}}$ is not uniformly lower-bounded, coerciveness and thus well-posedness of the minimisation problem are not guaranteed.

3.2. Discretization in space. Let $M > 0$ be given, and suppose $\Omega = [x_0, ..., x_{M+1}]$ is a multilevel discretization that results from several successive local or global refinements applied to the initial interval $[x_0, x_{M+1}]$. Such a hierarchy is necessary to apply the Newton multigrid method which is used to solve the two minimisation problems (Steps I and III above). The transport problem (Step II) is solved using an upwind finite difference scheme.

The Ritz-Galerkin approximation of both of the minimization problems, in the standard continuous piecewise-linear finite element space, can be written

$$\text{find} \quad \mathbf{u} \in \mathbb{R}^I \quad \text{s.t.} \quad \mathcal{J}(\mathbf{u}) \leq \mathcal{J}(\mathbf{v}), \qquad \forall \mathbf{v} \in \mathbb{R}^I, \tag{3.1}$$

where $I \in \mathbb{N}$. Here $\mathcal{J} : \mathbb{R}^I \longrightarrow \mathbb{R} \cup \{+\infty\}$ is a strictly convex, coercive, lower semi-continuous, but not necessarily smooth, nonlinear functional. Indeed, the obstacle in (2.4) and the case $m = 0$ in (2.9) lead to different nonsmooth terms in $\mathcal{J}_{SIA}$ and $\mathcal{J}_{SSA}$, respectively. Also, $\mathcal{J}_{SIA}$ and $\mathcal{J}_{SSA}$ are coercive in the finite dimensional spaces since all norms are equivalent.

Since Newton-type methods require smoothness, we use a truncated method, closely-related to the primal dual active set approach [**Kor94**, **Kor96**, **GK09**], to solve problems (3.1). Following the techniques that have been developed in [**GK09**, **GSS09**, **Grä11**] for linear and nonlinear obstacle problems, we use the Truncated Nonsmooth Newton MultiGrid (TNMMG) method, described next. By contrast, in [**Sch06a**] a regularisation parameter was introduced to deal with nonsmoothness for the approximation of the SSA when $m = 0$. Unfortunately, this approach might slow down the algebraic solver if the regularisation parameter is too small, or it might lead to a wrong solution if the parameter is too large.

Let $\mathcal{F} : \mathbb{R}^I \to \mathbb{R}^I$ be a nonlinear Gauß-Seidel smoother. More precisely, for a given iterate $\mathbf{u} \in \mathbb{R}^I$, $\mathcal{F}(\mathbf{u})$ provides the correction such the new iterate $\mathbf{u} + \mathcal{F}(\mathbf{u})$ minimizes successively $\mathcal{J}$ in each coordinate direction. Each scalar minimisation can be achieved by an inexact method. In practise, we use a bisection method since it does not require any smoothness. The TNNMG method [**Grä11**] then defines a sequence $\mathbf{u}^\nu$ by the following three steps per iteration:

$$\mathbf{u}^{\nu+\frac{1}{3}} = \mathbf{u}^\nu + \mathcal{F}(\mathbf{u}^\nu), \tag{3.2}$$

$$\mathbf{u}^{\nu+\frac{2}{3}} = \mathbf{u}^{\nu+\frac{1}{3}} - (\mathcal{J}''(\mathbf{u}^{\nu+\frac{1}{3}})_{\mathcal{I},\mathcal{I}})^{-1}\mathcal{J}'(\mathbf{u}^{\nu+\frac{1}{3}})_{\mathcal{I}}, \tag{3.3}$$

$$\mathbf{u}^{\nu+1} = \operatorname{argmin}_{\mathbf{w},\, \rho\in[0,1]}\{\mathcal{J}(\mathbf{w}); \quad \mathbf{w} = \rho\mathbf{u}^{\nu+\frac{1}{3}} + (1-\rho)\mathcal{P}_{Dom(\mathcal{J})}(\mathbf{u}^{\nu+\frac{2}{3}})\} \tag{3.4}$$

where the active index set is denoted

$$\mathcal{I} = \mathcal{I}(\mathbf{v}) = \{i, \text{ the subdifferential } \ \partial\mathcal{J}(v_i) \ \text{ is single-valued}\}. \tag{3.5}$$

Here $(\cdot)_{\mathcal{I}}$ and $(\cdot)_{\mathcal{I},\mathcal{I}}$ denote the truncation of vectors and matrices, respectively, to the index set $\mathcal{I}$; i.e. the i-th entry (and also the i-th column in the matrix case) is set to zero if i is not in $\mathcal{I}$. Also, $\mathcal{P}_{Dom(\mathcal{J})}$ denotes the projection onto the convex set $Dom(\mathcal{J}) = \{\mathbf{u}, \mathcal{J}(\mathbf{u}) < +\infty\}$.

Step (3.2) acts as a smoother of all nonlinearities. In particular, this step allows the set of active nodes $\mathcal{I}(\mathbf{v})$ to be modified. Step (3.3) consists of a Newton correction that applies in the smooth coordinate directions only. The set of active nodes $\mathcal{I}(\mathbf{v})$ is fixed during this step. Since the matrix $\mathcal{J}''(\mathbf{u}^{\nu+\frac{1}{2}})_{\mathcal{I},\mathcal{I}}$ is symmetric and positive definite on the subspace $\{\mathbf{u} \in \mathbb{R}^I, u_i = 0 \text{ if } i \notin \mathcal{I}(v)\}$ we apply a linear multigrid method for the inversion of $\mathcal{J}''(\mathbf{u}^{\nu+\frac{1}{2}})_{\mathcal{I},\mathcal{I}}$. More precisely, we implement a V-cycle type multigrid solver combined with a linear Gauß-Seidel smoother with 3 pre- and post-smoothing steps [**Hac85**]. Since the Newton correction (3.3) might act beyond the convex set $Dom(\mathcal{J})$, the third projection step (3.4) is constrains the Newton correction to remain in $Dom(\mathcal{J})$. Since the projected correction is not assured to have a lower energy than $\mathbf{u}^{\nu+\frac{1}{3}}$, a damping parameter ρ guarantees the monotonicity and the global convergence of the method [**Grä11**].

The implementation is based on the code *DUNE* (http://www.dune-project.org/) and its module *Dune-Tnnmg* [**GSS09**].

4. Numerical results

We consider the polynomial bedrock over the domain $\Omega = [-1000, 1000]$ km shown in Fig. 2. We initialize the ice geometry by a ten meter thick layer of ice (grounded and floating) on Ω and we apply a constant-in-time mass balance defined $a(x) = -1$ m y^{-1} if $x < -500$ km and $a = 0.3$ m y^{-1} elsewhere. We run our model until reaching a steady state shape. On the left-hand-side of the domain Ω, the chosen mass balance with an ablation area allows a free grounded margin to form, while there is an ice-water cliff on the right boundary of Ω. We use of the following parameters: $\rho_i = 900$ kg m^{-3}, $\rho_w = 1000$ kg m^{-3}, $g = 9.81$ m s^{-2} and $A = 4.6416 \times 10^{-24}$ Pa^{-3} s^{-1}. Two experiments are performed using different sliding parameterizations: (**a**) $m = 1/3$ and $C = 7.624 \times 10^6$ Pa m$^{-1/3}$ s$^{1/3}$ and (**b**) $m = 0$ and $C = 10^5$ Pa. Experiment (**a**) involves a power-type sliding law like in [**Sch07, PSP$^+$12**] while experiment (**b**) involves a Coulomb-type law [**Sch06b, Sch09**].

The domain Ω is uniformly meshed at a resolution of ~ 15 km and adaptively refined 5 km around the grounding line to a resolution of ~ 0.1 km in order to capture the high gradients of the solution expected in this area [**PSP$^+$12**]. Since we use an upwind finite difference scheme to solve the mass conservation equation, each time step is updated such that the CFL number never exceeds one. In practise, we start with a time step of 5 years, however, this time step decreases as low as ~ 0.1 year because of this stability criterion.

Fig. 2 displays the steady state shapes of our model with the corresponding SIA and SSA velocities for both sliding parameterizations. Both shapes were found after $\sim 10\,000$ years. The results lead to the following physical observations. First, as expected the SIA velocities are dominant in the ice sheet part while the SSA velocities are much larger in the ice shelf area for both sliding parameterizations. Second, power-law-type sliding allows SSA velocities to be nonzero everywhere, while Coulomb-type sliding allows nonzero SSA velocities only in the transition and floating areas. Third, the power-type sliding law induces a steeper gradient in surface elevation at the grounding line position compared to the Coulomb-type law. This geometric difference arises because, in the power law case, the basal shear stress can rise with increasing sliding velocity as the grounding line is approached from the grounded side, so as to balance higher driving stress from a steeper surface gradient. In the Coulomb case, by contrast, the basal stress is limited to the prescribed yield stress (i.e. $C = 10^5$ Pa in experiment (**b**)).

From a numerical point of view, the TNNMG method turns out to be a valuable tool to solve the two minimisation problems related to the SIA and the SSA. Indeed, Newton-type methods do not directly apply since $\mathcal{J}_{SIA}$ is nonsmooth because of the obstacle while $\mathcal{J}_{SSA}$ is nonsmooth in the $m = 0$ (Coulomb sliding) case. The TNNMG method based on truncation is expected to be faster than a classical regularized solver [**Sch06b**] since the truncated system in (3.3) is smaller, and an experimental comparison in the $m = 0$ case shows that TNMMG is $\sim 12\%$ faster compared to a Newton multigrid method with regularized nonsmooth parts. Of course, TNNMG does not depend on arbitrary regularisation parameters. Note that the TNNMG method converges quadratically in the iterations (3.2)-(3.4) and linearly when skipping the Newton correction (3.3), i.e. when applying a single Gauß-Seidel method. For instance, in our example the TNNMG method used only 13 iterations were needed against more than 500 for a single Gauß-Seidel method to compute one SSA velocity field with the same accuracy. It took around 10 minutes (CPU time) on a single 3.30GHz processor to compute the 15 000 time steps necessary to reach the steady state shape of the experiment (**b**). The CPU times for minimizing $\mathcal{J}_{SIA}$ and $\mathcal{J}_{SSA}$ were comparable while the time to compute the advection problem (Step II) was negligible.

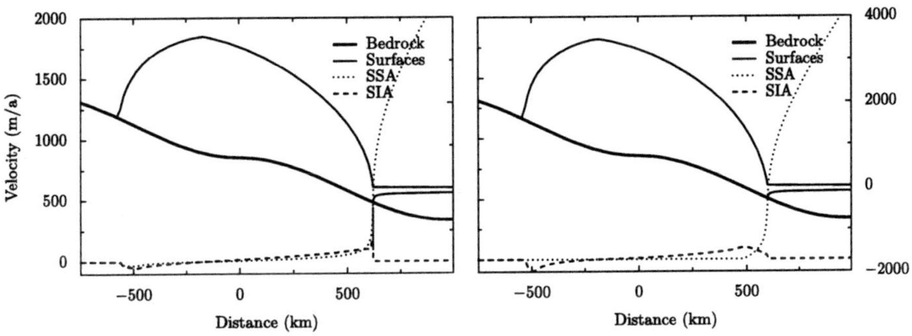

FIGURE 2. Steady state shapes of a two-dimensional idealized system ice sheet/shelf for (**a**; left) a power-type sliding law (**b**; right) a Coulomb-type sliding law. SIA and SSA velocities are displayed with dashed and dotted lines, respectively.

Combining a Newton correction, a truncation procedure to treat nonsmoothness, and a linear multigrid solver provides a fast and robust solver for variational inequality problems (2.4) and (2.9). The TNNMG method can be easily extended to three-dimensional ice sheet and ice shelf models. However, grid refinements around the grounding line, and numerical techniques to solve the mass conservation equation, require more attention. These issues will be addressed in future work.

References

[BB09] E. Bueler and J. Brown, *Shallow shelf approximation as a "sliding law" in a thermomechanically coupled ice sheet model*, Journal of Geophysical Research - Earth Surface **114** (2009), no. F3, F03008+.

[CDD+02] N. Calvo, J. I. Díaz, J. Durany, E. Schiavi, and C. Vázquez, *On a doubly nonlinear parabolic obstacle problem modelling ice sheet dynamics*, SIAM J. Appl. Math. **63** (2002), no. 2, 683–707 (electronic), DOI 10.1137/S0036139901385345. MR1951955 (2004g:86019)

[DGDF+09] G. Durand, O. Gagliardini, B. De Fleurian, T. Zwinger, and E. Le Meur, *Marine ice sheet dynamics: Hysteresis and neutral equilibrium*, Journal of Geophysical Research **114** (2009).

[DPF11] D. Docquier, L. Perichon, and Pattyn F., *Representing grounding line dynamics in numerical ice sheet models: Recent advances and outlook*, Surveys in Geophysics **32** (2011), 417–435.

[GB09] R. Greve and H. Blatter, *Dynamics of ice sheets and glaciers*, Springer Verlag, 2009.

[GHS09] D. Goldberg, D. M. Holland, and C. Schoof, *Grounding line movement and ice shelf buttressing in marine ice sheets*, Journal of Geophysical Research **114** (2009), no. F04026.

[GK09] C. Gräser and R. Kornhuber, *Multigrid methods for obstacle problems*, J. Comput. Math. **27** (2009), no. 1, 1–44. MR2493556 (2010b:35165)

[Glo03] R. Glowinski, *Finite element methods for incompressible viscous flow*, Handbook of numerical analysis, Vol. IX, Handb. Numer. Anal., IX, North-Holland, Amsterdam, 2003, pp. 3–1176. MR2009826

[Grä11] C. Gräser, *Convex minimization and phase field models*, Ph.D. thesis, FU Berlin, 2011.

[GSS09] C. Gräser, U. Sack, and O. Sander, *Truncated nonsmooth Newton multigrid methods for convex minimization problems*, Domain decomposition methods in science and engineering XVIII, Lect. Notes Comput. Sci. Eng., vol. 70, Springer, Berlin, 2009, pp. 129–136, DOI 10.1007/978-3-642-02677-5_12. MR2743965

[Hac85] W. Hackbusch, *Multi-grid methods and applications*, Springer series in computational mathematics, Springer, 1985.

[JB] G. Jouvet and E. Bueler, *Steady, shallow ice sheets as obstacle problems: well-posedness and finite element approximation*, SIAM Journal on Applied Mathematics **72** (2012), no. 4, 1292-1314.

[Kor94] R. Kornhuber, *Monotone multigrid methods for elliptic variational inequalities. I*, Numer. Math. **69** (1994), no. 2, 167–184. MR1310316 (96d:65201)

[Kor96] R. Kornhuber, *Monotone multigrid methods for elliptic variational inequalities. II*, Numer. Math. **72** (1996), no. 4, 481–499, DOI 10.1007/s002110050178. MR1376109 (96k:65081)

[PHDBDS06] F. Pattyn, A. Huyghe, S. De Brabander, and B. De Smedt, *Role of transition zones in marine ice sheet dynamics*, Journal of Geophysical Research **111** (2006), no. F2, 1–10.

[PSP+12] F. Pattyn, C. Schoof, L. Perichon, R. C. A. Hindmarsh, E. Bueler, B. de Fleurian, G. Durand, O. Gagliardini, R. Gladstone, D. Goldberg, G. H. Gudmundsson, V. Lee, F. M. Nick, A. J. Payne, D. Pollard, O. Rybak, F. Saito, and A. Vieli, *Results of the marine ice sheet model intercomparison project, MISMIP*, The Cryosphere Discussions **6** (2012), no. 1, 267–308.

[Sch06a] C. Schoof, *A variational approach to ice stream flow*, J. Fluid Mech. **556** (2006), 227–251, DOI 10.1017/S0022112006009591. MR2263447 (2008f:86019)

[Sch06b] C. Schoof, *Variational methods for glacier flow over plastic till*, J. Fluid Mech. **555** (2006), 299–320, DOI 10.1017/S0022112006009104. MR2265330 (2008f:86020)

[Sch07] C. Schoof, *Ice sheet grounding line dynamics: Steady states, stability, and hysteresis*, Journal of Geophysical Research **112** (2007).

[Sch09] C. Schoof, *Coulomb friction and other sliding laws in a higher-order glacier flow model*, Math. Models Methods Appl. Sci. **20** (2010), no. 1, 157–189, DOI 10.1142/S0218202510004180. MR2606247 (2011d:86014)

[Sch11] C. Schoof, *Marine ice sheet dynamics. Part 2. A Stokes flow contact problem*, J. Fluid Mech. **679** (2011), 122–155, DOI 10.1017/jfm.2011.129. MR2817693

[VA07] D. G. Vaughan and R. Arthern, *Why is it hard to predict the future of ice sheets ?*, Science **315** (2007), no. 5818, 1503–1504.

[VP05] A. Vieli and A. J. Payne, *Assessing the ability of numerical ice sheet models to simulate grounding line migration*, Journal of Geophysical Research **110** (2005), no. F1, 1–18.

[WMH+11] R. Winkelmann, M. A. Martin, M. Haseloff, T. Albrecht, E. Bueler, C. Khroulev, and A. Levermann, *The Potsdam Parallel Ice Sheet Model (PISM-PIK) Part 1: Model description*, The Cryosphere **5** (2011), 715–726.

INSTITUT FÜR MATHEMATIK, FREIE UNIVERSITÄT BERLIN, BERLIN, GERMANY
E-mail address: `guillaume.jouvet@fu-berlin.de`

DEPARTMENT OF MATHEMATICS AND STATISTICS, AND GEOPHYSICAL INSTITUTE, UNIVERSITY OF ALASKA FAIRBANKS, FAIRBANKS, ALASKA
E-mail address: `elbueler@alaska.edu`

INSTITUT FÜR MATHEMATIK, FREIE UNIVERSITÄT BERLIN, BERLIN, GERMANY
E-mail address: `graeser@math.fu-berlin.de`

INSTITUT FÜR MATHEMATIK, FREIE UNIVERSITÄT BERLIN, BERLIN, GERMANY
E-mail address: `ralf.kornhuber@fu-berlin.de`

Component mode synthesis for laminar viscous incompressible fluid flow

Mats G. Larson and Fredrik Bengzon

ABSTRACT. In this paper we develop a Component Modal Synthesis (CMS) model reduction technique for incompressible fluid flow. We extend the traditional CMS method, developed primarily for linear elasticity, by constructing a divergence free orthogonal modal basis for the fluid velocity. This allows elimination of the fluid pressure from the equations. For the reduced solution we derive a discrete a posteriori error estimate. The performance of our CMS method and a posteriori error estimate is illustrated in numerical experiments.

1. Introduction

Model reduction techniques are becoming increasingly important, since they often allow a complicated problem to be solved at a very affordable computational cost and reasonable accuracy. The approach is particularly important when a large number of problems should be solved, for instance, in time dependent problems, control problems, or parameter studies.

A common model reduction technique, originally developed for linear elasticity, is Component Mode Synthesis (CMS), which builds on a domain decomposition approach combined with modal reduction, see Hurty [**Hur65**], Craig and Bampton [**CB68**], Bourquin [**Bou90**] and d'Hennezel [**Bd92a, Bd92b**]. More precisely the reduced space is spanned by subdomain eigonmodes, with support in each subdomain, and so called interface eigenmodes. The actual reduction is obtained by choosing a limited number of eigen modes.

In this paper we extend CMS to incompressible fluid flow. The difficulty is that the subdomain and coupling modes, which represent the fluid velocity, must be divergence free, or solenoidal. We construct these modes by solving certain Stokes eigenvalue problems associated with the subdomains and the interface. In order to define the interface eigenvalue problem we first use harmonic extension from the interface to the domain based on the Stokes operator. Since the extension is divergence free we first need to construct a suitable space of functions on the interface that satisfies a conservation property for each subdomain. Our approach manufactures modes, which are divergence free, orthogonal with respect to the

---

1991 *Mathematics Subject Classification.* Primary 76D99, 65N30; Secondary 70J10, 65M15.

Key words and phrases. Model reduction, component modal synthesis, finite element method, computational fluid mechanics.

diffusion operator $-\Delta$, and allow for the formulation of CMS methods using modal truncation.

We derive a discrete energy norm a posteriori error estimate for the difference between the reduced and full finite element solutions. The estimate is based on the approach in Hannukainen, Stenberg, and Vohralik [**HSV10**], or Elman, Silvester, and Wathen [**ESW05**]. The main difference is that the frequency of the modes act like the usual mesh size parameter. Thus, error reduction comes from adding modes with higher and higher frequency to the modal basis. We use the a posteriori error estimate to design a simple adaptive algorithm, which automatically selects the appropriate number of modes in each subdomain and on the interface so that the error can be kept below a user-defined tolerance, see previous works by the authors [**JBL10, JBL**] on linear elasticity and thermo-elasticity.

The reminder of this paper is organized as follows; in Section 2 we review the Stokes equations and finite elements; in Section 3 we present our reduced method and the construction of the involved CMS spaces; in Section 4 we derive the a posteriori error estimate; finally, in Section 5 we present a some numerical experiments.

2. Stokes Equations

Let $\Omega \subset \mathbb{R}^2$ be a bounded domain with polygonal boundary $\partial\Omega$ and outward unit normal n. The domain is filled with a viscous incompressible fluid. We wish to find the velocity $u : \Omega \to \mathbb{R}^2$ and pressure $p : \Omega \to \mathbb{R}$ satisfying Stokes equations

$$-\nu\Delta u + \nabla p = f, \quad \text{in } \Omega, \tag{2.1a}$$

$$\nabla \cdot u = 0, \quad \text{in } \Omega, \tag{2.1b}$$

$$u = g, \quad \text{on } \partial\Omega, \tag{2.1c}$$

where $\nu > 0$ is a given viscosity parameter, $f : \Omega \to \mathbb{R}^2$ is a given force in, say, $[L^2(\Omega)]^2$, and $g : \partial\Omega \to \mathbb{R}^2$ a given no-slip boundary condition. For simplicity, we shall assume $g = 0$. The case $g \neq 0$ can be treated by modifying f.

Because the flow is enclosed the pressure p is only determined up to the constant hydrostatic pressure level.

2.1. Weak Form.

Let $V = [H_0^1(\Omega)]^2 = \{v \in [H^1(\Omega)]^2 : v|_{\partial\Omega} = 0\}$ and $Q = L_0^2(\Omega) = \{v \in L^2(\Omega) : (v, 1) = 0\}$. The weak form of (2.1) reads: find $(u, p) \in V \times Q$ such that

$$B((u,p),(v,q)) = (f, v), \quad \forall (v, q) \in V \times Q, \tag{2.2}$$

where the big bilinear form $B(\cdot,\cdot)$ is defined by

$$B((u,p),(v,q)) = a(u,v) + b(p,v) + b(q,u), \tag{2.3}$$

and the smaller bilinear forms $a(\cdot,\cdot)$ and $b(\cdot,\cdot)$ are defined by

$$a(u,v) = \nu(\nabla u : \nabla v), \tag{2.4}$$

$$b(p,v) = -(p, \nabla \cdot v). \tag{2.5}$$

The weak form (2.2) is a saddle point problem. It is well posed due to the inf-sup condition

$$C\|q\|_Q \leq \sup_{v \in V} \frac{b(q,v)}{\|v\|_V}, \quad \forall q \in Q, \tag{2.6}$$

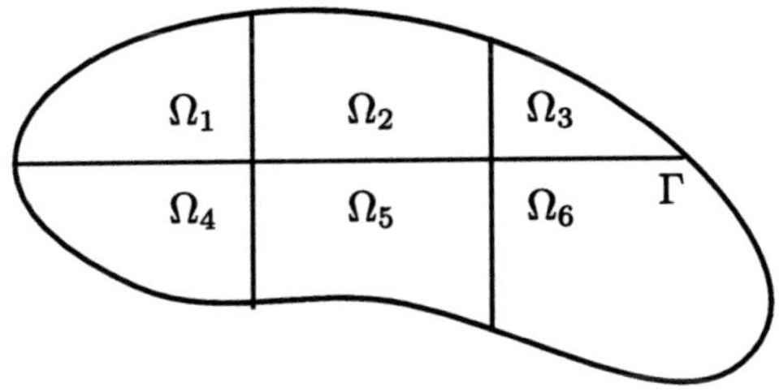

FIGURE 1. A domain Ω with a structured grid like partitioning into 6 subdomains Ω_i, $i = 1, \ldots, 6$, and interface Γ.

which holds for some constant $C > 0$. See Brezzi and Fortin [**FB91**] for details. Here, $\|\cdot\|_Q$ and $\|\cdot\|_V$ means the L^2 and energy norm $\sqrt{a(\cdot,\cdot)}$, respectively.

In this context a key observation is that on the null space $Z = \{v \in V : b(q, v) = 0,\ \forall q \in Q\}$ of $b(\cdot,\cdot)$ (i.e., the space of all divergence free velocities) the pressure disappears from the weak form (2.2), which reduces to: find $u \in Z$ such that

$$a(u, z) = (f, z), \quad \forall z \in Z. \tag{2.7}$$

2.2. Standard Galerkin Approximation. Let $\mathcal{K} = \{K\}$ be a mesh of Ω into shape regular triangle elements, and let (V^h, Q^h) be the inf-sup stable pair of discrete velocity and pressure spaces stemming from the non-conforming P_1/P_0 finite element on this mesh. We recall that the non-conforming approximation consists of linear velocity components and a constant pressure on each element. The pressure is discontinuous between adjacent triangles, whereas the velocities are continuous at the mid-points of the triangle edges. Indeed, the edge mid-points are the nodes or degrees of freedom for the velocity. This finite element is very popular in computational fluid dynamics, since it is elementwise solenoidal and easy to implement in software. The former property implies the important inclusion $Z^h \subset Z$ for the discrete null space Z^h of all discretely divergence free velocities.

Replacing V and Q with V^h and Q^h in the weak form (2.2) we obtain the standard Galerkin method: find $(u^h, p^h) \in V^h \times Q^h$ such that

$$B((u^h, p^h), (v, q)) = (f, v), \quad \forall (v, q) \in V^h \times Q^h. \tag{2.8}$$

As usual, any standard Galerkin method becomes computationally costly on a large mesh. One way of cutting this cost is to use component modal synthesis.

3. Component Modal Synthesis

Consider a partition $\bigcup_{i=1}^n \Omega_i$ of the domain Ω into n non-overlapping subdomains Ω_i, $i = 1, \ldots, n$. The subdomains are connected by the interface $\Gamma = \bigcup_{i,j=1, i\neq j}^n \Gamma_{I(i,j)}$ consisting of p interface segments $\Gamma_{I(i,j)} = \partial\Omega_i \cap \partial\Omega_j$. Here, $I(i, j)$ is an index function that maps an admissible pair (i, j) onto a unique integer $1 \leq k \leq p$. For simplicity, we shall assume that the partition of the domain is structured like a grid as illustrated in Figure 1.

A decomposition of the null space Z associated with the partition of the domain can be constructed as follows. For every subdomain Ω_i, define the subspaces $V_i =$

$[H_0^1(\Omega_i)]^2$ and $Q_i = L_0^2(\Omega_i)$, $i = 1, \ldots, n$. The subspaces Z_i of subdomain modes are then defined by

$$Z_i = \{z \in V_i : b(q, z) = 0,\ \forall q \in Q_i\}, \quad i = 1, \ldots, n. \tag{3.1}$$

Needless to say we have the orthogonality $a(u, v) = 0$, for $u \in Z_i$, and $v \in Z_j$, with $i \neq j$.

To couple the subdomain modes a subspace $Z_0 \subset Z$ associated with the interface Γ is also constructed. To this end, let $Z|_\Gamma$ be the trace space of Z associated with Γ. Further, let $Ez \in V$ be the harmonic extension of a function $z \in Z|_\Gamma$ to Ω, given by the solution to the problem: find $(Ez, p) \in V \times Q$, such that

$$B((Ez, p), (v, q)) = 0, \quad \forall (v, q) \in V_i \times Q_i, \quad i = 1, \ldots, n, \tag{3.2}$$

$$Ez|_\Gamma = z. \tag{3.3}$$

This construction makes the harmonic extension Ez divergence free. Thus, we have $Ez \in Z$. The subspace Z_0 of coupling modes, a-orthogonal to Z_i, $i = 1, \ldots, n$, is then defined by

$$Z_0 = \{Ez \in V : z \in Z|_\Gamma\}. \tag{3.4}$$

With Z_i, $i = 0, \ldots, n$, defined as above, it follows that an orthogonal decomposition of Z with respect to the inner product $a(\cdot, \cdot)$ is given by

$$Z = \bigoplus_{i=0}^{n} Z_i. \tag{3.5}$$

To construct a basis for Z, eigenvalue problems associated with each subspace Z_i are formulated: find $(\lambda_{i,j}, z_{i,j}) \in \mathbb{R} \times Z_i$ for $i = 0, \ldots, n$, such that

$$a(z_{i,j}, v) = \lambda_{i,j}(z_{i,j}, v), \quad \forall v \in Z_i, \quad i = 0, \ldots, n. \tag{3.6}$$

The solutions consist of eigenpairs $(\lambda_{i,j}, z_{i,j})$, where $0 < \lambda_{i,1} \leq \lambda_{i,2} \leq \cdots$, for $i = 0, \ldots, n$. Each set $\{z_{i,j}\}_{j=1}^{\infty}$ of eigenmodes forms an a-orthogonal basis in the associated subspace Z_i, $i = 0, \ldots, n$. Moreover, a basis for Z is given by the combined set of eigenmodes $\bigcup_{i=0}^{n}\{z_{i,j}\}_{j=1}^{\infty}$.

The harmonic extension Ez deserves special attention, since the problem (3.2) is ill-posed if the boundary condition $Ez|_\Gamma = z$ for some reason is not consistent with the incompressibility constraint $\nabla \cdot Ez = 0$. Due to the solenoidal nature of Ez it has zero flux across the boundary of anyone of the subdomains. That is, $(Ez, n)_{\partial\Omega_i} = 0$, $i = 1, \ldots, n$. Now, consider an interface segment Γ_k. On Γ_k we can write Ez (or any member of $Z|_\Gamma$) as $Ez = z_0 + c_k z_n$, where z_0 is a vector with zero flux, c_k a constant, and z_n a vector such that $(z_n, n)_{\Gamma_k} = 1$. Obviously, c_k is the magnitude of the net flux transported by Ez across Γ_k. However, since Γ_k is shared by the two subdomains Ω_i and Ω_j for which $k = I(i, j)$ this constant can not be arbitrary, but has to be determined from the requirement $(Ez, n)_\Gamma = \sum_{i=1}^{n} \sum_{\Gamma_k \in \partial\Omega_i} c_k = 0$. In the discrete setting this gives an under determined linear system of size $n \times p$ for c_k. Further, with a suitable choice of finite element the vectors z_0 and z_n are easy to represent on a mesh.

3.1. CMS Method. Now, let V_i^h, Q_i^h, and Z_i^h be the discrete counterparts of the spaces V_i, Q_i, and Z_i, respectively. Repeating the arguments from the previous section, and replacing all continuous spaces with their discrete ditto, it follows that an a-orthogonal decomposition of Z^h associated with the subdomains and the interface is given by $Z^h = \bigoplus_{i=0}^n Z_i^h$. Moreover, a modal basis $\{z_{i,j}\}_{j=1}^{n_i}$ for each of the subspaces Z_i^h, $i = 0, \dots, n$, is given by discrete eigenvalue problems similar to (3.6). Here, n_i is the dimension of Z_i (i.e., the number of velocity degrees of freedom not on the domain boundary). We assume that the mesh $\mathcal{K}$ respects the boundaries of the subdomains, and has no hanging nodes on the interface.

To reduce Z^h let us truncate the basis $\{z_{i,j}\}_{j=1}^{n_i}$ for each Z_i^h after m_i modes. More specific, let $Z^{h,m} \subset Z^h$ be defined by

$$Z^{h,m} = \bigoplus_{i=0}^{n} Z_i^{h,m_i}, \tag{3.7}$$

where $m = (m_i)_{i=0}^n$ is a multi-index with $m_i \leq n_i$, $i = 0, \dots, n$, and

$$Z_i^{h,m_i} = \text{span}\{z_{i,j}\}_{j=1}^{m_i}, \quad i = 0, \dots, n. \tag{3.8}$$

The modal truncation leads to the following CMS method: find $U^m \in Z^{h,m}$ such that

$$a(U^m, z) = (f, z), \quad \forall z \in Z^{h,m} \tag{3.9}$$

Here, we observe that the stiffness matrix is trivial to invert, since it is diagonal.

We remark that a pressure P^m corresponding to the velocity U^m can be reconstructed from the variational equation $b(P^m, w) = (f, w) - a(U^m, w)$ for all $w \in W$ with W a suitably chosen space. Indeed, for our particular finite element, P^m can be recovered at a very low cost, cf. Stevenson [**Ste98**].

3.2. Some Implementation Details. Let $Z^h|_\Gamma$ be the discrete trace space consisting of the restriction of Z^h to Γ. For the particular choice of the nonconforming P_1/P_0 finite element any discrete divergence free velocity $z \in Z^h|_\Gamma$ can on an interface segment Γ_k be written as $z = z_0 + c_k z_n$, where z_0 has zero flux and z_n unit flux. Recall that harmonic extension of z_0 and $c_k z_n$ gives, in principle, the coupling modes. Therefore, it is important to be able to represent z_0 and z_n as finite element functions on the mesh. Starting with z_0, it is possible to define two types of velocities with zero flux across Γ_k, namely:

- A velocity tangential to Γ_k at a node.
- Two oppositely directed normal velocities at any pair of nodes on Γ_k.

See Figure 2. The velocity at all other nodes on the interface must be zero for reasons of linear independence.

Continuing, to define $c_k z_n$ on Γ_k we use that we are at liberty to choose z_n more or less arbitrary as long as $(z_n, n_k)_{\Gamma_k} = 1$. Here, n_k is a fixed unit normal on Γ_k. A simple choice is to pick a node N on triangle edge $E \subset \Gamma_k$ and define z_n as the finite element function with value $z_n(N) = n_k/\text{meas}(E)$ at N and zero value at all other nodes. Further, to define the constant c_k specifying the net flux across Γ_k we use the fact that the flux of z is conserved in each subdomain. For example, consider the partition of the square domain Ω into a 3-by-3 grid with 9 subdomains and 12 interface segments shown in Figure 3. For this domain partition, the requirement $(z, n)_{\partial\Omega_i} = 0$ for $i = 1, \dots, 9$ leads to the under determined linear system $Gc = 0$,

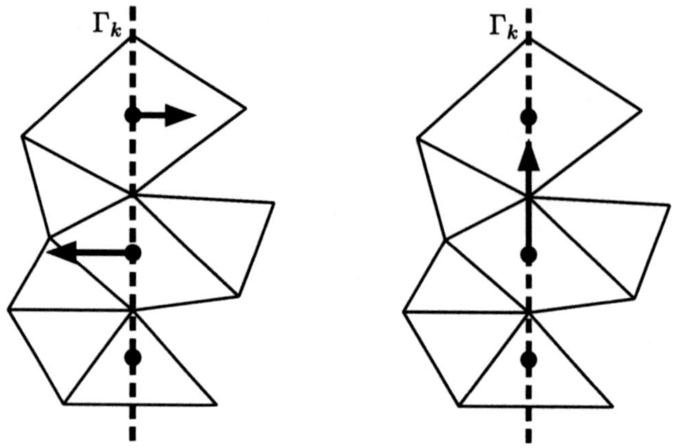

(a) Pair of normal velocities. (b) Tangential velocity.

FIGURE 2. Velocity types with zero net flux across an interface segment Γ_k on the mesh. The bullets (•) symbolize the location of the nodes.

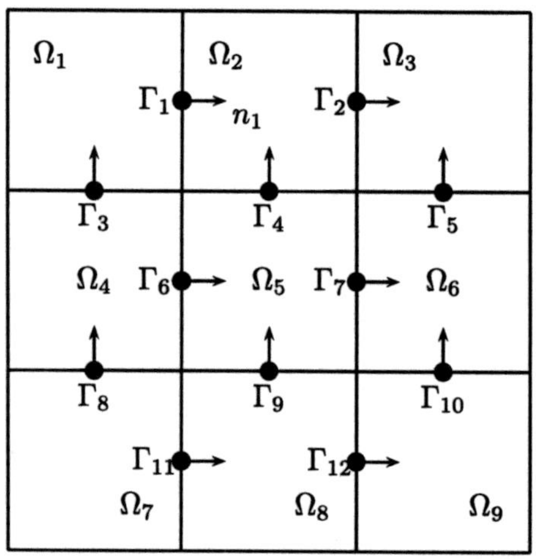

FIGURE 3. Domain partition of a square Ω into a 3-by-3 grid with 9 subdomains and 12 interface segments. Also shown are the normal directions.

where the 9×12 matrix G is given by

$$G = \begin{bmatrix} 1 & 0 & -1 & 0 & 0 & 0 & 0 & 0 & 0 & 0 & 0 & 0 \\ 1 & -1 & 0 & 1 & 0 & 0 & 0 & 0 & 0 & 0 & 0 & 0 \\ 0 & 1 & 0 & 0 & 1 & 0 & 0 & 0 & 0 & 0 & 0 & 0 \\ 0 & 0 & -1 & 0 & 0 & -1 & 0 & 1 & 0 & 0 & 0 & 0 \\ 0 & 0 & 0 & -1 & 0 & 1 & -1 & 0 & 1 & 0 & 0 & 0 \\ 0 & 0 & 0 & 0 & -1 & 0 & 1 & 0 & 0 & 1 & 0 & 0 \\ 0 & 0 & 0 & 0 & 0 & 0 & 0 & -1 & 0 & 0 & -1 & 0 \\ 0 & 0 & 0 & 0 & 0 & 0 & 0 & 0 & -1 & 0 & 1 & -1 \\ 0 & 0 & 0 & 0 & 0 & 0 & 0 & 0 & 0 & -1 & 0 & 1 \end{bmatrix} \tag{3.10}$$

The null space of G consist of four 12×1 vectors, and each of these vectors contain a set of admissible values of c_k, $k = 1, \ldots, 12$, on Γ_k.

Together, the tangential velocities, the node pairs of velocities, and the velocities from the null space of G form a basis for the discrete trace space $Z^h|_\Gamma$.

4. A Posteriori Error Estimate

Let us next turn to the question of accuracy. Suppose we wish to measure the error $e = u^h - U^m$ in the reduced solution relative to the finite element solution. By subtracting the CMS method (3.9) from the standard Galerkin method (2.8) we see that e satisfies the Galerkin orthogonality

$$a(e, z) = 0, \quad \forall z \in Z^{h,m} \tag{4.1}$$

Using Galerkin orthogonality and that $e \in Z^h$ we have

$$\|e\|_V = a(e, \|e\|_V^{-1} e) \leq \sup_{v \in Z^h} \frac{a(e,v)}{\|v\|_V} = \sup_{v \in Z^h} \sum_{i=0}^{n} \frac{a(e, v - \pi_i^{m_i} v)}{\|v\|_V} \tag{4.2}$$

$$= \sup_{v \in Z^h} \sum_{i=0}^{n} \frac{(f, v - \pi^{m_i} v) - a(U^m, v - \pi_i^{m_i} v)}{\|v\|_V} \tag{4.3}$$

$$= \sup_{v \in Z^h} \sum_{i=0}^{n} \frac{(R_i(U^m), v - \pi_i^{m_i} v)}{\|v\|_V}, \tag{4.4}$$

where $R_i(U^m)$, $i = 0, \ldots, n$ are the subspace residuals, defined by $(R_i(U^m), z) = (f, z) - a(U^m, z)$ for all $z \in Z_i$. The operator $\pi_i^{m_i} : Z^h \to Z^{h,m_i}$ denotes (Fourier) series expansion. That is, $\pi_i^{m_i} v = \sum_{j=1}^{m_i} (z_{i,j}, v) z_{i,j}$ for any $v \in Z^h$.

In the so-called error representation formula (4.4) the supremum is difficult to compute, and we would like to avoid it. To this end, we shall utilize the following approximation result.

LEMMA 4.1. *For any $v \in Z^h$ there holds the estimate*

$$\|v - \pi_i^{m_i} v\| \leq \frac{1}{\lambda_{i,m_i+1}} \|v\|_V, \quad i = 0, \ldots, n, \tag{4.5}$$

PROOF. For each $i = 0, \ldots, n$, the eigenvalue problem (3.6) admits a sequence $\{\lambda_{i,j}\}_{j=1}^{n_i}$ of positive eigenvalues of increasing magnitude, so that, by virtue of Parseval's identity and Cauchy-Schwarz inequality,

$$\|v - \pi_i^{m_i} v\|^2 = \sum_{j=m_i+1}^{n_i} (z_{i,j}, v)^2 \leq \frac{1}{\lambda_{i,m_i+1}^2} \sum_{j=m_i+1}^{n_i} \lambda_{i,j}^2 (z_{i,j}, v)^2 \tag{4.6}$$

$$\leq \frac{1}{\lambda_{i,m_i+1}^2} \sum_{j=1}^{n_i} \lambda_{i,j}^2 (z_{i,j}, z_{i,j})(v,v) \leq \frac{1}{\lambda_{i,m_i+1}^2} a(v,v) = \frac{1}{\lambda_{i,m_i+1}^2} \|v\|_V^2. \tag{4.7}$$

□

Now, using Lemma 4.1 and, again, Cauchy-Schwarz inequality we conclude

$$\|e\|_V \leq \sup_{v \in Z^h} \sum_{i=0}^{n} \frac{(R_i(U^m), v - \pi_i^{m_i} v)}{\|v\|_V} \leq \sum_{i=0}^{n} \frac{1}{\lambda_{i,m_i+1}} \|R_i(U^m)\|, \tag{4.8}$$

which is our a posteriori error estimate.

The a posteriori error estimate gives a bound on the error from the different subspaces Z_i^h and can therefore be used as a tool for adaptively selecting the dimension m_i of these spaces (i.e., the number of subdomain and coupling modes). This allows for the design of a CMS method with different accuracy in each subdomain and on the interface in order to keep the error e small. A basic adaptive algorithm for doing so takes the following form:

Algorithm 1 Adaptive CMS Method.

Start with a guess of the subspace dimensions m.
Choose a desired accuracy ϵ.
while $\|e\|_V > \epsilon$ **do**
  Solve (3.9) for U^m.
  for $i = 0$ to n **do**
    Compute the so-called error indicator

$$\eta_i = \frac{1}{\lambda_{i,m_i+1}} \|R_i(U^m)\|, \tag{4.9}$$

    and use it together with a refinement criterion to decide if subspace Z_i^h is eligible for refinement, and if so how much its dimension m_i should be increased.
  end for
end while

We use the simple refinement criterion $\eta_i > \beta \max_{0 \le j \le n} \eta_j$, with $0 \le \beta \le 1$, to classify the error indicators as large or small and to determine which subspaces are eligible for refinement. Other types of refinement criteria are of course also possible to use. In a refinement of a subspace we add three new modes.

5. Numerical Experiments

5.1. Fluid Flow due to a Gaussian Force. As a first numerical example we solve the Stokes' equations (2.1) on the square domain $\Omega = \{x : 0 \le x_1, x_2 \le 3\}$. Zero no-slip boundary conditions apply on the whole boundary $\partial\Omega$. The viscosity parameter $\nu = 0.1$, and the force $f = \exp(-cr^2)\,[1,1]^T$, with $c = 5$, and r the distance from the point $(1.5, 1.25)$. This causes a fluid flow inside the square with a one bigger and one smaller vortex. We solve this problem using a partition of Ω into the 3 by 3 grid with 9 subdomains Ω_i shown previously. The mesh is shown in Figure 4. On this mesh the number of possible divergence free velocities are $1873 = \dim(Z^h)$, so we may argue that reduction is called for. A few of the lowest subdomain and coupling modes are shown in Figure 5 and Figure 6, respectively. The resulting CMS solution U^m is shown in Figure 9. The two vortexes are clearly visible.

In order to study the qualitative performance of the a posteriori error estimate we have implemented the adaptive procedure of Algorithm 1. Stating from a uniform distribution with one mode (i.e., $m_i = 1$) in each subspace Z_i^h, we make 30 adaptive refinement loops. The distribution of modes at the end of the last refinement loop is shown in Table 1. We observe that the result is in agreement with intuition. Also, the solution looks seamless across the interface.

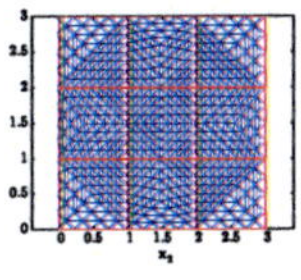

FIGURE 4. The mesh $\mathcal{K}$.

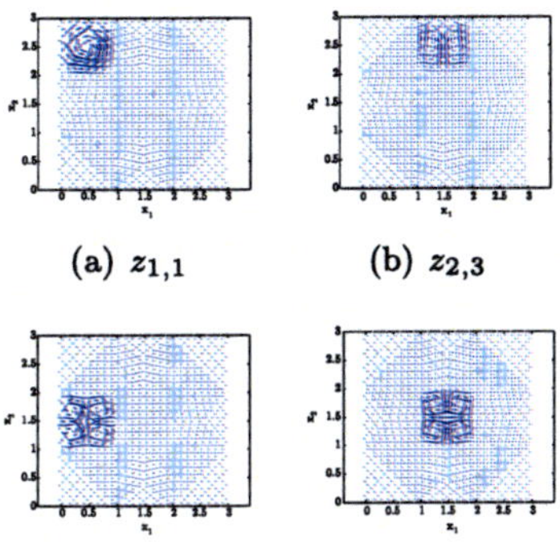

(a) $z_{1,1}$ (b) $z_{2,3}$

(c) $z_{4,4}$ (d) $z_{5,4}$

FIGURE 5. Glyph plots of $z_{i,j}$, the jth mode in subdomain Ω_i, for various i, j.

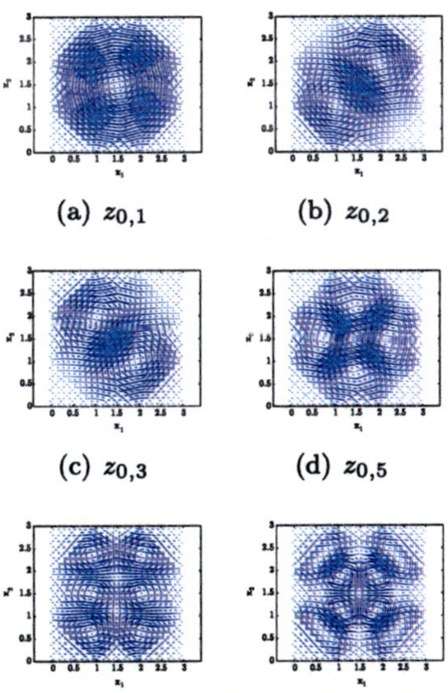

(a) $z_{0,1}$ (b) $z_{0,2}$

(c) $z_{0,3}$ (d) $z_{0,5}$

(e) $z_{0,7}$ (f) $z_{0,12}$

FIGURE 6. Glyph plots of $z_{0,j}$, the jth coupling mode, for $j = 1, 2, 3, 5, 7, 12$.

We remark that the a posteriori estimate is not suitable for quantitative error estimation, since the error is over estimated in general. This is clearly seen in Figure 8, which shows the total number of modes $|m| = m_0 + \ldots + m_n$, the energy norm

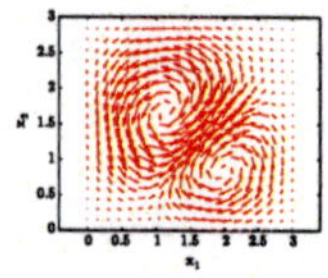

FIGURE 7. Glyph plot of CMS solution U^m.

TABLE 1. Distribution of modes at the end of the 30th refinement loop.

i	0	1	2	3	4	5	6	7	8	9
m_i	91	1	1	1	34	88	28	4	82	10

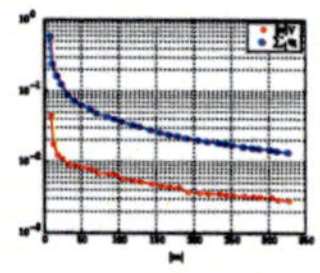

FIGURE 8. Number of modes $|m|$ versus error $\|e\|_V$ and bound $\sum_{i=0}^n \eta_i$.

error $\|e\|_V$, and the upper bound $\sum_{i=0}^n \eta_i$ on this error for each refinement loop. We observe that the convergence rate of the error is monotone, and drops fast in the beginning but slows towards the end. This is typical for these kind of reduced methods, which are designed to capture large scale effects of the solution only.

5.2. The Lid-driven Cavity. As a second numerical example we study the well know lid-driven cavity benchmark. Again the domain is the square $\Omega = [0,3]^2$. In this example we include the inertia term $(u \cdot \nabla)u$ and solve the full Navier-Stokes equations

$$-\nu\Delta u + (u \cdot \nabla)u + \nabla p = 0, \quad \text{in } \Omega, \tag{5.1a}$$

$$\nabla \cdot u = 0, \quad \text{in } \Omega, \tag{5.1b}$$

$$u = g, \quad \text{on } \partial\Omega, \tag{5.1c}$$

Here, g is zero on the cavity walls and bottom and $[g_0, 0]$ on the lid. This non-zero boundary condition is accounted for by adding the mode $z_{0,\infty} = g_0[e^{cx_2}/e^{3c},\ 0]$, with $g_0 = 33$ and $c = 50$, to the modal basis. The viscosity $\nu = 0.1$. For these settings the flow has a steady state with Reynolds number Re $= 1000$. The mesh and domain partition is the same as used previously. The non-linearity is handled by fixed point iteration. A small amount of streamline-diffusion stabilization is needed for convergence. We use 20 modes in each subdomain modes and 100 modes on the interface. The resulting CMS solution U^m is shown in Figure 9. The characteristic vortex is clearly visible.

References

[Bd92a] F. Bourquin and F. d'Hennezel, *Intrinsic component mode synthesis and plate vibrations*, Computers and Structures **44** (1992), no. 1–2, 315–324. MR1179203 (93f:73065)

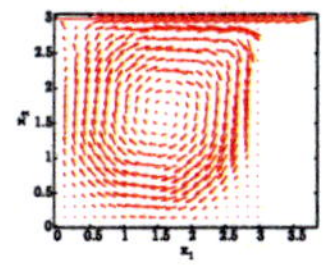

FIGURE 9. Glyph plot of CMS solution U^m.

[Bd92b] ______, *Numerical study of an intrinsic component mode synthesis method*, Computer Methods in Applied Mechanics and Engineering **97** (1992), 49–76. MR1182434 (93g:73082)

[Bou90] F. Bourquin, *Analysis and comparison of several component mode synthesis methods on one dimensional domains*, Numerische Mathematik **58** (1990), no. 1, 11–33. MR1069651 (92a:65323)

[CB68] R. R. Craig and M. C. C. Bampton, *Coupling of substructures for dynamic analysis*, AIAA Journal (1968), no. 6, 1313–1321.

[ESW05] H. Elman, D. Silvester, and A. Wathen, *Finite elements and fast iterative solvers*, Oxford University Press, 2005. MR2155549 (2006f:65002)

[FB91] M. Fortin and F. Brezzi, *Mixed and hybrid finite element methods*, Springer series in computational mathematics, Springer, 1991. MR1115205 (92d:65187)

[HSV10] A. Hannukainen, R. Stenberg, and M. Vohralik, *A unified framework for a posteriori error estimation for the Stokes problem*, Tech. Report A587, Helsinki University of Technology Institute of Mathematics Research Reports, 2010.

[Hur65] W. C. Hurty, *Dynamic analysis of structural systems using component modes*, AIAA Journal (1965), no. 4, 678–685.

[JBL] H. Jakobsson, F. Bengzon, and M. G. Larson, *Duality based adaptive model reduction for one-way coupled thermoelastic problems*, International Journal for Numerical Methods in Engineering, To appear.

[JBL10] ______, *Adaptive component mode synthesis in linear elasticity*, International Journal for Numerical Methods in Engineering **86** (2010), no. 7, 829–844. MR2829770

[Ste98] R. Stevenson, *A stable, direct post-processing procedure to compute the pressure for the Stokes equations*, Tech. Report 9810, Department of Mathematics, University of Nijmegen, April 1998.

DEPARTMENT OF MATHEMATICS, UMEÅ UNIVERSITY, SE-903 87 UMEÅ, SWEDEN
E-mail address: mats.larson@math.umu.se

DEPARTMENT OF MATHEMATICS, UMEÅ UNIVERSITY, SE-903 87 UMEÅ, SWEDEN
E-mail address: fredrik.bengzon@math.umu.se

Numerical analysis of wall adapted nonlinear filter models of turbulent flows

W. Layton and A. Takhirov

ABSTRACT. Stabilization using filters is intended to model and extract the energy lost to resolved scales due to nonlinearity breaking down resolved scales to unresolved scales. This process is highly nonlinear. We consider nonlinear filters which select eddies for damping (simulating breakdown) based on knowledge of how nonlinearity acts in real flow problems. The particular form of the nonlinear filter allows for easy incorporation of more knowledge into the filter process and its computational complexity is comparable to calculating a linear filter of similar form. Herein we show how to adapt nonlinear filters to the near wall region, give a convergence result for the wall-adapted method and give a test which shows eddy viscosity can be highly localized and produce an excellent result.

1. Introduction

Nonlinear filtering (NLF), recently introduced in [**20**], gives an approach to modeling and simulation of turbulent flows that (i) has a strong mathematical foundation, (ii) yields new turbulence models that can be evolved to greater accuracy and reliability and (iii) provides a modular implementation of any selected turbulence models within legacy codes, laminar flow codes and complex application codes. The physical idea behind nonlinear filtering is that a turbulence model should act as a proxy for the action of nonlinearity upon marginally resolved structures. Nonlinearity does not break down scales uniformly. Intermittence, nonuniformity, locality and backscatter occur. If nonlinearity breaks down a local structure, the model should strongly and locally damp the structure while if nonlinearity allows such a structure to persist the modeling terms should be negligible locally. If this physical idea is realized correctly, the NLF model immediately corrects through indicator functions the over damping of persistent, transitional, recirculation and other flows in eddy viscosity models. The examples given in [**20**] of indicator functions and their associated eddy viscosity models are all adapted to free turbulence (away from walls). Herein we show in Section 2.2 how to use the WALE construction of Nicoud and Ducros [**24**] to construct an indicator function for the nonlinear filter model that is adapted to the important near wall region.

2000 *Mathematics Subject Classification.* Primary 76F65; Secondary 65M12.
Key words and phrases. Turbulence, eddy viscosity, nonlinear filter.
The work of WL and AT was partially supported by NSF grant DMS-0810385.
This paper is in final form and no version of it will be submitted for publication elsewhere.

Consider the NSE in a domain $\Omega \subset \mathbb{R}^{2 \text{ or } 3}$

$$u_t + u \cdot \nabla u - \nu \Delta u + \nabla p = f(x,t), \tag{1.1}$$
$$\nabla \cdot u = 0\ ,\ u = 0 \text{ on } \partial\Omega \times (0,T], \text{ and } u(x,0) = u^0(x) \text{ in } \Omega.$$

Given a method for the NSE at low or moderate Reynolds numbers, e.g. (LegacyStep) below, the method (Modular NLFilter) adapts it to high Reynolds number flows by adding the modular steps (NLfilter) and (Relax) below. Suppressing the spacial discretization: we advance $u^n \simeq u(t^n)$ to u^{n+1} by:

ALGORITHM 1 (Modular NLFilter). *Pick* $\chi \in [0,1]$ *and* $\Delta t > 0$, *given* u^n, p^n
Step 1: *Find* w^{n+1}

$$\frac{w^{n+1} - u^n}{\Delta t} + w^{n+1} \cdot \nabla w^{n+1} - \nu \Delta w^{n+1} + \nabla p^{n+1} = f^{n+1} \tag{LegacyStep}$$
$$\nabla \cdot w^{n+1} = 0 \ \ and \ \ w^{n+1} = 0\ ,\ on\ \partial\Omega.$$

Step 2: *Nonlinear filter:* $w^{n+1} \rightarrow \overline{w^{n+1}}$ *by*

$$-\delta^2 \nabla \cdot \left(a(w^{n+1}) \nabla \overline{w^{n+1}} \right) + \overline{w^{n+1}} + \nabla \lambda = w^{n+1}, \tag{NLfilter}$$
$$\nabla \cdot \overline{w^{n+1}} = 0, \quad and \quad \overline{w^{n+1}} = w^{n+1}\ ,\ on\ \partial\Omega.$$

Step 3: *Relax:*

$$u^{n+1} := (1-\chi) w^{n+1} + \chi \overline{w^{n+1}}. \tag{Relax}$$

The nonlinear filtering in Step 2 requires solving one linear system. With common FEM discretizations of (NLfilter) and $\delta = O(\Delta x)$ the condition number of the $1,1$ block in the associated mixed linear system is $O(1)$. To specify (NLfilter), requires selecting the function $a(u, \nabla u)$ (abbreviated $a(u)$) which we call an *indicator function.*

DEFINITION 1.1 (Indicator function). $a = a(u, \nabla u)$ is a function with $0 \leq a(\cdot) \leq 1$, and

$$a(u(x)) \doteq 0 \text{ for laminar or persistent flow structures}$$
$$a(u(x)) \doteq 1 \text{ for rapidly decaying flow structures.}$$

Quite generally, Proposition 1, (Modular NLFilter) is stable and convergent:

$$\text{Global Error} = O(\Delta t + \frac{\chi}{\Delta t}\delta^2 + \text{Spacial Error}).$$

Related work. Linear filter stabilization was developed by Boyd [**4**] and Fischer and Mullen [**12**], [**26**] (who introduced relaxation in Step 3), used by Dunca [**9**] and Mathew et al [**25**], Garnier, Adams and Sagaut [**7**] and Visbal and Rizzetta [**6**] and analyzed in [**11**]. Numerical analysis of modular postprocessing algorithms was begun in [**11**], [**21**], culminating in the step of using nonlinear filters to adapt models in [**20**], and continuing to regularizations in [**2**]. Olshanskii and Xiong [**23**] have recently given a precise elaboration and analysis of the connection of (Modular NLFilter) with an eddy viscosity model.

2. Three Examples of Indicator Functions and Nonlinear Filters

The method (NLfilter) is no better than the indicator function selected. This limitation has been one intractable difficulty in previous eddy viscosity models: the EV used is over-sensitive to some persistent structures and the calculation is over-damped as a result. The idea of nonlinear filters is that different indicator functions with different sensitivities can be combined to advantage: since the *geometric average* of indicator functions is again an indicator function, given indicator functions $a_i(u), i = 1, \cdots, M$, we can choose in (NLfilter)

$$a(u) := (a_1(u) \cdot \dots \cdot a_M(u))^{1/M} \tag{2.1}$$

curing the over-sensitivity problem. Thus, *evolving the accuracy and reliability of the induced turbulence model means adding one function subroutine.* These $a_i(u)$ are obtained from theories of intermittence and eduction. We review three indicator functions in [**20**] then construct and test a wall adapted indicator from the WALE model of [**24**], the main contribution herein.

2.1. Three indicators tested in [20]. The Q criterion indicator. The Q criterion [**18**] marks persistent, coherent vortices where $Q > 0$ or local rigid body rotation dominates deformation:

$$\begin{aligned} Q(u,u) \quad &: \quad = \frac{1}{2}\left(\nabla^{ss}u : \nabla^{ss}u - \nabla^{s}u : \nabla^{s}u\right) \text{ where} \\ \nabla^{s}u \quad &: \quad = \frac{1}{2}\left(\nabla u + \nabla u^{tr}\right) \text{ and } \nabla^{ss}u := \frac{1}{2}\left(\nabla u - \nabla u^{tr}\right). \end{aligned}$$

Rescale so $Q > 0$ or $Q < 0$ corresponds to $a(u)$ close to 0 or 1, respectively.

DEFINITION 2.1. The Q-criterion indicator function is

$$a_Q(u) := \frac{1}{2} - \frac{1}{\pi}\arctan\left(\delta^{-1}\frac{Q(u,u)}{Q(u,u)+\delta^2}\right).$$

Vreman [**29**] constructs an eddy viscosity model based on a function $B(u)$ below. $B(u)$ is constructed to vanish for many coherent (non turbulent) flows:

$$\begin{aligned} |\nabla w|_F^2 \quad &= \quad \sum_{i,j=1,2,3}\left(\frac{\partial u_j}{\partial x_i}\right)^2, \beta_{ij} := \sum_{m=1,2,3}\frac{\partial u_i}{\partial x_m}\frac{\partial u_j}{\partial x_m}, \text{ and} \\ B(u) \quad &: \quad = \beta_{11}\beta_{22} - \beta_{12}^2 + \beta_{11}\beta_{33} - \beta_{13}^2 + \beta_{22}\beta_{33} - \beta_{23}^2. \end{aligned}$$

DEFINITION 2.2. The Vreman indicator function is

$$a_V(u) = \sqrt{B(u)/|\nabla u|_F^4}.$$

Relative helicity density. Let $\omega = \nabla \times u$. High helicity suppresses nonlinearity and thus breakdown by the NSE nonlinearity. (It may be a cause of intermittence, e.g., [**28**].) Indeed, helicity, $u \cdot \omega$, and the NSE nonlinearity, $u \times \omega$, are related by

$$\frac{\text{Helicity}^2 + |\text{NSE nonlinearity}|^2}{|u|^2|\omega|^2} = 1.$$

DEFINITION 2.3. The relative helicity indicator function is

$$a_H(u) := 1 - \left|\frac{u(x,t)\cdot\omega(x,t)}{|u(x,t)||\omega(x,t)| + \delta^2}\right|.$$

2.2. A Wall Adapted Indicator. Many models, including the ones induced by the above indicator functions, are optimized for turbulence away from walls. One exception is the WALE model of Nicoud and Ducros [**24**] which accounts for both strain and rotation of the smallest resolved scales and recovers the proper y^3 near-wall eddy viscosity scaling. Let $g_{ij} = \partial u_j/\partial x_i$ and $g_{ij}^2 := g_{ik}g_{kj}$ (summation convention). Consider the traceless symmetric part of g_{ij}^2:

$$S_{ij}^d = \frac{1}{2}(g_{ij}^2 + g_{ji}^2) - \frac{1}{3}\delta_{ij}g_{kk}^2,\ \delta_{ij} = \text{Kronecker } \delta.$$

Nicoud and Ducros [**24**] construct an eddy viscosity model beginning with

$$W(u) := \frac{(S^d(u) : S^d(u))^{3/2}}{(S^d(u) : S^d(u))^{5/4} + (D(u) : D(u))^{5/2}}. \tag{2.2}$$

We adapt $W(u)$ to an indicator function as follows.

DEFINITION 2.4. The wall adjusted indicator function is given by

$$a_{WALE}(u) := \frac{2.0}{\pi}\arctan\left(\frac{1}{\delta}\frac{|W(u)|}{\delta^2 + |W(u)|}\right).$$

The form of the arctangent function in $a(u)$ is one of many options. Considering the plot $a = a(W)$ (below for δ moderate), $a(W)$ transitions quickly from zero where $W(u) = 0$ to near 1 at all other values of $W(u)$.

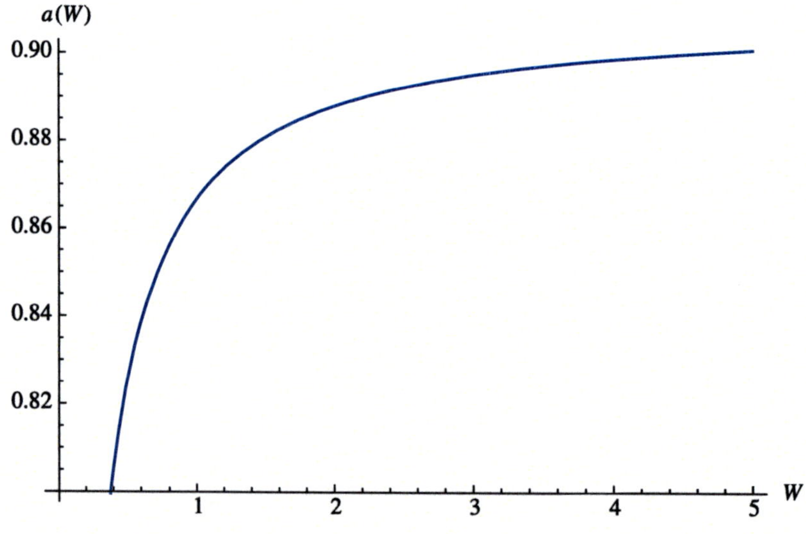

FIGURE 1. Indicator: $a(W)$ against W

3. Convergence of Nonlinear Filter Based Stabilization

There is a general stability and convergence theory for (Modular NLFilter) from [**20**] which implies convergence for the WALE indicator based model.

PROPOSITION 1 (Unconditional Stability and Convergence, [**20**]). *Let spacial discretization be by finite element methods with velocity-pressure finite element* *spaces satisfying the discrete inf-sup condition. The energy equality (implying stability) holds:*

$$\frac{1}{2}||u_h^l||^2 + \Delta t \sum_{n=0}^{l-1} \left\{ \frac{\Delta t}{2} || \frac{w_h^{n+1} - u_h^n}{\Delta t} ||^2 + \nu ||\nabla w_h^{n+1}||^2 \right\} +$$

$$\Delta t \sum_{n=0}^{l-1} \frac{\chi}{\Delta t} \left\{ \frac{2-\chi}{2} \left(w_h^{n+1} - \overline{w_h^{n+1}}^h, w_h^{n+1} \right) + \frac{\chi}{2} \left(w_h^{n+1} - \overline{w_h^{n+1}}^h, \overline{w_h^{n+1}}^h \right) \right\}$$

$$= \frac{1}{2}||u_h^0||^2 + \Delta t \sum_{n=0}^{l-1} (f^{n+1}, w_h^{n+1}), \quad \textit{for any } l > 0,$$

For $0 \le \chi \le 2$ *the model diffusion term is non-negative:*

$$\frac{\chi}{\Delta t} \left[\frac{2-\chi}{2} \left(w_h^{n+1} - \overline{w_h^{n+1}}, w^{n+1} \right) + \frac{\chi}{2} \left(w_h^{n+1} - \overline{w_h^{n+1}}, \overline{w_h^{n+1}} \right) \right] \ge 0.$$

Let $0 \le \chi \le 1$ *and suppose the velocity, pressure spaces contain piecewise polynomials of degree* $(k, k-1)$. *For* u, p, *and* f *sufficiently regular, the errors satisfy*

$$\|u(t^l) - w_h^l\| + \|u(t^l) - u_h^l\| + \left(\nu \Delta t \sum_{n=1}^{l} \|\nabla (u(t^n) - w_h^n)\|^2 \right)^{1/2}$$

$$\le C(u, p, data, \nu) \left[h^k + \Delta t + \frac{\chi}{\Delta t} (\gamma h^k + h^{k+1} + \delta h^k + \delta^2 \min\{\delta^{-1}, ||\nabla \cdot (a(w_h) \nabla u)||^2\}) \right].$$

The model dissipation induced by steps 2 and 3 is independent of Step 1 and is of eddy viscosity type

$$\text{Model Dissipation} = \frac{\chi}{\Delta t} \delta^2 \left(a(w) \nabla \overline{w}, \nabla w \right) + HOdissipationTs.$$

4. Numerical Experiments

Our tests of the WALE indicator in (Modular NLFilter) used *FreeFEM++* [**17**] and Taylor-Hood elements ($X_h = C^0$ piecewise quadratics, $Q_h = C^0$ piecewise linears).

Test of the convergence rate. First we test the predicted error and convergence rates for $a = a_{WALE}(\cdot)$ for the Green-Taylor vortex [**13**], [**14**]:

$$\begin{aligned} u_1(x,y,t) &= -\cos(\omega \pi x) \sin(\omega \pi y) \exp(-2\omega^2 \pi^2 t/\tau) \\ u_2(x,y,t) &= \sin(\omega \pi x) \cos(\omega \pi y) \exp(-2\omega^2 \pi^2 t/\tau) \\ p(x,y,t) &= -\frac{1}{4}(\cos(2\omega \pi x) + \cos(2\omega \pi y)) \exp(-2\omega^2 \pi^2 t/\tau) \end{aligned} \tag{4.1}$$

When $\tau = Re$, this is a solution of NSE with $f = 0$, consisting of $\omega \times \omega$ array of oppositely signed vortices that decay as $t \to \infty$. We take $\omega = 1$, $\tau = Re = 100$, $\Omega = (0,1)^2$, $\chi = \Delta t$, $\delta = \Delta x$ and $T = 0.1$. Table 1 shows the predicted convergence rates.

(Table 1)

Δt	h, δ	$\|\|\|u - u_h\|\|\|_{2,1}$	rate
0.005	$\frac{1}{4}$	0.0955	–
$\frac{0.005}{2}$	$\frac{1}{8}$	0.0264	1.86
$\frac{0.005}{4}$	$\frac{1}{16}$	0.0048	2.46
$\frac{0.005}{8}$	$\frac{1}{32}$	0.0012	2

Errors and convergence rates: (Modular NLFilter) using $a_{WALE}(\cdot)$

Flow over a step. The domain is a 40×10 rectangular channel with a 1×1 step five units into the channel. Boundary conditions are no-slip on the top, bottom boundaries, a parabolic inflow profile $(y(10-y)/25, 0)^T$, $\Delta t = 0.01$, $\nu = 1/600$ and do-nothing outflow. The correct behavior is a smooth velocity field away from the step and for eddies to periodically form and shed behind the step. *While not turbulent, this flow is a good test if a turbulence model over damps important dynamic structures. Generally, under damping and under resolved gives nonsensical solutions and over damping gives one large attached eddy that does not break up.* First we shown below velocity streamlines over speed contours at $T = 40$ for 2D flow over a step, found by an under resolved NSE solution (Step 1 without Steps 2 and 3, no filtering), on meshes with 1762 and 3226 dof (Figures 2 and 3). These are nonsensical compared to fully resolved simulations in [**5**].

Next compare these nonsensical solutions to solutions on the two same meshes using the modular nonlinear filtering model with the WALE indicator (Figures 4 and 5).

Compared to fine mesh results in [**5**], both velocities are accurate in major features and the eddy behind the step is starting to break up roll downstream. Outflow boundary noise is typical behavior. We next give plots of (the scalar) $a_{WALE}(u)$ for the two meshes (Figures 6 and 7).

As expected, stabilization is needed around the outflow boundary. It is remarkable that, away from the outflow region, the WALE indicator localized eddy viscosity to a very small area at the leading edge of the step. The WALE indicator shows that beyond a few mesh cells there, no extra eddy viscosity or stabilization is needed to get a good, coarse mesh approximation.

5. Conclusions

With *nonlinear* filtering, $w - \overline{w}$ and thus model dissipation is small in laminar regions and regions of persistent, coherent flow structures. It reduces both numerical and modeling errors and model dissipation *more closely mimics the exact physics of the energy cascade.* Nonlinear filtering reduces implementing a complex turbulence model in a flow code (possibly a legacy code of great length) to solving one well conditioned linear system each time step. Model accuracy is increased simply by providing additional function subroutines. The WALE indicator by itself and without combination of other indicators precise localizes the EV needed for the tested flow. We conjecture that this is because step flow is about flow-wall interactions for which WALE is an excellent tool and that for combinations of flow-wall

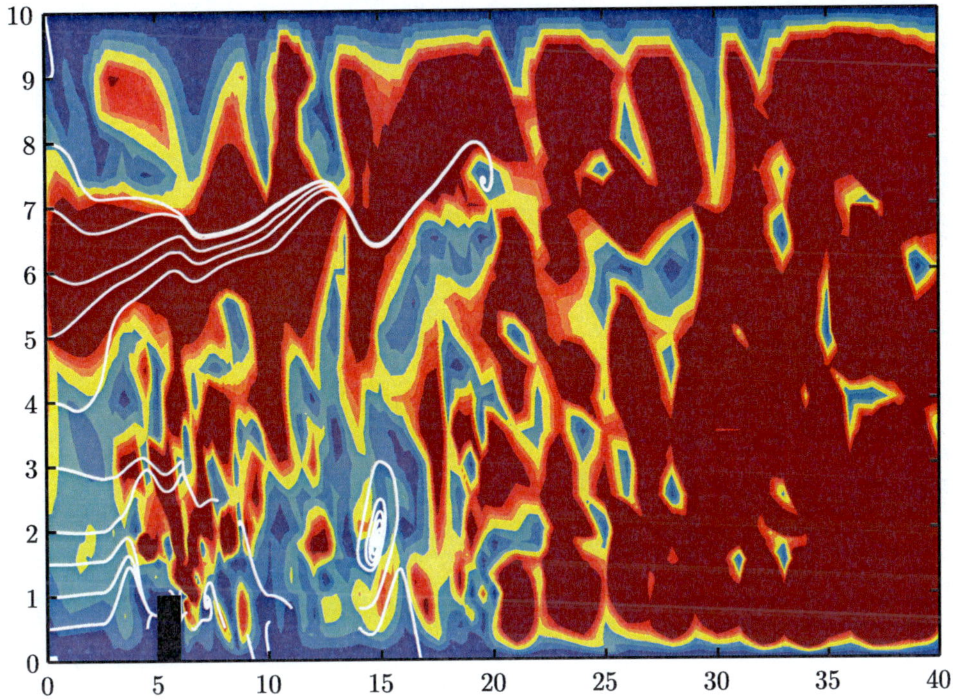

FIGURE 2. Velocity, $T = 40$, 2D step, Underresolved NSE solution, 1762 dof

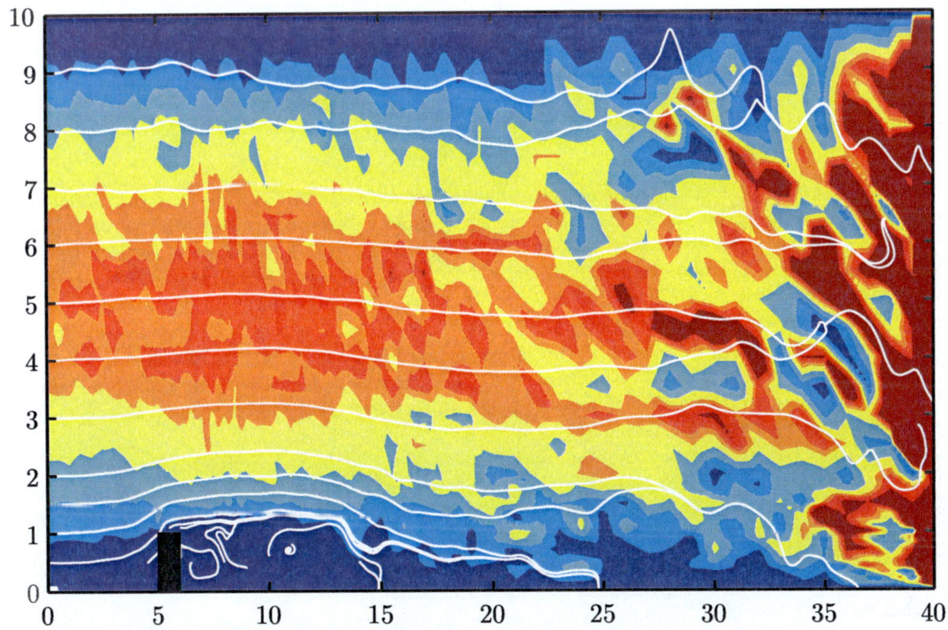

FIGURE 3. Velocity, $T = 40$, 2D step, Underresolved NSE solution, 3226 dof

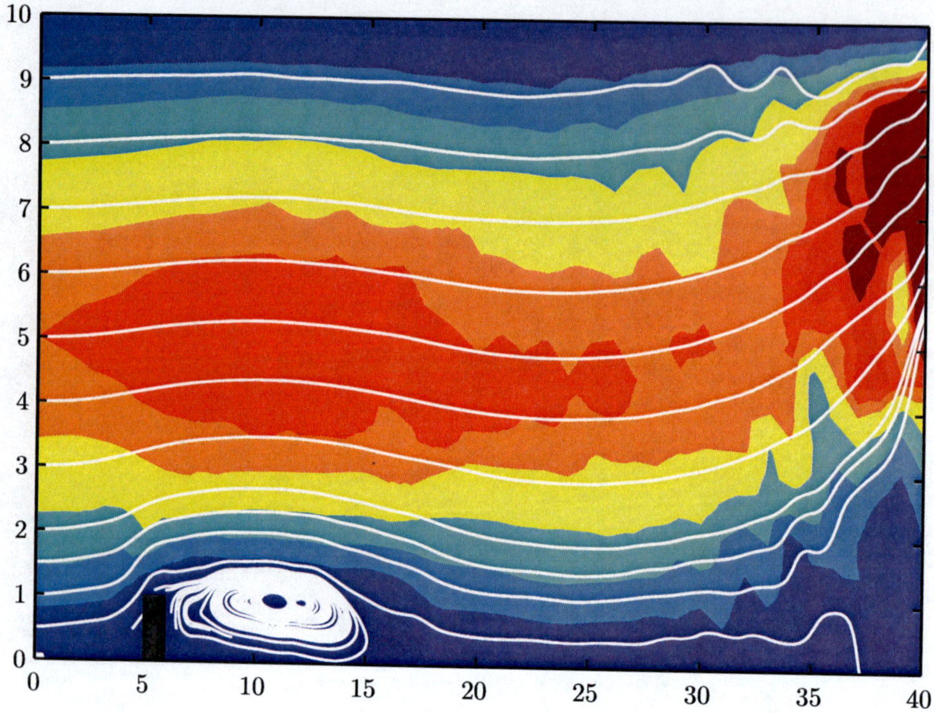

FIGURE 4. NLFiltering velocity, $T = 40$ using WALE indicator, 1762 dof

interactions with free turbulence and coherent structures, combinations of WALE with other indicators would produce improved results over WALE alone.

References

[1] L.C. Berselli, T. Iliescu and W. Layton, Large Eddy Simulation, Springer, Berlin, 2004

[2] A. Bowers, L. Rebholz, A. Takhirov, and C. Trenchea, Improved accuracy in regularization models of incompressible flow via adaptive nonlinear filtering, IJNMF, to appear, 2012.

[3] Jeff Borggaard, Traian Iliescu, and John Paul Roop, *A bounded artificial viscosity large eddy simulation model*, SIAM J. Numer. Anal. **47** (2008/09), no. 1, 622–645, DOI 10.1137/060656164. MR2475955 (2010b:76069)

[4] John P. Boyd, *Two comments on filtering (artificial viscosity) for Chebyshev and Legendre spectral and spectral element methods: preserving boundary conditions and interpretation of the filter as a diffusion*, J. Comput. Phys. **143** (1998), no. 1, 283–288, DOI 10.1006/jcph.1998.5961. MR1624716

[5] William Layton, Carolina C. Manica, Monika Neda, and Leo G. Rebholz, *Numerical analysis and computational testing of a high accuracy Leray-deconvolution model of turbulence*, Numer. Methods Partial Differential Equations **24** (2008), no. 2, 555–582, DOI 10.1002/num.20281. MR2382797 (2009b:76064)

[6] M.R. Visbal and D.P. Rizzetta, Large eddy simulation on curvilinear grids using compact differencing and filtering schemes, J. Fluids Eng. 124 (2002) 836-847.

[7] E. Garnier, N. Adams, and P. Sagaut, *Large eddy simulation for compressible flows*, Scientific Computation, Springer, Berlin, 2009. MR2542010 (2010g:76074)

[8] J. Connors and W. Layton, *On the accuracy of the finite element method plus time relaxation*, Math. Comp. **79** (2010), no. 270, 619–648, DOI 10.1090/S0025-5718-09-02316-3. MR2600537 (2011c:65187)

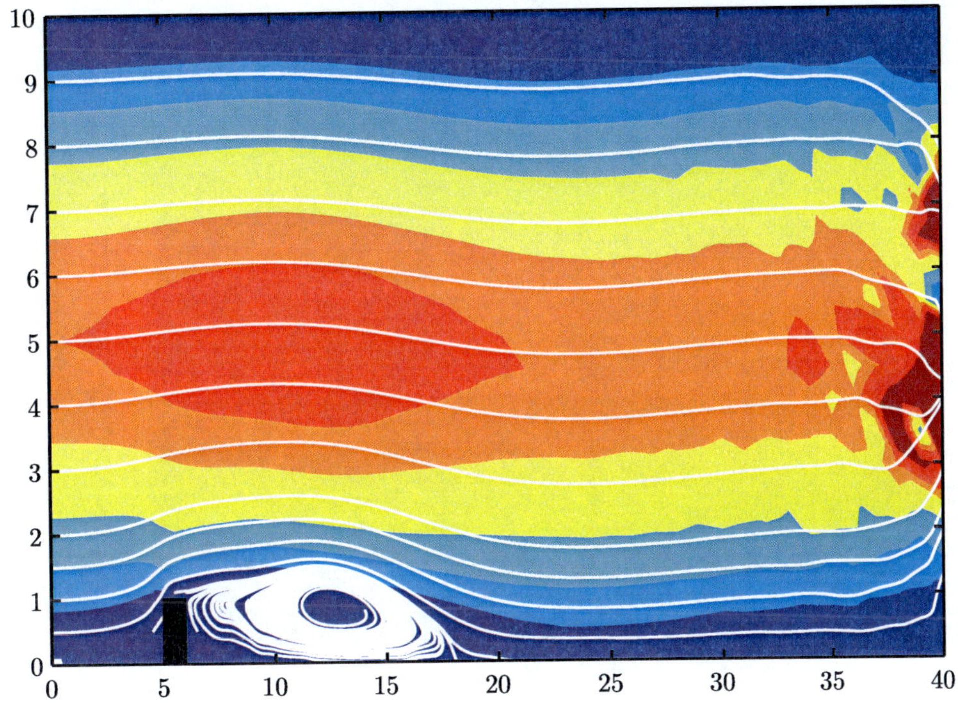

FIGURE 5. NLFiltering velocity, $T = 40$ using WALE indicator, 3226 dof

[9] A. Dunca, Investigation of a shape optimization algorithm for turbulent flows, report ANL/MCS-P1101-1003, Argonne National Lab, 2002.

[10] Vincent J. Ervin, William J. Layton, and Monika Neda, *Numerical analysis of a higher order time relaxation model of fluids*, Int. J. Numer. Anal. Model. **4** (2007), no. 3-4, 648–670. MR2344062 (2008j:76035)

[11] V. Ervin, W. Layton and M. Neda, Numerical analysis of filter based stabilization for evolution equations, technical report, submitted to SINUM 2009.

[12] Paul Fischer and Julia Mullen, *Filter-based stabilization of spectral element methods*, C. R. Acad. Sci. Paris Sér. I Math. **332** (2001), no. 3, 265–270, DOI 10.1016/S0764-4442(00)01763-8 (English, with English and French summaries). MR1817374 (2001m:65129)

[13] A.E. Green and G.I. Taylor, Mechanism of the production of small eddies from larger ones, Proc. Royal Soc. A, 158 (1937), 499-521.

[14] G.I. Taylor, On decay of vortices in a viscous fluid, Phil. Mag., 46, (1923), 671-674.

[15] Max D. Gunzburger, *Finite element methods for viscous incompressible flows*, Computer Science and Scientific Computing, Academic Press Inc., Boston, MA, 1989. A guide to theory, practice, and algorithms. MR1017032 (91d:76053)

[16] M. Germano, *Differential filters of elliptic type*, Phys. Fluids **29** (1986), no. 6, 1757–1758, DOI 10.1063/1.865650. MR845232 (87h:76075b)

[17] F. Hecht and O. Pironneau, FreeFEM++ , webpage: http://www.freefem.org.

[18] J.C. Hunt, A.A. Wray and P. Moin, Eddies stream and convergence zones in turbulent flows, CTR report CTR-S88, 1988.

[19] William Layton, *Superconvergence of finite element discretization of time relaxation models of advection*, BIT **47** (2007), no. 3, 565–576, DOI 10.1007/s10543-007-0142-z. MR2338532 (2009g:65127)

[20] William Layton, Leo G. Rebholz, and Catalin Trenchea, *Modular nonlinear filter stabilization of methods for higher Reynolds numbers flow*, J. Math. Fluid Mech. **14** (2012), no. 2, 325–354, DOI 10.1007/s00021-011-0072-z. MR2925112

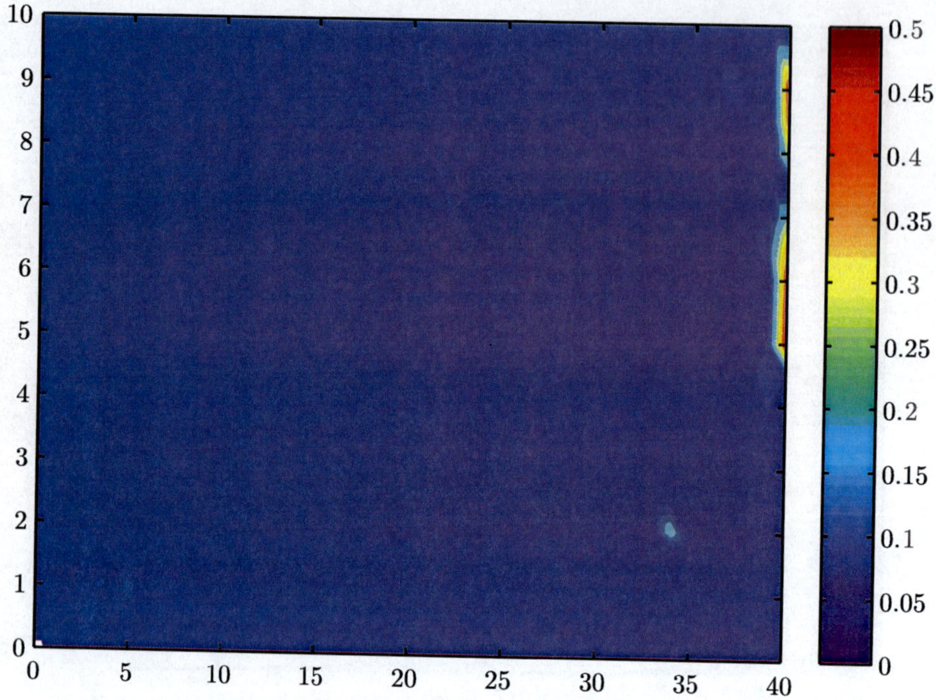

FIGURE 6. $T = 40$, WALE indicator for 1762 dof velocity

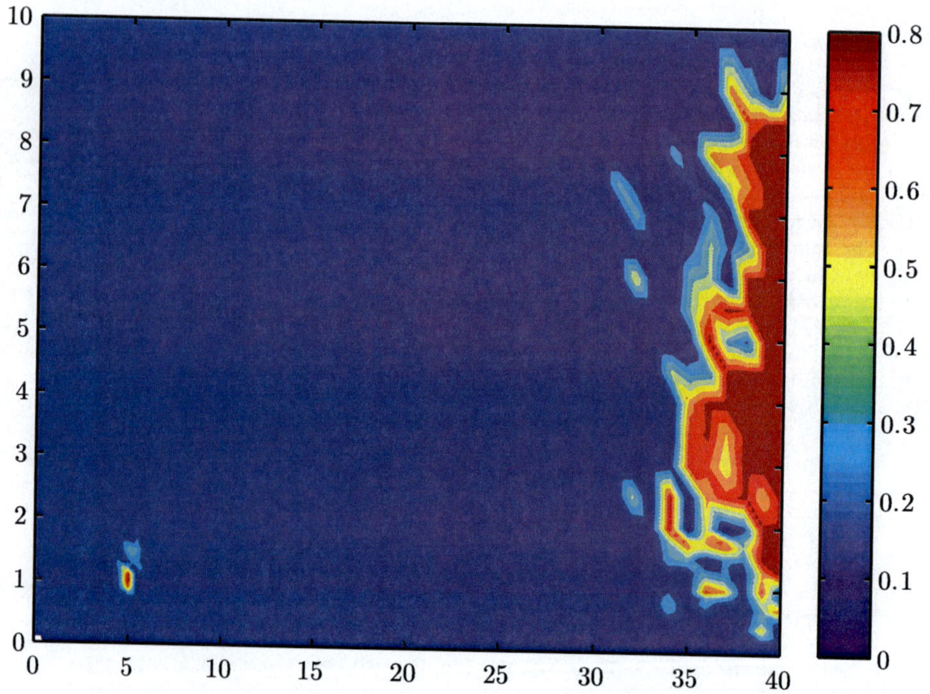

FIGURE 7. $T = 40$, WALE indicator for 3226 dof velocity

[21] W. Layton, L. Roehe and H. Tran, A modular, postprocessed induced VMS method, CMAME, to appear 2011.
[22] William Layton and Monika Neda, *Truncation of scales by time relaxation*, J. Math. Anal. Appl. **325** (2007), no. 2, 788–807, DOI 10.1016/j.jmaa.2006.02.014. MR2270051 (2008c:76049)
[23] M. Olshanskii and Xin Xiong, A connection between filter stabilization and eddy viscosity models, technical report 2012.
[24] F. Nicoud and F. Ducros, Subgrid-Scale Stress Modelling Based on the Square of the Velocity Gradient Tensor, Flow, Turbulence and Combustion, 62 (1999), 183-200.
[25] J. Mathew, R. Lechner, H. Foysi, J. Sesterhenn and R. Friedrich, An explicit filtering method for large eddy simulation of compressible flows, Physics of Fluids, 15 (2003), 2279-2289.
[26] Julie S. Mullen and Paul F. Fischer, *Filtering techniques for complex geometry fluid flows*, Comm. Numer. Methods Engrg. **15** (1999), no. 1, 9–18. MR1671621 (99i:76126)
[27] Pierre Sagaut, *Large eddy simulation for incompressible flows*, Scientific Computation, Springer-Verlag, Berlin, 2001. An introduction; With an introduction by Marcel Lesieur; Translated from the 1998 French original by the author. MR1815221 (2002f:76047)
[28] A. Tsinober and E. Levich, On the helical nature of 3-dimensional coherent structures in turbulent flows, Physics Letters 99A (1983) 321-324.
[29] A.W. Vreman, An eddy-viscosity subgrid-scale model for turbulent shear flow: algebraic theory and applications, Phys. Fluids 16 (2004), 3670-3681.

Department of Mathematics, University of Pittsburgh, Pittsburgh, Pennsylvania 15260
E-mail address: `WJL@pitt.edu`
URL: `http://www.math.pitt.edu/~wjl`

Current address: Department of Mathematics, University of Pittsburgh, Pittsburgh, Pennsylvania 15260
E-mail address: `AZT7@pitt.edu`
URL: `http://www.math.pitt.edu/~azt7`

Stability of partitioned methods for magnetohydrodynamics flows at small magnetic Reynolds number

William Layton, Hoang Tran, and Catalin Trenchea

ABSTRACT. MHD flows at small magnetic Reynolds number are governed by the Navier-Stokes equations coupled with the Ohm's law. The physical processes of fluid flows and electricity and magnetism are quite different and numerical simulations of non-model problems can require different meshes, time steps and methods. We introduce implicit-explicit (IMEX) methods where the MHD equations can be evolved in time by calls to the NSE and Ohm's law codes, each possibly optimized for the subproblem's respective physics.

1. Introduction

The MHD equations describe the motion of electrically conducting, incompressible flows in the presence of a magnetic field. If an electrically conducting fluid moves in a magnetic field, the magnetic field exerts forces which may substantially modify the flow. Conversely, the flow itself gives rise to a second, induced field and thus modifies the magnetic field. Initiated by Alfven in 1942 [**1**], MHD models occur in astrophysics, geophysics as well as engineering. Understanding these flows is central to many important applications, e.g., liquid metal cooling of nuclear reactors [**2, 8, 19**], sea water propulsion [**12**], process metallurgy [**3**].

The magnetic Reynolds number R_m is an important parameter in MHD, being indicative of the relative strength of induced magnetic field and imposed magnetic field:

$$R_m = \frac{\text{Induced field}}{\text{Applied field}} = \mu\sigma u L.$$

Here μ is the permeability of free space, σ is the electrical conductivity, u and L are the characteristic velocity and length scale correspondingly. Large values of velocity and length scale are unreachable in most industrial and laboratory flows. Consequently, MHD flows in terrestrial applications typically occur at small magnetic Reynolds number. While the magnetic field considerably alters the fluid motion, the induced field is usually found to be negligible by comparison with the imposed field [**4, 9, 17**]. Neglecting the induced magnetic field reduces MHD models to the system (RMHD) below, which is studied herein.

2010 *Mathematics Subject Classification.* Primary 65M12; Secondary 76W05.

The first author was partially supported by NSF grant DMS 0810385.

The second author was partially supported by Air Force grant FA 9550-09-1-0058 and by NSF grant DMS 0810385.

The third author was partially supported by Air Force grant FA 9550-09-1-0058.

Let Ω be a bounded, Lipschitz domain in $\mathbb{R}^d$ $(d = 3)$. With body force $\mathbf{f}$ and external imposed magnetic field $\mathbf{B}$ known, the electrically conducting flow can be completely described in terms of the fluid velocity $\mathbf{u}$, pressure p and electric potential ϕ. The reduced MHD (RMHD) equations are given by, see, e.g., [**16, 7, 21**]: Given $\mathbf{f}$, $\mathbf{B}$ and time $T > 0$, find $\mathbf{u}$, p and ϕ such that:

$$\text{(RMHD)} \qquad \begin{aligned} \frac{1}{N}(\mathbf{u}_t + \mathbf{u}\cdot\nabla\mathbf{u}) - \frac{1}{M^2}\Delta\mathbf{u} + \nabla p &= \mathbf{f} + (\mathbf{B}\times\nabla\phi + \mathbf{B}\times(\mathbf{B}\times\mathbf{u})), \\ \Delta\phi = \nabla\cdot(\mathbf{u}\times\mathbf{B}), &\text{ and } \nabla\cdot\mathbf{u} = 0. \end{aligned}$$

Here M, N are the Hartman number and interaction parameter given by

$$M = BL\sqrt{\frac{\sigma}{\rho\nu}}, \quad N = \sigma B^2\frac{L}{\rho u}$$

where B is the characteristic magnetic field, ρ is the density, and ν is the kinematic viscosity, all assumed constant. (RMHD) are supplemented by the homogeneous Dirichlet boundary conditions

$$\mathbf{u} = 0, \ \phi = 0 \text{ on } \partial\Omega\times[0,T]$$

and the initial data

$$\mathbf{u}(x,0) = \mathbf{u}_0(x), \ \phi(x,0) = \phi_0(x) \ \forall x\in\Omega. \tag{1.1}$$

Constant parameters and simple boundary conditions allows us to focus on the uncoupling of RMHD into physical subprocesses.

MHD flows involve different physical processes: the motion of fluid is governed by hydrodynamics equations and the electric potential is governed by electrodynamics equations. One approach to any coupled problem is monolithic methods. In these methods, the globally coupled problem is assembled at each time step and then solved iteratively. Partitioned methods, which solve the coupled problem by successively solving the sub-physics problems, are another attractive and promising approach for solving RMHD system. These allow us to employ the best NSE codes and best Ohm's law codes, each highly optimized for the sub-problems' respective physics.

The two partitioned methods we study herein include a first order, one step scheme and a second order, two steps scheme, both of which consist of implicit discretization of the subproblem terms and explicit discretization of coupling terms. We prove that these methods are stable over $0 \le t < \infty$. Particularly, the first order method IMEX1 is unconditionally stable. This stability result is surprising in that in combining implicit and explicit time discretizations one often sees the stability of the combination governed by that of the explicit method used.

For results on the steady-state MHD problems see [**20**] (2D), [**16**] (small R_m) and [**7**] (perfectly conducting walls). [**13, 14, 11, 15**] studied more boundary conditions that account for the electromagnetic interaction of the fluid with the outside world. For further discussions, see [**6, 5**]. For evolutionary MHD, see Schmidt [**18**] and for stability of fully coupled time discretization schemes, we refer to [**21**].

2. Notation and preliminaries

We denote the $L^2(\Omega)$ norms and corresponding inner products by $\|\cdot\|$ and $(\cdot,\cdot)$. Let $H^{-1}(\Omega)$ denote the dual space of $H_0^1(\Omega)$. The velocity, pressure and potential spaces are $X = (H_0^1(\Omega))^d$, $Q = L_0^2(\Omega)$ and $S = H_0^1(\Omega)$, respectively. The space of divergence free functions is given by

$$V = \{\mathbf{v} \in X : (\nabla \cdot \mathbf{v}, q) = 0 \;\; \forall q \in Q\}.$$

A weak formulation of (RMHD) is: Find $\mathbf{u} : [0,T] \to X$, $p : [0,T] \to Q$ and $\phi : [0,T] \to S$ for a.e. $t \in (0,T]$ satisfying

$$\begin{aligned} &\frac{1}{N}((\mathbf{u}_t, \mathbf{v}) + (\mathbf{u} \cdot \nabla \mathbf{u}, \mathbf{v})) + \frac{1}{M^2}(\nabla \mathbf{u}, \nabla \mathbf{v}) - (p, \nabla \cdot \mathbf{v}) \\ &\qquad + (\mathbf{u} \times \mathbf{B}, \mathbf{v} \times \mathbf{B}) - (\nabla \phi, \mathbf{v} \times \mathbf{B}) = (\mathbf{f}, \mathbf{v}) \;\; \forall \mathbf{v} \in X, \\ &(\nabla \cdot \mathbf{u}, q) = 0 \;\; \forall q \in Q, \\ &-(\nabla \phi, \nabla \psi) + (\mathbf{u} \times \mathbf{B}, \nabla \psi) = 0 \;\; \forall \psi \in S, \end{aligned} \tag{2.1}$$

with the initial condition (1.1) a.e. in Ω. Note that, setting $\mathbf{v} = \mathbf{u}$, $\psi = \phi$ and adding, the coupling terms exactly cancel in the monolithic sum and one verifies the stability of the continuous problem.

To make a spatial discretization of the RMHD system by the finite element method, we select finite element spaces

velocity: $X^h \subset X$, pressure: $Q^h \subset Q$, and potential: $S^h \subset S$

which are built on a conforming, edge to edge triangulation with maximum triangle parameter denoted by a subscript "h". We assume that $X^h \times Q^h$ satisfies the usual discrete inf-sup condition for the stability of the discrete pressure and X^h, Q^h, S^h satisfy approximation properties of piecewise polynomials on quasi-uniform meshes of local degrees $k, k-1, k$ respectively. The discretely divergence free velocity space is denoted by

$$V^h := X^h \cap \{\mathbf{v}_h : (q_h, \nabla \cdot \mathbf{v}_h) = 0, \;\text{ for all } q_h \in Q^h\}.$$

Also define the usual, explicitly skew symmetrized trilinear form

$$b(\mathbf{u}, \mathbf{v}, \mathbf{w}) = \frac{1}{2}((\mathbf{u} \cdot \nabla \mathbf{v}, \mathbf{w}) - (\mathbf{u} \cdot \nabla \mathbf{w}, \mathbf{v})).$$

The monolithic, semi-discrete approximation of (2.1) are maps $(\mathbf{u}_h, p_h, \phi_h) : [0,T] \to X^h \times Q^h \times S^h$ satisfying for all $\mathbf{v}_h \in X^h, q_h \in Q^h, \psi_h \in S^h$

$$\begin{aligned} &\frac{1}{N}((\mathbf{u}_{h,t}, \mathbf{v}_h) + b(\mathbf{u}_h, \mathbf{u}_h, \mathbf{v}_h)) + \frac{1}{M^2}(\nabla \mathbf{u}_h, \nabla \mathbf{v}_h) - (p_h, \nabla \cdot \mathbf{v}_h) \\ &\qquad + (\mathbf{u}_h \times \mathbf{B}, \mathbf{v}_h \times \mathbf{B}) - (\nabla \phi_h, \mathbf{v}_h \times \mathbf{B}) = (\mathbf{f}, \mathbf{v}_h), \\ &\qquad (\nabla \cdot \mathbf{u}_h, q_h) = 0, \\ &-(\nabla \phi_h, \nabla \psi_h) + (\mathbf{u}_h \times \mathbf{B}, \nabla \psi_h) = 0. \end{aligned} \tag{2.2}$$

2.1. The implicit-explicit partitioned schemes. The methods we propose and analyze herein have the coupling terms lagged or extrapolated, thus the system uncouples into two subproblem solves.

ALGORITHM 2.1 (First order IMEX scheme). *Given* $\mathbf{u}_h^n \in X^h, p_h^n \in Q^h, \phi_h^n \in S^h$, *find* $\mathbf{u}_h^{n+1} \in X^h, p_h^{n+1} \in Q^h, \phi_h^{n+1} \in S^h$ *satisfying*

$$\begin{aligned}
&\frac{1}{N}\left(\left(\frac{\mathbf{u}_h^{n+1}-\mathbf{u}_h^n}{\Delta t}, \mathbf{v}_h\right) + b(\mathbf{u}_h^{n+1}, \mathbf{u}_h^{n+1}, \mathbf{v}_h)\right) + \frac{1}{M^2}(\nabla \mathbf{u}_h^{n+1}, \nabla \mathbf{v}_h) \\
&\quad -(p_h^{n+1}, \nabla\cdot \mathbf{v}_h) + (\mathbf{u}_h^{n+1}\times \mathbf{B}, \mathbf{v}_h \times \mathbf{B}) - (\nabla\phi_h^n, \mathbf{v}_h\times\mathbf{B}) = (\mathbf{f}^{n+1}, \mathbf{v}_h), \\
&\text{(IMEX1)} \qquad (\nabla\cdot \mathbf{u}_h^{n+1}, q_h) = 0, \\
&\quad -(\nabla\phi_h^{n+1}, \nabla\psi_h) + (\mathbf{u}_h^n \times \mathbf{B}, \nabla\psi_h) = 0,
\end{aligned}$$

for all $\mathbf{v}_h \in X^h, q_h \in Q^h$ *and* $\psi_h \in S^h$.

The second scheme we consider employs second order, three level BDF discretization for the subproblem terms. The coupling terms are treated by two step extrapolation in Navier-Stokes equation and by implicit method in Ohm's law. Since one needs the updated value of u_h at current time level to compute ϕ_h, this method is uncoupled but sequential: $\phi_h^n \to \mathbf{u}_h^{n+1} \to \phi_h^{n+1}$. Nevertheless, solving the subproblems sequentially does not take considerably longer time, since for RMHD system, computing time for the nonlinear equation of $\mathbf{u}_h$ would dominate that for the linear equation of ϕ_h.

ALGORITHM 2.2 (Second order IMEX scheme). *Given* $\mathbf{u}_h^{n-1}, \mathbf{u}_h^n \in X^h, p_h^{n-1}, p_h^n \in Q^h, \phi_h^{n-1}, \phi_h^n \in S^h$, *find* $\mathbf{u}_h^{n+1} \in X^h, p_h^{n+1} \in Q^h, \phi_h^{n+1} \in S^h$ *satisfying*

$$\begin{aligned}
&\frac{1}{N}\left(\left(\frac{3\mathbf{u}_h^{n+1} - 4\mathbf{u}_h^n + \mathbf{u}_h^{n-1}}{2\Delta t}, \mathbf{v}_h\right) + b(\mathbf{u}_h^{n+1}, \mathbf{u}_h^{n+1}, \mathbf{v}_h)\right) + \frac{1}{M^2}(\nabla \mathbf{u}_h^{n+1}, \nabla \mathbf{v}_h) \\
&\quad -(p_h^{n+1}, \nabla\cdot \mathbf{v}_h) + (\mathbf{u}_h^{n+1}\times \mathbf{B}, \mathbf{v}_h \times \mathbf{B}) - (\nabla(2\phi_h^n - \phi^{n-1}), \mathbf{v}_h\times\mathbf{B}) = (\mathbf{f}^{n+1}, \mathbf{v}_h), \\
&\text{(IMEX2)} \qquad (\nabla\cdot \mathbf{u}_h^{n+1}, q_h) = 0, \\
&\quad -(\nabla\phi_h^{n+1}, \nabla\psi_h) + (\mathbf{u}_h^{n+1} \times \mathbf{B}, \nabla\psi_h) = 0,
\end{aligned}$$

for all $\mathbf{v}_h \in X^h, q_h \in Q^h$ *and* $\psi_h \in S^h$.

3. Stability of the two partitioned methods

In this section, we establish stability of the approximations in Algorithms IMEX1 and IMEX2.

THEOREM 3.1 (Unconditional stability of Algorithm IMEX1). *Let* $(\mathbf{u}_h^n, p_h^n, \phi_h^n) \in X^h \times Q^h \times S^h$ *satisfy* (IMEX1) *for each* $n \in \{1, 2, ..., \frac{T}{\Delta t}\}$. *Then*

$$\begin{aligned}
&\frac{1}{N}\|\mathbf{u}_h^n\|^2 + \frac{1}{N}\sum_{j=0}^{n-1}\|\mathbf{u}_h^{j+1} - \mathbf{u}_h^j\|^2 + \Delta t\|\nabla\phi_h^n\|^2 + \Delta t\|\mathbf{B}\times\mathbf{u}_h^n\|^2 + \frac{\Delta t}{M^2}\sum_{j=0}^{n-1}\|\nabla \mathbf{u}_h^{j+1}\|^2 \\
&\quad + \Delta t\sum_{j=0}^{n-1}\left(\| - \nabla\phi_h^j + \mathbf{u}_h^{j+1}\times\mathbf{B}\|^2 + \| - \nabla\phi_h^{j+1} + \mathbf{u}_h^j \times \mathbf{B}\|^2\right) \\
&\qquad \le \frac{1}{N}\|\mathbf{u}_h^0\|^2 + \Delta t\|\nabla\phi_h^0\|^2 + \Delta t\|\mathbf{B}\times\mathbf{u}_h^0\|^2 + M^2\Delta t\sum_{j=0}^{n-1}\|\mathbf{f}^{j+1}\|_{-1}^2.
\end{aligned}$$

PROOF. In (IMEX1), setting $\mathbf{v}_h = \mathbf{u}_h^{j+1}, q_h = p_h^{j+1}, \psi_h = \phi_h^{j+1}$, we have

$$\begin{aligned}
&\frac{1}{2\Delta t}\cdot\frac{1}{N}\left(\|\mathbf{u}_h^{j+1}\|^2 - \|\mathbf{u}_h^j\|^2 + \|\mathbf{u}_h^{j+1} - \mathbf{u}_h^j\|^2\right) + \frac{1}{M^2}\|\nabla\mathbf{u}_h^{j+1}\|^2 + \|\mathbf{B}\times\mathbf{u}_h^{j+1}\|^2 \\
&\qquad = \left(\mathbf{u}_h^{j+1}\times\mathbf{B}, \nabla\phi_h^j\right) + (\mathbf{f}^{j+1}, \mathbf{u}_h^{j+1}), \qquad (3.1)\\
&\|\nabla\phi_h^{j+1}\|^2 = \left(\mathbf{u}_h^j\times\mathbf{B}, \nabla\phi_h^{j+1}\right).
\end{aligned}$$

Applying polarization identity $ab = \frac{1}{2}a^2 + \frac{1}{2}b^2 - \frac{1}{2}(a-b)^2$, it gives

$$\begin{aligned}
\left(\mathbf{u}_h^{j+1}\times\mathbf{B}, \nabla\phi_h^j\right) &= \frac{1}{2}\|\mathbf{u}_h^{j+1}\times\mathbf{B}\|^2 + \frac{1}{2}\|\nabla\phi_h^j\|^2 - \frac{1}{2}\|-\nabla\phi_h^j + \mathbf{u}_h^{j+1}\times\mathbf{B}\|^2, \\
\left(\mathbf{u}_h^j\times\mathbf{B}, \nabla\phi_h^{j+1}\right) &= \frac{1}{2}\|\mathbf{u}_h^j\times\mathbf{B}\|^2 + \frac{1}{2}\|\nabla\phi_h^{j+1}\|^2 - \frac{1}{2}\|-\nabla\phi_h^{j+1} + \mathbf{u}_h^j\times\mathbf{B}\|^2.
\end{aligned} \qquad (3.2)$$

Inserting (3.2) into (3.1), adding (3.1), then multiplying by $2\Delta t$ and summing from $j = 0$ to $n-1$ give

$$\begin{aligned}
&\frac{1}{N}\|\mathbf{u}_h^n\|^2 + \frac{1}{N}\sum_{j=0}^{n-1}\|\mathbf{u}_h^{j+1} - \mathbf{u}_h^j\|^2 + \Delta t\|\nabla\phi_h^n\|^2 + \Delta t\|\mathbf{B}\times\mathbf{u}_h^n\|^2 \\
&+ \frac{2\Delta t}{M^2}\sum_{j=0}^{n-1}\|\nabla\mathbf{u}_h^{j+1}\|^2 + \Delta t\sum_{j=0}^{n-1}\left(\|-\nabla\phi_h^j + \mathbf{u}_h^{j+1}\times\mathbf{B}\|^2 + \|-\nabla\phi_h^{j+1} + \mathbf{u}_h^j\times\mathbf{B}\|^2\right) \\
&\qquad = \frac{1}{N}\|\mathbf{u}_h^0\|^2 + \Delta t\|\nabla\phi_h^0\|^2 + \Delta t\|\mathbf{B}\times\mathbf{u}_h^0\|^2 + 2\Delta t\sum_{j=0}^{n-1}(\mathbf{f}^{j+1}, \mathbf{u}_h^{j+1}).
\end{aligned}$$

Applying Young's inequality yields the result. □

Algorithm IMEX2 is stable over $0 \le t < \infty$ with a condition on time step.

THEOREM 3.2 (Stability of Algorithm IMEX2). *Let* $(\mathbf{u}_h^n, p_h^n, \phi_h^n) \in X^h\times Q^h\times S^h$ *satisfy* (IMEX2) *for each* $n \in \{1, 2, ..., \frac{T}{\Delta t}\}$. *Under the time step restriction*

$$\Delta t < \frac{1}{2N\|\mathbf{B}\|_{L^\infty}^2(M^2C_P^2\|\mathbf{B}\|_{L^\infty}^2 + 1)}$$

Algorithm IMEX2 *is uniformly in time stable*

$$\begin{aligned}
&\frac{1}{2N}\|\mathbf{u}_h^n\|^2 + \frac{1}{2N}\|2\mathbf{u}_h^n - \mathbf{u}_h^{n-1}\|^2 + \frac{\Delta t}{2M^2}\sum_{j=1}^{n-1}\|\nabla\mathbf{u}_h^{j+1}\|^2 \\
&+ \Delta t\sum_{j=1}^{n-1}\|-\nabla(2\phi_h^j - \phi_h^{j-1}) + \mathbf{u}_h^{j+1}\times\mathbf{B}\|^2 \\
&\qquad + \Delta t\sum_{j=1}^{n-1}\|-\nabla(2\phi_h^j - \phi_h^{j-1}) + (2\mathbf{u}_h^j - \mathbf{u}_h^{j-1})\times\mathbf{B}\|^2 \\
&\le \frac{1}{2N}\|\mathbf{u}_h^1\|^2 + \frac{1}{2N}\|2\mathbf{u}_h^1 - \mathbf{u}_h^0\|^2 + 2\Delta t M^2\sum_{j=1}^{n-1}\|\mathbf{f}^{j+1}\|_{-1}^2.
\end{aligned}$$

For a detailed proof we refer to [**10**].

4. Numerical experiments

First, using the exact solutions introduced in [**21**], we verify the convergence rates of our methods. Second, we consider large values of M and N and test the stability established in Section 3.

4.1. Test 1. The domain is $\Omega = [0, \pi]^2$ and true solution $(\mathbf{u}, p, \phi)$ from [**21**] is given by

$$\begin{aligned}\mathbf{u}(x, y, t) &= (2\cos(2x)\sin(2y), -2\sin(2x)\cos(2y))e^{-5t},\\ p(x, y, t) &= 0,\\ \phi(x, y, t) &= (\cos(2x)\cos(2y) + x^2 - y^2)e^{-5t}.\end{aligned}$$

Take the time interval $0 \le t \le 1$ and set $M = 20, N = 16$. The imposed magnetic field is $\mathbf{B} = (0, 0, 1)$. We utilize piecewise quadratic for velocity and piecewise linear for pressure for the Navier-Stokes equation and continuous piecewise quadratic finite elements for the Ohm's law. The boundary condition on the problem is inhomogeneous Dirichlet: $\mathbf{u}_h = \mathbf{u}$ on $\partial\Omega$. The initial data and source terms are chosen to correspond the exact solution.

We denote $\|\cdot\|_\infty = \|\cdot\|_{L^\infty(0,T;L^2(\Omega))}$ and $\|\cdot\|_2 = \|\cdot\|_{L^2(0,T;L^2(\Omega))}$. From the tables, IMEX1 is first order and IMEX2 is second order.

TABLE 1. The convergence performance for Algorithm IMEX1.

h	Δt	$\\|\mathbf{u}-\mathbf{u}_h\\|_\infty$	$\\|\nabla\mathbf{u}-\nabla\mathbf{u}_h\\|_2$	$\\|\nabla\cdot(p-p_h)\\|_\infty$	$\\|\phi-\phi_h\\|_\infty$	$\\|\nabla\phi-\nabla\phi_h\\|_2$
1/5	1/40	1.047e+0	2.921e+0	1.956e+0	5.760e-1	9.764e-1
1/10	1/80	7.406e-1	2.062e+0	1.005e+0	3.913e-1	6.764e-1
1/20	1/160	4.338e-1	1.214e+0	5.094e-1	2.277e-1	3.952e-1
1/40	1/320	2.348e-1	6.522e-1	2.564e-1	1.237e-1	2.137e-1
1/80	1/640	1.223e-1	3.374e-1	1.286e-1	6.459e-2	1.113e-1
Rate of conv.		0.7853	0.7889	0.9825	0.7975	0.7928

TABLE 2. The convergence performance for Algorithm IMEX2.

h	Δt	$\\|\mathbf{u}-\mathbf{u}_h\\|_\infty$	$\\|\nabla\mathbf{u}-\nabla\mathbf{u}_h\\|_2$	$\\|\nabla\cdot(p-p_h)\\|_\infty$	$\\|\phi-\phi_h\\|_\infty$	$\\|\nabla\phi-\nabla\phi_h\\|_2$
1/5	1/40	3.217e-1	1.108e+0	2.837e-1	1.634e-1	3.206e-1
1/10	1/80	4.633e-2	3.670e-1	6.676e-2	2.374e-2	6.380e-2
1/20	1/160	8.973e-3	8.325e-2	1.614e-2	4.694e-3	1.519e-2
1/40	1/320	2.081e-3	1.533e-2	4.003e-3	1.096e-3	3.806e-3
1/80	1/640	5.118e-4	3.104e-3	1.001e-3	2.698e-4	9.577e-4
Rate of conv.		2.1747	2.1541	2.0353	2.2922	2.0841

4.2. Test 2. Many important applications of MHD in laboratory and industry involve large Hartmann number and interaction parameter, see, e.g., [**17, 4**]. The time step condition for stability of IMEX2 looks pessimistic in these cases. We compare our methods for such flows, confirming the unconditional stability of IMEX1. We observe that IMEX2 is stable for much larger time steps than predicted by Theorem 3.2.

Let $\Omega = [0, 10^{-1}]^2$ and $\mathbf{B} = (0, 0, 1)$. We consider the flow of liquid aluminium at 700°C:

$$\sigma = 4.1 \cdot 10^6 \text{ mho/m}, \qquad \rho = 2400 \text{ kg/m}^3,$$
$$\nu = 6 \cdot 10^{-7} \text{ m}^2/\text{s}, \qquad \eta = 1.94 \cdot 10^{-1} \text{ m}^2/\text{s}.$$

We take the characteristic values of length, velocity and magnetic field to be $L = 0.1\text{m}$, $u = 0.1\text{m/s}$, $B = 1\text{T}$, typical for laboratory and industrial flows. The Reynolds number, magnetic Reynolds number, Hartmann number and interaction parameter are then $Re = 16667$, $R_m = 0.051496$, $M = 5336$, $N = 1708$ correspondingly.

We take the source term $\mathbf{f}$ and the boundary condition to be 0 and the initial condition is given by

$$\mathbf{u}_0(x, y) = (10\pi \cos(10\pi x)\sin(10\pi y), -10\pi \sin(10\pi x)\cos(10\pi y)),$$
$$\phi_0(x, y) = (\cos(10\pi x)\cos(10\pi y) + x^2 - y^2).$$

For a system lacking of external energy exchange and body forces, the system energy decays over time. The energy $E^j = \|\mathbf{u}_h^j\|^2 + \|\phi_h^j\|^2$ is computed using two different methods studied herein, on $h = 1/10$. For each algorithm, the time step is chosen purposely to give us an estimate of practical restriction on time step for the stability of the method. The results are showed in Figure 1.

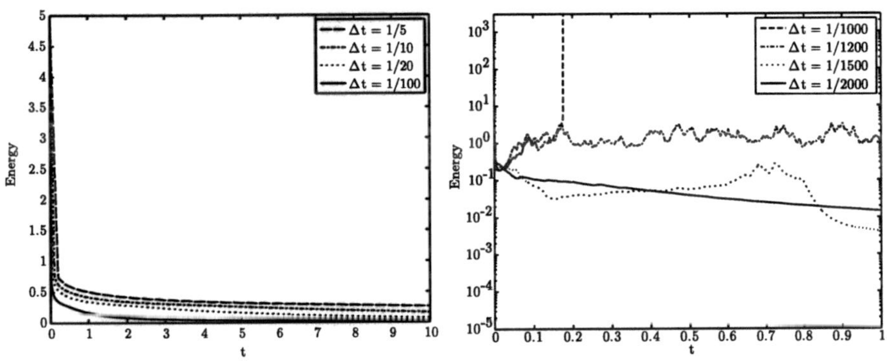

FIGURE 1. The decay of system energy computed by IMEX1 (left) and IMEX2 (right) with several different time steps chosen.

Figure 1 confirms the unconditional stability of IMEX1 established in Theorem 3.1. It also indicates that the experimental stability condition for IMEX2 is $\Delta t \lesssim 1/1500$, which, while restrictive, is significantly better than the time step restriction in Theorem 3.2.

References

1. H. Alfvén, *Existence of electromagnetic-hydrodynamic waves*, Nature **150** (1942), 405.
2. L. Barleon, V. Casal, and L. Lenhart, *MHD flow in liquid-metal-cooled blankets*, Fusion Engineering and Design (1991), no. 14, 401–412.
3. P. A. Davidson, *Magnetohydrodynamics in material processing*, Annu. Rev. Fluid Mech. **31** (1999), 273–300.
4. ______, *An introduction to magnetohydrodynamics*, Cambridge Texts in Applied Mathematics, Cambridge University Press, Cambridge, 2001. MR1825486 (2002c:76103)

5. M. Discacciati, *Numerical approximation of a steady MHD problem*, Domain decomposition methods in science and engineering XVII, no. 60, Springer, Berlin, 2008, Lect. Notes Comput. Sci. Eng., pp. 313–320. MR2436096
6. J. L. Guermond and P. D. Minev, *Mixed finite element approximation of an MHD problem involving conducting and insulating regions: The 3D case*, Numer. Methods Partial Differential Equations **19** (2003), no. 6, 709–731. MR2009590 (2004h:65099)
7. M.D. Gunzburger, A.J. Meir, and J. Peterson, *On the existence, uniqueness, and finite element approximation of solutions of the equations of stationary, incompressible magnetohydrodynamics*, Math. Comp. **56** (1991), no. 194, 523–563. MR1066834 (91m:76127)
8. H. Hashizume, *Numerical and experimental research to solve MHD problem in liquid-blanket system*, Fusion Engineering and Design **81** (2006), no. 8–14, 1431–1438.
9. L. D. Landau and E. M. Lifshitz, *Electrodynamics of continuous media*, Course of Theoretical Physics, Vol. 8. Translated from the Russian by J. B. Sykes and J. S. Bell, Pergamon Press, Oxford, 1960. MR0121049 (22:11796)
10. W. Layton, H. Tran, and C. Trenchea, *Numerical analysis of two partitioned time stepping methods for MHD flows at low magnetic Reynolds numbers*, (2012), in preparation.
11. W. J. Layton, A. J. Meir, and P. G. Schmidt, *A two-level discretization method for the stationary MHD equations*, Electron. Trans. Numer. Anal. **6** (1997), 198–210 (electronic). MR1615165 (99c:76067)
12. T. F. Lin, J. B. Gilbert, and R. Kossowsky, *Sea water magnetohydrodynamic propulsion for next-generation undersea vehicles*, ONR annual report for period from Feb. 1, 1989 to Jan. 31, 1990, under contract no. N00014-89-J-1693, NTIS AD Number: A218318 (1990).
13. A. J. Meir, *The equations of stationary, incompressible magnetohydrodynamics with mixed boundary conditions*, Comput. Math. Appl. **25** (1993), no. 12, 13–29. MR1216012 (94d:76107)
14. A. J. Meir and P. G. Schmidt, *Variational methods for stationary MHD flow under natural interface conditions*, Nonlinear Anal. **26** (1996), no. 4, 659–689. MR1362743 (96m:76073)
15. ———, *Analysis and numerical approximation of a stationary MHD flow problem with nonideal boundary*, SIAM J. Numer. Anal. **36** (1999), no. 4, 1304–1332. MR1701784 (2000f:76140)
16. J. Peterson, *On the finite element approximation of incompressible flows of an electrically conducting fluid*, Numer. Methods Partial Differential Equations **4** (1988), no. 1, 57–68. MR1012474 (90i:65231)
17. P. H. Roberts, *An introduction to magnetohydrodynamics*, Elsevier, USA, 1967.
18. P. G. Schmidt, *A Galerkin method for time-dependent MHD flow with nonideal boundaries*, Commun. Appl. Anal. **3** (1999), no. 3, 383–398. MR1696344 (2000b:65191)
19. S. Smolentsev, R. Moreau, L. Bühler, and C. Mistrangelo, *MHD thermofluid issues of liquid-metal blankets: Phenomena and advances*, Fusion Engineering and Design **85** (2010), no. 7–9, 1196–1205.
20. N. Winowich and W. Hughes, *A finite element analysis of two dimensional MHD flow*, Liquid-Metal Flows and Magnetohydrodynamics, AIAA, New York, 1983.
21. G. Yuksel and R. Ingram, *Numerical analysis of a finite element, Crank-Nicolson discretization for MHD flow at small magnetic Reynolds number*, Tech. report, University of Pittsburgh, 2011.

Department of Mathematics, University of Pittsburgh, Pittsburgh, Pennsylvania 15260

E-mail address: `wjl@pitt.edu`

Department of Mathematics, University of Pittsburgh, Pittsburgh, Pennsylvania 15260

E-mail address: `hat25@pitt.edu`

Department of Mathematics, University of Pittsburgh, Pittsburgh, Pennsylvania 15260

E-mail address: `trenchea@pitt.edu`

LNG_FEM: Graded Meshes on Domains of Polygonal Structures

Hengguang Li and Victor Nistor

ABSTRACT. We develop LNG_FEM, a software package for graded mesh generation and for solving elliptic equations. LNG_FEM generates user-specified graded meshes on arbitrary 2D domains with straight edges for different boundary conditions. We shall focus on a detailed exposition on the implementation of the software. In addition, we demonstrate that LNG_FEM is equipped with advanced algorithms and data structures to perform efficiently in numerical tests. We hope that LNG_FEM can broaden the use and understanding of graded mesh in the finite element approximation of singular solutions.

We develop a software package entitled LNG_FEM, which comes from "Linear Graded Finite Element Method". LNG_FEM is a free software package, written in C, for the generation of graded meshes in general 2D domains with polygonal structures [**6, 7, 11**], and for the construction of linear finite element solutions of elliptic boundary value problems. As a by-product of our research on numerical approximations of singular functions, LNG_FEM is a fast, memory-efficient, user-friendly package that can handle different boundary conditions. LNG_FEM is designed for studying and demonstrating grading algorithms, as well as educating students on the finite element method.

This expository article shall mainly present instructions on the use of LNG_FEM (Section 2). In Section 3, we also briefly describe several features on the algorithms and data structures, which make LNG_FEM a reliable and efficient software package feasible for various problems. We keep the package up-to-date. Suggestions to improve the software in any aspect are welcome.

1. Graded meshes

It is well known that elliptic boundary value problems may have singular solutions, even when the given data is smooth. Some typical situations that lead to singular solutions include: non-smooth domains, changes in boundary conditions, interfaces (especially the non-smooth ones) in transmission problems, and differential operators with non-smooth coefficients. Various numerical schemes have been developed to improve the convergence rate of the finite element approximations of

---

2010 *Mathematics Subject Classification.* Primary 65Y04; Secondary 65N30.

H. Li was partially supported by the NSF Grant DMS-1158839. V. Nistor was partially supported by the NSF Grants OCI-0749202 and DMS-1016556. LNG_FEM is freely available at http://hengguangli.blogspot.com/2009/04/lngfem.html.

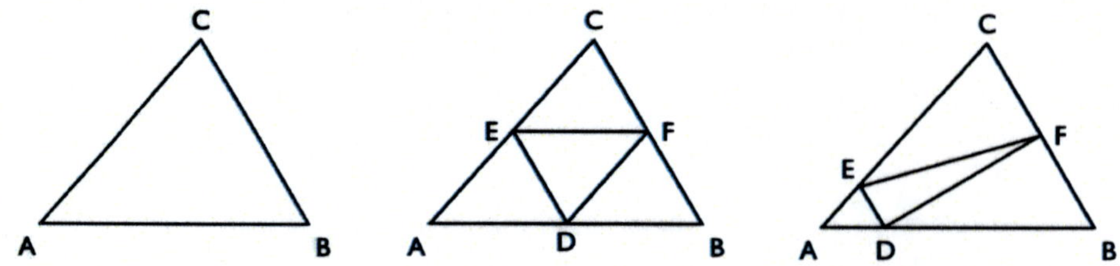

FIGURE 1. The initial triangle ABC (left); the uniform refinement, $\kappa = 0.5$ (center); the κ-refinement with $\kappa < 0.5$ for vertex A (right), $\kappa = \frac{|AD|}{|AB|} = \frac{|AE|}{|AC|} = \frac{|DE|}{|BC|}$.

these singular solutions. Among these numerical schemes, based on *a priori* estimates of the equation in special function spaces [**5, 9, 8, 13**], mesh grading has proved to be a powerful technique [**1, 4, 3, 2, 10, 11, 12, 14**].

Consider the following model problem in a polygonal domain Ω with the mixed boundary condition,

$$(1.1) \qquad -\Delta u = f \quad \text{in } \Omega, \qquad \partial_\nu u = 0 \quad \text{on } \partial\Omega_N, \qquad u = 0 \quad \text{on } \partial\Omega_D.$$

Assuming a smooth f, let $S \subset \bar{\Omega}$ be the set of singular points near which the solution u is not locally in H^2. Then, there is a systematic construction of graded meshes [**3, 11, 14**] to deal with the lack of regularity of the solution at those points.

DEFINITION 1.1. Let $\mathcal{T}$ be a triangulation of Ω. We require that every point in S be a vertex in $\mathcal{T}$ and no two singular points belong to the same triangle. For each point in S, define a grading parameter $\kappa \in (0, 1/2]$. Then *the κ-refinement of $\mathcal{T}$*, denoted by $\kappa(\mathcal{T})$ is obtained by dividing each edge AB of $\mathcal{T}$ in two parts as follows. If neither A nor B is a singular point, then we divide AB into two equal parts. Otherwise, if $A \in S$, we divide AB into AC and CB such that $|AC| = \kappa|AB|$. This procedure will divide each triangle of $\mathcal{T}$ into four triangles.. Let $\mathcal{T}_0$ be an initial triangulation with the above properties. Then, we define by induction $\mathcal{T}_{n+1} = \kappa(\mathcal{T}_n)$.

For each point in the singular set S, there is an optimal range for the grading parameter κ [**3, 11**], determined by the regularity estimate of the solution u in weighted Sobolev spaces. Once κ is chosen within that range, the graded mesh $\mathcal{T}_n$ yields finite element approximations of equation (1.1) with optimal convergence rates. (This result applies to general uniformly strongly elliptic equations mentioned above with various singular solutions.)

2. Instructions

One of the purposes of the software package LNG_FEM is to encourage a broader understanding and use of graded meshes for the finite element approximation of singular solutions of elliptic PDEs. Therefore, in addition to ensuring the reliability and efficiency of the algorithms, we also try to maintain a friendly user interface. The main features of LNG_FEM include:

*1. **Modularization.*** Related algorithms and data structures are grouped in different modules for easy updates and modification; input files and outputs of the program are bundled in two directories (/Sourcefiles and /Results, respectively), to simplify the initialization of the program and the analysis of the results afterwards.

*2. **Generality.*** The user is allowed to set up various parameters such as the

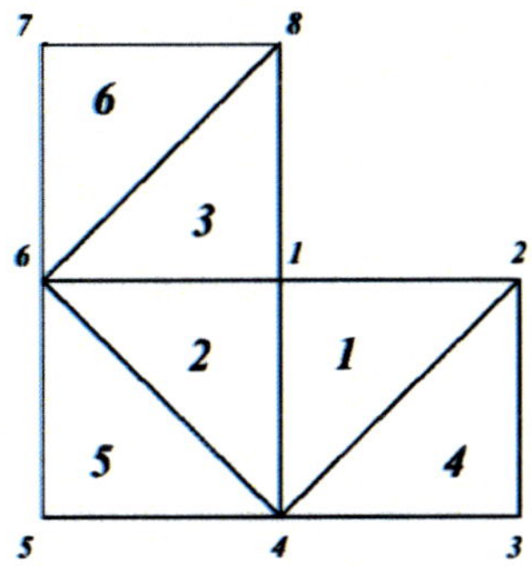

FIGURE 2. An initial triangulation of the L-shape domain with the Dirichlet boundary condition: the numbering of nodes and triangles.

computational domain, boundary conditions, and the grading parameter for each specific singular point, by customizing the input files. The current model problem in the package is equation (1.1) with $f = 1$. It is possible to work on more general equations by updating the corresponding module in the package.

3. ***Efficiency.*** LNG_FEM is equipped with advanced algorithms that perform efficiently in terms of storage and speed. We thus managed to minimize the time for mesh generation and matrix assembling. Details will be discussed in Section 3.

4. ***Analysis of the result.*** LNG_FEM can either compute the numerical solution on the current mesh or compare the current numerical solution with the solution from the previous mesh to provide the convergence rate. The mesh and the solution can be visualized easily in MATLAB with embedded commands in the package.

We now provide a detailed instruction on the implementation of the package.

2.1. Source files. After unzipping the downloaded file, we shall have created the directory /LNG_FEM, including two sub-directories (/Sourcefiles and /Results) and several other files. Note that the the executable file is .out. In particular, /Sourcefiles contains information on the initial triangulation, boundary conditions, and grading parameters. We elaborate on the settings in /Sourcefiles by taking the mesh in Figure 2 as an example.

Below are the rules for the initial triangulation.

1. Any vertex or singular point of the domain is an initial node.
2. An initial triangle cannot contain more than one singular point of the domain.
3. Any node cannot have more than six adjacent triangles.
4. Suppose that there are n initial triangles. The numbering of the triangles can be any one of the $n!$ *permutations of the set* $\{i,\ 1 \le i \le n\}$. *Figure 2 shows only one possible numbering. Suppose there are l singular points and m non-singular nodes in the initial nodes. Then, the numbering of the singular nodes can be any one of the* $l!$ *permutations of the set* $\{i,\ 1 \le i \le l\}$; *the numbering of the non-singular nodes can be any one of the* $m!$ *permutations of the set* $\{i,\ l+1 \le i \le l+m\}$. *For example, "1" has to be assigned to the node on the re-entrant corner in Figure 2, because it is the only singular point. The numbering of the other seven nodes can be any one of the* 7! *permutations of the set* $\{i,\ 2 \le i \le 8\}$.

More precisely, as in Figure 3, for **LNG_Initialnode.txt**, the integer in the first row indicates the number of nodes (eight nodes in Figure 2). Starting from the second row, the ith row, $i \ge 2$, lists the coordinates (x, y) of the $(i-1)$st initial node. (The second node is $(1, 0)$, for example.) **LNG_Initialtriangle.txt** contains

```
8
0 0
1 0
1 -1
0 -1
-1 -1
-1 0
-1 1
0 1
```

```
8 8
1 2 2 3 3 4 4 5 5 6 6 7 7 8 8 1
```

```
6
1 2 4
1 4 6
1 6 8
2 3 4
4 5 6
6 7 8
```

```
1
0.2
```

FIGURE 3. Files (in the directory /Sourcefiles) for the triangulation in Figure 2 with the grading parameter $\kappa = 0.2$: LNG_Initialnode.txt (left), LNG_Dedge.txt (top), LNG_Initialtriangle.txt (right), LNG_Ratiocontrol.txt (bottom).

information on initial triangles. The first number in the file stands for the number of initial triangles, while the three-tuple in the ith row, $i \geq 2$, identifies the vertices of the $(i-1)$st triangle with the numbering of vertices in the *ascending* order. For example, since the third triangle in Figure 2 has vertices $1, 6$, and 8, the 4th row of LNG_Initialtriangle.txt is 1 6 8 in Figure 3. With these two source files, it suffices to pass all geometric information of the domain to the program.

LNG_Dedge.txt is to specify boundary conditions, namely, the Dirichlet and Neumann edges, respectively. The first integer in the first row represents the number of non-duplicate endpoints (each point counted only once) of Dirichlet edges; the second integer is the number of Dirichlet edges. The integers in the second row are the numberings of the endpoints of Dirichlet edges. For example, LNG_Dedge.txt in Figure 3 implies that there are eight Dirichlet nodes and eight Dirichlet edges. The Dirichlet edges are $1-2$, $2-3$, $\cdots$, and $8-1$, imposing the pure Dirichlet condition on the L-shape domain. Mixed boundary conditions can be imposed similarly.

LNG_Ratiocontrol.txt controls the grading parameter $\kappa \in (0, 1/2]$ (Figure 1) for specified singular nodes. The first integer of the file is the number of singular nodes, while the decimal in the ith row, $i \geq 2$, is the grading parameter for the $(i-1)$st singular node. As in LNG_Ratiocontrol.txt (Figure 3), there is one singular node on the L-shape domain (the first node) and the grading parameter is $\kappa = 0.2$.

Source files for some typical domains and boundary conditions (cracks, mixed boundary conditions, multiple singular nodes, etc.) can be found in LNG_Demo, which is downloadable on our web page. To implement these files, one can simply replace the original files in /Sourcefiles by the source files given in LNG_Demo.

2.2. Outputs. With all the source files ready, one can open a command terminal and enter the directory /LNG_FEM. To implement, type ./LNG_FEM.out in the terminal and follow the on-screen instructions (Figure 4).

Note that if we choose to compare consecutive solutions, it calculates the H^1-error between the current numerical solution and the solution from last implementation. Therefore, to compare solutions on the third-level mesh and on the 4th-level mesh, we need to compute the solution on the third level first and make another run on the 4th level for the comparison.

All the outputs are automatically placed into the directory /Results, with a MATLAB-recognizable format for the visualization of solutions. We prepared two .m files LNG_MESH.m and LNG_PLOT.m in the package, for graphing graded

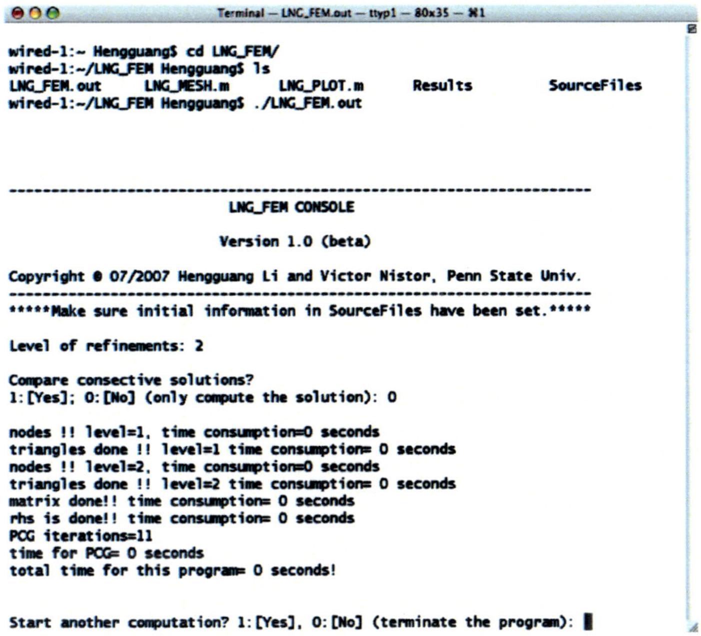

FIGURE 4. The interface of LNG_FEM.

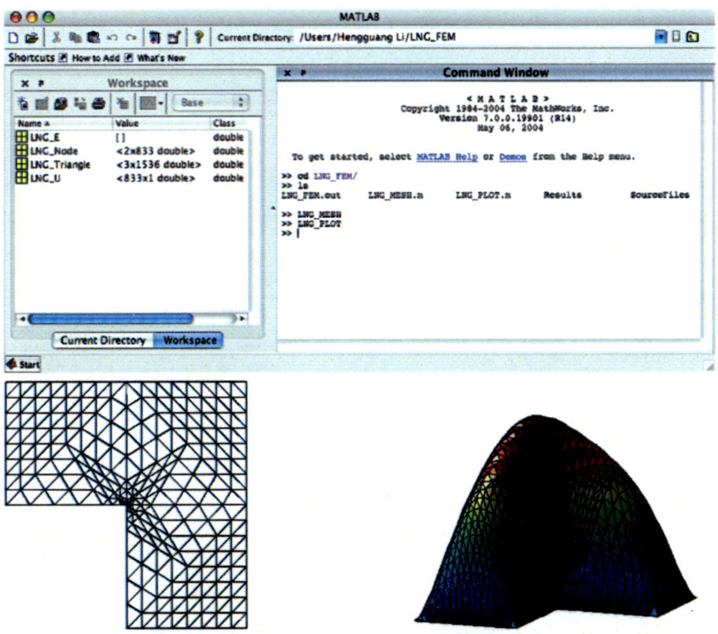

FIGURE 5. MATLAB graphs the 4th-level graded mesh (left) and the corresponding solution (right) on the L-shape domain from the initial triangulation in Figure 2.

meshes and solutions, respectively. After launching MATLAB, set the current directory to be /LNG_FEM. Figure 5 shows a screen shot of MATLAB and the

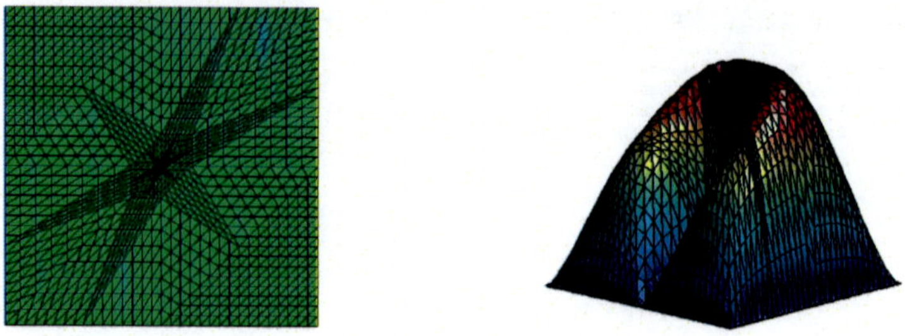

FIGURE 6. The graded mesh toward the tip of crack at the center of the domain after four refinements (left), $\kappa = 0.2$; the corresponding numerical solution of equation (1.1) (right).

resulting pictures of these commands. As a reminder, if we choose to compare solutions, LNG_FEM does not export the current mesh or the current solution.

The files in /Results are: LNG_Node.txt including the coordinates of nodes; LNG_Triangle.txt containing the vertex numbering of triangles; LNG_U.txt being the finite element solution; LNG_Rtp.txt specifying the triangles a node belongs to. LNG_E.txt is for graphing and LNG_Pre.txt is the number of nodes in the mesh.

2.3. The processing flow and examples. We provide a concise diagram below for the working procedure of LNG_FEM. In addition, besides for the L-shape domain, we include other examples (Figures 6 – 8) from the package for illustrations.

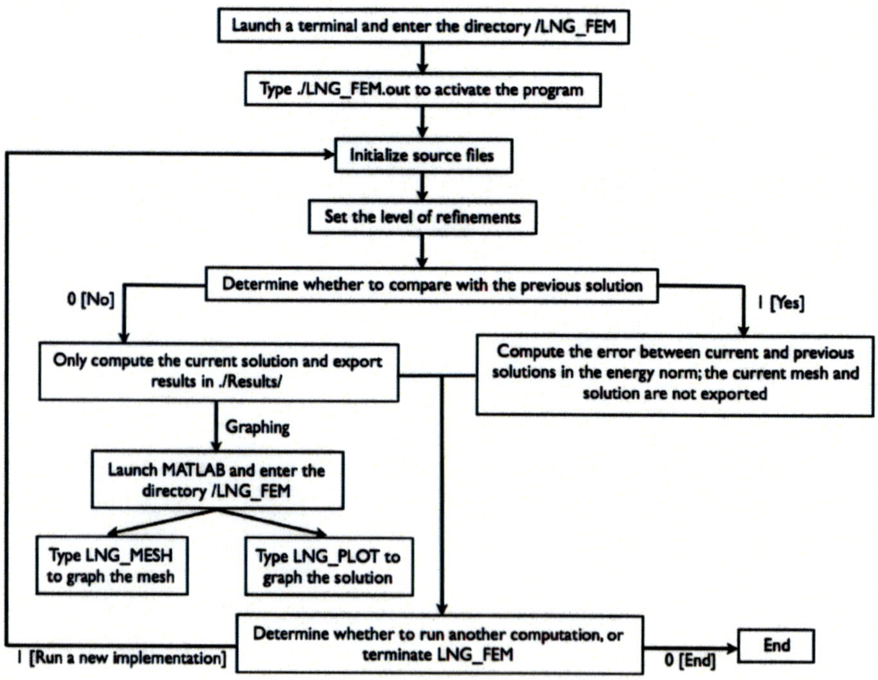

3. Algorithms

We used a compact format [**15**] to store sparse matrices and vectors. Pointers were used to assign and release vectors dynamically to minimize the use of the memory. In particular, the memory needed for LNG_FEM to generate meshes,

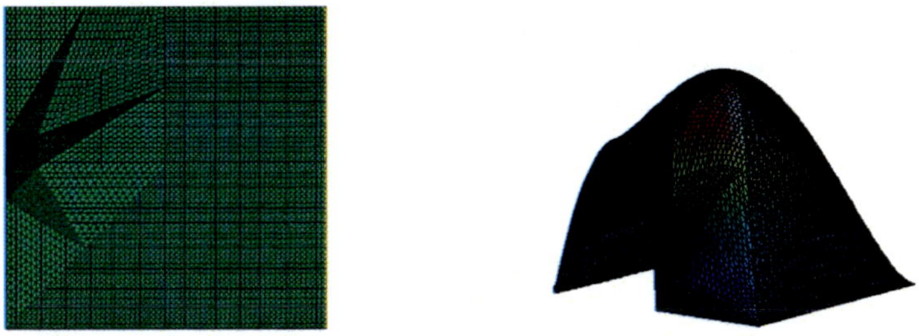

FIGURE 7. The graded mesh toward the point where the boundary condition changes, $\kappa = 0.2$, level=5 (left); the corresponding numerical solution of equation (1.1) (right).

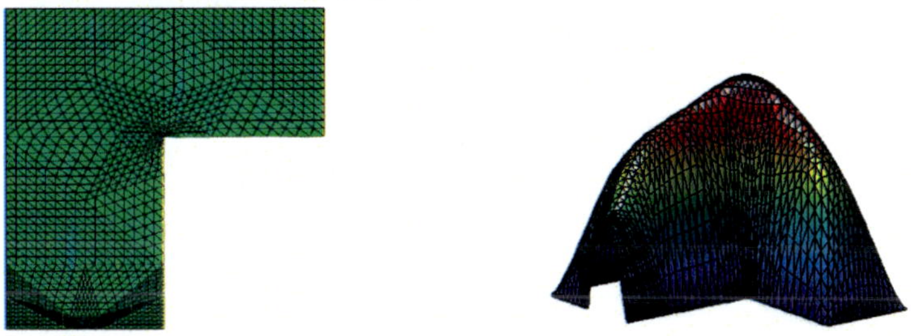

FIGURE 8. The graded mesh after four refinements for distinct singular points: re-entrant corner, $\kappa = 0.3$, and the point where the boundary condition changes, $\kappa = 0.2$ (left); the corresponding numerical solution for equation (1.1) (right).

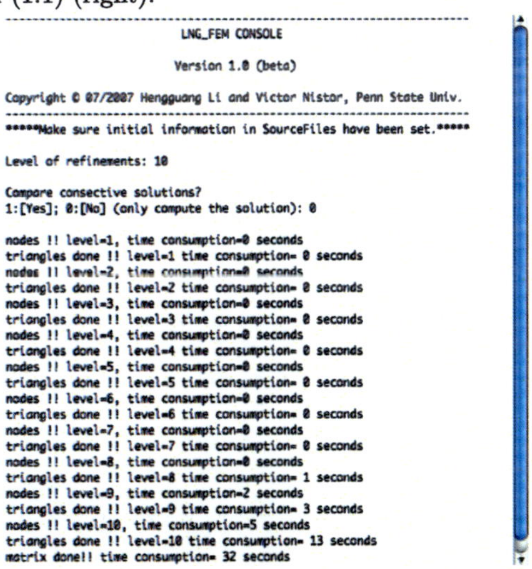

```
---------------------------------------------------------------------------
                              LNG_FEM CONSOLE

                            Version 1.0 (beta)

Copyright © 07/2007 Hengguang Li and Victor Nistor, Penn State Univ.
---------------------------------------------------------------------------
*****Make sure initial information in SourceFiles have been set.*****

Level of refinements: 10

Compare consective solutions?
1:[Yes]; 0:[No] (only compute the solution): 0

nodes !! level=1, time consumption=0 seconds
triangles done !! level=1 time consumption= 0 seconds
nodes !! level=2, time consumption=0 seconds
triangles done !! level=2 time consumption= 0 seconds
nodes !! level=3, time consumption=0 seconds
triangles done !! level=3 time consumption= 0 seconds
nodes !! level=4, time consumption=0 seconds
triangles done !! level=4 time consumption= 0 seconds
nodes !! level=5, time consumption=0 seconds
triangles done !! level=5 time consumption= 0 seconds
nodes !! level=6, time consumption=0 seconds
triangles done !! level=6 time consumption= 0 seconds
nodes !! level=7, time consumption=0 seconds
triangles done !! level=7 time consumption= 0 seconds
nodes !! level=8, time consumption=0 seconds
triangles done !! level=8 time consumption= 1 seconds
nodes !! level=9, time consumption=2 seconds
triangles done !! level=9 time consumption= 3 seconds
nodes !! level=10, time consumption=5 seconds
triangles done !! level=10 time consumption= 13 seconds
matrix done!! time consumption= 32 seconds
```

FIGURE 9. The processing time for LNG_FEM.

assemble matrices, solve the system of equations, and to compare solutions, is linearly dependent of the problem size. For example, starting with eight initial triangles, as for the domain with a crack (Figure 6), LNG_FEM needs 1.7GB of memory to refine the mesh 10 times, which generates $2^{23} \approx 8.4 \times 10^6$ triangles; and it needs 430MB of memory for the 9th refinement, with $2^{21} \approx 2.1 \times 10^6$ triangles. Therefore, we can easily go up to the 10th level on regular desktops and more on relatively powerful machines.

The algorithms for the mesh generation and matrix assembling were carefully designed, such that the computational cost in the final triangulation almost linearly depends on the number of triangles. We timed the program for 10 consecutive refinements on the original domain with eight triangles for the crack problem (Linux Redhat 9.0 with two 2.8GHz Intel Xeon processors and 2GB of memory, Figure 9). It takes unnoticeable little time for LNG_FEM to generate $2^{17} \approx 1.3 \times 10^5$ triangles, 6 seconds to generate 2.1 million triangles, and 24 seconds to generate 8.4 million triangles. Assembling the matrix takes a little longer, namely 32 seconds.

In fact, the most time consuming part is solving the system of equations. With the built-in PCG solver, it takes about 20 minutes for the 9th level. In fact, it is the only module that is not optimized in LNG_FEM. We are working on a multigrid solver which will definitely lead to a speed boost for the program.

References

1. T. Apel, A.-M. Sändig, and J. Whiteman, *Graded mesh refinement and error estimates for finite element solutions of elliptic boundary value problems in non-smooth domains*, Math. Methods Appl. Sci. **19** (1996), no. 1, 63–85.
2. I. Babuška, R.B. Kellogg, and J. Pitkäranta, *Direct and inverse error estimates for finite elements with mesh refinements*, Numer. Math. **33** (1979), no. 4, 447–471.
3. C. Băcuţă, V. Nistor, and L. Zikatanov, *Improving the rate of convergence of 'high order finite elements' on polygons and domains with cusps*, Numer. Math. **100** (2005), no. 2, 165–184.
4. S.C. Brenner, J. Cui, T. Gudi, and L.-Y. Sung, *Multigrid algorithms for symmetric discontinuous Galerkin methods on graded meshes*, Numer. Math. **119** (2011), no. 1, 21–47.
5. M. Costabel, M. Dauge, and C. Schwab, *Exponential convergence of hp-FEM for Maxwell equations with weighted regularization in polygonal domains*, Math. Models Methods Appl. Sci. **15** (2005), no. 4, 575–622.
6. M. Dauge, *Elliptic boundary value problems on corner domains*, Lecture Notes in Mathematics, vol. 1341, Springer-Verlag, Berlin, 1988.
7. P. Grisvard, *Singularities in boundary value problems*, Research Notes in Applied Mathematics, vol. 22, Springer-Verlag, New York, 1992.
8. V.A. Kondrat′ev, *Boundary value problems for elliptic equations in domains with conical or angular points*, Transl. Moscow Math. Soc. **16** (1967), 227–313.
9. V.A. Kozlov, V. Mazya, and J. Rossmann, *Spectral problems associated with corner singularities of solutions of elliptic equations*, Mathematical Surveys and Monographs, vol. 5, AMS, Rhode Island, 2001.
10. H. Li, *A-priori analysis and the finite element method for a class of degenerate elliptic equations*, Math. Comp. **78** (2009), 713–737.
11. H. Li, A. Mazzucato, and V. Nistor, *Analysis of the finite element method for transmission/mixed boundary value problems on general polygonal domains*, Electron. Trans. Numer. Anal. **37** (2010), 41–69.
12. H. Li and V. Nistor, *Analysis of a modified Schrödinger operator in 2D: regularity, index, and FEM*, J. Comput. Appl. Math. **224** (2009), no. 1, 320–338.
13. S. Nicaise, *Polygonal interface problems*, Lang, Peter Publishing, Incorporated, 1993.
14. G. Raugel, *Résolution numérique par une méthode d'éléments finis du problème de Dirichlet pour le laplacien dans un polygone*, C. R. Acad. Sci. Paris Sér. A-B **286** (1978), no. 18, A791–A794.
15. L. Zikatanov, *A note on FE assembly*, 2006.

Hengguang Li, Department of Mathematics, Wayne State University, Detroit, MI 48202, USA. E-mail address: hli@math.wayne.edu

Victor Nistor, Department of Mathematics, The Pennsylvania State University, University Park, PA 16802, USA. E-mail address: nistor@math.psu.edu

Homogenized discontinuous Galerkin method for Maxwell's equations in periodic structured dispersive media

Jichun Li and Jiajia Waters

ABSTRACT. For the first time, the discontinuous Galerkin method is coupled with the homogenization technique to solve the time-dependent Maxwell's equations in dispersive media with periodic microstructures. To illustrate the effectiveness of our method, we present three 2-D numerical examples: one solves a time-harmonic Maxwell's equations with a rapidly oscillating coefficient; one uses our DG method to form the homogenized coefficient matrix; the last one solves the time-dependent Maxwell's equations in dispersive media with highly oscillatory coefficient.

1. Introduction

The popularity of the discontinuous Galerin (DG) method is attributable to its great flexibility in mesh construction, its easily handling of complex geometries, as well as its efficiency in parallel implementation. Because of these nice features, the DG method has been used to solve various partial differential equations [**7, 11**], including Maxwell's equations in free space (e.g.,[**8, 9, 10, 12**]). However, application of DG method to Maxwell's equations in dispersive media is still quite limited [**13, 14, 18, 20, 21**]. A dispersive medium is characterized by a frequency-dependent permittivity and/or permeability. Note that dispersive media are ubiquitous and examples include biological tissue, ionosphere, water, soil, snow, ice, plasma, optical fibers and radar absorbing materials. Hence the study of wave propagation in dispersive media plays an important role in many applications. Furthermore, the dispersive media in real applications are often formed by composite materials with periodic microstructures. When the period of the microstructure is small compared to the wavelength, the coefficients in Maxwell's equations become highly oscillatory. Numerically solving Maxwell's equations with rapidly changing coefficients is quite challenging and even impossible due to the fine mesh requirement for resolving the fine scale of those physical microstructures. In this case, a powerful tool is to use homogenization technique [**3**] to greatly reduce the computation time and memory requirements of numerical simulations.

There are many excellent works on homogenization of Maxwell's equations (cf. [**3, 2, 4**] and references therein). The early work of Bensoussan, Lions, and

2000 *Mathematics Subject Classification.* Primary 65F10, 35Q60, 78M40.

Key words and phrases. Maxwell's equations, discontinuous Galerkin method, homogenization, multiscale asymptotic expansion, composite materials.

Papanicolaou [3] established convergence results for Maxwell type equations with rapidly oscillating coefficients. In 2010, a multiscale finite element method using edge elements is developed in [6] to solve the time-harmonic Maxwell's equations in composite materials. In 2006, Banks et al. [1] used the periodic unfolding method to homogenize the parameters in a composite dispersive medium with periodic microstructures. Inspired by the works [6, 1] and our previous work on DG method [14, 16], in this paper we develop a homogenized DG method for solving the time-dependent Maxwell's equations in dispersive media with periodic microstructures.

The rest of the paper is organized as follows. In Sect.2, we first present the governing equations, then describe the homogenization technique for time-harmonic Maxwell's equations, and finally develop a DG scheme for the corrector equation on a unit cell and a fully discrete DG scheme for the Maxwell's equations in periodic structured dispersive media. Three numerical examples are presented in Sect.3 to demonstrate the effectiveness of our scheme. The paper is concluded in Sect.4.

2. The homogenized DG method

2.1. The governing equations. The governing equations for modeling electromagnetic wave propagation in isotropic non-magnetized cold electron plasma are [14]:

$$\boldsymbol{E}_{tt} + \nabla \times (c_v^2 \nabla \times \boldsymbol{E}) + \omega_p^2 \boldsymbol{E} - \boldsymbol{J}(\boldsymbol{E}) = \mathbf{f} \quad \text{in } \Omega \times (0,T), \tag{2.1}$$

where $\boldsymbol{E}$ is the electric field, c_v is the speed of light, ω_p is the plasma frequency, $\mathbf{f}$ is a fixed source, and $\boldsymbol{J}$ is the polarization current density represented as

$$\boldsymbol{J}(\boldsymbol{E}) \equiv \boldsymbol{J}(\boldsymbol{x}, t; \boldsymbol{E}) = \nu \omega_p^2 \int_0^t e^{-\nu(t-s)} \boldsymbol{E}(\boldsymbol{x}, s) ds. \tag{2.2}$$

Here $\nu \geq 0$ is the electron-neutron collision frequency. To describe the wave propagation in a medium with periodic microstructures, we assume that $c_v = c_v(\frac{\boldsymbol{x}}{\epsilon})$ is a periodically varying function in space, where ϵ denotes the relative size of the microstructure. More specifically, we are interested in the vector equations:

$$\boldsymbol{E}_{tt} + \nabla \times (c_v^2(\frac{\boldsymbol{x}}{\epsilon}) \nabla \times \boldsymbol{E}) + \omega_p^2 \boldsymbol{E} - \boldsymbol{J}(\boldsymbol{E}) = \mathbf{f} \quad \text{in } \Omega \times (0,T), \tag{2.3}$$

supplemented by perfect conducting boundary condition

$$\boldsymbol{n} \times \boldsymbol{E} = \boldsymbol{0} \quad \text{on } \partial\Omega \times (0,T), \tag{2.4}$$

and the initial conditions

$$\boldsymbol{E}(\boldsymbol{x}, 0) = \boldsymbol{E}_0(\boldsymbol{x}), \quad \boldsymbol{E}_t(\boldsymbol{x}, 0) = \boldsymbol{E}_1(\boldsymbol{x}), \tag{2.5}$$

where $\boldsymbol{n}$ denotes the unit outward normal to $\partial\Omega$, and $\boldsymbol{E}_0(\boldsymbol{x})$ and $\boldsymbol{E}_1(\boldsymbol{x})$ are some given functions.

2.2. The homogenization technique. Before we homogenize the time-dependent problem (2.3), let us consider a simpler time-independent problem:

$$\begin{aligned} &\nabla \times (A^\epsilon(\boldsymbol{x}) \nabla \times \boldsymbol{u}_\epsilon(\boldsymbol{x})) - \omega^2 \boldsymbol{u}_\epsilon(\boldsymbol{x}) = f(\boldsymbol{x}), \quad \text{in } \Omega, \\ &\nabla \cdot \boldsymbol{u}_\epsilon = 0, \quad \text{in } \Omega, \\ &\boldsymbol{n} \times \boldsymbol{u}_\epsilon = 0, \quad \text{on } \partial\Omega, \end{aligned} \tag{2.6}$$

where $\Omega \in R^d$ (d=2 or 3) is a bounded Lipschitz polygonal convex domain or a smooth domain with a microstructure, ω denotes the angular frequency, and the matrix $A^\epsilon(\boldsymbol{x}) = (a_{ij}^\epsilon(\boldsymbol{x})) = (a_{ij}(\frac{\boldsymbol{x}}{\epsilon}))$ is symmetric positive definite. Furthermore, we assume that $a_{ij}(\frac{x}{\epsilon})$ are 1-periodic functions. Let $\xi = \frac{x}{\epsilon}$, and the reference unit cell $Q = \{\xi = (\xi_1, \xi_2, \xi_3) : \ 0 < \xi_i < 1, \ i = 1, 2, 3\}$.

Following [**6**], the first order multiscale asymptotic expansion for the solution of problem (2.6) can be constructed as:

$$\boldsymbol{u}_\epsilon(\boldsymbol{x}) = \boldsymbol{u}^0(\boldsymbol{x}) + \epsilon\boldsymbol{\theta}(\xi)\nabla \times \boldsymbol{u}^0(\boldsymbol{x}), \tag{2.7}$$

where $\boldsymbol{u}^0$ is the solution to the homogenized problem (2.11) given below, and the matrix function $\boldsymbol{\theta}(\xi) = (\boldsymbol{\theta}^1, \boldsymbol{\theta}^2, \boldsymbol{\theta}^3)$. The corrector functions $\boldsymbol{\theta}^k$ are solutions to the corrector equations [**6**]:

$$\begin{aligned} &\nabla_\xi \times (A(\xi)\nabla_\xi \times \boldsymbol{\theta}^k(\xi)) = -\nabla_\xi \times (A(\xi)\mathbf{e}_k), \quad \xi \in Q, \\ &\nabla_\xi \cdot \boldsymbol{\theta}^k(\xi) = 0, \quad \xi \in Q, \\ &\boldsymbol{\theta}^k(\xi) \times \boldsymbol{n} = 0, \quad \xi \in \partial Q, \end{aligned} \tag{2.8}$$

where e_k is the canonical basis of R^d, i.e., we have $e_1 = (1,0,0)'$, $e_2 = (0,1,0)'$, $e_3 = (0,0,1)'$ when $d = 3$.

The homogenized coefficient matrix $\widehat{A}$ can be computed by two methods [**3**, pp.142-145]. The first method is to solve for components $\chi^j(\xi), j = 1,2,3$, which satisfy the equation:

$$\nabla_\xi \cdot (A^{-1}(\xi)\nabla_\xi \chi^j(\xi)) = -\nabla_\xi \cdot (A^{-1}(\xi)\mathbf{e}_j). \tag{2.9}$$

Then we form a matrix function $\boldsymbol{\chi}(\xi) = (\chi^1(\xi), \chi^2(\xi), \chi^3(\xi))$ and construct

$$\widehat{A} = (\mathcal{M}(A^{-1}(\xi)(I_3 + \nabla_\xi \boldsymbol{\chi}(\xi))))^{-1},$$

where $\mathcal{M}$ is the averaging operator defined as $\mathcal{M}v = \int_Q v(\xi)\, d\xi$.

The second method is to set

$$\widehat{A} = \mathcal{M}(A(\xi) + A(\xi)\nabla_\xi \times \boldsymbol{\theta}(\xi)). \tag{2.10}$$

With $\widehat{A}$, we can define the solution $\boldsymbol{u}^0(\boldsymbol{x})$ for the homogenized problem of (2.6) as

$$\begin{aligned} &\nabla \times (\widehat{A}\nabla \times \boldsymbol{u}^0) - \omega^2 u^0 = f(\boldsymbol{x}), \quad \boldsymbol{x} \in \Omega, \\ &\nabla \cdot \boldsymbol{u}^0 = 0, \quad \boldsymbol{x} \in \Omega, \\ &\boldsymbol{u}^0 \times \boldsymbol{n} = 0, \quad \boldsymbol{x} \in \partial\Omega. \end{aligned} \tag{2.11}$$

We like to remark that in 2D, all the above derivations hold true, except that we treat $A(\xi)$ as a scalar function, and the 2D curl operators become as: $\nabla_\xi \times \chi(\xi) = (\frac{\partial \chi}{\partial \xi_2}, -\frac{\partial \chi}{\partial \xi_1})'$ for a scalar function $\chi(\xi)$; while $\nabla_\xi \times \chi(\xi) = \frac{\partial \chi_2}{\partial \xi_1} - \frac{\partial \chi_1}{\partial \xi_2}$ for a vector function $\chi = (\chi_1, \chi_2)'$. More specifically, in 2D, $A(\xi)\mathbf{e}_k$ in (2.8) is simplified to a scalar function $A(\xi)$.

2.3. The homogenized DG schemes. Now we want to construct a homogenized DG method for solving (2.3). To do this, we have to solve two sets of equations: one for the corrector functions on the unit cell Q, and one for the homogenized problem over domain Ω.

2.3.1. *A DG scheme for the corrector functions in the unit cell Q.* We consider a shape-regular mesh T_{h_0} that partitions Q into disjoint tetrahedral elements K (or triangles in 2D) such that $\overline{Q} = \bigcup_{K\in T_{h_0}} K$. We denote the diameter of K by h_K, the mesh size h_0 by $h_0 = \max_{K\in T_{h_0}} \{h_K\}$, the set of all interior faces by $F^I_{h_0}$, the set of all boundary faces by $F^B_{h_0}$, and the set of all faces by $F_{h_0} = F^I_{h_0} \bigcup F^B_{h_0}$.

We assume that the finite element space $\boldsymbol{V}_{h_0}$ is given by

$$\boldsymbol{V}_{h_0} = \{\boldsymbol{v} \in \boldsymbol{L}^2(Q) : \boldsymbol{v}|_K \in (P_l(K))^d, K \in T_{h_0}\}, \quad d = 2,3, \quad l \geq 1, \tag{2.12}$$

where $P_l(K)$ denotes the space of polynomials of total degree at most l on K. To construct the DG scheme, we introduce a bilinear form $a_{h_0}(\cdot,\cdot))$ defined on $\boldsymbol{V}_{h_0} \times \boldsymbol{V}_{h_0}$:

$$\begin{aligned} a_{h_0}(\boldsymbol{u},\boldsymbol{v}) = &\sum_{K\in T_{h_0}} \int_K A(\xi)\nabla\times\boldsymbol{u}\cdot\nabla\times\boldsymbol{v}d\boldsymbol{x} - \sum_{f\in F_{h_0}} \int_f [[\boldsymbol{u}]]_T\cdot\{\{A(\xi)\nabla\times\boldsymbol{v}\}\}dA \\ &- \sum_{f\in F_{h_0}} \int_f [[\boldsymbol{v}]]_T\cdot\{\{A(\xi)\nabla\times\boldsymbol{u}\}\}dA + \sum_{f\in F_{h_0}} \int_f a[[\boldsymbol{u}]]_T\cdot[[\boldsymbol{v}]]_T dA, \end{aligned}$$

where $[[\boldsymbol{v}]]_T$ and $\{\{\boldsymbol{v}\}\}$ denote the tangential jump and average of $\boldsymbol{v}$ across an interior face $f = \partial K^+ \bigcap \partial K^-$ between two neighboring elements K^+ and K^-, i.e.,

$$[[\boldsymbol{v}]]_T = \boldsymbol{n}^+\times\boldsymbol{v}^+ + \boldsymbol{n}^-\times\boldsymbol{v}^-, \quad \{\{\boldsymbol{v}\}\} = (\boldsymbol{v}^+ + \boldsymbol{v}^-)/2,$$

where $\boldsymbol{v}^+$ and $\boldsymbol{v}^-$ denote the traces of $\boldsymbol{v}$ from K^+ and K^-, respectively, and $\boldsymbol{n}^+$ and $\boldsymbol{n}^-$ are the unit outward normal vectors to ∂K^+ and ∂K^-, respectively. On a boundary face $f = \partial K \bigcap \partial\Omega$, we define $[[\boldsymbol{v}]]_T = \boldsymbol{n}\times\boldsymbol{v}$ and $\{\{\boldsymbol{v}\}\} = \boldsymbol{v}$. Finally, a is a penalty function, which is defined on each face $f \in F_{h_0}$ as

$$a|_f = \gamma A(\xi) h_0^{-1},$$

where $h_0|_f = min\{h_{K^+}, h_{K^-}\}$ for an interior face $f = \partial K^+ \bigcap \partial K^-$, and $h_0|_f = h_k$ for a boundary face $f = \partial K \bigcap \partial Q$. The penalty parameter γ is a positive constant and has to be chosen sufficiently large so in order to guarantee the coercivity of $a_{h_0}(\cdot,\cdot)$.

To construct a fully discrete DG scheme for the corrector equations (2.8), we multiply both sides of (2.8) by $\boldsymbol{v} \in \boldsymbol{V}_{h_0}$ and integrate over Q:

$$\begin{aligned} &\sum_{K\in T_{h_0}} \int_K A(\xi)\nabla_\xi\times\boldsymbol{\theta}^p_{h_0}(\xi)\cdot\nabla_\xi\times\boldsymbol{v}d\xi - \sum_{f\in F_{h_0}} \int_f [[\boldsymbol{\theta}^p_{h_0}(\xi)]]_T\cdot\{\{A(\xi)\nabla_\xi\times\boldsymbol{v}\}\}dA \\ &- \sum_{f\in F_{h_0}} \int_f [[\boldsymbol{v}]]_T\cdot\{\{A(\xi)\nabla_\xi\times\boldsymbol{\theta}^p_{h_0}(\xi)\}\}dA + \sum_{f\in F_{h_0}} \int_f a[[\boldsymbol{\theta}^p_{h_0}(\xi)]]_T\cdot[[\boldsymbol{v}]]dA \\ &= - \sum_{K\in T_{h_0}} \int_K (A(\xi)\mathbf{e}_p)\cdot\nabla_\xi\times\boldsymbol{v}, \quad p = 1,2,3. \end{aligned} \tag{2.13}$$

2.3.2. *A fully-discrete homogenized DG scheme.* Following [**3**], it is easy to see that the solution $\boldsymbol{E}^0$ of the homogenized problem for (2.3) shall satisfy the equation:

$$\begin{aligned} &\boldsymbol{E}^0_{tt} + \nabla\times(\widehat{A}\nabla\times\boldsymbol{E}^0) + \omega_p^2\boldsymbol{E}^0 - \boldsymbol{J}(\boldsymbol{E}^0) = \mathbf{f} \quad \text{in } \Omega\times(0,T), \\ &\boldsymbol{n}\times\boldsymbol{E} = 0 \quad on \quad \partial\Omega\times(0,T), \end{aligned} \tag{2.14}$$

where $\widehat{A}$ is the homogenized coefficient matrix of $A(\xi) = c_v^2(\frac{\boldsymbol{x}}{\epsilon})$ and can be computed as mentioned in sect.2.2.

To define a fully discrete scheme for solving (2.14), we divide the time interval $(0,T)$ into M uniform subintervals by points $0 = t_0 < t_1 < t_2 < \dots < t_M = T$, where $t_k = k\tau$ and $\tau = \frac{T}{M}$. Moreover, let $\boldsymbol{u}^k = \boldsymbol{u}(.,t_k)$ for $0 \le k \le M$, and define the following difference operators:

$$\overline{\boldsymbol{u}}^k = (\boldsymbol{u}^{k+1} + \boldsymbol{u}^{k-1})/2, \quad \delta_\tau^2 \boldsymbol{u}^k = (\boldsymbol{u}^{k+1} - 2\boldsymbol{u}^k + \boldsymbol{u}^{k-1})/\tau^2.$$

Now we can formulate an implicit scheme for solving (2.14): For any $1 \le k \le M-1$, find $\boldsymbol{E}_h^{0,k+1} \in \boldsymbol{V}_h$ such that [**14**]:

$$(2.15)\quad (\delta_\tau^2 \boldsymbol{E}_h^{0,k}, \boldsymbol{v}) + a_h(\overline{\boldsymbol{E}}_h^{0,k}, \boldsymbol{v}) + \omega_p^2(\overline{\boldsymbol{E}}_h^{0,k}, \boldsymbol{v}) - (\boldsymbol{J}_h^{0,k}, \boldsymbol{v}) = (\overline{\boldsymbol{f}}^k, \boldsymbol{v}), \quad \forall \boldsymbol{v} \in \boldsymbol{V}_h.$$

Here for simplicity we use $\boldsymbol{V}_h$ to denote the same space as $\boldsymbol{V}_{h_0}$ defined in (2.12), but now our mesh is on Ω and has a mesh size h.

The initial approximations of (2.5) are taken to be

$$(2.16)\qquad \boldsymbol{E}_h^{0,0} = P_h \boldsymbol{E}_0^0, \quad \boldsymbol{E}_h^{0,1} = P_h(\boldsymbol{E}_0^0 + \tau \boldsymbol{E}_1^0 + \frac{\tau^2}{2}\boldsymbol{E}_{tt}^0(0)),$$

where P_h is the standard L^2 projection operator onto $\boldsymbol{V}_h$, and $\boldsymbol{E}_{tt}^0(0)$ can be obtained by setting $t = 0$ in (2.14). Moreover, $\boldsymbol{J}_h^{0,k}$ in (2.15) is defined by a recursive formula [**14**]:

$$(2.17)\qquad \boldsymbol{J}_h^{0,0} = 0, \quad \boldsymbol{J}_h^{0,k} = e^{-\nu\tau}\boldsymbol{J}_h^{0,k-1} + \frac{\nu\omega_p^2}{2}\tau(e^{-\nu\tau}\boldsymbol{E}_h^{0,k-1} + \boldsymbol{E}_h^{0,k}), \quad k \ge 1.$$

With the corrector $\boldsymbol{\theta}_{h_0}$ and the homogenized solution $\boldsymbol{E}_h^{0,k+1}$, we can use the multiscale asymptotic expansion (2.7) to form the approximate solution $\boldsymbol{E}_{h,h_0}^{k+1}$ of (2.3) as follows:

$$(2.18)\qquad \boldsymbol{E}_{h,h_0}^{k+1}(\boldsymbol{x}) = \boldsymbol{E}_h^{0,k+1}(\boldsymbol{x}) + \epsilon\boldsymbol{\theta}_{h_0}(\xi)\nabla \times \boldsymbol{E}_h^{0,k+1}(\boldsymbol{x}).$$

3. Numerical results

The implementation of our homogenized DG scheme for solving (2.3) in periodic micro-structures can be broken into three steps:

- Step 1. Solve (2.13) on a reference cell $Q = (0,1)^2$ for for the corrector function θ_{h_0}.
- Step 2. Form the homogenized coefficient matrix $\widehat{A}_{h_0}$, then solve (2.15)-(2.17) on the whole domain Ω.
- Step 3. Construct the approximate multiscale asymptotic expansion (2.18).

To validate our algorithm, below we provide three 2D numerical examples. For simplicity, we choose $\Omega = Q = (0,1)^2$, the basis function used is the discontinuous linear element, and the mesh is composed of uniformly refined triangular elements.

EXAMPLE 3.1. This example is used to check how our DG method works for the time-harmonic Maxwell's equations (2.6) with a rapidly oscillating coefficient:

$$A(\frac{\boldsymbol{x}}{\epsilon}) = \frac{20}{(2 + 1.5\sin(2\pi(\frac{x}{\epsilon}) + 0.75))(2 + 1.5\sin(2\pi(\frac{y}{\epsilon}) + 0.75))},$$

which is a 2D simplification of Example 5.1 of [**6**]. In our tests, we choose $\epsilon = \frac{1}{100}$, $\omega^2 = 1$ and the source function $\boldsymbol{f} = (30,30)^T$ in (2.6).

The numerical solutions $\boldsymbol{E}_h$ obtained by (2.7) using various mesh sizes are plotted in Figures 1-3. Though there is no exact solution to be compared, these figures show that our numerical solution converges as the mesh size reduces.

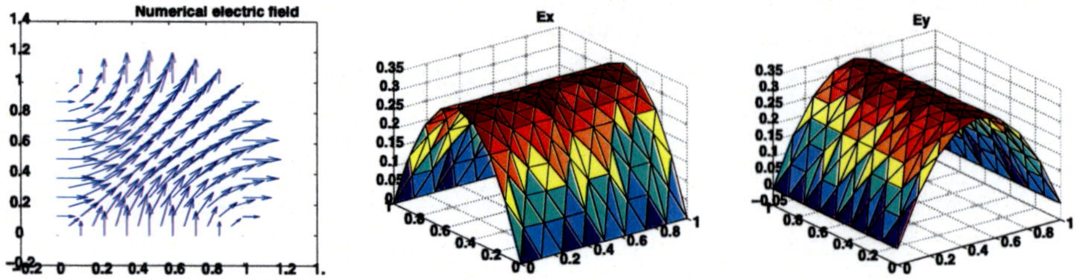

(A) The electric field $\boldsymbol{E}_h$.

(B) Contour plot of the x component of $\boldsymbol{E}_h$

(C) Contour plot of the y component of $\boldsymbol{E}_h$

FIGURE 1. Example 1. Computed electric fields using unit cell mesh size $1/2^7$ and domain Ω mesh size $1/2^3$.

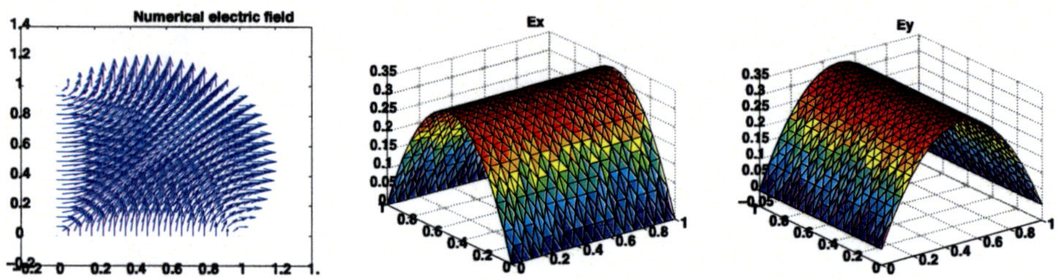

(A) The electric field $\boldsymbol{E}_h$.

(B) Contour plot of the x component of $\boldsymbol{E}_h$

(C) Contour plot of the y component of $\boldsymbol{E}_h$

FIGURE 2. Example 1. Computed electric fields using unit cell mesh size $1/2^7$ and domain Ω mesh size $1/2^4$.

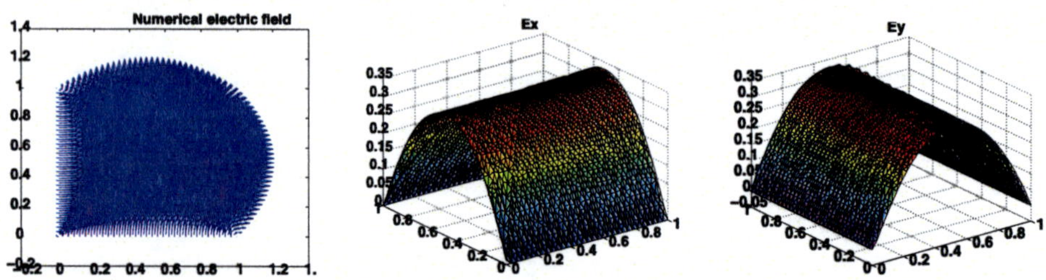

(A) The electric field $\boldsymbol{E}_h$.

(B) Contour plot of the x component of $\boldsymbol{E}_h$

(C) Contour plot of the y component of $\boldsymbol{E}_h$

FIGURE 3. Example 1. Computed electric fields using unit cell mesh size $1/2^7$ and domain Ω mesh size $1/2^5$.

EXAMPLE 3.2. This example is used to evaluate how our DG method performs on evaluating the homogenized coefficient matrix. Here we compared the homogenized DG method (2.10) with the classic Maxwell-Garnett (M-G) formula, which is given by:

$$\varepsilon_{eff} = \varepsilon_e \left[1 + \frac{3\delta(\varepsilon_i - \varepsilon_e)}{\varepsilon_i + 2\varepsilon_e - \delta(\varepsilon_i - \varepsilon_e)} \right],$$

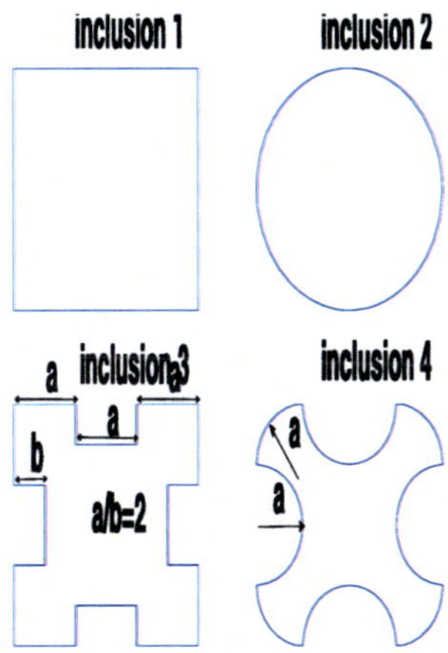

FIGURE 4. Geometry of the 2D inclusions used in Example 3.2

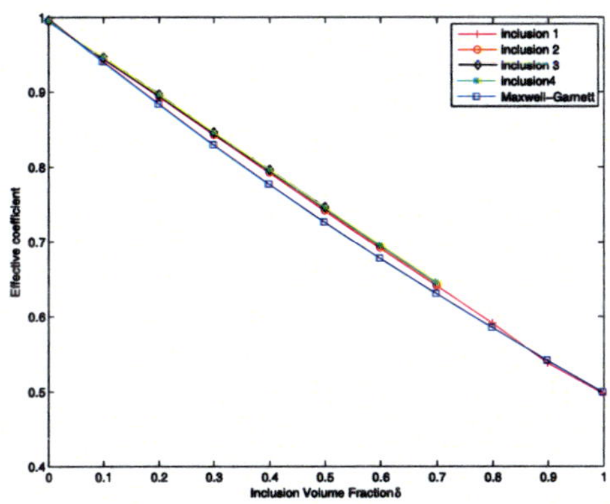

FIGURE 5. Computed effective coefficients $\widehat{A}$ for inclusions in Fig.4. From (2.1), $a(\xi) = 0.5$ inside each inclusion. The host medium is free space and the mesh size is $h_0 = 1/2^7$.

where ε_{eff} is the effective dielectric constant of the composite medium, ε_i is the one of the inclusions, ε_e is the one of the host medium, and δ is the volume fraction of the embedded material.

Following **[19]**, we consider various microstructures (see Fig.4) with $c_v^2 = 0.5$ inside the inclusions and $c_v^2 = 1$ elsewhere (i.e., the host medium is free space). In Fig.5, homogenized effective parameters c_v^2 are plotted against the volume fraction. From Fig.5, we find that our homogenized DG method produces the effective parameters which very close to those obtained by the M-G formula.

EXAMPLE 3.3. This example is to apply our homogenized DG method to solve problem (2.3) with a highly oscillatory periodic coefficient. In this problem we consider the circular inclusion shown in Fig.4 and choose $c_v^2 = 0.5$ inside the inclusion, and $c_v^2 = 1$ in the host medium. The relative size of the microstructure ϵ is

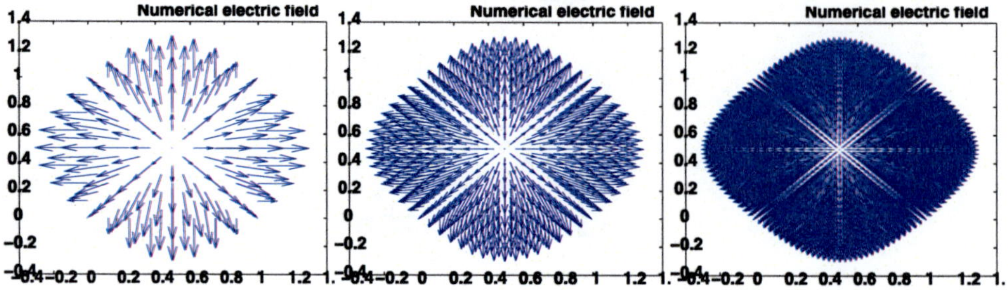

FIGURE 6. Example 3.3. The numerical electric fields $\boldsymbol{E}_h$ at time $T = 1$ obtained with $\tau = 0.02$ and different meshes $h = \frac{1}{8}$ (Left), $h = \frac{1}{16}$ (Middle), $h = \frac{1}{32}$ (Right).

chosen to be 10^{-5}, and the volume fraction is chosen to be 0.4. Moreover, we set $\omega_p = \nu = 1, \boldsymbol{f} = \boldsymbol{0}$ in the equation (2.3).

The 2-D problem (2.3) is solved with initial conditions:

$$\boldsymbol{E}_0 = \boldsymbol{E}(x, y, 0) = (\cos(\pi x)\sin(\pi y), \sin(\pi x)\cos(\pi y))^T,$$
$$\boldsymbol{E}_1 = \boldsymbol{E}_t(x, y, 0) = e^{-\tau}(\cos(\pi x)\sin(\pi y), \sin(\pi x)\cos(\pi y))^T.$$

Note that for both $i = 0, 1$, $\nabla \cdot \boldsymbol{E}_i = 0$ in Ω and $\boldsymbol{n} \times \boldsymbol{E}_i = 0$ on $\partial\Omega$. For our tests, we fix the final time $T = 1$, the time step size $\tau = 0.02$, the unit cell mesh size $h_0 = \frac{1}{2^7}$, but vary the mesh sizes h. Since the exact solution is unknown, we plot the numerical solution $\boldsymbol{E}_h$ at time $T = 1$ obtained with different meshes $h = \frac{1}{8}, \frac{1}{16}, \frac{1}{32}$ in Fig.6, which show clearly that the numerical solution is convergent as the mesh size $h \to 0$.

4. Concluding remarks

In this paper, we present a homogenized discontinuous Galerkin method for solving the time-dependent Maxwell's equations in dispersive media with periodic microstructures. The effectiveness of our algorithm is demonstrated through three numerical examples. In the future, we plan to extend this work to simulating wave propagation in metamaterials [**17**] constructed from periodic microstructures [**5, 15**]. Rigorous error analysis of our algorithm will be investigated also.

References

[1] H. Banks, V. Bokil, D. Cioranescu, N. Gibson, G. Griso, and B. Miara. Homogenization of periodically varying coefficients in electromagnetic materials. *J. Sci. Comp.*, 28(2):191–221, 2006. MR2272629 (2007i:78015)

[2] G. Barbatis and I. G. Stratis. Homogenization of maxwell's equations in dissipative bianisotropic media. *Math. Methods Appl. Sci.*, 26(14):1241–1253, 2003. MR2002980 (2004j:78007)

[3] A. Bensoussan, J.L. Lions, and G. Papanicolaou. *Asymptotic Analysis for Periodic Structures.* North-Holland, Amsterdam, 1978. MR503330 (82h:35001)

[4] A. Bossavit, G. Griso, and B. Miara. Modelling of periodic electromagnetic structures bianisotropic materials with memory effects. *J. Math. Pures Appl.*, 84(7):819–850, 7 2005. MR2144646 (2006g:35261)

[5] G. Bouchitté and B. Schweizer. Homogenization of maxwell's equations in a split ring geometry. *Multiscale Model. Simul.*, 8:717–750, 2010. MR2609637 (2011i:35008)

[6] L. Cao, Y. Zhang, W. Allegretto, and Y. Lin. Multiscale asymptptoc method for maxwell's equations in composite materials. *SIAM J. Numer. Anal.*, 47(6):4257–4289, 2009. MR2585187 (2011d:65276)

[7] B. Cockburn, G. Karniadakis, and C.-W. Shu (eds.). *Discontinuous Galerkin Methods: Theory, Computation and Applications.* Springer, New York, 2000. MR1842160 (2002b:65004)

[8] V. Dolean, H. Fahs, L. Fezoui, and S. Lanteri. Locally implicit discontinuous galerkin method for time domain electromagnetics. *J. Comp. Phys.*, 229:512–526, 2010. MR2565614 (2010j:78018)

[9] L. Fezoui, S. Lanteri, S. Lohrengel, and S. Piperno. Convergence and stability of a discontinuous galerkin time-domain method for the 3d heterogeneous maxwell equations on unstructured meshes. *Model. Math. Anal. Numer.*, 39(6):1149–1176, 2005. MR2195908 (2007a:65149)

[10] M.J. Grote, A. Schneebeli, and D. Schötzau. Interior penalty discontinuous galerkin method for maxwell's equations: optimal l^2-norm error estimates. *IMA J. Numer. Anal.*, 28:440–468, 2008. MR2433208 (2009g:65122)

[11] J.S. Hesthaven and T. Warburton. *Nodal Discontinuous Galerkin Methods: Algorithms, Analysis, and Applications.* Springer, New York, 2008. MR2372235 (2008k:65002)

[12] P. Houston, I. Perugia, A. Schneebeli, and D. Schötzau. Interior penalty method for the indefinite time-harmonic maxwell equations. *Numer. Math.*, 100:485–518, 2005. MR2194528 (2006k:65323)

[13] Y. Huang and J. Li. Interior penalty discontinuous galerkin method for maxwell's equations in cold plasma. *J. Sci. Comp.*, 41(3):321–340, 2009. MR2556468 (2010j:65187)

[14] Y. Huang, J. Li, and W. Yang. Interior penalty dg methods for maxwell's equations in dispersive media. *J. Comp. Phys.*, 230(12):4559–4570, 2011. MR2795970 (2012a:78006)

[15] R.V. Kohn and S.P. Shipman. Magnetism and homogenization of microresonators. *Multiscale Model. Simul.*, 7:62–92, 2008. MR2399538 (2009d:78025)

[16] J. Li. Optimal l^2 error estimates for the interior penalty dg method for maxwell's equations in cold plasma. *Commun. Comput. Phys.*, 11:319–334, 2012. MR2850897 (2012m:65340)

[17] J. Li and Y. Huang. *Time-Domain Finite Element Methods for Maxwell's Equations in Metamaterials.* Springer, New York, 2013. Springer Series in Computational Mathematics, vol.43,.

[18] T. Lu, P. Zhang, and W. Cai. Discontinuous galerkin methods for dispersive and lossy maxwell's equations and pml boundary conditions. *J. Comp. Phys.*, 200(2):549–580, 2004. MR2095277 (2005e:78025)

[19] O. Ouchetto, C.-W. Qiu, S. Zouhdi, L.-W. Li, and A. Razek. Homogenization of 3-d periodic bianisotropic metamaterials. *IEEE Trans. Microw. Theory Tech.*, 54(11):3893–3898, 2006.

[20] C. Scheid and S. Lanteri. Convergence of a discontinuous galerkin scheme for the mixed time-domain maxwell's equations in dispersive media. *IMA J. Numer. Anal.*, 2012.

[21] B. Wang, Z. Xie, and Z. Zhang. Error analysis of a discontinuous galerkin method for maxwell equations in dispersive media. *J. Comp. Phys.*, 229(22):8552–8563, 2010. MR2719188 (2011f:78011)

Department of Mathematical Sciences, University of Nevada Las Vegas, Las Vegas, Nevada 89154-4020

E-mail address: `jichun@unlv.nevada.edu`

Department of Mathematical Sciences, University of Nevada Las Vegas, Las Vegas, Nevada 89154-4020

E-mail address: `wangj16@unlv.nevada.edu`

Immersed finite element method of lines for moving interface problems with nonhomogeneous flux jump

Tao Lin, Yanping Lin, and Xu Zhang

ABSTRACT. This article presents an immersed finite element method of lines for solving parabolic moving interface problems with a non-homogeneous flux jump. The immersed finite elements are used for spatial discretization, which allow the material interface to be embedded in the interior of elements in the mesh. This feature makes it possible to employ the method of lines for solving a moving interface problem over a fixed solution mesh. Numerical experiments are provided to show features of this new method.

1. Introduction

We consider the following parabolic moving interface problem:

$$\begin{cases} u_t - (\beta u_x)_x = f(t,x), & x \in \Omega,\ t \in (0, T_{end}], \\ u(t,x) = g(t,x), & x \in \partial\Omega,\ t \in (0, T_{end}], \\ u(0,x) = u_0(x), & x \in \overline{\Omega}. \end{cases} \tag{1.1}$$

Here $\Omega \subset \mathbb{R}$ is an open interval, and for the sake of simplicity, we assume that there is only one interface point $\alpha(t)$ moving inside Ω. This point separates Ω into two sub-intervals $\Omega^-(t)$ and $\Omega^+(t)$ such that $\Omega = \Omega^-(t) \cup \Omega^+(t) \cup \{\alpha(t)\}$ with $\Omega^-(t)$ on the left hand side of $\alpha(t)$. The diffusivity coefficient $\beta(t,x)$ is discontinuous across the interface $\alpha(t)$. Without loss of generality, we assume that it is a piece-wise constant function defined as follows:

$$\beta(t,x) = \begin{cases} \beta^-, & x \in \Omega^-(t), \\ \beta^+, & x \in \Omega^+(t). \end{cases} \tag{1.2}$$

Consequently, the solution u is assumed to satisfy the following nonhomogeneous interface jump conditions:

$$[u]|_{\alpha(t)} = 0, \tag{1.3}$$

$$[\beta u_x]|_{\alpha(t)} = Q(t). \tag{1.4}$$

2010 *Mathematics Subject Classification.* Primary 65M20, 65M60; Secondary 35R05.

Key words and phrases. Immersed finite element, method of lines, moving interface, nonhomogeneous.

This work is partially supported by NSF grant DMS-1016313, GRF grant of Hong Kong (Project No. PolyU 501709), AMA-JRI of PolyU, Polyu grant No. 5020/10P and NSERC (Canada).

The model problem (1.1)-(1.4) appears in many applications, for instance, the Stefan problem [**CDH67**] for simulating temperature distribution undergoing a phase transition such as ice melting into water. The well-known Stefan condition requires the flux jump to be proportional to the velocity of the moving front, which is nonzero in general. Other interesting applications include the simulation of the electric potential [**Coo75**], the modeling of the water flow in porous media with a source at the interface [**Miy06**], and the modeling of air-vapor-heat transport through textile materials [**ZS11**], to name just a few.

It is well known that finite element (FE) methods can be used for solving interface problems provided that solution meshes are aligned with the material interfaces. Consequently, each element in a mesh contains essentially only one type of materials. However, the basis functions on each element are generic polynomials independent of the interface jump conditions. The recently developed immersed finite element (IFE) methods [**LLW03, LLLR04, GLL07, AL09, He09, HLL11, WLL11, XLQ11, WWY12**] employ an alternative approach to solve interface problems. IFE methods can use meshes independent of the interface location, but their basis functions are specially constructed to satisfy pertinent interface jump conditions.

For moving interface problems, using IFEs in spatial discretization has a major advantage over traditional FEs in the sense that a fixed mesh can be used throughout the whole simulation so that popular methods such as the method of lines (MoL) can be employed to reduce a moving interface problem to an ordinary differential equation (ODE) system that can be solved numerically by a preferred ODE solver. Fully discrete Crank-Nicolson type IFE schemes [**HLLZ12**] and the IFE MoL [**LLZ12**] have been developed for solving moving interface problems with a homogeneous flux jump, *i.e.*, $Q(t) = 0$. In this article, we extend IFE MoL to problems with a nonhomogeneous flux jump.

The rest of this article is organized as follows. In Section 2, we develop an IFE MoL based on a semi-discrete formulation using linear IFEs. In Section 3, we present several numeral results to demonstrate features of this new scheme. Brief conclusions are summarized in Section 4.

2. IFE Method of Lines

2.1. Linear IFE Functions for Nonhomogeneous Flux Jump. Let $\mathcal{T}_h$ be a uniform partition of Ω with a mesh size h. An element in this mesh is an interface element if $\alpha(t)$ is in its interior. Let $\mathcal{T}_h^{i,t}$, and $\mathcal{T}_h^{n,t}$ be the collections of interface elements and non-interface elements at time t, respectively. Also let $\mathcal{N}_h$ be the set of nodes in $\mathcal{T}_h$.

We use $\phi_j^t(x)$ to denote the global linear IFE basis function [**AL09**] associated with the node $x_j \in \mathcal{N}_h$. The superscript t on the basis function emphasizes the fact that it is either a linear IFE basis function or a standard linear FE basis function depending on whether $\alpha(t)$ is in the elements adjacent to the node x_j. Since $\alpha(t)$ changes with respect to time, $\phi_j^t(x)$ is time dependent even though the mesh is time independent [**HLLZ12**]. Then, we let $S_h^t(\Omega) = \text{span}\{\phi_j^t : x_j \in \mathcal{N}_h\}$ be the linear IFE space at the time t defined on the mesh $\mathcal{T}_h$. Note that in [**HLL11**], the authors enriched the IFE space by introducing an additional IFE function in order to handle the nonhomogeneous flux jump for elliptic interface problems. Our effort here is to extend this idea to the moving interface problem (1.1)-(1.4).

Assume, at time t, the interface point $\alpha(t)$ moves into the element $T = (x_j, x_{j+1})$, *i.e.*, T becomes an interface element such that $\alpha(t)$ separates T into $T^- = (x_j, \alpha)$, and $T^+ = (\alpha, x_{j+1})$. We construct an enrichment linear IFE function $\phi_J^t(x)$ for treating the nonhomogeneous flux jump as follow **[HLL11]**:

$$\phi_J^t(x) = \begin{cases} c_1(t)(x - x_j), & \text{for } x \in T^-, \\ c_2(t)(x_{j+1} - x), & \text{for } x \in T^+, \\ 0, & \text{otherwise.} \end{cases} \tag{2.1}$$

Note that ϕ_J^t is also time dependent, and $c_1(t)$, and $c_2(t)$ are determined by the following nonhomogeneous interface jump conditions

$$[\phi_J^t(\alpha)] = 0, \text{ and } [\beta\phi_{J,x}^t(\alpha)] = 1.$$

It is easy to show that $c_1(t)$ and $c_2(t)$ are uniquely determined by these conditions with the following formulas,

$$c_1(t) = \frac{x_{i+1} - \alpha}{\beta^-(\alpha - x_{i+1}) + \beta^+(x_i - \alpha)}, \quad c_2(t) = \frac{\alpha - x_i}{\beta^-(\alpha - x_{i+1}) + \beta^+(x_i - \alpha)}. \tag{2.2}$$

2.2. An IFE Method of Lines. We note that each IFE basis function $\phi_j^t(x)$ is associated with a fixed node $x_j \in \mathcal{N}_h$ even though $\phi_j^t(x)$ may changes with respect to t. Therefore, following the basic idea of the method of lines, we can look for a semi-discrete IFE solution to the moving interface problem in the following form:

$$u_h(t, x) = \tilde{u}_h(t, x) + \sum_{T \in \mathcal{T}_h^{i,t}} Q(t)\phi_J^t(x), \tag{2.3}$$

where $\tilde{u}_h(t, \cdot) \in S_h^t(\Omega)$ is the homogenized solution such that

$$\tilde{u}_h(t, x) = \sum_{x_j \in \mathcal{N}_h} u_j(t)\phi_j^t(x). \tag{2.4}$$

First, it is easy to see that $u_h(t, x)$ satisfies the jump conditions (1.3) and (1.4). Then, we determine the unknown coefficients $u_j(t)$ through a set of ODEs in t. Consider the weak form of the problem. Multiply $v \in H_0^1(\Omega)$ on both side of (1.1) and integrate on each Ω^s, $s = +, -$:

$$\int_{\Omega^s} u_t v \mathrm{d}x + \int_{\Omega^s} (\beta u_x)_x v \mathrm{d}x = \int_\Omega f v \mathrm{d}x, \ \forall\, v \in H_0^1(\Omega),$$

Applying integration by parts, summing over $s = +, -$, and using the relation (1.4), we obtain the following weak form:

$$\sum_{T \in \mathcal{T}_h} \int_T u_t v \mathrm{d}x + \sum_{T \in \mathcal{T}_h} \int_T \beta u_x v_x \mathrm{d}x = \int_\Omega f v \mathrm{d}x - Q(\alpha)v(\alpha), \ \forall\, v \in H_0^1(\Omega).$$

This weak form suggests the following IFE equations for determining $u_h \in S_h^t(\Omega)$:

$$\sum_{T \in \mathcal{T}_h} \int_T \frac{\partial u_h}{\partial t} v_h \mathrm{d}x + \sum_{T \in \mathcal{T}_h} \int_T \beta u_{hx} v_{hx} \mathrm{d}x = \int_\Omega f v_h \mathrm{d}x - Q(\alpha)v_h(\alpha), \ \forall\, v_h \in S_h^{t,0}(\Omega),$$

where $S_h^{t,0}(\Omega) = \{\phi_j \in S_h^t(\Omega) : x_j \in \mathcal{N}_h^0\}$, and $\mathcal{N}_h^0 \subset \mathcal{N}_h$ is the collection of nodes inside Ω. Taking the time derivative of u_h yields

$$\frac{\partial u_h}{\partial t} = \sum_{x_j \in \mathcal{N}_h} \frac{du_j}{dt}\phi_j^t + \sum_{x_j \in \mathcal{N}_h^{i,t}} u_j \frac{\partial \phi_j^t}{\partial t} + \sum_{T \in \mathcal{T}_h^{i,t}} \frac{dQ}{dt}\phi_J^t + \sum_{T \in \mathcal{T}_h^{i,t}} Q\frac{\partial \phi_J^t}{\partial t}. \tag{2.5}$$

Using (2.3)-(2.5) in the IFE equation above and substituting v_h by the IFE basis functions ϕ_i^t, we obtain an ODE system of $u_j(t)$:

$$\sum_{x_j\in\mathcal{N}_h} u_j'(t)\int_\Omega \phi_i^t\phi_j^t\mathrm{d}x + \sum_{x_j\in\mathcal{N}_{h,t}^i} u_j(t)\int_\Omega \phi_i^t\left(\frac{\partial}{\partial t}\phi_j^t\right)\mathrm{d}x + \sum_{x_j\in\mathcal{N}_h} u_j(t)\int_\Omega \beta\phi_{ix}^t\phi_{jx}^t\mathrm{d}x$$

$$(2.6)\quad = \int_\Omega f\phi_i^t\mathrm{d}X - Q(\alpha)\phi_i^t(\alpha) - \sum_{T\in\mathcal{T}_h^{i,t}}\int_T \phi_i^t\frac{\partial Q}{\partial t}\phi_J^t\mathrm{d}x - \sum_{T\in\mathcal{T}_h^{i,t}}\int_T \phi_i^t Q\frac{\partial\phi_{J,T}^t}{\partial t}\mathrm{d}x$$

$$- \sum_{T\in\mathcal{T}_h^{i,t}}\int_T \beta\phi_{i,x}^t(Q\phi_{J,x}^t)\mathrm{d}x,\qquad \forall\phi_j^t\in S_h^{t,0}(\Omega).$$

The matrix form of this IFE method of lines is as follows:

$$(2.7)\qquad M(t)\mathbf{u}'(t) + (K(t)+A(t))\mathbf{u}(t) = \mathbf{rhs}_f(t) - \mathbf{rhs}_Q(t),$$

where

- $M(t)$, $A(t)$ are mass and stiffness matrices associated to the first and third terms on the left hand side of (2.6).
- $K(t)$ is corresponding to the second term on the left hand side of (2.6).
- $\mathbf{rhs}_f(t)$ is the source term vector associated to the first term on the right hand side of (2.6).
- $\mathbf{rhs}_Q(t)$ is related to the last four terms of (2.6).

REMARK 2.1. If we solve a moving interface problem with a homogeneous flux jump condition, *i.e.*, $Q(t)=0$, then $\mathbf{rhs}_Q(t)=\mathbf{0}$ in (2.7); if we solve a problem with a static interface, *i.e.*, $\alpha(t)=\alpha$, then $K(t)=0$ in (2.7).

2.3. Some Implementation Issues. In this subsection, we briefly discuss some implementation issues for the IFE method of lines.

Matrices $M(t)$, $A(t)$ and vector $\mathbf{rhs}_f(t)$: Standard IFE packages developed for the problem with a fixed interface can be directly used to assemble $M(t)$, $A(t)$, and $\mathbf{rhs}_f(t)$. The only thing one needs to pay attention is to update the interface location for a given value of the time variable t in the computation.

Matrix $K(t)$: This matrix involves integrals of the time derivative of IFE basis functions $\frac{\partial\phi_i^t}{\partial t}$. We would like to mention that $K(t)$ is much sparser than $M(t)$ or $A(t)$; hence, it costs little time to get it assembled. The complete procedure for assembling $K(t)$ is provided in [**HLLZ12**].

Vector $\mathbf{rhs}_Q(t)$: The assembling of this vector involves the integrals of ϕ_J^t, $\phi_{J,x}^t$, $\frac{\partial\phi_J^t}{\partial t}$. From (2.1) and (2.2) we can easily obtain

$$(2.8)\qquad \frac{\partial\phi_J^t(x)}{\partial t} = \begin{cases} \dfrac{\beta^+(x_{i+1}-x_i)\alpha'(t)}{(\beta^-(\alpha-x_{i+1})+\beta^+(x_i-\alpha))^2}(x-x_i), & \text{in } T^-,\\ \dfrac{-\beta^-(x_{i+1}-x_i)\alpha'(t)}{(\beta^-(\alpha-x_{i+1})+\beta^+(x_i-\alpha))^2}(x_{i+1}-x), & \text{in } T^+,\end{cases}$$

$$(2.9)\qquad \phi_{J,x}(x) = \begin{cases} \dfrac{x_{i+1}-\alpha(t)}{\beta^-(\alpha(t)-x_{i+1})+\beta^+(x_i-\alpha(t))}, & \text{in } T^-,\\ \dfrac{x_i-\alpha(t)}{\beta^-(\alpha(t)-x_{i+1})+\beta^+(x_i-\alpha(t))}, & \text{in } T^+.\end{cases}$$

We emphasize again that assembling vector $\mathbf{rhs}_Q(t)$ costs little time. This is because we only need to calculate related integrals locally over the interface element. As a consequence, this vector is also very sparse.

3. Numerical Experiments

In this section, we present numerical examples to demonstrate features of this IFE MoL for solving moving interface problem with a nonhomogeneous flux jump.

We consider the interface problem defined on the solution domain $\Omega \times [0, T_{end}]$, where $\Omega = (0,1)$ and $T_{end} = 1$. The interface $\alpha(t)$ is a moving point which separates Ω into two sub-domains $\Omega^-(t) = (0, \alpha(t))$ and $\Omega^+ = (\alpha(t), 1)$. The exact solution $u(t,x)$ is chosen as follows

$$
(3.1) \qquad u(x,t) = \begin{cases} \left((x-\alpha(t))^2 + \frac{1}{\beta^-}\right) e^x - \frac{1}{\beta^-} e^{\alpha(t)}, & x \in (0, \alpha(t)), \\ \left((x-\alpha(t))^2 + \frac{2}{\beta^+}\right) e^x - \frac{2}{\beta^+} e^{\alpha(t)}, & x \in (\alpha(t), 1). \end{cases}
$$

Simple calculations show $Q(t) = e^{\alpha(t)} > 0$.

EXAMPLE 3.1. **General ODE Solver.** In this example, we test some general ODE solvers with fixed time step. We denote the step size by τ, and let $t_n = n\tau$, with $n = 0, 1, \cdots, N$. We assume the moving interface is governed by

$$
\alpha(t) = \alpha_1(t) = \left(\frac{\pi}{5} - \frac{1}{3}\right) t + \frac{1}{3}.
$$

The ODE system in the IFE MoL (2.7) is equivalent to the standard form:

$$
\mathbf{u}' = F(t, \mathbf{u}), \qquad \mathbf{u}(0) = \mathbf{u}_0,
$$

where

$$
F(t, \mathbf{u}) = M(t)^{-1}\Big(- (K(t) + A(t))\mathbf{u} + \mathbf{rhs}_f(t) - \mathbf{rhs}_Q(t)\Big),
$$

and $\mathbf{u}_0 = (u_0(x_j))$, $x_j \in \mathcal{N}_h$. Popular ODE solvers such as Runge Kutta methods and multi-step methods can be employed to solve this ODE system. We have tested several second order ODE solvers in our experiments because they have the order of accuracy comparable to that of the IFE space used in the spacial discretization, and we select $h = \tau$ in these experiments.

Numerical results produced by the implicit midpoint scheme and second order backward difference (BDF) scheme **[AP98]** are listed in Tables 1 and 2. Errors of numerical solutions in L^∞, L^2, and semi-H^1 norms are calculated at the final time level $t = 1$. Linear regression of these errors with different mesh size h yields the convergence order for each Sobolev norm. Both small coefficient jump $(\beta^-, \beta^+) = (1, 3)$, and large coefficient jump $(\beta^-, \beta^+) = (1, 1000)$ are tested.

TABLE 1. Errors of Linear IFE MoL at $t = 1$ with $\beta^- = 1$, $\beta^+ = 3$

	Implicit Midpoint			BDF2		
h	$\\|\cdot\\|_{L_\infty}$	$\\|\cdot\\|_{L_2}$	$\|\cdot\|_{H_1}$	$\\|\cdot\\|_{L_\infty}$	$\\|\cdot\\|_{L_2}$	$\|\cdot\|_{H_1}$
1/20	$1.42E-4$	$1.20E-3$	$7.98E-2$	$1.55E-4$	$1.23E-3$	$7.98E-2$
1/40	$3.02E-5$	$2.95E-4$	$4.06E-2$	$4.43E-5$	$2.86E-4$	$4.06E-2$
1/80	$3.95E-5$	$8.22E-5$	$2.01E-2$	$2.31E-5$	$7.83E-5$	$2.01E-2$
1/160	$5.00E-6$	$1.98E-5$	$1.00E-2$	$5.65E-6$	$1.97E-5$	$1.00E-2$
1/320	$3.12E-6$	$4.71E-6$	$5.02E-3$	$4.60E-6$	$4.52E-6$	$5.02E-3$
1/640	$3.86E-7$	$1.25E-6$	$2.51E-3$	$7.06E-7$	$1.18E-6$	$2.51E-3$
1/1280	$3.53E-7$	$2.53E-7$	$1.25E-3$	$1.71E-7$	$2.23E-7$	$1.25E-3$
Order	1.51	2.02	1.00	1.56	2.04	1.00

TABLE 2. Errors of Linear IFE MoL at $t = 1$ with $\beta^- = 1$, $\beta^+ = 1000$

	Implicit Midpoint			BDF2		
h	$\|\cdot\|_{L_\infty}$	$\|\cdot\|_{L_2}$	$\|\cdot\|_{H_1}$	$\|\cdot\|_{L_\infty}$	$\|\cdot\|_{L_2}$	$\|\cdot\|_{H_1}$
1/20	$1.96E{-4}$	$9.68E{-4}$	$6.78E{-2}$	$6.40E{-5}$	$9.96E{-4}$	$6.77E{-2}$
1/40	$7.10E{-5}$	$2.62E{-4}$	$3.81E{-2}$	$2.11E{-5}$	$2.64E{-4}$	$3.81E{-2}$
1/80	$2.00E{-5}$	$6.40E{-5}$	$1.74E{-2}$	$8.79E{-6}$	$6.41E{-5}$	$1.74E{-2}$
1/160	$5.84E{-6}$	$1.55E{-5}$	$8.57E{-3}$	$1.22E{-6}$	$1.57E{-5}$	$8.56E{-3}$
1/320	$2.87E{-6}$	$3.94E{-6}$	$4.45E{-3}$	$8.83E{-7}$	$3.87E{-6}$	$4.45E{-3}$
1/640	$8.94E{-7}$	$9.52E{-7}$	$2.16E{-3}$	$1.68E{-7}$	$9.63E{-7}$	$2.16E{-3}$
1/1280	$2.04E{-7}$	$2.51E{-7}$	$1.07E{-3}$	$4.84E{-8}$	$2.50E{-7}$	$1.07E{-3}$
Order	1.61	2.00	1.01	1.73	2.00	1.01

The data in Table 1 and 2 suggest that errors in L^2 and H^1 norms obey optimal rates. However, errors in L^∞ norm has a sub-optimal convergence order, and the reason for this sub-optimal convergence is an interesting topic for future research. Other ODE solvers such as the Crank-Nicolson, diagonally implicit Runge Kutta (DIRK2) schemes, etc., are also tested. The performances are similar, so we omit the corresponding data in this manuscript.

EXAMPLE 3.2. **Adaptive ODE Solver.** One attractive feature of the method of lines is that the adaptivity in time variable can be easily realized. Hence, with a suitably chosen adaptive ODE solver, the IFE MoL can efficiently produce a reliable numerical solution to a parabolic initial boundary value problem whose interface changes with respect to time at a variable rate.

To demonstrate this feature, we consider another example which is basically the same as that in the previous example but its interface is governed by:

$$\alpha(t) = \alpha_2(t) = \frac{1}{30}\left(\log(10t + 0.25) + \frac{1}{500}e^{6t+3} + 4\right).$$

Obviously, this interface function changes with respect to time at a variable rate, see the plot on the left in Figure 1. We use the standard embedded DIRK45 scheme in [**HW96**] to solve these two moving interface problems. The plot on the right in Figure 1 displays the time step sizes used for producing a numerical solution to the second moving interface problem. Comparing this plot with the derivative of $\alpha(t)$ in the plot on the left, we can see that our IFE MoL with this adaptive ODE solver uses larger time step sizes in the region where $|\alpha'(t)|$ has smaller values while it uses smaller time step sizes in the region where $|\alpha'(t)|$ is large. This obviously indicates that this IFE MoL can efficiently handle the change of the interface with respect to time via the adaptivity feature in DIRK45.

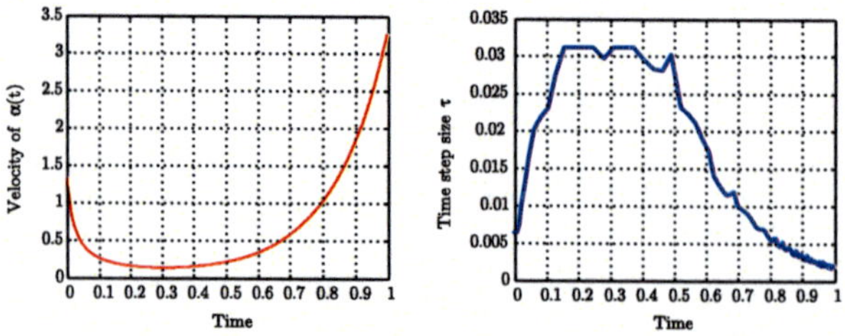

FIGURE 1. left plot: the graph of $|\alpha'(t)|$; right plot: time step sizes used in the adaptive IFE MoL on a uniform mesh with $h = 1/160$.

As demonstrated by the data in Table 3, this adaptive feature enables the IFE MoL to produce reliable numerical results by strategically choosing the step size in the time integration. The mesh sizes used in this group of numerical experiments are listed in the first column of Table 3. The second and third columns contain error data of the IFE solutions to the two moving interface problems. The numbers below N are the total numbers of time iterations. We note that the interface in the second moving interface problem changes more complicatedly than the first. Consequently, the adaptive IFE MoL uses less time iterations for the first problem. More importantly, applying linear regression to the data in this table, we can see that the adaptive IFE MoL produces numerical solutions to both problems with optimal convergence rates in the L^2 and H^1 norms.

Figure 2 displays the curve for the L^∞ norm errors in the IFE MoL solution generated on a mesh of $h = 1/160$ with an adaptive ODE solver at a sequence of time levels. This adaptive IFE MoL solution uses 108 time steps. The other two curves are for L^∞ norm errors in numerical solutions produced by IFE methods with a uniform time step size $\tau = 1/108$. It is clear that the adaptive IFE MoL can maintain the error in its solution below a certain level almost uniformly over the whole time interval while errors in those IFE methods based a uniform time step size grow along with the time. These numerical results further demonstrate the reliability of the IFE MoL combined with a good adaptive ODE solver.

TABLE 3. Errors of Adaptive DIRK45 at $t = 1$ with $\beta^- = 1$, $\beta^+ = 1000$

	DIRK45: $\alpha(t) = \alpha_1(t)$				DIRK45: $\alpha(t) = \alpha_2(t)$			
h	N	$\|\cdot\|_{L_\infty}$	$\|\cdot\|_{L_2}$	$\lvert\cdot\rvert_{H_1}$	N	$\|\cdot\|_{L_\infty}$	$\|\cdot\|_{L_2}$	$\lvert\cdot\rvert_{H_1}$
1/20	8	$3.27E{-5}$	$9.87E{-4}$	$6.77E{-2}$	16	$2.27E{-3}$	$1.98E{-3}$	$1.68E{-1}$
1/40	13	$1.42E{-5}$	$2.66E{-4}$	$3.81E{-2}$	30	$6.85E{-4}$	$5.31E{-4}$	$4.90E{-2}$
1/80	25	$2.60E{-6}$	$6.35E{-5}$	$1.74E{-2}$	57	$1.13E{-4}$	$1.03E{-4}$	$1.76E{-2}$
1/160	49	$1.09E{-6}$	$1.58E{-5}$	$8.56E{-3}$	108	$3.10E{-5}$	$2.89E{-5}$	$7.74E{-3}$
1/320	95	$8.01E{-7}$	$4.03E{-6}$	$4.45E{-3}$	217	$2.22E{-5}$	$9.13E{-6}$	$3.76E{-3}$
1/640	190	$2.59E{-7}$	$1.04E{-6}$	$2.16E{-3}$	432	$3.99E{-6}$	$2.29E{-6}$	$1.88E{-3}$
1/1280	379	$7.86E{-8}$	$2.68E{-7}$	$1.07E{-3}$	864	$8.45E{-7}$	$6.32E{-7}$	$9.38E{-4}$
Order		1.41	1.98	1.01		1.83	1.93	1.22

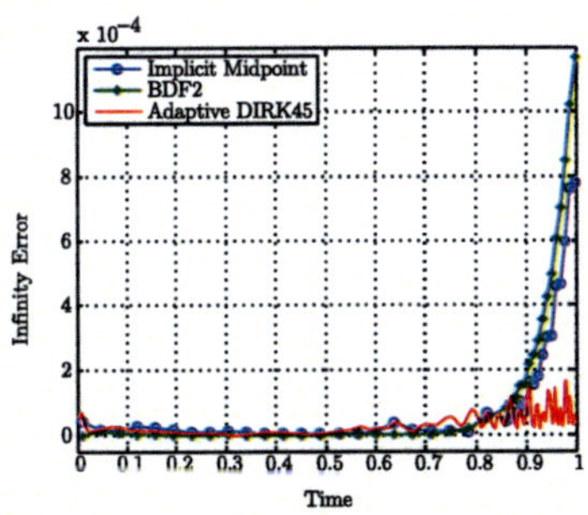

FIGURE 2. Curves for L^∞ errors of IFE solutions generated on a uniform mesh with $h = 1/160$. The adaptive IFE MoL uses 108 times steps determined according to error control strategy, but the other two IFE methods use a uniform time step size $\tau = 1/108$.

4. Conclusion

In this article, we develop an IFE MoL for the one dimensional parabolic moving interface problem with a nonhomogeneous flux jump condition. An enrichment linear IFE function is constructed to homogenize the original problem. Abundant choice of efficient ODE solvers allows this IFE MoL to solve this kind of moving interface problems efficiently.

References

[AL09] S. Adjerid and T. Lin, *A p-th degree immersed finite element for boundary value problems with discontinuous coefficients*, Appl. Numer. Math. **59** (2009), no. 6, 1303–1321. MR2510495 (2009m:65123)

[AP98] U. Ascher and L. Petzold, *Computer methods for ordinary differential equations and differential-algebraic equations*, Society for Industrial and Applied Mathematics (SIAM), Philadelphia, PA, 1998. MR1638643 (99k:65052)

[CDH67] J. Cannon, J. Douglas, and C. Hill, *A multi-boundary Stefan problem and the disappearance of phases*, J. Math. Mech. **17** (1967), 21–33. MR0269999 (42:4892)

[Coo75] D. Cook, *The theory of electromagnetic field*, Prentice-Hall Physics Series, Prentice Hall College Div, 1975.

[GLL07] Y. Gong, B. Li, and Z. Li, *Immersed-interface finite-element methods for elliptic interface problems with nonhomogeneous jump conditions*, SIAM J. Numer. Anal. **46** (2007), no. 1, 472–495. MR2377272 (2008m:65319)

[He09] X. He, *Bilinear immersed finite elements for interface problems*, Ph.D. thesis, Virginia Tech, 2009.

[HLL11] X. He, T. Lin, and Y. Lin, *Immersed finite element methods for elliptic interface problems with non-homogeneous jump conditions*, Int. J. Numer. Anal. Model. **8** (2011), no. 2, 284–301. MR2740492 (2011h:65224)

[HLLZ12] X. He, T. Lin, Y. Lin, and X. Zhang, *Immersed finite element methods for parabolic equations with moving interface*, Numer. Methods Partial Differential Equations (2012), to appear.

[HW96] E. Hairer and G. Wanner, *Solving ordinary differential equations*, Springer Series in Computational Mathematics, 14, vol. II, Springer-Verlag, Berlin, 1996. MR1439506 (97m:65007)

[LLLR04] Z. Li, T. Lin, Y. Lin, and R. Rogers, *An immersed finite element space and its approximation capability*, Numer. Methods Partial Differential Equations **20** (2004), no. 3, 338–367. MR2046521 (2005f:65153)

[LLW03] Z. Li, T. Lin, and X. Wu, *New cartesian grid methods for interface problems using the finite element formulation*, Numer. Math. **96** (2003), no. 1, 61–98. MR2018791 (2005c:65104)

[LLZ12] T. Lin, Y. Lin, and X. Zhang, *A method of lines based on immersed finite elements for parabolic moving interface problems*, Adv. Appl. Math. Mech. (2012), to appear.

[Miy06] T. Miyazaki, *Water flow in soils*, CRC Press, 2006.

[WLL11] C. Wu, Z. Li, and M. Lai, *Adaptive mesh refinement for elliptic interface problems using the non-conforming immersed finite element method*, Int. J. Numer. Anal. Model. **8** (2011), no. 3, 466–483. MR2805671 (2012g:65267)

[WWY12] K. Wang, H. Wang, and X. Yu, *An immersed Eulerian-Lagrangian localized adjoint method for transient advection-diffusion equations with interfaces*, Int. J. Numer. Anal. Model. **9** (2012), no. 1, 29–42. MR2871300 (2012k:65118)

[XLQ11] H. Xie, Z. Li, and Z. Qiao, *A finite element method for elasticity interface problems with locally modified triangulations*, Int. J. Numer. Anal. Model. **8** (2011), no. 2, 189–200. MR2740487 (2011h:74113)

[ZS11] Q. Zhang and W. Sun, *A numerical study of air-vapor-heat transport through textile materials with a moving interface*, J. Comput. Appl. Math. **236** (2011), no. 5, 819–833. MR2853507

Department of Mathematics, Virginia Tech, Blacksburg, Virginia 24061
E-mail address: tlin@vt.edu

Department of Applied Mathematics, Hong Kong Polytechnic University, Hung Hom, Hong Kong, and Department of Mathematical and Statistics Science, University of Alberta, Edmonton AB, T6G 2G1, Canada
E-mail address: malin@polyu.edu.hk and yanlin@ualberta.ca

Department of Mathematics, Virginia Tech, Blacksburg, Virginia 24061
E-mail address: xuz@vt.edu

B-spline Gaussian collocation software for 1D parabolic PDEs

Paul H. Muir

ABSTRACT. This paper surveys B-spline Gaussian collocation software for the numerical solution of one-dimensional (1D) parabolic partial differential equations (PDEs). In this approach, the approximate solution is represented as a linear combination of known spatially dependent B-spline basis functions with unknown time-dependent coefficients. These coefficients are determined by imposing the boundary conditions (in some form) and the collocation conditions: the approximate solution is required to satisfy the PDE at selected points (collocation points) within the problem domain. The collocation points are the images of the Gauss points on each subinterval of a mesh which partitions the spatial domain. The collocation conditions represent a system of time-dependent ordinary differential equations (ODEs) which must be solved, together with the boundary conditions, to obtain the time-dependent B-spline basis coefficients. When the boundary conditions are differentiated, additional ODEs are obtained, and the coupled collocation/boundary condition system is a system of ODEs; when the boundary conditions are treated directly, algebraic equations are obtained, and the coupled collocation/boundary condition system is a system of differential algebraic equations (DAEs). We survey the evolution of B-spline Gaussian collocation software for 1D parabolic PDEs over the last three and a half decades. The transition from software with no spatial adaptivity to software featuring spatial adaptivity and error control and the transition from the use of ODE software to the use of DAE software will be highlighted. Recent work involving the development of software packages from this family that use new efficient spatial error estimation schemes will also be reviewed.

1. Introduction

In this paper, we survey the evolution of B-spline Gaussian collocation software for the numerical solution of systems of one-dimensional (1D) time-dependent parabolic partial differential equations (PDEs). The general form for the problem class we consider is

$$u_t(x,t) = f\left(x, t, u(x,t), u_x(x,t), u_{xx}(x,t)\right), \tag{1}$$

for $x \in (a,b)$, $t \in (t_0, t_{out})$, with $u : R \times R \to R^n$, $f : R \times R \times R^n \times R^n \times R^n \to R^n$, where n is the number of PDEs. The initial conditions and separated boundary

2000 *Mathematics Subject Classification.* Primary 65M70; Secondary 65M20.

Funding support from NSERC, MITACS, Saint Mary's University.

conditions are given by

$$u(x,t_0) = u_0(x), \quad b_L\left(t, u(a,t), u_x(a,t)\right) = 0, \quad b_R\left(t, u(b,t), u_x(b,t)\right) = 0,$$

where $u_0 : R \to R^n$, $x \in [a,b]$, and $b_L, b_R : R \times R^n \times R^n \to R^n$, $t \in (t_0, t_{out})$.

B-spline Gaussian collocation methods represent an important class of spatial discretization schemes for 1D PDEs. The approximate solution is represented as a linear combination of known spatial basis functions, B-splines [**deBo78**], with unknown time-dependent coefficients. Time-dependent ordinary differential equations (ODEs) involving these unknown coefficients are then obtained by requiring the collocation solution to satisfy the PDE at a set of collocation points in the spatial domain. In Gaussian collocation (or orthogonal spline collocation), the spatial domain is partitioned by a mesh of points and the collocation points are chosen to be the Gauss points (see, e.g., [**AMR95**]) mapped onto each spatial mesh subinterval.

The first approach commonly used for treatment of the boundary conditions involved asking the user for sufficient information about the boundary conditions to allow the time-differentiated form of these conditions to be obtained. These additional time-dependent ODEs, together with the ODEs arising from the collocation conditions, were solved using software for the numerical solution of ODEs. A better approach, now used in current software from this family, features the direct treatment of the boundary conditions; the collocation solution is required to satisfy the boundary conditions, yielding a set of algebraic equations. These equations, coupled with the ODEs arising from the collocation conditions, represent a system of differential-algebraic equations (DAEs) which are solved using DAE software. In either case, once the time-dependent B-spline coefficients are obtained for a given time, the corresponding collocation solution can be constructed as a linear combination of the B-spline basis functions and evaluated at any point in the spatial domain. The time-stepping scheme implemented within the time-integration software adaptively controls an estimate of the local error of the B-spline coefficients.

Another dimension in which the capability of B-spline Gaussian collocation software has progressed quite significantly is in spatial mesh adaptivity and spatial error control. The first codes developed within this family employed only a fixed spatial mesh provided at the beginning of the computation. Later members of this family computed a high order estimate of the spatial error and adapted the spatial mesh in order to control this estimate with respect to a user-provided tolerance. The most recently developed members of this family make use of new interpolation based spatial error estimation to improve the efficiency of the codes.

While this survey focuses on B-spline Gaussian collocation codes, there are of course many other types of software for the numerical solution of 1D PDEs. These include codes based on other types of collocation; an important example is the MOVCOL package [**HuRu96**]. See also the excellent survey by Fairweather and Meade [**FaMe89**].

This paper is organized as follows. In Section 2, we briefly review 1D B-spline Gaussian collocation. The remaining sections provide a survey of a family of B-spline Gaussian collocation software packages that have been developed over the last few decades. Section 3 discusses the two earliest packages which employ fixed spatial meshes and a differentiated form of the boundary conditions so that an ODE time-integrator can be employed. Section 4 discusses two more recently developed packages that provide adaptive control of a spatial error estimate and treat the boundary conditions directly, using DAE time integration software. This section also briefly discusses several papers that present numerical comparisons of some of the software packages considered in this survey. Section 5 discusses recently developed software packages, based on the codes described in Section 4, that use new efficient interpolation-based spatial error estimation schemes. Section 6 provides a summary and suggestions for future work.

2. B-spline Gaussian Collocation

The B-spline Gaussian collocation algorithm begins by assuming the existence of a spatial mesh $\{x_i\}_{i=0}^{N}$ that partitions the spatial domain (with $x_0 = a, x_N = b$.) Given an integer p and a requirement for C^1-continuity at the internal mesh points, we define on the given mesh a set of piecewise polynomials of degree p on each subinterval, known as a B-spline basis. We denote the B-spline basis functions by $\{B_j(x)\}_{j=1}^{NC}$, where $NC = N(p-1)+2$. The B-spline basis is constructed so that on any subinterval, $[x_{i-1}, x_i), i = 1, \ldots, N$, at most $p+1$ of the B-spline basis functions, namely, $\{B_m(x)\}_{m=(i-1)(p-1)+1}^{i(p-1)+2}$, are non-zero.

The collocation solution, $U(x,t)$, is expressed in the form

$$U(x,t) = \sum_{j=1}^{NC} w_j(t)B_j(x), \tag{2}$$

where $w_j(t)$ is the vector ($w_j(t) : R \to R^n$) of time-dependent B-spline coefficients multiplying the jth B-spline basis function, $B_j(x)$. For $x \in [x_{i-1}, x_i)$, (since most of the B-splines basis functions are zero) (2) can be expressed as

$$U(x,t) = \sum_{m=(i-1)(p-1)+1}^{i(p-1)+2} w_m(t)B_m(x). \tag{3}$$

Next, letting $\{\rho_j\}_{j=1}^{p-1}$ be the Gaussian points mapped onto $[0,1]$, we can define the $p-1$ collocation points on the ith subinterval, $[x_{i-1}, x_i]$, to be $\xi_l = x_{i-1} + h_i\rho_j$, where $h_i = x_i - x_{i-1}$ and $l = 1 + (i-1)(p-1) + j$, $j = 1, \cdots, p-1$. (Note that $\xi_1 = a$ and $\xi_{NC} = b$ are associated with the boundary conditions.)

The Gaussian collocation algorithm requires the approximate solution, (2), to satisfy the PDEs at the collocation points, $\{\xi_l\}_{l=2}^{NC-1}$. Substituting the collocation solution and its derivatives, evaluated at the collocation points, $\xi_l, l = 2, \ldots, NC-1$, into the PDE gives the collocation conditions

$$U_t(\xi_l, t) = f(\xi_l, t, U(\xi_l, t), U_x(\xi_l, t), U_{xx}(\xi_l, t)). \tag{4}$$

Expressing the approximate solution in the form (3) allows us to express the collocation conditions in a form that shows the explicit dependence on the time-dependent B-spline coefficients. When $\xi_l \in [x_{i-1}, x_i]$, we have

$$\sum_{m=(i-1)(p-1)+1}^{i(p-1)+2} w_m'(t)B_m(\xi_l) =$$

$$f\left(\xi_l, t, \sum_{m=(i-1)(p-1)+1}^{i(p-1)+2} w_m(t)B_m(\xi_l), \sum_{m=(i-1)(p-1)+1}^{i(p-1)+2} w_m(t)B_m'(\xi_l),\right.$$

$$(5) \qquad \left. \sum_{m=(i-1)(p-1)+1}^{i(p-1)+2} w_m(t) B''_m(\xi_l) \right) .$$

The above conditions represent a system of $(NC-2)n$ ODEs that depend on the unknown time-dependent B-spline coefficients, $w_j(t), j = 1, \ldots, NC$.

3. PDECOL, EPDCOL: fixed spatial meshes and ODE software

The B-spline package [**deBo78**], a popular collection of software routines that provided a stable implementation of the B-spline basis, was released by de Boor in 1978 and shortly afterward Madsen and Sincovec used this package to implement a B-spline basis within a Gaussian collocation package for the numerical solution of 1D PDEs called PDECOL [**MaSi79**]. Given an initial spatial mesh and a value for p, this software employed the B-spline Gaussian collocation algorithm described in Section 2 to discretize the PDEs, yielding a system of $(NC-2)n$ ODEs in time. The user was required to write subroutines that provided sufficient information about the boundary conditions and their derivatives to allow the code to obtain the time derivatives of the boundary conditions, yielding an additional $2n$ ODEs in time. These ODEs, together with the ODEs (5) arising from the collocation process, represented $NC \cdot n$ time-dependent ODEs that were solved to obtain the $NC \cdot n$ time-dependent B-spline coefficients, $\{w_j(t)\}_{j=1}^{NC}$, appearing in (2).

This ODE system was solved in PDECOL using the ODE solver, STIFIB (a modification of GEARIB [**Hind77**]). STIFIB was based on two families of multistep methods known as Adams methods and backward differentiation formulas (BDFs). The code used an approximate Newton-type algorithm for the solution of the nonlinear systems that arose and thus Jacobians of these nonlinear systems were required. Since most of the B-spline basis functions on a given spatial subinterval are zero, the Jacobian matrix had nonzero elements only in a region near the diagonal. Accordingly, PDECOL made use of an option provided by STIFIB to employ a banded matrix data structure and associated algorithms to handle the linear algebra computations that arose.

However, a closer examination of the sparsity structure of the matrices revealed that they possessed a structure known as almost block diagonal (ABD); see, e.g., [**DFK83**]. Based on this observation, Keast and Muir developed an improved version of PDECOL called EPDCOL [**KeMu91**]. The key modification involved introducing a package specifically designed for the factorization and solution of ABD linear systems called COLROW [**DFK83**]. The authors showed that the use of COLROW approximately doubled the speed of the computation.

The PDECOL and EPDCOL packages provided temporal control of the B-spline coefficients through the adaptive time-stepping and method order adaptivity implemented within STIFIB but since the spatial mesh was fixed, no control of the spatial error was possible. As well, since these codes made use of an ODE solver, the boundary conditions could not be treated directly; as indicated above, they were differentiated and added to the collection of ODEs resulting from the collocation process. This implied that the boundary conditions were satisfied only to within the tolerance used for control of the time integration.

4. BACOL, BACOLR: spatial error control and DAE software

Slightly more than a decade after the release of EDPCOL, Wang, Keast, and Muir developed a new B-spline Gaussian collocation code that addressed the two major issues associated with PDECOL/EPDCOL. The new code, called BACOL [**WKM04a, WKM04b**], also employed the B-spline Gaussian collocation spatial discretization algorithm described in Section 2 but treated the boundary conditions directly. The ODE system arising from the collocation process, coupled with the boundary conditions, was solved using a modified version of the DAE solver, DASSL [**BCP89**]. DASSL employed BDFs and used adaptivity in time step size and formula order to control an estimate of the local temporal error. It was modified to allow COLROW to be used for the treatment of the linear systems.

The second major feature of the BACOL package was the introduction of spatial adaptivity and error control. The spatial error estimate was obtained by computing a second collocation solution, $\bar{U}(x,t)$, based on a B-spline basis consisting of C^1-continuous piecewise polynomials *of degree* $p+1$ (on the same spatial mesh used for the computation of $U(x,t)$).

The papers [**DoDu74, CePa76**] show that the collocation solution, $U(x,t)$, is of order $p+1$, i.e.,

$$|U(x,t) - u(x,t)| \sim O(h^{p+1}),$$

where $u(x,t)$ is the exact solution and h is the maximum mesh spacing. Then,

$$|U(x,t) - \bar{U}(x,t)| = |(U(x,t) - u(x,t)) - (\bar{U}(x,t) - u(x,t))| = |C_p h^{p+1}| + O(h^{p+2}),$$

where C_p is the coefficient of h^{p+1} in the leading order term in the collocation error for $U(x,t)$. Thus the difference between the two collocation solutions gives, asymptotically, an estimate of the error in $U(x,t)$.

In BACOL, two normalized spatial error estimates were computed. (We describe these for the case $n = 1$; generalizations for larger n values are straightforward.) The first was

$$E(t) = \sqrt{\int_a^b \left(\frac{U(x,t) - \bar{U}(x,t)}{ATOL + RTOL|U(x,t)|} \right)^2 dx}, \tag{6}$$

where $ATOL$ and $RTOL$ are the absolute and relative spatial error tolerances supplied by the user. $U(x,t)$ was accepted at time t if $E(t) \leq 1$. The second set of N error estimates had the form,

$$\hat{E}_i(t) = \sqrt{\int_{x_{i-1}}^{x_i} \left(\frac{U(x,t) - \bar{U}(x,t)}{ATOL + RTOL|U(x,t)|} \right)^2 dx}, \quad i = 1, \ldots, N. \tag{7}$$

In addition to considering the acceptability of $U(x,t)$ as indicated above, BACOL also determined if the mesh was sufficiently equidistributed with respect to the error estimates, $\hat{E}_i(t)$. If not, based on these error estimates, BACOL attempted to construct a new mesh that would have a sufficient number of points to yield a collocation solution whose spatial error estimate, $E(t)$, would satisfy the user tolerances. Also, the mesh points were positioned so that the spatial error estimates, $\hat{E}_i(t)$, were approximately equidistributed over the subintervals of the new mesh. See [**WKM04a**] for further details.

In the paper [**WKM04c**], BACOL was compared with a number of similar codes, including EPDCOL, D03PPF [**NAG**], TOMS731 [**BlZe1994**], MOVCOL, and HPNEW [**Moor01**], with respect to accuracy and efficiency. The authors showed that the performance of BACOL was superior to that of the other codes, especially for problems whose solutions exhibited sharp, time-dependent layers and when high accuracy was required.

Three years later, a new code, BACOLR, based on a modification of BACOL, was developed by Wang, Keast, and Muir [**WKM07**]. Because the higher order BDFs employed by DASSL were multistep methods, it was necessary, in BACOL, to interpolate past solution values from the previous spatial mesh to the new spatial mesh, whenever a spatial remeshing was performed, thus detracting from the efficiency and accuracy of the computations. Furthermore, it is well known that the stability regions of the higher order BDFs do not include the imaginary axis and thus problems that lead to DAE systems that have Jacobians for which the eigenvalues are complex or nearly complex cannot be treated by DASSL, using the higher order BDFs. The paper [**WKM07**] showed that BACOL failed on a problem of this type due to the failure of DASSL. The BACOLR code was obtained by modifying BACOL to replace DASSL with DAE solver RADAU5 [**HaWa96**], which employed a one-step, A-stable, Runge-Kutta method. The one-step property of the Runge-Kutta method meant that BACOLR did not need to interpolate past values whenever a remeshing was performed and the A-stability of the Runge-Kutta method ensured that the code did not have difficulty with problems leading to DAE systems whose Jacobians had eigenvalues near the imaginary axis. On the other hand, there was substantially more work per step associated with the implementation of the Runge-Kutta method employed by RADAU5. Numerical comparisons of BACOL and BACOLR showed that the two codes had similar performance on several standard test problems and that BACOLR had much superior performance on problems for which the stability of the BDFs was an issue [**WKM07**].

The above findings were in contrast to those of [**Moor95**] in which a code called HPDASSL (a predecessor to HPNEW) was compared with a code called HPSIRK. The two codes used the same Galerkin-type spatial discretization and featured spatial mesh and spatial order adaptivity. For the time integration, HPDASSL used DASSL while HPSIRK used a singly-implicit Runge-Kutta method. The author found that HPDASSL was superior to HPSIRK.

A more recent paper [**GrPa07**] considered a comparison of EPDCOL, BACOL, and BACOLR for the numerical solution of mathematical models of fixed bed and simulated moving bed chromatography. For the simulated moving bed model, the authors presented error results and CPU time results; these showed that BACOL and BACOLR were able to obtain more accurate solutions, with better efficiency, than EPDCOL. Their results also allow a limited comparison of BACOL and BACOLR. For the case of significant mass transfer resistance, the results show that BACOLR, while slightly more accurate than BACOL (for a given tolerance), required somewhat more CPU time. For negligible mass transfer resistance, the results show that the two codes had comparable accuracy but that BACOLR was more efficient.

5. BACOLI, BACOLRI: new efficient spatial error estimation schemes

It was clear that the error estimation algorithm used by BACOL/BACOLR, involving the computation of a second, independent collocation solution, essentially doubled the cost of the computation. Arsenault, Smith, Muir, and Pew looked at two alternate approaches for obtaining a second low cost solution approximation against which the collocation solution could be compared to obtain the required spatial error estimates.

The first approach, called the SuperConvergent Interpolant (SCI) scheme, involved replacing the computation of the higher order collocation solution, $\bar{U}(x,t)$, with an interpolant which was based on superconvergent values of the collocation solution, $U(x,t)$. Arsenault, Smith, and Muir [**ASM2009**] reported on work involving the modification of BACOL to introduce the SCI error estimation scheme. The papers [**DoDu74, CePa76**] proved the existence of more accurate, i.e., superconvergent, solution and derivative approximations for a collocation solution evaluated at the spatial mesh points. In addition, corresponding results in the area of boundary value ODEs (BVODEs) suggested that other superconvergent solution values, at points internal to each subinterval, might also be present. Theorem 5.140/Corollary 5.142 of [**AMR95**] stated that (letting $u(x)$ be the exact solution of the BVODE and $U(x)$ be the collocation solution of the BVODE)

$$(8) \qquad u(x) - U(x) = u^{(k+2)}(x_i) P_k\left(\frac{x - x_i}{h_i}\right) h_i^{k+2} + O(h_i^{k+3}) + O(h^{2k}),$$

where $x_i < x < x_{i+1}$, $i = 0, \ldots, N-1$, $k = p-1$, is the number of collocation points per subinterval, $h_i = x_{i+1} - x_i$, $h = \max_{i=1}^{N} h_i$, $u^{(k+2)}(x_i)$ is the $(k+2)$ derivative of $u(x)$ evaluated at x_i, and the polynomial

$$(9) \qquad P_k(\xi) = \frac{1}{k!}\int_0^{\xi} (t-\xi) \prod_{r=1}^{k} (t - \rho_r) dt,$$

where $\{\rho_r\}_{r=1}^{k}$ are the images of the Gauss points on (0,1). To our knowledge, this result has not been generalized to the PDE case. However, Arsenault, Smith, Muir, and Keast [**ASMK11**] demonstrated experimentally, for the 1D PDE case, the existence of superconvergent points within each subinterval that corresponded to the zeros of the polynomial, P_k (9).

The key idea in the SCI scheme was to evaluate $U(x,t)$, generally of order $p+1$, at a sufficient number of superconvergent points (where $U(x,t)$ is of order $p+2$) in order to obtain enough superconvergent data to construct a C^1-continuous piecewise polynomial interpolant of order $p+2$ over the spatial domain. On each subinterval, the interpolant was a polynomial that interpolated (i) the superconvergent solution and derivatives values at each end of the subinterval, (ii) the $p-1$ superconvergent solution values available within the subinterval, and (iii) the two closest external superconvergent values immediately adjacent to the subinterval. This gave a sufficient number of superconvergent values so that the data error of the interpolated superconvergent values dominated the interpolation error. Since solution and derivative values were interpolated, a Hermite-Birkhoff representation for the interpolant was employed. The paper [**ASM2009**] provided experimental results that showed that it was usually the case that the SCI error estimates were comparable to the original BACOL error estimates. However, an issue for the SCI scheme was that because two of the interpolated values were taken from outside the subinterval on which each Hermite-Birkhoff interpolant was obtained, the interpolation error depended on the ratio of the size of that subinterval to the size of its immediately neighboring subintervals. When either of these ratios was large, the authors observed that SCI scheme would significantly overestimate the error.

The second alternative error estimation approach, considered by Arsenault, Smith, Muir, and Pew, was called the Lower Order Interpolant (LOI) scheme. In this scheme, $U(x,t)$, was replaced by an interpolant that was based on selected values of the higher order collocation solution, $\bar{U}(x,t)$. *The LOI was constructed so that it had the property that its interpolation error was asymptotically equivalent to the collocation error of* $U(x,t)$. (An earlier example of this approach is [**Moor01**].) In [**ASMP12**], Arsenault, Smith, Muir, and Pew described the LOI, a C^1-continuous piecewise polynomial, represented by a Hermite-Birkhoff interpolant on each subinterval. The interpolated values were (i) the mesh point values of $\bar{U}(x,t)$ and $\bar{U}_x(x,t)$ and (ii) the $p-1$ values of $\bar{U}(x,t)$ internal to each subinterval that corresponded to the internal superconvergent points associated with $U(x,t)$. The authors presented numerical results comparing the error estimates from the LOI scheme with those from the SCI scheme and the original BACOL error estimate. These results showed good agreement between the LOI and the BACOL error estimates, and generally good agreement with the SCI error estimates (with exceptions, as noted above, when the SCI scheme overestimated the spatial error.)

Recent work by Pew, Li, and Muir involved the development of a new version of BACOL, called BACOLI, that computed only one collocation solution and controlled the spatial error in that collocation solution using either the SCI or LOI error estimation scheme. Numerical results showed that BACOLI was almost twice as fast as BACOL. Similar very recently completed work by Pew and Muir involved the development of BACOLRI, a new version of BACOLR, in which only one collocation solution was computed and either the SCI or LOI error estimation scheme was used to compute the spatial error estimate. Preliminary numerical results showed that this new code was almost twice as fast as the original BACOLR.

6. Summary and Future Work

In this paper, we have provided a survey of the evolution of a family of 1D PDE software packages that employ B-spline Gaussian collocation as the spatial discretization scheme. Two important improvements were the move from ODE solvers to DAE solvers as the time-integrators (allowing for the direct treatment of the boundary conditions), and the move from fixed spatial mesh, non-adaptive codes, to codes that employ spatial adaptivity to control a high order estimate of the spatial error. The most recent codes in this family have demonstrated the effectiveness of efficient new interpolation-based spatial error estimation schemes.

This survey suggests a number of directions for future work. Further comparison of the SCI and LOI error estimation schemes should be performed to determine which of the two should become the primary spatial error estimation scheme in the latest codes in this family. The extension of the codes to include order adaptivity might also be worthwhile; this may require an extension to a more general linear algebra package. A major area of future work is to generalize the approach to 2D PDEs. Some work in this direction has been reported in [**LiMu12**].

References

ASM2009. T. Arsenault, T. Smith, P.H. Muir, Superconvergent interpolants for efficient spatial error estimation in 1D PDE collocation solvers, Can. Appl. Math. Q., 17, 2009, 409–431. MR2848313 (2012j:65352)

ASMK11. T. Arsenault, S. Smith, P.H. Muir, and P. Keast, Efficient interpolation-based error estimation for 1D time-dependent PDE collocation codes, Technical Report 2011_001, Department of Mathematics and Computing Science, Saint Mary's University, Halifax, N.S., 2011.

ASMP12. T. Arsenault, T. Smith, P.H. Muir, J. Pew, Asymptotically exact interpolation-based error estimates for collocation solutions of 1D PDEs, to appear, Can. Appl. Math. Q., 2012. MR2848313 (2012j:65352)

AMR95. U.M. Ascher, R.M.M. Mattheij, and R.D. Russell, *Numerical Solution of Boundary Value Problems for Ordinary Differential Equations*, Classics in Applied Mathematics, 13, Society for Industrial and Applied Mathematics, Philadelphia, PA, 1995. MR1351005 (96f:65075)

BlZe1994. J.G. Blom and P.A. Zegeling, Algorithm 731: A moving-grid interface for systems of one-dimensional time-dependent partial differential equations, ACM Trans. Math. Softw., 20, 1994, 194–214.

deBo78. C. de Boor, *A Practical Guide to Splines*, Applied Mathematical Sciences, 27, Springer-Verlag, New York, 1978. MR507062 (80a:65027)

BCP89. K.E. Brenan, S.L. Campbell, and L.R. Petzold, *Numerical Solution of Initial-Value Problems in Differential-Algebraic Equations*, Society for Industrial and Applied Mathematics, Philadelphia, PA, 1989. MR1101809 (92e:65001)

CePa76. J.H. Cerutti and S.V. Parter, Collocation methods for parabolic partial differential equations in one space dimension, Numer. Math., 26, 1976, 227–254. MR0433922 (55:6892)

DFK83. J.C. Diaz, G. Fairweather, and P. Keast, FORTRAN packages for solving certain almost block diagonal linear systems by modified alternate row and column elimination, ACM Trans. Math. Softw., 9, 1983, 358–375. MR791972

DoDu74. J. Douglas, Jr. and T. Dupont, *Collocation Methods for Parabolic Equations in a Single Space Variable*, Lecture Notes in Mathematics, 385, Springer-Verlag, Berlin, 1974. MR0483559 (58:3551)

FaMe89. G. Fairweather and D. Meade, A survey of spline collocation methods for the numerical solution of differential equations, J.C. Diaz (Ed.), Mathematics for Large Scale Computing, Lecture Notes in Pure and Applied Mathematics, 120, Marcel Dekker, New York, 1989, 297–341. MR1014015 (90m:65004)

GrPa07. N.S. Graca and L.S. Pais, Modelling and simulation of fixed-bed and simulated moving bed chromatographic process using PDECOL, EPDCOL, BACOL, and BACOLR software packages, CMNE/CILAMCE 2007, Oporto, Portugal, 2007.

HaWa96. E. Hairer and G. Wanner, *Solving Ordinary Differential Equations. II*, Springer Series in Computational Mathematics, 14, 2nd Ed., Springer-Verlag, Berlin, 1996. MR1439506 (97m:65007)

Hind77. A.C. Hindmarsh, GEARIB: Solution of ordinary differential equations having banded Jacobians, Technical Report UCID-30059, Lawrence Livermore National Laboratory, 1977.

HuRu96. W. Huang and R.D. Russell, A moving collocation method for solving time dependent partial differential equations, Appl. Numer. Math., 20, 1996, 101–116. MR1385237

KeMu91. P. Keast and P.H. Muir, Algorithm 688. EPDCOL: A more efficient PDECOL code, ACM Trans. Math. Softw., 17, 1991, 153–166.

LiMu12. Z. Li and P.H. Muir, B-Spline collocation software for two-dimensional, time-dependent parabolic PDEs, to appear in Advances in Applied Mathematics and Mechanics, 2012.

MaSi79. N.K. Madsen and R.F. Sincovec, Algorithm 540: PDECOL, general collocation software for partial differential equations, ACM Trans. Math. Softw., 5, 1979, 326–351.

Moor95. P.K. Moore, Comparison of adaptive methods for one dimensional parabolic systems, Appl. Numer. Math., 16, 1995, 471–488. MR1325260 (95m:65170)

Moor01. P.K. Moore, Interpolation error-based a posteriori error estimation for two-point boundary value problems and parabolic equations in one space dimension, Numer. Math., 90, 2001, 149–177. MR1868766 (2002h:65111)

NAG. D03PPF, NAG Fortran library, Mark 16A. The Numerical Algorithms Group, Oxford, U.K.

WKM04a. R. Wang, P.Keast, and P.H. Muir, A high-order global spatially adaptive collocation method for 1D parabolic PDEs, Appl. Numer. Math., 50, 2004, 239–260. MR2066739

WKM04b. R. Wang, P.Keast, and P.H. Muir, BACOL: B-spline Adaptive COLlocation software for 1D parabolic PDEs, ACM Trans. Math. Softw., 30, 2004, 454–470. MR2124439

WKM04c. R. Wang, P.Keast, and P.H. Muir, A comparison of adaptive software for 1D parabolic PDEs, J. Comput. Appl. Math., 169, 2004, 127–150. MR2071265

WKM07. R. Wang, P.Keast, and P.H. Muir, Algorithm 874: BACOLR—spatial and temporal error control software for PDEs based on high-order adaptive collocation, ACM Trans. Math. Softw., 34, 2008, Art. 15, 28. MR2444073

MATH. AND COMP. SCI., SAINT MARY'S UNIVERSITY, HALIFAX, N.S., CANADA B3H 3C3
E-mail address: muir@smu.ca

The central-upwind finite-volume method for atmospheric numerical modeling

Ramachandran D. Nair and Kiran K. Katta

ABSTRACT. A semi-discretized central-upwind finite-volume (CFV) scheme has been developed for atmospheric modeling applications. The non-oscillatory property of the scheme is achieved by employing high-order weighted essentially non-oscillatory (WENO) reconstruction method, and time integration relies on explicit Runge-Kutta method. The WENO reconstruction is fifth-order accurate and implemented in a dimension-split manner, and a fully 2D fourth-order reconstruction is also considered for comparison. The CFV scheme is computationally efficient and employs a compact non-staggered computational stencil with an optional positivity-preserving filter. The scheme has been validated for benchmark advection tests on the cubed-sphere. A global shallow-water model and a 2D non-hydrostatic Euler solver are also developed based on the same central finite-volume scheme.

1. Introduction

Finite-volume (FV) discretization has become a method of choice for many new generation atmospheric models, due to its inherent conservation properties and geometric flexibility to adapt various grid structures. A large class of FV methods for solving hyperbolic conservation laws are based on high-order extensions of the Godunov scheme [**G**], collectively known as the Godunov-type schemes. The central-upwind finite-volume (CFV) schemes [**KNP**, **KL**] are a subset of Godunov-type methods for solving hyperbolic conservation laws, and which combines the nice features of the classical upwind and central FV [**NT**] methods. In contrast with the upwind methods, the CFV schemes do not require characteristic decomposition of the hyperbolic system or expensive Riemann solvers. Semi-discrete formulation of CFV schemes avoid staggered grids and are relatively easy to implement, for practical applications. Our focus is on the application of CFV scheme for solving three major building blocks of a complex 3D atmospheric model: global linear transport equation with positivity preservation, shallow-water equations on the sphere which mimics the horizontal aspect of atmospheric dynamics, and the compressible 2D Euler equations in x-z plane for the vertical dynamics.

1991 *Mathematics Subject Classification.* Primary 86A10, 86-08; Secondary 76U99.

Key words and phrases. Central-upwind, finite-volume, shallow-water model, cubed-sphere, WENO method, non-hydrostatic Euler solver.

The first author was supported in part by DOE BER Program #DE-SC0001658. The National Center for Atmospheric Research is sponsored by NSF.

2. Semi-Discrete CFV Schemes

In order to describe the 2D CFV scheme, we first consider the following conservation law on (x, y) Cartesian plane with a source term S, as follows:

$$\frac{\partial U}{\partial t} + \nabla \cdot \mathbf{F}(U) = S(U) \text{ in } D \times (0, T], \tag{2.1}$$

where $U = U(x, y, t)$ is a conservative quantity with initial value $U_0 = U(x, y, 0)$, $\mathbf{F}(U)$ is a generic the flux function. The domain D is assumed to be rectangular and doubly periodic with non-overlapping rectangular cells $I_{ij} = [x_{i-1/2}, x_{i+1/2}] \otimes [y_{i-1/2}, y_{i+1/2}]$, with the grid-spacings Δx and Δy. Following [**KNP, KY**], the semi-discrete form on a rectangular cell I_{ij} with boundary Γ_k can be written as follows:

$$\frac{d\bar{U}_{ij}}{dt} = \frac{-1}{\Delta x \Delta y}\left[\sum_{k=1}^{4} \int_{\Gamma_k} \hat{\mathbf{F}} \cdot \mathbf{n}\right] + \bar{S}_{ij} \tag{2.2}$$

where $\bar{S}_{ij}$ is the the cell-averaged source term and, $\bar{U}_{ij}$ is the cell average which evolves in time and is subject to the following conservation constraint,

$$\bar{U}_{ij} = \frac{1}{\Delta x\, \Delta y} \int_{y_{j-1/2}}^{y_{j+1/2}} \int_{x_{i-1/2}}^{x_{i+1/2}} P_{ij}(x, y)\, dxdy. \tag{2.3}$$

In (2.3), $P_{ij}(x, y)$ is the piecewise polynomial function on I_{ij}, which is approximated by a suitable reconstruction function of targeted order of accuracy. We consider a fifth-order Weighted Essentially Non-Oscillatory (WENO5) 1D reconstruction [**BL**] with a dimension-split approach, and a fully 2D fourth-order reconstruction [**KY**]. For CFV with WENO5 reconstruction one flux-point on each cell walls are used (Fig. 1 left panel), resulting a simple FV discretization. However, with a fully 2D reconstruction each cell wall requires three flux evaluation (Fig. 1, right panel). For example, the flux integral in (2.2) on the east wall Γ_E for the fully 2D reconstruction is approximated by the Simpson's rule,

$$\int_{\Gamma_E} \hat{\mathbf{F}} \cdot \mathbf{n} \approx \frac{\Delta y}{6}\left[\hat{F}_{i+1/2,j-1/2} + 4\hat{F}_{i+1/2,j} + \hat{F}_{i+1/2,j+1/2}\right],$$

where $\hat{F}_{i+1/2,:}$ is the one-sided central-upwind flux formula [**KY**] which is dependent on the local speed $\alpha_{i+1/2,:}$ as follows (for convenience dependence on j is suppressed):

$$\hat{F}_{i+1/2} = \frac{\alpha^+_{i+1/2} F(U^-_{i+1/2}) - \alpha^-_{i+1/2} F(U^+_{i+1/2})}{\alpha^+_{i+1/2} - \alpha^-_{i+1/2}} + \frac{\alpha^+_{i+1/2}\alpha^-_{i+1/2}}{\alpha^+_{i+1/2} - \alpha^-_{i+1/2}}\left[U^+_{i+1/2} - U^-_{i+1/2}\right]$$

The local speed is given by the eigenvalues λ_ℓ of the flux Jacobian at the left ($^-$) and right ($^+$) sides of the cell interface such that

$$\alpha^+_{i+1/2} = \max\left[\lambda_\ell\left(\frac{\partial F}{\partial U}\right), 0\right], \quad \alpha^-_{i+1/2} = \min\left|\lambda_\ell\left(\frac{\partial F}{\partial U}\right), 0\right|$$

The final form of the semi-discrete CFV scheme reduces to the following ODE:

$$\frac{d\bar{U}_{ij}}{dt} = L(\bar{U}_{ij}), \tag{2.4}$$

which can be solved using a high-order explicit strong stability-preserving Runge-Kutta method [**GS, KNK**].

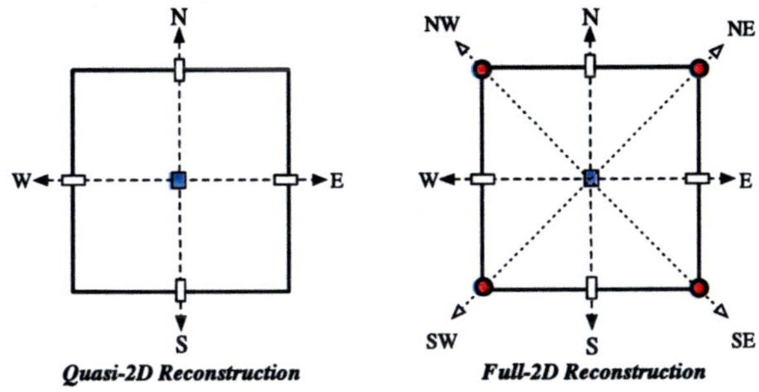

FIGURE 1. Schematic of reconstruction on CFV cells, where marked points on the cell walls denote the flux points on which fluxes are evaluated. The left panel shows a cell requiring two 1D reconstruction (WENO5) in each coordinate direction. The right panel shows a cell requiring fully 2D reconstruction for which 3 flux points on each wall.

3. The shallow water model on the cubed-sphere

The physical domain is a sphere $\mathcal{S}$, based on the cubed-sphere geometry [**SA**]. The cubed-sphere geometry consists of partitioning $\mathcal{S}$ into six identical regions which are obtained by the equiangular central projection (gnomonic) projection [**N**] of the faces of the inscribed cube onto the surface of $\mathcal{S}$. Each of the local coordinate systems is free of singularities, and creates a non-orthogonal curvilinear coordinate system on $\mathcal{S}$. However, the edges of the cubed-sphere faces are not continuous. The local coordinates (or central angles of the projection) for each face are $x^1 = x^1(\lambda, \theta)$, $x^2 = x^2(\lambda, \theta)$ such that $x^1, x^2 \in [-\pi/4, \pi/4]$, where λ and θ are the longitude and latitude, respectively, of a sphere with radius R. The metric tensor, G_{ij} associated with the transformation is

$$G_{ij} = \frac{R^2}{\rho^4 \cos^2 x^1 \cos^2 x^2} \begin{bmatrix} 1 + \tan^2 x^1 & -\tan x^1 \tan x^2 \\ -\tan x^1 \tan x^2 & 1 + \tan^2 x^2 \end{bmatrix},$$

where $i, j \in \{1, 2\}$ and $\rho^2 = 1 + \tan^2 x^1 + \tan^2 x^2$. The Jacobian of the transformation (the metric term) is $\sqrt{G} = [\det(G_{ij})]^{1/2}$.

The flux-form shallow water (SW) model developed on the cubed-sphere relies on non-orthogonal curvilinear coordinates [**SA**]. The SW equations are treated in tensorial form with covariant (u_i) and contravariant (u^i) wind vectors, which are related through $u_i = G_{ij}u^j$, $u^i = G^{ij}u_j$ and $G^{ij} = G_{ij}^{-1}$; where $i, j \in \{1, 2\}$. The orthogonal components of the spherical wind vector $\mathbf{v}(\lambda, \theta) = (u, v)$ can be expressed in terms of contravariant vectors (u^1, u^2) as follows,

$$\begin{bmatrix} u \\ v \end{bmatrix} = A \begin{bmatrix} u^1 \\ u^2 \end{bmatrix}, \quad A^T A = G_{ij}.$$

The details of the local transformation laws and A for each face of the cubed-sphere can be found in [**N**].

The governing equations for inviscid shallow-water flow (on a rotating sphere) of a thin layer of fluid (2D) are the horizontal momentum and the continuity equations for the height h. Here, h is the depth of the fluid and it is related to the free surface geopotential height (above sea level) $\Phi = g\,(h_s + h)$, where h_s denotes the height

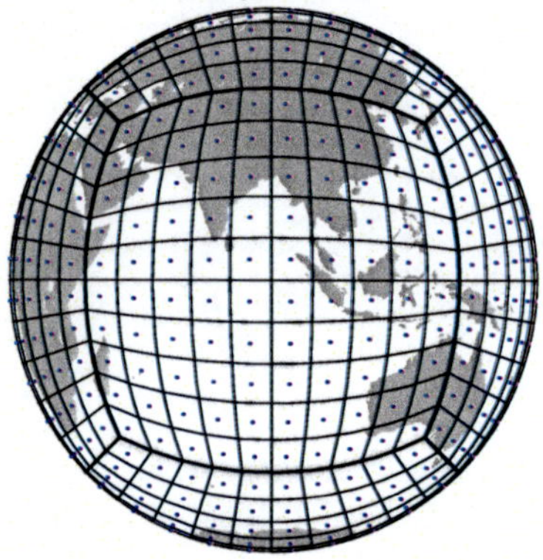

FIGURE 2. A cubed-sphere with $10 \times 10 \times 6$ FV cells (volumes), which span the entire surface of the sphere $\mathcal{S}$. On the computational domain the grid cells are uniformly spaced ($\Delta x^1 = \Delta x^2$). The halo regions required for CFV stencil at the cubed-sphere face edges are created with 1D interpolation along the grid lines.

of the underlying mountains and g is the gravitational acceleration. Thus the prognostic variables are u_1, u_2 and h, and the shallow-water equations on $\mathcal{S}$ can be written in a compact form following the formulation [**NTL**]:

$$\frac{\partial}{\partial t}\mathbf{U} + \frac{\partial}{\partial x^1}\mathbf{F}_1(\mathbf{U}) + \frac{\partial}{\partial x^2}\mathbf{F}_2(\mathbf{U}) = \mathbf{S}(\mathbf{U}), \tag{3.1}$$

where the state vector $\mathbf{U}$ and the flux vectors $\mathbf{F}_1, \mathbf{F}_2$ are defined by

$$\mathbf{U} = \left[u_1, u_2, \sqrt{G}\,h\right]^T, \quad \mathbf{F}_1 = \left[E, 0, \sqrt{G}hu^1\right]^T, \quad \mathbf{F}_2 = \left[0, E, \sqrt{G}hu^2\right]^T,$$

$E = \Phi + \frac{1}{2}(u_1\,u^1 + u_2\,u^2)$ is the energy term. The source terms $\mathbf{S}$ is a function of relative vorticity ζ, Coriolis term $f = 2\omega \sin\theta$, and the contravariant wind vectors (u^1, u^2), defined as:

$$\mathbf{S}(\mathbf{U}) = \left[\sqrt{G}u^2(f+\zeta), -\sqrt{G}\,u^1(f+\zeta), 0\right]^T, \qquad \zeta = \frac{1}{\sqrt{G}}\left[\frac{\partial u_2}{\partial x^1} - \frac{\partial u_1}{\partial x^2}\right] \tag{3.2}$$

3.1. Numerical Experiments.

3.1.1. *Conservative transport on the cubed-sphere.* The transport problem has fundamental importance for atmospheric modeling. Therefore the first test we consider is a linear transport problem, a solid-body rotation of cosine profile (cosine-bell) on the sphere, which is a benchmark test suggested in [**W**]. The transport equation on the cubed-sphere for a scalar h can be written as [**N**]:

$$\frac{\partial \psi}{\partial t} + \frac{\partial u^1\psi}{\partial x^1} + \frac{\partial u^2\psi}{\partial x^2} = 0, \tag{3.3}$$

where $\psi = \sqrt{G}\,h$. Note that Eq.(3.3) may be considered as a special case of (3.1) with the prescribed velocity fields.

The non-divergent velocity filed is defined to be $u = u_0(\cos\theta + \sin\lambda\sin\theta)/\sqrt{2}$, $v = -(u_0 \sin\lambda)/\sqrt{2}$. The parameter $u_0 = 2\pi R/(12 \text{ days})$, for a sphere with radius R, scaled in such a way that the wind field translates the initial cosine-bell in the

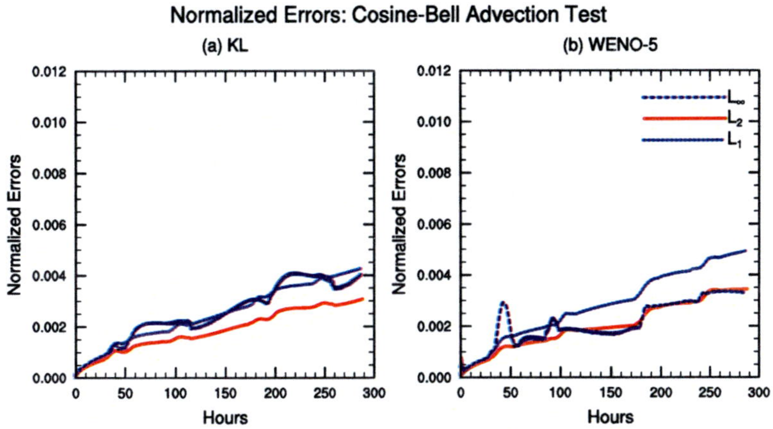

FIGURE 3. Time traces of normalized l_1, l_2, l_∞ errors for the solid-body rotation on the sphere for a complete revolution (12 days).

north-east direction, with a period of 12 days for a complete revolution. Since the exact solution is known, time tracers of the error measures (l_1, l_2, l_∞) can be computed. Initially the cosine-bell $h(\lambda, \theta, t)$ is centered at $(\lambda = 3\pi/2, \theta = 0)$ with a base radius $r_0 = R/3$

$$h(\lambda, \theta, t = 0) = \begin{cases} 500\,[1 + \cos(\pi r_d/r_0)] & \text{if } r_d < r_0 \\ 0 & \text{if } r_d \geq r_0 \end{cases},$$

where r_d is the great-circle distance between (λ, θ) and the centre of the cosine-bell.

The CFV schemes employing a fully 2D fourth-order reconstruction and a dimension split fifth-order Weighted Essentially Non-Oscillatory (WENO5) reconstruction methods are used for the transport equation (3.3). The height of the cosine-bell upon initialization is $h \in [984.2, 0]$, after a full revolution with the positivity-preserving filter [**KNK**] the height $h \in [978.6, 0]$. The resolution is $45 \times 45 \times 6$ ($\approx 2°$) with CFL 0.25. Figure 3 shows the time evolution of l_1, l_2 and l_∞ errors. The fourth-order scheme scores slightly better than the split WENO5 in terms of accuracy. However, as far as the computational efficiency and ease of implementation on the cubed-sphere are concerned, the WENO5 turned out to be a better option. Hereafter, for the numerical experiments, we only report the results with the CFV scheme employing WENO5 reconstruction.

3.1.2. *Zonal flow over an isolated mountain.* This test case is particularly useful for studying the effectiveness of the numerical scheme in conserving integral invariants such as mass and total energy. For this benchmark test, the flow filed is highly nonlinear and no analytic solution is available, a complete description of this test is given in [**W**]. The initial velocity $(u, v) = (u_0 \cos\theta, 0)$ and height field is given by $g\,h = g\,h_0 - \frac{u_0}{2}(2\,a\,\omega + u_0)\sin^2\theta$, where R and ω are the earth's radius and angular velocity, respectively; $u_0 = 2\pi R/(12 \text{ days})$, and $gh_0 = 2.94 \times 10^4$ m^2/s^2. The *mountain* (circular cone) is centered at $(\lambda_c = 3\pi/2, \theta_c = \pi/6)$ with height $h_s = 2000\,(1 - r/a)$ m, where $a = \pi/9$ and $r^2 = \min[a^2, (\lambda - \lambda_c)^2 + (\theta - \theta_c)^2]$. The mean equivalent depth of the atmosphere is set to be $h_0 = 5960$ m. Figure 4 shows simulated results (WENO5) after 2 and 15 days, where an approximate resolution of $2°$, and CFL ≈ 0.5 were used.

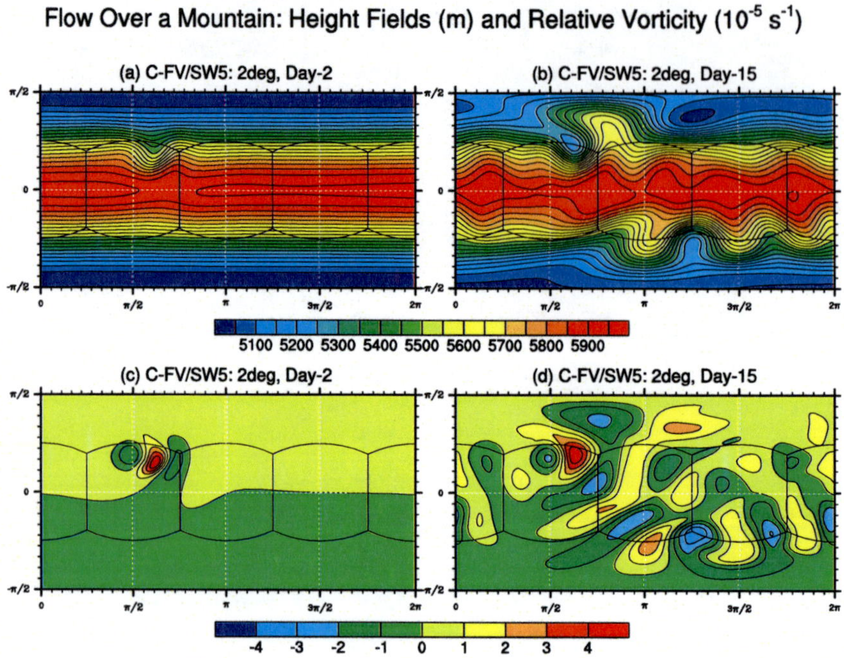

FIGURE 4. Height (top) and vorticity (bottom) fields after 2 and 15 days.

4. CFV Non-Hydrostatic 2D Model

Here we demonstrate the application of CFV for a non-hydrostatic (NH) atmospheric model, in a simple 2D (x-z) Cartesian setup. We have developed a compressible Euler solver based on CFV, which employs the following atmospheric adaptation of compressible Euler system [**GR, NN**]

$$\frac{\partial}{\partial t}\mathbf{U} + \frac{\partial}{\partial x}\mathbf{F}(\mathbf{U}) + \frac{\partial}{\partial z}\mathbf{G}(\mathbf{U}) = \mathbf{S}(\mathbf{U}) \tag{4.1}$$

where

$$\mathbf{U} = \begin{bmatrix} \rho' \\ \rho u \\ \rho w \\ \rho\Theta' \end{bmatrix}, \quad \mathbf{F} = \begin{bmatrix} \rho u \\ \rho u^2 + p' \\ \rho uw \\ \rho u\Theta \end{bmatrix}, \quad \mathbf{G} = \begin{bmatrix} \rho w \\ \rho wu \\ \rho w^2 + p' \\ \rho w\Theta \end{bmatrix}, \quad \mathbf{S} = \begin{bmatrix} 0 \\ 0 \\ -\rho' g \\ 0 \end{bmatrix}.$$

In the above system, ρ is the density of fluid, u is the velocity component in the x-direction (horizontal), w is the velocity component in the z-direction (vertical), p is the pressure, Θ is the potential temperature and $\mathbf{S}$ is the source term. The variables ρ, Θ and p are decomposed as the sum of mean state $\bar{(.)}$ and perturbation $(.)'$; $\rho = \bar{\rho} + \rho'$, $p = \bar{p} + p'$ and $\Theta = \bar{\Theta} + \Theta'$, such that the mean-state holds hydrostatic relation $d\bar{p}/dz = -\bar{\rho}g$.

The potential temperature Θ and the real temperature T follow the Exner relation $\Theta = T(p_0/p)^{R_d/c_p}$. The system (4.1) is closed by the equation of state, $p = C_0(\rho\Theta)^\gamma$ where $C_0 = R_d^\gamma p_0^{-R_d/c_v}$. The reference surface pressure $p_0 = 10^5$ Pa and the other thermodynamic constants are: $\gamma = c_p/c_v$, R_d =287 J kg^{-1} K^{-1}, c_p = 1004 J kg^{-1} K^{-1}, c_v = 717 J kg^{-1} K^{-1}.

Figure 4 shows a rising bubble [**WS**, **NN**] in a convectively neutral atmosphere simulated with the CFV scheme with WENO5 reconstruction procedure. The convective thermal uses a hydrostatic balance based on a constant potential temperature with zero initial wind, and the bubble is *triggered* by perturbing the potential temperature. The model domain is $[0, 20]$ km $\times[0, 10]$ km, and grid-spacing is uniformly set to $\Delta x = \Delta z = 133$ m such that CFL ≈ 0.65, with an explicit third-order Ruge-Kutta integrator [**GS**]. Top panels in Fig. 4 shows the initial and final (1000s) thermal bubble (Θ'), and lower panels show the horizontal and vertical wind fields. The CFV Euler solver simulates bubble structure quite well as compared to other high-order model results [**WS**, **NN**].

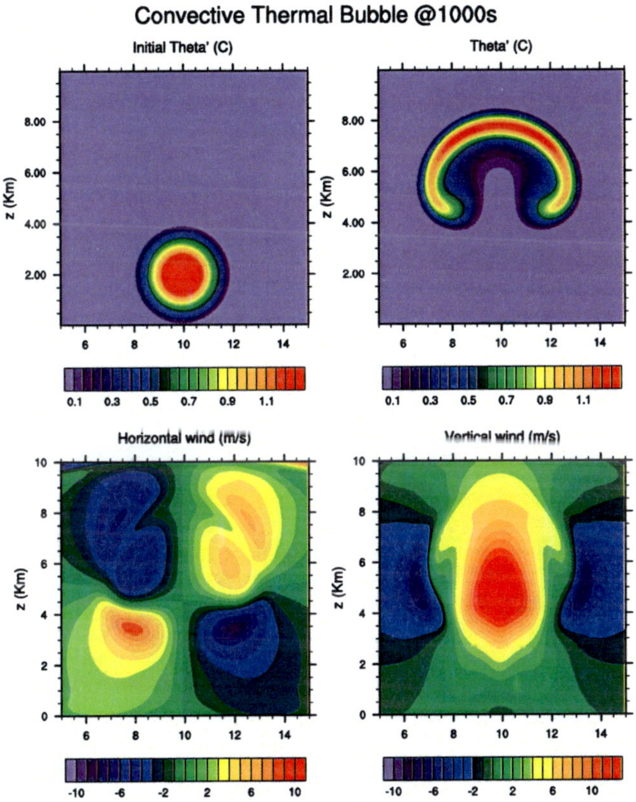

FIGURE 5. Convection in a neutral atmosphere simulated by the CFV Euler solver. Initial potential temperature perturbation (Θ') is shown in the top left panel and top right panel shows the convective bubble after 1000 s of simulation, lower left and right panels show the wind fields u and w at 1000 s, respectively.

5. Summary

We have developed a central-upwind finite volume (CFV) scheme for atmospheric modeling applications. A novel feature of this method is its simplicity, in terms of implementation. The CFV approach does not rely on characteristic decomposition or expensive Riemann solver. The CFV scheme combined with the fifth-order WENO reconstruction employed in a dimension-split manner. The high-order dimension-split approach on the cubed-sphere did not create any significant accuracy issue as compared to a fourth-order fully 2D reconstruction. The tracer transport with CFV has the positivity-preserving option. The nonlinear shallow-water results are comparable to those with high-order conventional upwind based Godunov-type FV schemes [**CX**, **UJ**]. The Euler solver based on CFV results are promising. The 2D CFV schemes considered herein will be further extended to a 3D non-hydrostatic atmospheric model in future.

References

[BL] Steve Bryson and Doron Levy, *On the total variation of high-order semi-discrete central schemes for conservation laws*, J. Sci. Comput. **27** (2006), no. 1-3, 163–175, DOI 10.1007/s10915-005-9046-8. MR2285773 (2008c:65197)

[CX] Chungang Chen and Feng Xiao, *Shallow water model on cubed-sphere by multi-moment finite volume method*, J. Comput. Phys. **227** (2008), no. 10, 5019–5044, DOI 10.1016/j.jcp.2008.01.033. MR2414846 (2009e:86001)

[GR] F. X. Giraldo and M. Restelli, *A study of spectral element and discontinuous Galerkin methods for the Navier-Stokes equations in nonhydrostatic mesoscale atmospheric modeling: equation sets and test cases*, J. Comput. Phys. **227** (2008), no. 8, 3849–3877, DOI 10.1016/j.jcp.2007.12.009. MR2403870 (2009c:86009)

[GS] Sigal Gottlieb, Chi-Wang Shu, and Eitan Tadmor, *Strong stability-preserving high-order time discretization methods*, SIAM Rev. **43** (2001), no. 1, 89–112 (electronic), DOI 10.1137/S003614450036757X. MR1854647 (2002f:65132)

[G] S. K. Godunov, *A difference method for numerical calculation of discontinuous solutions of the equations of hydrodynamics*, Mat. Sb. (N.S.) **47 (89)** (1959), 271–306 (Russian). MR0119433 (22 #10194)

[KNK] K. Katta, R. D. Nair, and V. Kumar, *High-order central finite-volume schemes for linear transport problems on the cubed-sphere*, Quart. J. Roy. Meteor. Soc., 2012, Submitted.

[KNP] Alexander Kurganov, Sebastian Noelle, and Guergana Petrova, *Semidiscrete central-upwind schemes for hyperbolic conservation laws and Hamilton-Jacobi equations*, SIAM J. Sci. Comput. **23** (2001), no. 3, 707–740 (electronic), DOI 10.1137/S1064827500373413. MR1860961 (2003a:65065)

[KL] Alexander Kurganov and Doron Levy, *Central-upwind schemes for the Saint-Venant system*, M2AN Math. Model. Numer. Anal. **36** (2002), no. 3, 397–425, DOI 10.1051/m2an:2002019. MR1918938 (2003d:76115)

[KY] A. Kurganov and Y. Liu, *New adaptive artificial viscosity method for hyperbolic systems of conservation laws*, J. of Comput. Phys, 2012, submitted.

[N] R. D. Nair, S. J. Thomas, and R. D. Loft, *A discontinuous Galerkin transport scheme on the cubed-sphere*, Mon. Wea. Rev., **133**, 2005, pp. 814–828.

[NTL] R. D. Nair, S. J. Thomas, and R. D. Loft, *A discontinuous Galerkin global shallow water model*, Mon. Wea. Rev., **133**, pp. 876–888.

[NN] Matthew R. Norman, Ramachandran D. Nair, and Fredrick H. M. Semazzi, *A low communication and large time step explicit finite-volume solver for non-hydrostatic atmospheric dynamics*, J. Comput. Phys. **230** (2011), no. 4, 1567–1584, DOI 10.1016/j.jcp.2010.11.022. MR2753379 (2012c:86008)

[NT] Haim Nessyahu and Eitan Tadmor, *Nonoscillatory central differencing for hyperbolic conservation laws*, J. Comput. Phys. **87** (1990), no. 2, 408–463, DOI 10.1016/0021-9991(90)90260-8. MR1047564 (91i:65157)

[SA] R. Sadourny, *Conservative finite-difference approximations of the primitive equations on quasi-uniform spherical grids*, Mon. Wea. Rev., **100**, 1972, pp. 136–144.

[UJ] Paul A. Ullrich, Christiane Jablonowski, and Bram van Leer, *High-order finite-volume methods for the shallow-water equations on the sphere*, J. Comput. Phys. **229** (2010), no. 17, 6104–6134, DOI 10.1016/j.jcp.2010.04.044. MR2657861 (2011d:76069)

[W] David L. Williamson, John B. Drake, James J. Hack, Rüdiger Jakob, and Paul N. Swarztrauber, *A standard test set for numerical approximations to the shallow water equations in spherical geometry*, J. Comput. Phys. **102** (1992), no. 1, 211–224, DOI 10.1016/S0021-9991(05)80016-6. MR1177513 (93d:86006)

[WS] L. J. Wicker and W. C. Skamarock, *A time-splitting scheme for the elastic equations incorporating second-order Runge-Kutta time differencing* Mon. Wea. Rev., **126**, pp. 1992–1999

Institute for Mathematics Applied to Geosciences, National Center for Atmospheric Research, 1850 Table Mesa Drive, Boulder, Colorado 80305
E-mail address: rnair@ucar.edu

The University of Texas at El Paso, 500 W University Avenue, El Paso, Texas 79902
E-mail address: kkkatta@miners.utep.edu

Sensitivity computations of the Leray-α model

Monika Neda, Faranak Pahlevani, and Jiajia Waters

ABSTRACT. We present a computational study of the sensitivity of the Leray-α model with respect to the filter width α on the 2D Cavity problem. The goal of the study is to compute and compare the sensitivity of the model using two different techniques, forward finite difference method and sensitivity equation method, and identify the range of α values for which the velocity approximation via Leray-α model is considered accurate while being computationally efficient. Sensitivity for the 3D Eithier-Steinman problem was investigated as well.

1. Introduction

The study of sensitivity has become an important tool in analyzing the fluid behavior. Even when a fluid flow model has performed well in practice, the reliability of the model is often not addressed [**1**]. The reliability can be affected if the model displays sensitivity to certain parameters. Sensitivity analysis eliminates the arising uncertainties and provides a reliable interval for the parameters to be chosen from, e.g. see [**12, 13**] . Over the years, there have been investigations on the sensitivity topics, see [**1, 3, 8, 12, 13, 14, 15**]. Two methods could be used in order to calculate the sensitivities: Forward Finite Difference method (FFD) or the Sensitivity Equation Method (SEM), see [**16, 12**].

Accurate direct numerical simulation (DNS) of the Navier–Stokes equations (NSE) at high Reynolds numbers is well known to be a formidable computational challenge. For this reason, regularization models of the NSE have been studied since these models allow for more efficient simulations of high Reynolds-number flows by resolving only large-scale flow structures. In 1934, J. Leray [**11**] introduced the following regularization of the NSE (currently known as the Leray model) as a theoretical tool

$$(1.1) \qquad \mathbf{u}_t + \overline{\mathbf{u}} \cdot \nabla \mathbf{u} - \nu \Delta \mathbf{u} + \nabla p = \mathbf{f} \text{ and } \nabla \cdot \mathbf{u} = 0\,, \quad \text{in } \Omega \times (0,T).$$

Here, $\mathbf{u}$ and p represent velocity vector and pressure respectively, ν represents the viscosity and $\mathbf{f}$ represents the body. Initially, Leray chose $\overline{\mathbf{u}} = g_\alpha \star \mathbf{u}$, where g_α is a Gaussian associated with a length scale α, and proved existence and uniqueness of strong solutions to (1.1). The good theoretical properties have brought a re-examination of the Leray model (1.1) with the modification that the Gaussian filter is replaced by a less expensive differential filter,

$$(1.2) \qquad -\alpha^2 \Delta \overline{\mathbf{u}} + \overline{\mathbf{u}} = \mathbf{u}.$$

2010 *Mathematics Subject Classification.* Primary.

Through the years, the theory and computations of the model were studied in [**7, 5, 18, 10, 4**] and their references.

Herein, we explore the sensitivity of a Leray-α model with respect to the filter width α using both FFD and SEM for the 2D Cavity problem and also present the sensitivity computation for the 3D Eithier-Steinman problem. Similar sensitivity computational studies of cavity problem using an eddy viscosity type of models are presented in [**12, 13**].

2. Numerical Scheme

The approach of SEM yields the following system of equations by implicitly differentiating the Leray-α model with respect to the parameter α,

$$
\begin{aligned}
\mathbf{s}_t + \mathbf{w}\cdot\nabla\mathbf{u} + \overline{\mathbf{u}}\cdot\nabla\mathbf{s} - \nu\Delta\mathbf{s} + \nabla q &= 0, \quad \text{in } \Omega\times(0,T),\\
\nabla\cdot\mathbf{s} &= 0, \quad \text{in } \Omega\times(0,T),\\
\mathbf{s}(x,0) &= 0, \quad \text{in } \Omega.
\end{aligned} \tag{2.1}
$$

where $\mathbf{s} = \frac{\partial\mathbf{u}}{\partial\alpha}$ is the sensitivity of velocity, $q = \frac{\partial p}{\partial\alpha}$ is the sensitivity of pressure, and $\mathbf{w} = \frac{\partial\overline{\mathbf{u}}}{\partial\alpha}$ is the sensitivity of filtered velocity. Here, $\mathbf{w}$ satisfies sensitivity equation of filter problem, i.e.

$$
-\alpha^2\Delta\mathbf{w} + \mathbf{w} = -\frac{2}{\alpha}(\mathbf{u} - \overline{\mathbf{u}}) + \mathbf{s}. \tag{2.2}
$$

This section presents the algorithm for solving numerically (1.1) and (1.2), coupled with (2.1) and (2.2). The finite element spaces are $(\mathbf{X}^h, Q^h) \subset (H_0^1(\Omega), L_0^2(\Omega))$ where Ω is the domain and

$$
H_0^1(\Omega) := \{v \in H^1(\Omega) : v|_{\partial\Omega} = 0\}, \text{ and } L_0^2(\Omega) := \{q \in L^2(\Omega) : \int_\Omega q = 0\}.
$$

Also, trilinear $b^*(\cdot,\cdot,\cdot) : X \times X \times X \to R$ form is defined as $b^*(\mathbf{u},\mathbf{v},\mathbf{w}) := \frac{1}{2}(\mathbf{u},\nabla\mathbf{v},\mathbf{w}) - \frac{1}{2}(\mathbf{u},\nabla\mathbf{w},\mathbf{v})$, see [**9**] for more information.

Discrete approximation solutions of the Leray-α model in equations (1.1)-(1.2) on the time interval $(0,T]$, read as following: Given $(\mathbf{u}_h^n, p_h^n)$, find $(\mathbf{u}_h^{n+1}, p_h^{n+1}) \in (X_h, Q_h)$, for $n = 1,2,3,\ldots,N_T$, such that

$$
\begin{aligned}
\frac{1}{\Delta t}(\mathbf{u}_h^{n+1} - \mathbf{u}_h^n, \mathbf{v}) + b^*(\overline{\mathbf{u}}_h^{n+1/2}, \mathbf{u}_h^{n+1/2}, \mathbf{v}) + \nu(\nabla\mathbf{u}_h^{n+1/2}, \nabla\mathbf{v}) - (p_h^{n+1}, \nabla\cdot\mathbf{v})&\\
= (\mathbf{f}_h^{n+1/2}, \mathbf{v}), &\quad \forall\mathbf{v}\in\mathbf{X}_h,\\
(\nabla\cdot\mathbf{u}_h^{n+1/2}, q) = 0, &\quad \forall q\in Q_h,\\
\alpha^2(\overline{\mathbf{u}}_h^n, \nabla\mathbf{v}) + (\overline{\mathbf{u}}_h^n, \mathbf{v}) - (\mathbf{u}_h^n, \mathbf{v}) = 0, &\quad \forall\mathbf{v}\in\mathbf{X}_h.
\end{aligned} \tag{2.3}
$$

Similarly, for sensitivity equations (2.1)-(2.2) on the time interval $(0,T]$ we have: Given $(\mathbf{s}_h^n, q_h^n)$, we seek $(\mathbf{s}_h^{n+1}, q_h^{n+1}) \in (\mathbf{X}_h, Q_h)$, for $n = 1,2,3,\ldots,N_T$, satisfying

$$
\begin{aligned}
\frac{1}{\Delta t}(\mathbf{s}_h^{n+1} - \mathbf{s}_h^n, \mathbf{v}) + b^*(\mathbf{w}_h^{n+1/2}, \nabla\mathbf{u}_h^{n+1/2}, \mathbf{v}) + b^*(\overline{\mathbf{u}}_h^{n+1/2}, \nabla\mathbf{s}_h^{n+1/2}, \mathbf{v})&\\
+\nu(\nabla\mathbf{s}_h^{n+1/2}, \nabla\mathbf{v}) - (q_h^{n+1/2}, \nabla\cdot\mathbf{v}) = 0, &\quad \forall\mathbf{v}\in\mathbf{X}_h,\\
(\nabla\cdot\mathbf{s}_h^{n+1/2}, r) = 0, &\quad \forall r\in Q_h,\\
\alpha^2(\nabla\mathbf{w}_h^n, \nabla v) + (\mathbf{w}_h^n, v) - (\mathbf{s}_h^n, \mathbf{v}) + \frac{2}{\alpha}(\mathbf{u}_h^n - \overline{\mathbf{u}}_h^n, \mathbf{v}) = 0, &\quad \forall\mathbf{v}\in\mathbf{X}_h.
\end{aligned} \tag{2.4}
$$

3. Sensitivity Computations

In this section, we present the following computations regarding our sensitivity studies for the 2D Cavity problem and 3D Either-Steinman problem. The first experiment illustrates a comparison of the sensitivity computation using two different techniques, SEM and FFD. SEM uses the discretized sensitivity equations (2.4) and FFD is obtained using $\frac{\mathbf{u}(\alpha+\Delta\alpha)-\mathbf{u}(\alpha)}{\Delta\alpha}$, by computing $\mathbf{u}$ from (2.3) for two different filter lengths $\alpha + \Delta\alpha$ and α. To carry out the computations we used the software FreeFem++, [**17**].

3.1. 2D Cavity Problem. The flow domain Ω is $[0,1] \times [0,1]$ with upper boundary condition $\mathbf{u} = (16x^2(1-x)^2, 0)^T$ and zero everywhere else. The initial data is $\mathbf{u}(0,x,y) = (3y^2 - 2y, 0)^T$ in Ω. Since initial and boundary conditions have no dependence on α, they are set to zero for the sensitivity $\mathbf{s}$.

All the computations are carried out with a fixed mesh size $h = \frac{1}{36}$, and a uniform time step $\Delta t = 0.01$ for different viscosities corresponding to Reynolds numbers of 5000, 10000, and 50000. Let s_{FFD} and s_{SEM} represent the sensitivity computations using FFD and SEM, respectively. Figures 1-3 present $\|s_{SEM}(t)\|_{L^2(\Omega)}$ and $\|s_{FFD}(t)\|_{L^2(\Omega)}$ for $\alpha = kh$, where $k = \frac{1}{4}, \frac{1}{2}, 1, 2, 4$ with $\Delta\alpha = 0.001$ at times $t = 0.1$, and 1.

One observes that for all α, the computed sensitivity via FFD is overall larger than the one computed via SEM in all the tested Reynolds numbers. In both methods, sensitivity norm increases as α takes on values closer to 0 demonstrating that the average velocity $\mathbf{u}$ is more sensitive with respect to smaller values of α. A larger scale of sensitivity is obtained as time has progressed to $t = 1$. Larger Reynolds numbers show larger sensitivities especially at $t = 1$.

Table 1 presents the difference between $\|s_{SEM}\|_{L^2(\Omega)}$ and $\|s_{FFD}\|_{L^2(\Omega)}$ for different values of Reynolds number at $t = 1$. Sensitivity values for $\alpha = 2h$, and $4h$ are apart up to 0.25 in all cases and there is an increase in the difference as α decreases. For high Reynolds number, i.e. $Re = 10000$, and 50000, the difference in sensitivity norms is noticeable for $\alpha = \frac{1}{2}h$, and $\frac{1}{4}h$.

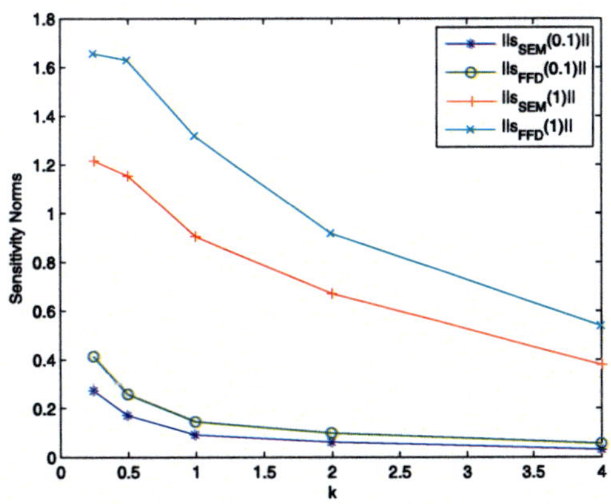

FIGURE 1. Sensitivity norms via SEM and FFD for $Re = 5000$

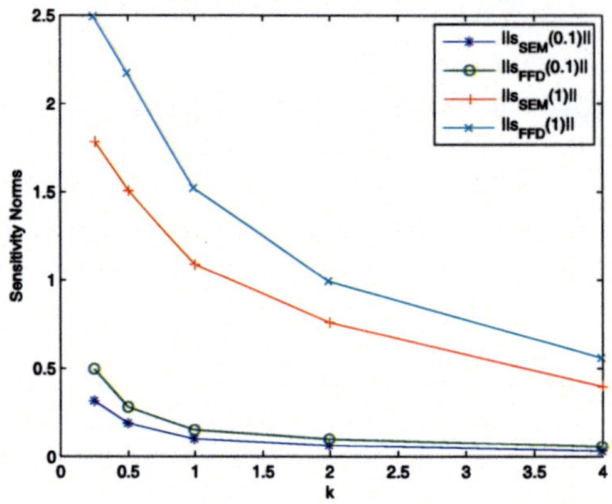

FIGURE 2. Sensitivity norms via SEM and FFD for $Re = 10000$

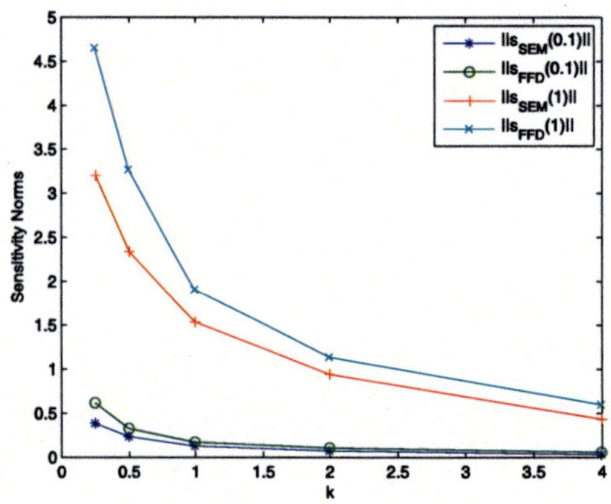

FIGURE 3. Sensitivity norms via SEM and FFD for $Re = 50000$

TABLE 1. Difference between $\|s_{SEM}\|$ and $\|s_{FFD}\|$ at $t = 1$ with $h = \frac{1}{36}$

α	$Re = 5000$	$Re = 10000$	$Re = 50000$
$4h$	0.1586	0.1603	0.1586
$2h$	0.2458	0.2335	0.1955
h	0.4118	0.4315	0.3685
$\frac{1}{2}h$	0.4732	0.6689	0.9310
$\frac{1}{4}h$	0.4394	0.7133	1.4550

The sensitivity of the average velocity $\mathbf{u}$ with respect to the variations of parameter α for different Reynolds numbers is tested by computing $\|\mathbf{s}\|_{l^2(0,T;L^2)}$ (note

$\|u\|_{l^2(0,T;L^2)} = \left[\Delta t \sum_{i=0}^{N} \|u(i\Delta t)\|_{L^p}^q\right]^{1/q}$) using sensitivity equations in (2.4) for final time $T = 1$. The numerical results obtained from these computations is shown in Figure 4. Note that by selecting large values for the parameter α, e.g. larger than $4h$, all the velocity scales that are less or equal to α are filtered. Hence, an approximated average velocity $\mathbf{u}$ corresponding to large values of α is not considered as accurate as the approximated velocity using smaller values of this parameter and therefore not reliable. Specially in the case of high Reynolds numbers where the flow structure contains a lot of small scales maintaining a reliable solution versus computational feasibility of the solution becomes important. Given that $\mathbf{u}$ depends on α implicitly, $\mathbf{u}(0)$ indicates the true solution to Navier-Stokes equations while $\mathbf{u}(\alpha)$ for $0 < \alpha < 1$ denotes the approximated average velocity from (2.3). Using Taylor expansion, $\mathbf{u}(0) = \mathbf{u}(\alpha) - \alpha\mathbf{s} + O(\alpha^2)$, the reliable α values are restricted to values with $O(\alpha^2)$ less than 0.01. Considering the fact that a higher sensitivity for smaller values of parameter α indicates the rise of computational complexity, the numerical results in Figure 4 suggest the following interval of α values as the optimal choice for the tested Reynolds numbers.

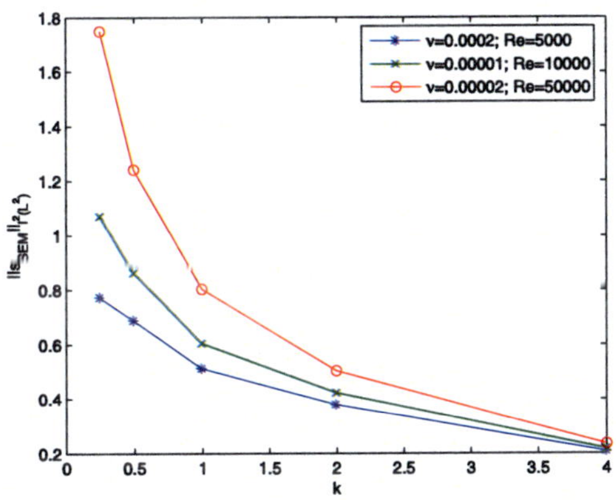

FIGURE 4. Sensitivity over the time interval of $[0, 1]$

TABLE 2. The interval of optimal values for parameter α

Re	Interval of α values
5000	$\frac{1}{4}h \leq \alpha \leq 4h$
10000	$\frac{1}{2}h \leq \alpha \leq 4h$
50000	$h \leq \alpha \leq 4h$

As indicated in Table 2, the interval of optimal values of α for the best accuracy and computational complexity is smaller for higher Reynolds number.

REMARK 3.1. In this experiment, we chose α values for which $\|s\|_{l^2(0,1;L^2)} < 1$ as the optimal values leading to a less expensive computation. The smaller α values with $\|s\|_{l^2(0,1;L^2)} \geq 1$, the more precise calculations of the approximated velocity, $\mathbf{u}$, becomes. However, including more velocity scales by selecting α values very small results in increased run time. The user must therefore consider the trade-off between increased precision and the corresponding longer run times and thus the computational complexity.

The following Figure 5 shows the computed sensitivity solution using (2.4) for $Re = 10000$ at $t = 0.1$, and 1 with $\alpha = h$.

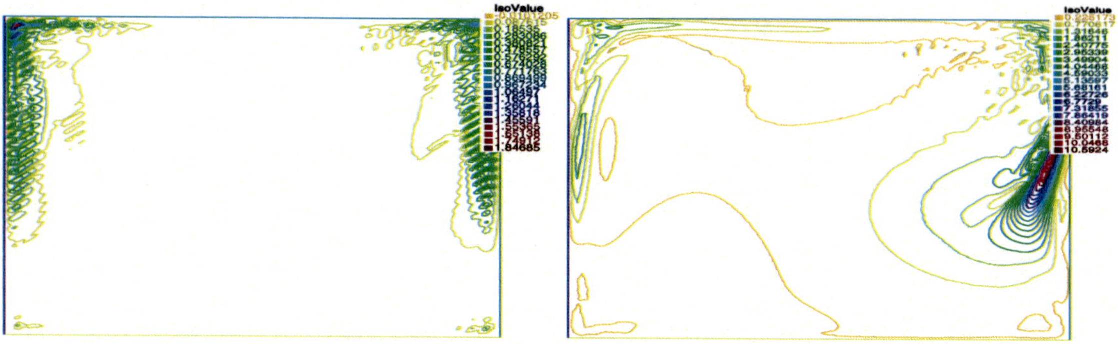

FIGURE 5. Sensitivity at $t = 0.1$ and 1, respectively (from left to right).

3.2. 3D Eithier-Steinman problem. The exact NSE solution for this problem [**6**], is given by

$$
\begin{aligned}
u1 &= -\,a(e^{ax}sin(ay+dz) + e^{az}cos(ax+dy))e^{-\nu d^2 t}\\
u2 &= -\,a(e^{ay}sin(az+dx) + e^{ax}cos(ay+dz))e^{-\nu d^2 t}\\
u3 &= -\,a(e^{az}sin(ax+dy) + e^{ay}cos(az+dx))e^{-\nu d^2 t}\\
p &= -\,\frac{a^2}{2}(e^{2ax} + e^{2ay} + e^{2az} + 2sin(ax+dy)cos(az+dx)e^{a(y+z)}\\
&\quad + 2sin(ay+dz)cos(ax+dy)e^{a(z+x)}\\
&\quad + 2sin(az+dx)cos(ay+dz)e^{a(x+y)})e^{-\nu d^2 t}.
\end{aligned}
$$

The following parameters are used $a = 1.25$, $d = 1$ at final time $T = 0.1$ with time step $\Delta t = 0.01$ on the $(-1,1)^3$ domain. We compute the sensitivity via SEM using Taylor-Hood finite elements for $Re = 1$ and 10000. The mesh size is $h = 2/13$, and $\alpha = h$. The complex flow structure is seen in the streamribbons in the box, and the velocity streamlines and speed contours on the sides in Figure 6, where the exact velocity solution is plotted at $Re = 10000$. Figure 7 represents the numerical approximation of the velocity and its sensitivity at $Re = 10000$. Based on Table 3, we see that the sensitivity increases as we go to higher Reynolds number, the same behavior as in the 2D cavity case.

4. Conclusions and future directions

Sensitivity of the Leray-α with respect to the filter width α on a two dimensional cavity problem was computed via SEM and FFD methods. For all values of α, the computed sensitivity via FFD is overall larger than the one computed via SEM in all the tested Reynolds numbers. In both methods, sensitivity norm increases as α takes values closer to 0, meaning that $\mathbf{u}$ is more sensitive with respect to smaller

TABLE 3. Sensitivity $\|s_{SEM}\|$ for different Re values.

α	$Re = 1$	$Re = 10000$
$\frac{1}{2}h$	0.00971594	0.101695
h	0.0328125	0.180056
$2h$	0.0459474	0.187778

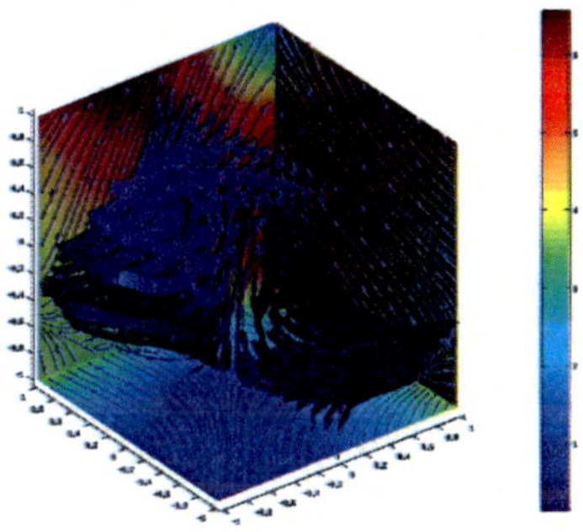

FIGURE 6. Exact velocity solution at $T = 0.1$.

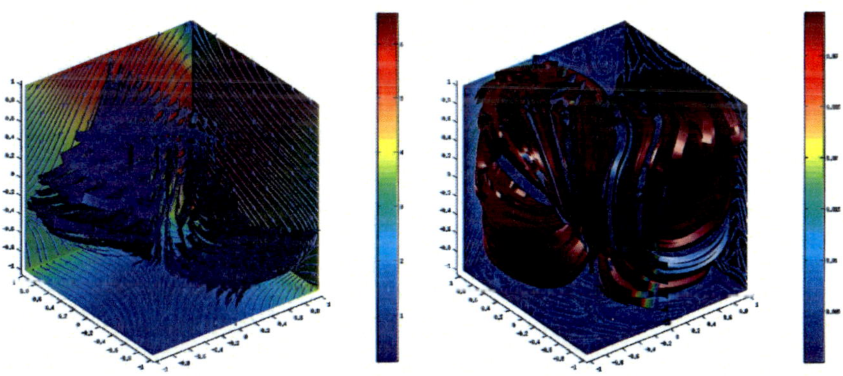

FIGURE 7. Numerical approximation of velocity for the Eithier-Steinman problem and its sensitivity at $T = 0.1$, respectively (from left to right).

values of α. The sensitivity value increased as time has progressed to $t = 1$. Also, sensitivities are larger at higher Reynolds number. A smaller interval of reliable α values is obtained for larger values of Re. Regarding the 3D Either-Steinman problem, the numerical solution was successfully obtained and its sensitivity too. The results show that the sensitivity is higher for larger Re. The further studies will include stochastic finite element discretization, which should give more insights into the parameter sensitivity. Other benchmark problems will be investigated as well.

References

1. M. Anitescu and W. J. Layton, Sensitivities in Large Eddy Simulation and Improved Estimates of Turbulent Flow Functionals *J.C.P*, 1782001, 391-426.
2. M. Anitescu, F. Pahlevani and W. J. Layton, Implicit for local effects and explicit for nonlocal effects is unconditionally stable *Electronic Trnasactions of Numerical Analysis*, 18(2004), 174-187. MR2150767 (2006g:65095)
3. J. Borggaard, D. Pelletier and E. Turgeon, Sensitivity and uncertainty analysis for variable property flows, *in Proceedings of the 39th AIAA Aerospace Sciences Meeting and Exhibit*, 0140, 1993.
4. A. Bowers and L. Rebholz, Increasing accuracy and efficiency in FE computations of the Leray-deconvolution model, To appear in *Numerical Methods for Partial Differential Equations.* MR2879803
5. A. Cheskidov, D.D. Holm, E. Olson and E.S. Titi, On a Leray-α model of turbulence, *Royal Society London, Proceedings, Series A, Mathematical, Physical and Engineering Sciences*, 461(2005), 629-649. MR2121928 (2006b:76041)
6. C. Eithier and D. Steinman, Exact fully 3d Navier-Stokes solutions for benchmarking, *International Journal for Numerical Methods in Fluids*, 19 (1994), 369-375.
7. B.J. Geurts and D.D. Holm, Leray and LANS-alpha modeling of turbulent mixing, *J. of Turbulence*, 7(2006), 1-33. MR2207369 (2006i:76047)
8. M. Gunzburger, Sensitivities, adjoints and flow optimization, *Int. Jour. Num. Meth. Fluids*, 31(1999), 53-78. MR1714510 (2000g:76035)
9. W. Layton, *Introduction to Finite Element Methods for Incompressible, Viscous Flows*, SIAM publications, 2008.
10. W. Layton, C. Manica, M. Neda and L. Rebholz, Numerical Analysis and Computational Testing of a high-order Leray-deconvolution turbulence model, *Numerical Methods for Partial Differential Equations*, 24(2008), 555-582. MR2382797 (2009b:76064)
11. J. Leray, Essay sur les mouvements plans d'une liquide visqueux que limitent des parois, *J. math. pur. appl.*, Paris Ser. IX, 13(1934), 331-418.
12. F. Pahlevani, Sensitivity Computations of Eddy Viscosity Models with an Application in Drag Computation, *International Journal for Numerical Methods in Fluids*, 52(2006), 381-392. MR2262632 (2007f:76145)
13. F. Pahlevani and L. Davis, Parameter Sensitivity of an Eddy Viscosity Model: Analysis, Computation and its Application to Quantifying Model Reliability, To appear in *International Journal of Uncertainty Quantification*, 2012.
14. P. Sagaut and T.Lê, Some investigations of the sensitivity of large eddy simulation, *Tech. Rep.*, 1997-12, ONERA.
15. L. Stanley and D. Stewart, Design sensitivity analysis: Computational issues of sensitivity equation methods, *Frontiers in Mathematics, SIAM*, Philadelphia, 2002. MR1946496 (2003j:49003)
16. M. Gunzburger, Perspectives in Flow Control and Optimization, *Frontiers in Mathematics, SIAM*, Philadelphia, 2003. MR1946726 (2003k:93001)
17. F. Hecht, O. Pironneau and K. Oshtsuka, *Software Freefem++*, http://www.freefem++.org, 2003.
18. M.I. Vishik, E.S. Titi and V.V. Chepyzhov, Trajectory attractor approximations of the 3d Navier-Stokes system by the Leray-alpha model, *Russian Math Dokladi*, 71(2005), 91-95.

Department of Mathematics, University of Nevada, Las Vegas, Nevada 89154
E-mail address: `monika.neda@unlv.edu`

Division of Science and Engineering, Penn State Abington, Abington, Pennsylvania 19001
E-mail address: `fxp10@psu.edu`

Department of Mathematics, University of Nevada, Las Vegas, Nevada 89154
E-mail address: `wangj16@unlv.nevada.edu`

A sparse multiresolution stochastic approximation for uncertainty quantification

D. Schiavazzi, A. Doostan, and G. Iaccarino

ABSTRACT. The present work proposes a novel sampling-based uncertainty propagation framework in which solutions are represented using a multiresolution dictionary. The coefficients of such an expansion are evaluated using greedy methodologies within the Compressive Sampling framework. The effect of various sampling strategies is investigated. The proposed methodology is verified on the Kraichnan-Orszag problem with one and two random initial conditions.

1. Introduction

Most physical systems are inevitably affected by uncertainties due to natural variabilities or incomplete knowledge about their governing laws. To achieve predictive computer simulations of such systems, a major task is, therefore, to study the impact of these uncertainties on response quantities of interest. Within the probabilistic framework, uncertainties may be represented in the form of random variables/processes. Several computational strategies may then be applied to *propagate* these uncertainty in order to obtain the statistics of quantities of interest.

Monte Carlo (MC) sampling methods have been widely used for this purpose. Besides their generally slow convergence, they offer a blend of simplicity, robustness, and efficiency for high dimensional problems. Recently, there has been an increasing interest in developing more efficient alternative numerical methods. Most notably, stochastic approximation schemes based on Wiener-Askey polynomial chaos expansions **[7, 15]** or interpolations on sparse grids **[14]** have been applied successfully to a variety of problems with random inputs. For sufficiently smooth solutions, these methods achieve as high as exponential convergence rate in the mean-squares sense. For non-smooth responses such as those exhibiting sharp gradients or discontinuities these methods, however, may result in poor approximations. To address this shortcoming, various methodologies including adaptive multiresolution **[10]**, multi-element (generalized) polynomial chaos **[12]**, adaptive sparse grid collocation

---

2010 *Mathematics Subject Classification.* Primary 60H35, 35R60; Secondary 76D06.

G.I. and D.S. s work is supported under Subcontract No. B597952, with Lawrence Livermore National Security under Prime Contract No. DE-AC52-07NA27344 from the Department of Energy National Nuclear Security Administration for the management and operation of the Lawrence Livermore National laboratory.

A.D.' s work is supported by the Department of Energy under Advanced Scientific Computing Research Early Career Award DE- SC0006402.

[11], rational approximation **[4]**, and simplex stochastic collocation **[13]**, have been suggested in recent years.

In this work, we present a multiresolution approach based on sparse multiwavelet expansions for uncertainty propagation. Unlike the work in **[10]** where the multiwavelet coefficients are computed via Galerkin projections (typically requiring modifications of deterministic physics solvers), we propose a Compressive Sampling (CS) strategy in which physics solvers are treated as "black boxes". CS is a new direction in signal processing that enables (up to) exact reconstruction of signals admitting sparse representations, in some suitable basis, using samples obtained at a sub-Nyquist rate **[3, 2]**. In **[5]**, the use of CS in the approximation of sparse polynomial chaos solutions to stochastic PDEs was first introduced. In the present study, we demonstrate the efficiency of CS in recovering stochastic functions having sparse expansions in multiwavelet bases. At the core of the proposed CS framework are the application of a greedy basis pursuit technique together with an Importance Sampling strategy that enable accurate reconstruction of non-smooth stochastic functions. To the authors' best of knowledge, no previous study is available in literature where the combination of the three elements of multiresolution approximations, CS, and Importance Sampling have been applied for the purpose of uncertainty propagation.

1.1. Problem Setup.

Consider a probability space $(\Omega, \mathcal{F}, \mathcal{P})$ in which Ω is the sample space, $\mathcal{F}$ is the σ-algebra of events, and $\mathcal{P}$ defines a probability measure on $\mathcal{F}$. A vector of independent and identically distributed random variables with joint probability density function $\rho(\mathbf{y}) : \mathbb{R}^d \to \mathbb{R}_{\geq 0}$ is indicated by $\mathbf{y} = (y_1, \ldots, y_d)$ with $y_i : \Omega \to \mathbb{R}$, $i = 1, \ldots, d$, and $d \in \mathbb{N}$. We state our problem as an evolution in time of $\mathbf{u}(\mathbf{y}, t) : \Omega \times [0, T] \to \mathbb{R}^q$, $q \in \mathbb{N}$, such that

$$\frac{\partial}{\partial t}\mathbf{u}(\mathbf{y}, t) = \mathbf{f}(\mathbf{u}, t), \quad t \in [0, T], \mathbf{y} \in \Omega, \quad \text{with} \quad \mathbf{u}(\mathbf{y}, t = 0) = \mathbf{u}_0(\mathbf{y}) \tag{1.1}$$

hold $\mathcal{P}$-a.s. in Ω. Here we assume the well-posedness (in $\mathcal{P}$-a.s. sense) of (1.1) with respect to the choices of the forcing and boundary functions $\mathbf{f}$ and $\mathbf{u}_0$, respectively. In the present study, we seek to construct a multiresolution representation of $\mathbf{u}(\mathbf{y}, t)$ at a fixed time $t_a \in [0, T]$ by using samples $\{\mathbf{u}(\mathbf{y}^{(k)}, t_a) : k = 1, \ldots, M\}$ corresponding to M realizations $\{\mathbf{y}^{(k)} : k = 1, \ldots, M\}$ of the random inputs $\mathbf{y}$. To simplify the notation and presentation, we henceforth drop the time variable t_a and describe our approach for a scalar, univariate solution $u(y)$, i.e., $q = d = 1$.

The remainder of this paper is organized as follows. In Section 2, we overview the multiresolution representation of a function $u(y) \in \mathbf{L}^2([0, 1])$ in multiwavelet bases. We describe our approach – within the Compressive Sampling framework – to obtain such a representation in Sections 3 and 4, respectively. Finally, in Section 5, we apply the proposed framework to a non-linear ODE problem.

2. Multiresolution Analysis and Multiwavelet Approximation

A *multiresolution* approximation of $\mathbf{L}^2([0, 1])$ is expressed by means of a nested sequence of closed subspaces $\mathbf{V}_0 \subset \mathbf{V}_1 \subset \cdots \subset \mathbf{V}_j \subset \cdots \subset \mathbf{L}^2([0, 1])$, where each $\mathbf{V}_j = \text{span}\{\phi_{j,k}(y) : k = 0, \ldots, 2^j - 1\}$ and

$$\phi_{j,k}(y) = 2^{j/2}\phi(2^j y - k) \tag{2.1}$$

are generated by dilations and translations of a *scaling* function $\phi(y) : [0,1] \to \mathbb{R}$. The scaling function $\phi(y)$ is such that the closure of the union of $\mathbf{V}_j$, i.e., $\overline{\bigcup_{k=1}^{\infty} \mathbf{V}_k}$, is dense in $\mathbf{L}^2([0,1])$. Let the *wavelet* subspace $\mathbf{W}_j$ denote the orthogonal complement of $\mathbf{V}_j$ in $\mathbf{V}_{j+1}$, that is $\mathbf{V}_{j+1} = \mathbf{V}_j \oplus \mathbf{W}_j$ and $\mathbf{V}_j \perp \mathbf{W}_j$. It can be shown that $\mathbf{W}_j = \operatorname{span}\{\varphi_{j,k}(y) : k = 0, \ldots, 2^j - 1\}$ where $\varphi_{j,k}(y)$ is generated from dilation and translation of a *mother* wavelet function $\varphi(y) : [0,1] \to \mathbb{R}$, i.e.,

$$\varphi_{j,k}(y) = 2^{j/2} \varphi(2^j y - k). \tag{2.2}$$

By the construction of the wavelet spaces $\mathbf{W}_j$, it is straightforward to see that $\mathbf{V}_j = \mathbf{V}_0 \oplus (\bigoplus_{k=0}^{j} \mathbf{W}_k)$, and consequently $\mathbf{V}_0 \oplus (\bigoplus_{k=0}^{\infty} \mathbf{W}_k) = \mathbf{L}^2([0,1])$. Therefore, any function $u(y) \in \mathbf{L}^2([0,1])$ admits an orthogonal decomposition of the form

$$u(y) = \tilde{\alpha}_{0,0} \phi_{0,0}(y) + \sum_{j=0}^{\infty} \sum_{k=0}^{2^j - 1} \alpha_{j,k} \varphi_{j,k}(y), \tag{2.3}$$

where $\tilde{\alpha}_{0,0} = \langle u, \phi_{0,0} \rangle_{\mathbf{L}^2([0,1])}$ and $\alpha_{j,k} = \langle u, \varphi_{j,k} \rangle_{\mathbf{L}^2([0,1])}$. To simplify the notation, we rewrite (2.3) in the form

$$u(y) = \sum_{i=1}^{\infty} \alpha_i \psi_i(y), \tag{2.4}$$

in which we establish a one-to-one correspondence between elements of the basis sets $\{\psi_i : i = 0, \ldots, \infty\}$ and $\{\phi_{0,0}, \varphi_{j,k} : k = 0, \ldots, 2^j - 1, j = 0, \ldots, \infty\}$.

In the present study we adopt the slightly more complicated multiresolution of Alpert [**1**] where multiple scaling functions $\{\phi_i(y) : i = 0, \ldots, m-1\}$ are used to construct $\mathbf{V}_0$. Specifically, we choose $\phi_i(y)$ as the Legendre polynomial of degree i defined on the interval $[0,1]$. An orthonormal basis $\{\varphi_i(y) : i = 0, \ldots, m-1\}$ for $\mathbf{W}_0$ is also established. More precisely, let $\mathbf{U}_m = \{u(y) \in \mathbf{L}^2([0,1]) : \int_{[0,1]} u(y)\, y^m\, dy = 0\}$ represent the subspace of functions in $\mathbf{L}^2([0,1])$ with vanishing moment of order m. We then construct $\varphi_i \in \mathbf{U}_j$, $j = 0, \ldots, i+m-1$, with the orthonormality constraint $\langle \varphi_i, \varphi_j \rangle_{\mathbf{L}^2[(0,1)]} = \delta_{ij}$, $i,j = 0, \ldots, m-1$, where δ_{ij} is the Kronecker delta. The *multiwavelet* basis functions $\varphi_{j,k}$ are then generated by dilations and translations of $\{\varphi_i(y) : i = 0, \ldots, m-1\}$. The resulting basis is unique (up to the sign) and provides a generalization of Legendre and Haar representations. In particular, Legendre polynomials can be obtained by stopping the expansion at the resolution $j = 0$, while Haar wavelets are obtained for $m = 0$. If expanded in the Alpert multiwavelet basis, sparse representations are likely to be observed for piecewise smooth functions. Sharp gradients, bifurcations or discontinuities, for example in hyperbolic problems, motivate the use of such dictionaries as multiwavelet with the ability of capturing these local *features* for which global polynomials may not be adequate, see, e.g., [**10**]. In addition to several numerical advantages, the orthogonality property of Alpert multiwavelets is also desirable allowing first and second order statistics of u to be evaluated directly from the expansion coefficients. We refer the interested reader to [**1**] for an in-depth derivation of the Alpert multiwavelet basis. The construction of a multiwavelet basis in $\mathbf{L}^2([0,1]^d)$ with $d \geq 2$ – employed in the numerical example of Section 5 – is also discussed in [**1**].

3. Compressive Sampling of Multiwavelet Expansions

3.1. Rudiments of Compressive Sampling. Compressive Sampling (CS) is a new direction in signal processing that breaks the traditional limits of the Shannon-Nyquist sampling rate for reconstruction of sparse signals. Consider a vector of measurements $\mathbf{u} = (u(y^{(1)}), \dots, u(y^{(M)}))^T \in \mathbb{R}^M$ of $u \in \mathbf{L}^2([0,1])$. Assuming that u admits a multiwavelet expansion of the form (2.4) with some finite m and resolution j, $\mathbf{u}$ can be represented as $\mathbf{u} = \boldsymbol{\Psi}\,\boldsymbol{\alpha}$, where the so-called *measurement* matrix $\boldsymbol{\Psi} \in \mathbb{R}^{M\times P}$ contains the realization of the multiwavelet basis $\{\psi_i(y)\}$ corresponding to $\mathbf{u}$ and $\boldsymbol{\alpha} \in \mathbb{R}^P$ is the vector of unknown expansion coefficients. Here, P is the cardinality of the truncated multiwavelet basis. Then u has a sparse multiwavelet representation if $\|\boldsymbol{\alpha}\|_0 = \#\{\alpha_i : \alpha_i \neq 0\} \ll P$. For a *sufficiently* sparse u, CS recovers u exactly using some $M \ll P$ measurements by solving an optimization problem of the form

$$(P_s) \qquad \min_{\boldsymbol{\alpha}\in\mathbb{R}^P} \|\boldsymbol{\alpha}\|_s \quad \text{subject to} \quad \mathbf{u} = \boldsymbol{\Psi}\,\boldsymbol{\alpha}.$$

The sparsest solution $\boldsymbol{\alpha}$ to (P_s) corresponds to $s = 0$, i.e., minimizing the ℓ_0 quasi-norm $\|\boldsymbol{\alpha}\|_0$, which is generally NP-hard to compute. To break this complexity several heuristics based on greedy pursuit, e.g., Orthogonal Matching Pursuit (OMP), and convex relaxation via ℓ_1-minimization, i.e., $s = 1$, have been proposed, among other approaches. Moreover, several metrics such as the *mutual coherence*, [**2**], or the *restricted isometry property*, [**3**], have been introduced to provide guarantees on the uniqueness of the sparsest solution to (P_s) as well as the ability of the heuristic approaches in recovering the solution. In particular, the mutual coherence of $\boldsymbol{\Psi}$ (e.g., see [**2**]) is defined as

$$\mu(\boldsymbol{\Psi}) = \max_{i\neq j} \frac{|\boldsymbol{\psi}_i^T \boldsymbol{\psi}_j|}{\|\boldsymbol{\psi}_i\|_2 \, \|\boldsymbol{\psi}_j\|_2}, \tag{3.1}$$

where $\boldsymbol{\psi}_i \in \mathbb{R}^M$ is the i-th column of $\boldsymbol{\Psi}$. Note that $\mu(\boldsymbol{\Psi}) \in [0,1]$ in general, and that it is strictly positive for $M < P$. Depending on the sparsity level $\|\boldsymbol{\alpha}\|_0$, the mutual coherence provides a sufficient condition on the number M of measurements for a successful recovery of $\boldsymbol{\alpha}$ from P_s, as shown in [**2**].

3.2. Sparse Multiwavelet Approximation Using Tree-based OMP. In the present study, we extend the greedy Tree-based OMP (TOMP) approach of [**9, 6**] to solve (P_0) for sparse multiwavelet coefficients $\boldsymbol{\alpha}$. Due to the local nature of the multiwavelet basis, it may occur, for example, that none of the realizations of the random input y – from which the measurement vector $\mathbf{u}$ is generated – belong to the support of the i-th basis function ψ_i, resulting in $\boldsymbol{\psi}_i = \mathbf{0}$. In the interest of brevity, only a short description of the adopted greedy algorithms is provided here; detailed expositions are provided in [**9, 6**]. Both the TOMP and OMP algorithms are iterative two-staged procedures. In the first *sensing* stage, basis functions are selected based on the correlation of their corresponding columns in $\boldsymbol{\Psi}$ with the residual vector. While only the most correlated basis is selected in OMP, few additional candidates with high correlations are considered in TOMP. During the second *reconstruction* stage, each candidate in turn, *with associated ancestors*, is temporarily added to the active basis set and the corresponding coefficients are evaluated using a standard least squares. Only the candidate (and ancestors) generating the minimum residual is permanently added to the index set and the process is re-iterated. The TOMP algorithm in [9] was originally developed for scalar tree representations. We, however, extend that work to account for vector multiwavelet trees, as needed in this work.

4. TOMP with Importance Sampling

A natural way to obtain the measurements $\mathbf{u}$ is to generate random realizations of the input y and evaluate the corresponding solution $u(y)$. However, for situation where u exhibits, for instance, sharp gradients or discontinuities, such sampling strategy may not necessarily lead to accurate approximation. This is because the higher resolution basis functions needed to capture the local structure of u may not be sampled enough to constitute a well-conditioned measurement matrix $\mathbf{\Psi}$. To achieve a local accumulation of samples, where needed, we here propose an Importance Sampling approach that is discussed next.

Importance Sampling strategy. Importance Sampling is a well known variance reduction methodology in Monte Carlo estimations. Sampling is performed according to a modified distribution, which promotes the important regions of the input variables and the quantity of interest whose expectation is sought.

An insight on the typical wavelet structure of piecewise smooth functions is given in [6]. In particular, the wavelet coefficients of piecewise smooth functions tend to form connected subtrees within wavelet trees. Additionally, a large wavelet coefficient (in magnitude) generally indicates the presence of a local singularity or sharp gradient. The above considerations form the basis of our sampling strategy. The idea is to concentrate samples at locations where large multiwavelet coefficients are observed while preconditioning the basis to maintain orthogonality. The proposed importance sampling consists of a number of steps that are applied iteratively:

(1) A multiwavelet approximation up to a given m and resolution j is obtained by solving (P_0).
(2) The coefficients α_i are sorted in decreasing order, based on the quantity $|\alpha_i|/|\text{supp}(\psi_i)|$, where $|\text{supp}(\psi_i)|$ is the size of the support of ψ_i.
(3) A sample is drawn in $\text{supp}(\alpha_i)$ according to a uniform distribution only if $|\alpha_i| > \alpha_{tol}$ ($\alpha_{tol} = 1.0 \times 10^{-3}$ is used in the present study).

Preconditioning the basis. Assuming y is uniformly distributed on $[0,1]$, i.e., $\rho(y) : [0,1] \to 1$, the direct application of the above modified sampling leads to measurement matrices $\mathbf{\Psi}$ with large mutual coherence $\mu(\mathbf{\Psi})$. This is because the multiwavelets are orthogonal with respect to the measure $\rho(y)$, i.e. $\int_0^1 \psi_i(y)\,\psi_j(y)\,dy = \delta_{ij}$. A correction, therefore, is needed to retain orthogonality for sufficiently large M. Let $\gamma(y) : [0,1] \to \mathbb{R}_{\geq 0}$ denote the density function according to which the (independent) modified samples $y^{(k)}$, $k = 1, \dots, M$, are distributed and $\hat{\psi}_i(y) = \psi_i(y)/\sqrt{\gamma(y)}$ be the scaled multiwavelet basis. Then,

$$\text{(4.1)} \qquad \frac{1}{M}\sum_{k=1}^{M} \hat{\psi}_i(y^{(k)})\,\hat{\psi}_j(y^{(k)}) \xrightarrow{\text{a.s.}} \int_0^1 \frac{\psi_i(y)}{\sqrt{\gamma(y)}}\,\frac{\psi_j(y)}{\sqrt{\gamma(y)}}\,\gamma(y)\,dy = \delta_{ij},$$

as a result of the strong law of large numbers.

In the CS framework, this translates in sampling according to $\gamma(y)$ and using a modified measurement matrix $\hat{\mathbf{\Psi}} = \mathbf{W\Psi}$ and the data $\hat{\mathbf{u}} = \mathbf{Wu}$ with the *preconditioner* matrix $\mathbf{W} = \text{diag}(1/\sqrt{\gamma(y^{(i)})})$, $i = 1, \dots, M$.

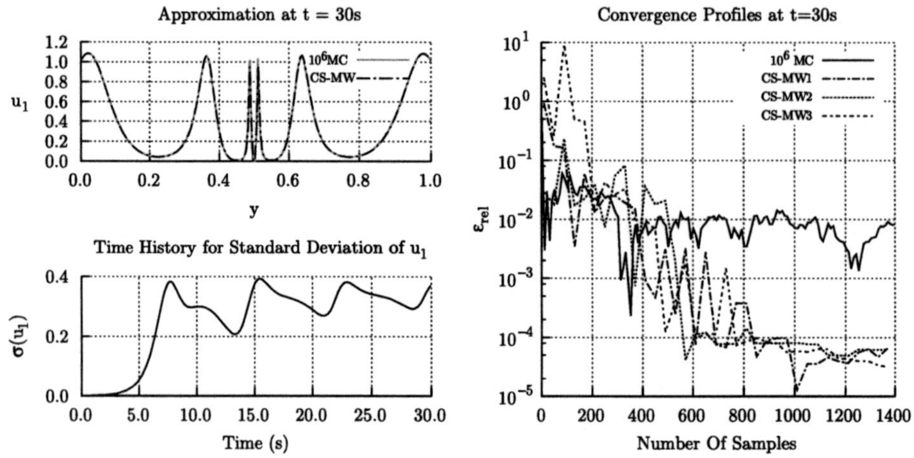

FIGURE 1. Results for the 1D KO problem. Monte Carlo (MC) estimates with 10^6 samples are compared to various independent runs of the proposed (CS-MW) strategy.

We now discuss how a piecewise constant measure $\gamma(y)$ can be defined on partitions of $[0,1]$ associated with a (truncated) multiwavelet representation. We focus on establishing a one-to-one relationship between a scalar wavelet tree and a partition of $[0,1]$. Vector trees (whose vertices are arrays of numbers) are used to store multiwavelet representations while scalar trees are usually adopted for wavelets. A partition of $[0,1]$ is build by first forming a scalar connected subtree $\mathcal{T}$, obtained by pruning all vertices with coefficients α_i with $|\alpha_i| < \alpha_{tol}$. The *leaves* ($L$ in total) of $\mathcal{T}$ are identified, their supports form a set of disjoint intervals $\{\mathcal{B}_i : i = 1, \dots, L\}$ which result in the desired partition of $[0,1]$.

Using the coefficient-driven sampling discussed above, $\gamma(y)$ is defined as a piecewise constant distribution. In particular, a set of probability masses $p_i = 2^{-j}(M_i/M)$ for every box $\mathcal{B}_i$ is considered in which M_i is the number of samples within the interval $\mathcal{B}_i$ (with $|\mathcal{B}_i| = 2^{-j}$), j is the resolution level of the associated leaf, and M is the total number of available samples.

5. Numerical Test: Kraichnan-Orszag Problem

The Kraichnan-Orszag (KO) problem is derived from simplified inviscid Navier-Stokes equations [**8**], and is expressed as a coupled system of non-linear ODEs. We here adopt a rotated version of the original KO problem

$$\frac{du_1}{dt} = u_1\,u_3, \quad \frac{du_2}{dt} = -u_2\,u_3, \quad \frac{du_3}{dt} = -u_1^2 + u_2^2, \tag{5.1}$$

with initial conditions specified below.

In [**12**], the KO problem is used as a benchmark and analytical solutions are provided in terms of Jacobi's elliptic functions. If the set of initial conditions is chosen such that the bifurcation point $(u_1, u_2, u_3) = (\sqrt{2}, 0, 1)$ is consistently crossed, it is shown that the accuracy of the global polynomial approximations (at the stochastic level) deteriorates rapidly with time.

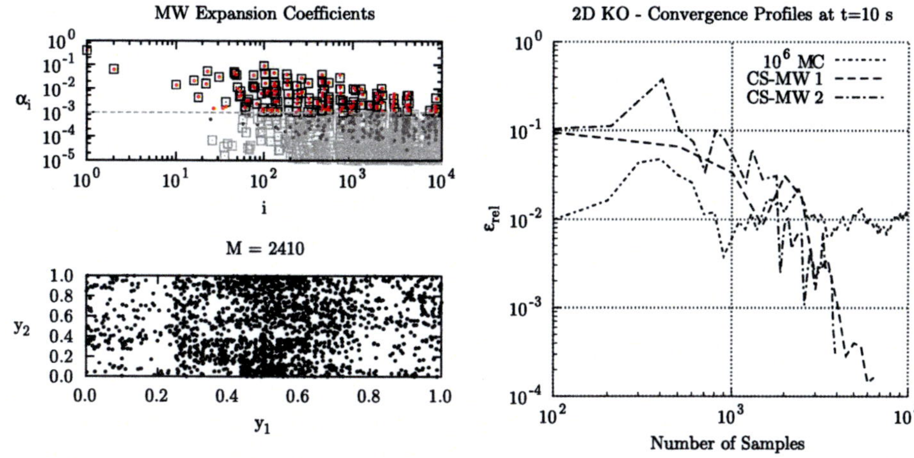

FIGURE 2. Results for the 2D KO problem. Convergence profiles for the Monte Carlo (MC) method and two independent runs of the proposed (CS-MW) approach are presented. Higher densities of samples can be observed close to $y_1 = 1/2$, as expected. Expansion coefficients obtained with CS-MW (dots) are compared to least-squares solutions based on 9×10^4 samples (squares).

5.1. Results for 1D KO Problem. Initial conditions for (5.1) are assumed to be uncertain and specified as

$$u_1(t=0) = 1 \quad ; \quad u_2(t=0) = 0.2\,y - 0.1 \quad ; \quad u_3(t=0) = 0, \tag{5.2}$$

where y is uniformly distributed on $[0,1]$. The stochastic response is evaluated at $t = 20s$ and $t = 30s$ using a multiwavelet dictionary with $m = 3$ and a resolution up to $j = 7$. The OMP solver was used with a relative tolerance $\epsilon = 1.0 \times 10^{-4}$. The time history of the standard deviation for variable u_1 together with the reconstructed response at $t = 30s$ and convergence graphs are illustrated in Figure 1. The error metric $\epsilon_{rel} = |\sigma_{CSMW} - \hat{\sigma}|/\hat{\sigma}$ is also evaluated where σ_{CSMW} is the estimate for the standard deviation calculated with the CS-based multiwavelet expansion and $\hat{\sigma}$ the corresponding exact value.

5.2. Results for 2D KO Problem at $t = 10s$. The initial conditions of the Kraichnan-Orszag problem are again assumed to be uncertain but this time are functions of two random variables

$$u_1(t=0) = 1 \quad ; \quad u_2(t=0) = 0.2\,y_1 - 0.1 \quad ; \quad u_3(t=0) = 2\,y_2 - 1, \tag{5.3}$$

where y_1 and y_2 are independent and uniformly distributed on $[0,1]$. A two dimensional multiwavelet measurement matrix is generated with $m = 2$ and resolution up to $j = 4$, resulting in a basis of cardinality $P = 9216$. Figure 2 shows results in terms of sampling distribution, multiwavelet coefficients, and convergence in statistics for the system's response. As noted from Figure 2, the expansion coefficients produced by the proposed strategy with about $M = 2400$ samples is comparable to those obtained using a multiwavelet least squares approximation (where coefficients are evaluated as $\boldsymbol{\alpha}_{LS} = (\boldsymbol{\Psi}^T\boldsymbol{\Psi})^{-1}\boldsymbol{\Psi}^T\mathbf{u}$ with 9×10^4 samples, demonstrating the efficiency of the CS-based reconstruction.

6. Conclusions

A novel framework for non-intrusive uncertainty propagation is proposed, which combines the ability of a multiresolution approximation in capturing piecewise smooth stochastic responses via Compressive Sampling. Within this framework, an adaptive Importance Sampling strategy is applied in order to improve converge rates of approximating responses exhibiting sharp gradients or discontinuities. Finally, the convergence of the method is demonstrated through its application to the Kraichnan-Orszag problem – a non-linear ODE system – with random initial conditions.

References

1. B.K. Alpert, *A class of bases in l^2 for the sparse representation of integral operators*, Siam J. Math. Anal **24** (1993), 246. MR1199538 (93k:65104)
2. A.M. Bruckstein, D.L. Donoho, and M. Elad, *From sparse solutions of systems of equations to sparse modeling of signals and images*, SIAM review **51** (2009), no. 1. MR2481111 (2010d:94012)
3. E.J. Candes and T. Tao, *Decoding by linear programming*, Information Theory, IEEE Transactions on **51** (2005), no. 12, 4203–4215. MR2243152 (2007b:94313)
4. T. Chantrasmi, A. Doostan, and G. Iaccarino, *Padé–legendre approximants for uncertainty analysis with discontinuous response surfaces*, Journal of Computational Physics **228** (2009), no. 19, 7159–7180. MR2568588 (2010j:76090)
5. A. Doostan and H. Owhadi, *A non-adapted sparse approximation of pdes with stochastic inputs*, Journal of Computational Physics **230** (2011), no. 8, 3015 – 3034. MR2774328 (2012f:60239)
6. M.F. Duarte, M.B. Wakin, and R.G. Baraniuk, *Fast reconstruction of piecewise smooth signals from incoherent projections*, SPARS'05 (2005).
7. R.G. Ghanem and P.D. Spanos, *Stochastic finite elements: a spectral approach*, Dover Pubns, 2003.
8. R.H. Kraichnan, *Direct-interaction approximation for a system of several interacting simple shear waves*, Physics of Fluids **6** (1963), 1603.
9. C. La and M.N. Do, *Tree-based orthogonal matching pursuit algorithm for signal reconstruction*, Image Processing, 2006 IEEE International Conference on, IEEE, 2006, pp. 1277–1280.
10. O.P. Le Maıtre, H.N. Najm, R.G. Ghanem, and O.M. Knio, *Multi-resolution analysis of wiener-type uncertainty propagation schemes*, Journal of Computational Physics **197** (2004), no. 2, 502–531. MR2063905 (2005b:65142)
11. X. Ma and N. Zabaras, *An adaptive hierarchical sparse grid collocation algorithm for the solution of stochastic differential equations*, Journal of Computational Physics **228** (2009), no. 8, 3084–3113. MR2509309 (2010c:65013)
12. X. Wan and G.E. Karniadakis, *An adaptive multi-element generalized polynomial chaos method for stochastic differential equations*, Journal of Computational Physics **209** (2005), no. 2, 617–642. MR2151997 (2006e:65007)
13. J.A.S. Witteveen and G. Iaccarino, *Simplex stochastic collocation with random sampling and extrapolation for nonhypercube probability spaces*, SIAM Journal on Scientific Computing **34** (2012), A814. MR2914305
14. D. Xiu and J.S. Hesthaven, *High-order collocation methods for differential equations with random inputs*, SIAM Journal on Scientific Computing **27** (2006), no. 3, 1118. MR2199923 (2006i:65010)
15. D. Xiu and G.E. Karniadakis, *The wiener–askey polynomial chaos for stochastic differential equations*, SIAM J. Sci. Comput. **24** (2002), no. 2, 619–644. MR1951058 (2003m:60174)

Dipartimento di Matematica, Università di Padova
E-mail address: danschi@stanford.edu

Aerospace Engineering Sciences Department, University of Colorado, Boulder, Colorado

Department of Mechanical Engineering, Stanford University, Stanford, California

A mixed finite element method for a time-dependent incompressible MHD problem

Dominik Schötzau and Xiaoxi Wei

ABSTRACT. We propose a mixed finite element method for the numerical solution of the equations of incompressible and resistive magnetohydrodynamics. We combine the spatial discretization developed in Greif et al. (2010) for stationary problems with an implicit time-stepping method. The resulting fully discrete scheme yields exactly divergence-free velocities, discretely divergence-free magnetic fields, and is energy-stable. We present a numerical test to demonstrate the feasibility of this approach.

1. Introduction

Incompressible magnetohydrodynamics (MHD) describes the flow of viscous, incompressible and electrically conducting fluids whose electric resistivity cannot be neglected. We consider a standard form of an incompressible MHD system as derived in [**AS96, GBL06, GMP91**]. That is, we neglect phenomena involving high frequency as well as the convection current, and consider a non-polarizable, non-magnetizable and homogeneous medium. The governing equations are then of the form

$$\begin{aligned} \partial_t \mathbf{u} - \nu\,\Delta\mathbf{u} + (\mathbf{u}\cdot\nabla)\mathbf{u} + \nabla p - \kappa\,(\nabla\times\mathbf{b})\times\mathbf{b} &= \mathbf{f} && \text{in } \Omega\times J, \\ \kappa\partial_t\mathbf{b} + \kappa\nu_m\,\nabla\times(\nabla\times\mathbf{b}) - \kappa\,\nabla\times(\mathbf{u}\times\mathbf{b}) &= \mathbf{0} && \text{in } \Omega\times J, \\ \nabla\cdot\mathbf{u} = \nabla\cdot\mathbf{b} &= 0 && \text{in } \Omega\times J. \end{aligned} \tag{1.1}$$

Here, Ω is a simply-connected Lipschitz polyhedron in $\mathbb{R}^3$ with a connected boundary, and $J = (0,T)$ a time interval. The unknowns are the velocity field $\mathbf{u}$, the magnetic field $\mathbf{b}$, and the pressure p. The right-hand side $\mathbf{f}$ is a given source term. The dimensionless parameters ν^{-1}, ν_m^{-1}, and κ are the hydrodynamic Reynolds number, the magnetic Reynolds numbers, and the coupling number, respectively. We consider the homogeneous Dirichlet boundary conditions

$$\mathbf{u} = \mathbf{0} \quad \text{and} \quad \mathbf{n}\times\mathbf{b} = \mathbf{0} \qquad \text{on } \partial\Omega\times J, \tag{1.2}$$

1991 *Mathematics Subject Classification.* Primary 65M60; Secondary 65M20.

Key words and phrases. Incompressible magnetohydrodynamics, mixed finite element methods, curl-conforming elements, exactly divergence-free elements.

This work was supported in part by the Natural Sciences and Engineering Research Council of Canada (NSERC).

with $\mathbf{n}$ denoting the unit outward normal on $\partial\Omega$, and impose the initial conditions

$$\mathbf{u}|_{t=0} = \mathbf{u}_0 \quad \text{and} \quad \mathbf{b}|_{t=0} = \mathbf{b}_0 \qquad \text{in } \Omega, \tag{1.3}$$

for given divergence-free fields $\mathbf{u}_0$ and $\mathbf{b}_0$.

In our recent work **[GLSW10]**, we derived and analyzed a new mixed finite element method for a stationary variant of (1.1). The approximation of the magnetic field is based on curl-conforming Nédélec elements of the first kind. This choice allows one to correctly capture the strongest magnetic singularities in non-convex domains, in contrast to straightforwardly applied C^0-finite element approximations, which are known converge to incorrect solutions. The fluid discretization is based on a mixed discontinuous Galerkin (DG) approach with divergence-conforming velocity and discontinuous pressure elements for the fluid unknowns, as originally developed in **[CKS07]** for the incompressible Navier-Stokes equations. The resulting finite element scheme yields exactly divergence-free velocity approximations, which ensures discrete energy-stability without any additional modifications. We proved nearly optimal error estimates in the mesh size for the method, and showed that singular magnetic solution components are correctly resolved.

The purpose of this paper is to apply the spatial discretization of **[GLSW10]** to the time-dependent MHD model (1.1). We first review its main theoretical and stability properties in the semi-discrete setting. Then we combine the spatial finite element method with the backward Euler method in time. As in **[AS96, Pro08]**, the nonlinear terms are discretized semi-implicitly. Hence, in each time step a linearized (but coupled) MHD system then needs to be solved. We present a numerical test to demonstrate that this approach is feasible for transient problems.

We point out that in the stationary setting of **[GLSW10]**, we introduced a scalar Lagrange multiplier associated with the divergence constraint on the magnetic field. The sole purpose of this additional unknown is to ensure the stability of the discretization. This multiplier is not necessary in the time-dependent case.

2. Weak formulation

We introduce the standard Sobolev spaces

$$\mathbf{V} = H_0^1(\Omega)^3, \qquad \mathbf{C} = H_0(\mathrm{curl};\Omega), \qquad Q = L^2(\Omega)/\mathbb{R},$$

and consider the following variational formulation for (1.1)–(1.3): find $\mathbf{u}(t) \in \mathbf{V}$, $\mathbf{b}(t) \in \mathbf{C}$, and $p(t) \in Q$, such that $\mathbf{u}(0) = \mathbf{u}_0$, $\mathbf{b}(0) = \mathbf{b}_0$ and

$$\begin{aligned} (\partial_t \mathbf{u}, \mathbf{v}) + A(\mathbf{u}, \mathbf{v}) + O(\mathbf{u}; \mathbf{u}, \mathbf{v}) + C(\mathbf{b}; \mathbf{v}, \mathbf{b}) + B(\mathbf{v}, p) &= (\mathbf{f}, \mathbf{v}), \\ B(\mathbf{u}, q) &= 0, \\ \kappa(\partial_t \mathbf{b}, \mathbf{c}) + M(\mathbf{b}, \mathbf{c}) - C(\mathbf{b}; \mathbf{u}, \mathbf{c}) &= 0, \end{aligned} \tag{2.1}$$

for all $(\mathbf{v}, \mathbf{c}, q) \in \mathbf{V} \times \mathbf{C} \times Q$ and $t \in J$.

Here, we denote by $(\cdot,\cdot)$ the L^2-inner product. The forms appearing in (2.1) are given by

$$\begin{aligned} A(\mathbf{u}, \mathbf{v}) &= \nu(\nabla \mathbf{u}, \nabla \mathbf{v}), & O(\mathbf{w}; \mathbf{u}, \mathbf{v}) &= ((\mathbf{w} \cdot \nabla)\mathbf{u}, \mathbf{v}), \\ M(\mathbf{b}, \mathbf{c}) &= \kappa\nu_m (\nabla \times \mathbf{b}, \nabla \times \mathbf{c}), & C(\mathbf{d}; \mathbf{v}, \mathbf{b}) &= \kappa(\mathbf{v} \times \mathbf{d}, \nabla \times \mathbf{b}), \\ B(\mathbf{u}, q) &= -(\nabla \cdot \mathbf{u}, q). \end{aligned}$$

The well-posedness of this formulation follows with the same arguments as those used in [**Sch04**] for stationary problems; see also [**Pro08**].

Let us point out the following properties of the variational formulation (2.1). First, by taking $\mathbf{c} = \nabla s$ for $s \in H_0^1(\Omega)$ as a test function in the third equation of (2.1), we obtain

$$\frac{d}{dt}(\mathbf{b}, \nabla s) = 0 \qquad \forall s \in H_0^1(\Omega),\ \forall t \in J. \tag{2.2}$$

This implies that the incompressibility constraint on the magnetic field $\mathbf{b}$ is satisfied for all times provided that $\mathbf{b}_0$ is divergence-free. Second, by taking $\mathbf{v} = \mathbf{u}$, $q = p$ and $\mathbf{c} = \mathbf{b}$ in (2.1), observing that $O(\mathbf{u};\mathbf{u},\mathbf{u}) = 0$, and adding the resulting equations, we readily conclude that the following energy identity holds:

$$\frac{1}{2}\frac{d}{dt}\left(\|\mathbf{u}\|^2 + \kappa\|\mathbf{b}\|^2\right) + \nu\|\nabla\mathbf{u}\|^2 + \kappa\nu_m\|\nabla\times\mathbf{b}\|^2 = (\mathbf{f},\mathbf{u}), \tag{2.3}$$

with $\|\cdot\|$ denoting the L^2-norm. Our mixed finite element method will satisfy discrete versions of (2.2) and (2.3).

3. Discretization in space

We consider a family of regular and quasi-uniform triangulations $\mathcal{T}_h$ of mesh size h that partition the domain Ω into simplices $\{K\}$. For $k \geq 1$, we now look for approximations $\mathbf{u}_h(t)$, $\mathbf{b}_h(t)$, and $p_h(t)$ in the finite element spaces

$$\mathbf{u}_h(t) \in \mathbf{V}_h = \{\, \mathbf{v} \in H_0(\mathrm{div};\Omega) \,:\, \mathbf{v}|_K \in \mathbb{P}_k(K)^3,\ K \in \mathcal{T}_h \,\},$$

$$\mathbf{b}_h(t) \in \mathbf{C}_h = \{\, \mathbf{c} \in \mathbf{C} \,:\, \mathbf{c}|_K \in \mathbb{P}_{k-1}(K)^3 \oplus \mathbb{R}_k(K),\ K \in \mathcal{T}_h \,\},$$

$$p_h(t) \in Q_h = \{\, q \in Q \,:\, q|_K \in \mathbb{P}_{k-1}(K),\ K \in \mathcal{T}_h \,\}.$$

The space $\mathbb{P}_k(K)$ denotes all polynomials of total degree at most k on element K, and $\mathbb{R}_k(K)$ is the space of homogeneous vector polynomials of total degree k which are orthogonal to the position vector.

The velocity space $\mathbf{V}_h$ is the divergence-conforming Brezzi-Douglas-Marini (BDM) space; it has degrees of freedom specified for the normal components of vector fields along faces. The space $\mathbf{C}_h$ represents the first family of curl-conforming Nédélec elements; its degrees of freedom are defined for the tangential components of functions along faces. We notice that the finite element spaces $\mathbf{C}_h$ and Q_h are conforming subspaces of $\mathbf{C}$ and Q, respectively, while the velocity space $\mathbf{V}_h$ is non-conforming with respect to $\mathbf{V}$.

Then, we consider the following semi-discrete approximation of (2.1): find $\mathbf{u}_h(t) \in \mathbf{V}_h$, $\mathbf{b}_h(t) \in \mathbf{C}_h$ and $p_h(t) \in Q_h$, such that $\mathbf{u}_h(0) = \mathbf{u}_{0,h}$, $\mathbf{b}_h(0) = \mathbf{b}_{0,h}$ and

$$\begin{aligned} (\partial_t \mathbf{u}_h, \mathbf{v}) + A_h(\mathbf{u}_h, \mathbf{v}) + O_h(\mathbf{u}_h; \mathbf{u}_h, \mathbf{v}) + C(\mathbf{b}_h; \mathbf{v}, \mathbf{b}_h) + B(\mathbf{v}, p_h) &= (\mathbf{f}, \mathbf{v}), \\ B(\mathbf{u}_h, q) &= 0, \\ \kappa(\partial_t \mathbf{b}_h, \mathbf{c}) + M(\mathbf{b}_h, \mathbf{c}) - C(\mathbf{b}_h; \mathbf{u}_h, \mathbf{c}) &= 0, \end{aligned} \tag{3.1}$$

for all $(\mathbf{v}, \mathbf{c}, q) \in \mathbf{V}_h \times \mathbf{C}_h \times Q_h$ and $t \in J$.

The discrete initial data $\mathbf{u}_{0,h}$ and $\mathbf{b}_{0,h}$ are suitable approximations to $\mathbf{u}_0$ and $\mathbf{b}_0$ (to be specified later). The forms B, M, and C are the same as in the continuous problem (2.1), whereas the forms A_h and O_h are defined in a mesh-dependent manner.

The form A_h will be chosen as the standard symmetric interior penalty form

$$A_h(\mathbf{u},\mathbf{v}) = \sum_{K\in\mathcal{T}_h}\int_K \nu\nabla\mathbf{u}:\nabla\mathbf{v}\,d\mathbf{x} - \sum_{F\in\mathcal{F}_h}\int_F \{\!\{\nu\nabla\mathbf{u}\}\!\}:[\![\mathbf{v}]\!]\,ds$$
$$- \sum_{F\in\mathcal{F}_h}\int_F \{\!\{\nu\nabla\mathbf{v}\}\!\}:[\![\mathbf{u}]\!]\,ds + \sum_{F\in\mathcal{F}_h}\frac{a_0\nu}{h_F}\int_F [\![\mathbf{u}]\!]:[\![\mathbf{v}]\!]\,ds.$$

Here, we denote by $\mathcal{F}_h$ the set faces of $\mathcal{T}_h$. The diameter of face F is denoted by h_F. We use the standard notation $\{\!\{\cdot\}\!\}$ and $[\![\cdot]\!]$ for averages and jumps over interelemental faces of piecewise smooth functions (with obvious modification for boundary faces). The parameter $a_0 > 0$ is the interior penalty stabilization parameter. It has to be chosen larger than a threshold value which is independent of ν and the mesh size h. For the convection term, we take O_h to be the original discontinuous Galerkin upwind form; see [**GLSW10**] and the references therein. We note that other choices for A_h and O_h are possible as well [**CKS07**].

4. Theoretical properties

We now review the stability and well-posedness properties of the spatial discretization of (3.1); see [**GLSW10**] for details.

4.1. Divergence constraints. We start by discussing how the divergence constraints on $\mathbf{u}$ and $\mathbf{b}$ are enforced in (3.1). For the velocity variable, we introduce the discrete kernel

$$\mathbf{J}_h = \{\,\mathbf{v}\in\mathbf{V}_h\,:\,B(\mathbf{v},q)=0\;\;\forall q\in Q_h\,\}.$$

From the second equation in (3.1), it follows directly that $\mathbf{u}_h(t)\in\mathbf{J}_h$ for all $t\in J$. For our particular choice of finite element spaces, we have $\nabla\cdot\mathbf{V}_h\subseteq Q_h$. Hence, the condition $B(\mathbf{u}_h,q)=0$ for all $q\in Q_h$ readily implies that the approximate velocity field $\mathbf{u}_h$ is exactly divergence-free:

$$\nabla\cdot\mathbf{u}_h(t)\equiv 0\qquad\forall t\in J. \tag{4.1}$$

For the magnetic unknown, we use the discrete and L^2-orthogonal Helmholtz decomposition

$$\mathbf{C}_h = \mathbf{X}_h\oplus\nabla S_h,$$

with the discrete spaces S_h and $\mathbf{X}_h$ given by

$$S_h = \{\,s\in H_0^1(\Omega)\,:\,s|_K\in\mathbb{P}_k(K),\;K\in\mathcal{T}_h\,\},$$
$$\mathbf{X}_h = \{\,\mathbf{v}\in\mathbf{C}_h\,:\,(\mathbf{v},\nabla s)=0\;\forall s\in S_h\,\}.$$

Testing the third equation in (3.1) with ∇s for $s\in S_h$ yields

$$\frac{d}{dt}(b_h,\nabla s)=0\qquad\forall s\in S_h,\;\forall t\in J.$$

Hence, if $\mathbf{b}_{0,h}\in\mathbf{X}_h$, then the approximate magnetic field is discretely divergence-free for all times:

$$\mathbf{b}_h(t)\in\mathbf{X}_h\qquad\forall t\in J. \tag{4.2}$$

4.2. Boundedness and coercivity. To discuss the stability properties of the spatial discretization in (3.1), we introduce the product forms

$$\mathcal{A}_h(\mathbf{w}, \mathbf{d}; \mathbf{u}, \mathbf{b}; \mathbf{v}, \mathbf{c}) = A_h(\mathbf{u}, \mathbf{v}) + M(\mathbf{b}, \mathbf{c}),$$
$$\mathcal{O}_h(\mathbf{w}, \mathbf{d}; \mathbf{u}, \mathbf{b}; \mathbf{v}, \mathbf{c}) = O_h(\mathbf{w}; \mathbf{u}, \mathbf{v}) + C(\mathbf{d}; \mathbf{v}, \mathbf{b}) - C(\mathbf{d}; \mathbf{u}, \mathbf{c}).$$

Problem (3.1) can then be written as follows: find $(\mathbf{u}_h(t), \mathbf{b}_h(t), p_h(t)) \in \mathbf{V}_h \times \mathbf{C}_h \times Q_h$ such that

$$\begin{aligned}(\partial_t \mathbf{u}_h, \mathbf{v}) + \kappa(\partial_t \mathbf{b}_h, \mathbf{c}) + \mathcal{A}_h(\mathbf{u}_h, \mathbf{b}_h; \mathbf{v}, \mathbf{c}) \\ + \mathcal{O}_h(\mathbf{u}_h, \mathbf{b}_h; \mathbf{u}_h, \mathbf{b}_h; \mathbf{v}, \mathbf{c}) + B(\mathbf{v}, p_h) - B(\mathbf{u}_h, q) = (\mathbf{f}, \mathbf{v}),\end{aligned} \tag{4.3}$$

for all $(\mathbf{v}, \mathbf{c}, q) \in \mathbf{V}_h \times \mathbf{C}_h \times Q_h$ and $t \in J$.

In what follows, we shall use the broken H^1-norm

$$\|\mathbf{u}\|_{1,h}^2 = \sum_{K \in \mathcal{T}_h} \|\nabla \mathbf{u}\|_{L^2(K)}^2 + \sum_{F \in \mathcal{F}_h} h_F^{-1} \|[\![\mathbf{u}]\!]\|_{L^2(F)}^2,$$

as well as the product norm

$$|\!|\!|(\mathbf{u}, \mathbf{b})|\!|\!|^2 = \nu \|\mathbf{u}\|_{1,h}^2 + \kappa \nu_m \|\mathbf{b}\|_{H(\mathrm{curl};\Omega)}^2.$$

Then the forms $\mathcal{A}_h$ and B are bounded over the discrete spaces:

$$|\mathcal{A}_h(\mathbf{u}, \mathbf{b}; \mathbf{v}, \mathbf{c})| \le C_{\mathcal{A}} \, |\!|\!|(\mathbf{u}, \mathbf{b})|\!|\!| \, |\!|\!|(\mathbf{v}, \mathbf{c})|\!|\!|,$$
$$B(\mathbf{v}, q)| \le C_B \|v\|_{1,h} \, \|q\|,$$

for all $\mathbf{u}, \mathbf{v} \in \mathbf{V}_h$, $\mathbf{b}, \mathbf{c} \in \mathbf{C}_h$, and $q \in Q_h$, with constants $C_{\mathcal{A}}, C_B$ independent of h, ν, κ, and ν_m. For the form $\mathcal{O}_h$, we have the continuity property

$$\begin{aligned}&|\mathcal{O}_h(\mathbf{w}_1, \mathbf{d}_1; \mathbf{u}, \mathbf{b}; \mathbf{v}, \mathbf{c}) - \mathcal{O}_h(\mathbf{w}_2, \mathbf{d}_2; \mathbf{u}, \mathbf{b}; \mathbf{v}, \mathbf{c})| \\ &\qquad \le \frac{C_{\mathcal{O}} \, \bar{\kappa}}{\bar{\nu}^{\frac{3}{2}}} \, |\!|\!|(\mathbf{w}_1 - \mathbf{w}_2, \mathbf{d}_1 - \mathbf{d}_2)|\!|\!| \, |\!|\!|(\mathbf{u}, \mathbf{b})|\!|\!| \, |\!|\!|(\mathbf{v}, \mathbf{c})|\!|\!|\end{aligned}$$

for all $\mathbf{w}_1, \mathbf{w}_2 \in \mathbf{V}_h$, $\mathbf{d}_1, \mathbf{d}_2 \in \mathbf{X}_h$, $\mathbf{u}, \mathbf{v} \in \mathbf{V}_h$, and $\mathbf{b}, \mathbf{c} \in \mathbf{C}_h$, with a constant $C_{\mathcal{O}}$ independent of h, ν, κ and ν_m. Here, we have set

$$\bar{\nu} = \min\{\nu, \kappa \nu_m\}, \qquad \bar{\kappa} = \max\{1, \kappa\}.$$

Moreover, the forms $\mathcal{A}_h$ and $\mathcal{O}_h$ satisfy the coercivity properties:

$$\mathcal{A}_h(\mathbf{u}, \mathbf{b}; \mathbf{u}, \mathbf{b}) \ge C_C \, |\!|\!|(\mathbf{u}, \mathbf{b})|\!|\!|^2, \qquad \mathbf{u} \in \mathbf{V}_h, \ \mathbf{b} \in \mathbf{X}_h, \tag{4.4}$$

$$\mathcal{O}_h(\mathbf{w}, \mathbf{d}; \mathbf{u}, \mathbf{b}; \mathbf{u}, \mathbf{b}) \ge 0, \qquad \mathbf{w} \in \mathbf{J}_h, \ \mathbf{u} \in \mathbf{V}_h, \ \mathbf{b}, \mathbf{d} \in \mathbf{C}_h, \tag{4.5}$$

with a constant $C_C > 0$ independent of h, ν, κ and ν_m. Finally, the form B is inf-sup stable

$$\inf_{q \in Q_h} \sup_{\mathbf{v} \in \mathbf{V}_h} \frac{B(\mathbf{v}, q)}{\|\mathbf{v}\|_{1,h} \|q\|} = \lambda_h \ge C > 0,$$

where the constant C is independent of h, ν, κ, and ν_m.

4.3. Energy-stability. It is now easy to establish a discrete version of the energy bound (2.3). Indeed, by choosing $\mathbf{v} = \mathbf{u}_h$, $\mathbf{c} = \mathbf{b}_h$ and $q = p_h$ in (3.1), by adding the resulting equations and by using the coercivity results in (4.4)–(4.5), we obtain the discrete energy bound

$$\frac{1}{2}\frac{d}{dt}\left(\|\mathbf{u}_h\|^2 + \kappa\|\mathbf{b}_h\|^2\right) + C_C|||(\mathbf{u}_h, \mathbf{b}_h)|||^2 \leq (\mathbf{f}, \mathbf{u}_h). \tag{4.6}$$

Let us remark here that we have strongly used property (4.1), i.e., the fact that the approximate velocity $\mathbf{u}_h$ is exactly divergence-free.

4.4. Linear stationary systems. The theoretical properties reviewed in Section 4.2 were used in [**GLSW10**] to derive nearly optimal error estimates in the mesh size h for similar discretizations of stationary but non-linear versions of (1.1). The loss in convergence rates is due to the use of inverse inequalities in the analysis. It is not observed numerically and probably an artefact of the techniques of proof. The suboptimality can be avoided theoretically if our spatial discretization is applied to linear stationary systems of the form

$$\begin{aligned} -\nu\,\Delta\mathbf{u} + \alpha\mathbf{u} + (\mathbf{w}\cdot\nabla)\mathbf{u} + \nabla p - \kappa\,(\nabla\times\mathbf{b})\times\mathbf{d} &= \mathbf{f} && \text{in } \Omega,\\ \nabla\cdot\mathbf{u} &= 0 && \text{in } \Omega,\\ \kappa\nu_m\,\nabla\times(\nabla\times\mathbf{b}) + \beta\mathbf{b} - \kappa\,\nabla\times(\mathbf{u}\times\mathbf{d}) &= \mathbf{g} && \text{in } \Omega, \end{aligned} \tag{4.7}$$

for positive parameters α, β, and divergence-free data $\mathbf{w} \in H_0^1(\Omega)^3$, $\mathbf{d} \in H_0(\mathrm{curl};\Omega)$ and $\mathbf{g} \in H(\mathrm{div};\Omega)$. Note that problems of this type appear naturally after implicit time discretization in (1.1).

5. Time discretization

We shall now discretize (3.1) in time using the backward Euler scheme. To that end, we introduce the time step $\Delta t = T/N$, and set $t_n = n\Delta t$ for $0 \leq n \leq N$. As in [**AS96, Pro08**], we take the nonlinear convection and coupling terms in a semi-implicit manner. If we define the difference quotient $Du^n = \frac{u^n - u^{n-1}}{\Delta t}$, we arrive at the fully discrete scheme: find approximations $(\mathbf{u}_h^n, \mathbf{b}_h^n, p_h^n) \in \mathbf{V}_h \times \mathbf{C}_h \times Q_h$ such that $\mathbf{u}_h^0 = \mathbf{u}_{0,h} \in \mathbf{J}_h$, $\mathbf{b}_h^0 = \mathbf{b}_{0,h} \in \mathbf{X}_h$, and

$$\begin{aligned} (D\mathbf{u}_h^n, \mathbf{v}) + A_h(\mathbf{u}_h^n, \mathbf{v}) + O_h(\mathbf{u}_h^{n-1}; \mathbf{u}_h^n, \mathbf{v}) + C(\mathbf{b}_h^{n-1}; \mathbf{v}, \mathbf{b}_h^n) + B(\mathbf{v}, p_h^n) &= (\mathbf{f}^n, \mathbf{v}),\\ B(\mathbf{u}_h^n, q) &= 0,\\ \kappa(D\mathbf{b}_h^n, \mathbf{c}) + M(\mathbf{b}_h^n, \mathbf{c}) - C(\mathbf{b}_h^{n-1}; \mathbf{u}_h^n, \mathbf{c}) &= 0, \end{aligned}$$

for all $(\mathbf{v}, \mathbf{c}, q) \in \mathbf{V}_h \times \mathbf{C}_h \times Q_h$ and $1 \leq n \leq N$, where $\mathbf{f}^n = \mathbf{f}(t_n)$.

This scheme amounts to solving a linear (but still coupled) MHD system in each time-step. It can be readily seen that the approximate solutions uniquely exist. Moreover, the properties discussed in Section 4 still hold. The approximate velocity fields are exactly divergence-free: $\nabla\cdot\mathbf{u}_h^n \equiv 0$ in Ω for all $1 \leq n \leq N$. The approximate magnetic fields are discretely divergence-free: $\mathbf{b}_h^n \in \mathbf{X}_h$ for all $1 \leq n \leq N$, provided that $\mathbf{b}_h^0 \in \mathbf{X}_h^n$ is discretely divergence-free. Finally, we obtain the fully discrete stability bound

$$\frac{1}{2}D\left(\|\mathbf{u}_h^n\|^2 + \kappa\|\mathbf{b}_h^n\|^2\right) + \frac{\Delta t}{2}\left(\|D\mathbf{u}_h^n\|^2 + \kappa\|D\mathbf{b}_h^n\|^2\right) + C_C|||(\mathbf{u}_h^n, \mathbf{b}_h^n)|||^2 \leq (\mathbf{f}^n, \mathbf{u}_h^n).$$

We refer to [**Pro08**] for results regarding the convergence of closely related fully discrete schemes with discrete energy principle.

6. Numerical test: Couette channel flow

We present a computational experiment using the lowest-order elements for the two-dimensional version of the MHD system (1.1). The computations have been carried out using MATLAB. We also refer to [**GLSW10**] for an exhaustive series of numerical experiments for the steady-state problem. The tests include problems with smooth and singular solutions, as well as three-dimensional channel and driven cavity flows. In particular, the results there indicate that convergence is achieved in non-convex domains where straightforwardly applied nodal elements for $\mathbf{b}$ are known to fail.

Here, we consider the two-dimensional Couette channel flow problem computed in [**Wei11**]. It involves an unidirectional flow in the channel $\Omega = (0,5) \times (0,1)$ with a moving top wall, under the influence of the constant transverse magnetic field $\mathbf{b}_D = (0,1)$. The two-dimensional version of the MHD system (1.1) then has a solution of the form

$$\mathbf{u}(x,y,t) = (u(y,t),0), \qquad \mathbf{b}(x,y,t) = (b(y,t),1), \qquad p(x,y,t) = p(y,t). \tag{6.1}$$

We impose no-slip boundary conditions on the top and bottom of the channel, inhomogeneous Neumann conditions at the inflow and outflow parts of Ω, and the magnetic boundary conditions $\mathbf{n}\times\mathbf{b} = \mathbf{n}\times\mathbf{b}_D$ on Γ. We take zero initial conditions. An exact solution can be found by means of Fourier series, cf. [**Wei11**] for details.

In Figures 1 and 2, we show the evolution of the computed velocity and magnetic fields, respectively. The parameters are chosen as $\nu = \nu_m = 1$, and $\kappa = 25$. The interior penalty parameter a_0 is taken as 10. We observe that the approximate fields reach their steady-states fairly quickly, as is expected. In Figure 2, the induced magnetic field $\mathbf{b} = (b(y,t),1)$ is normalized over the computational domain such that the largest magnitude of each component is one.

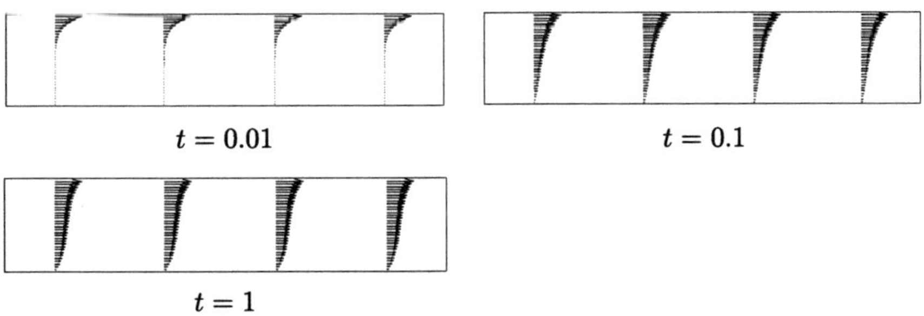

FIGURE 1. Velocity approximation.

7. Concluding remarks

We have presented a fully discrete scheme for a time-dependent MHD problem, where the space discretization is based on $H(\mathrm{div})$-conforming BDM elements for the fluid unknowns and on $H(\mathrm{curl})$-conforming Nédélec elements of the first kind for the

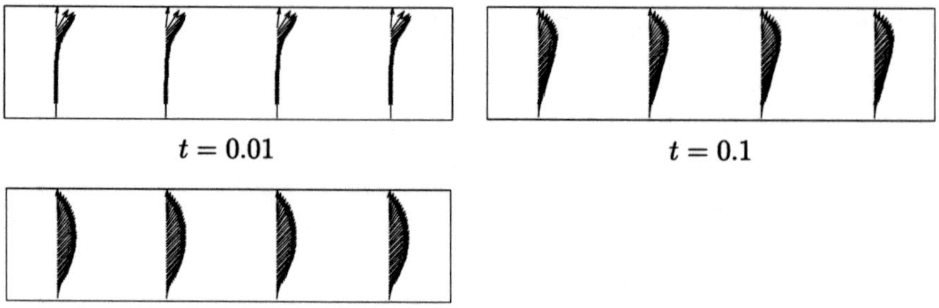

$t = 0.01$ $t = 0.1$

$t = 1$

FIGURE 2. Magnetic approximation.

magnetic unknowns. A detailed numerical analysis of the spatial discretization for stationary problems can be found in [**GLSW10**]. The feasibility of our approach has been validated numerically. To make it more efficient, fast iterative solvers and preconditioners for the resulting linear systems are needed. Developing such solution strategies is the subject of ongoing research.

References

[AS96] F. Armero and J.C. Simo, *Long-term dissipativity of time-stepping algorithms for an abstract evolution equation with applications to the incompressible MHD and Navier-Stokes equations*, Comput. Methods Appl. Mech. Engrg. **131** (1996), 41–90. MR1393572 (97g:76050)

[CKS07] B. Cockburn, G. Kanschat, and D. Schötzau, *A note on discontinuous Galerkin divergence-free solutions of the Navier-Stokes equations*, J. Sci. Comp. **31** (2007), 61–73. MR2304270 (2008f:76109)

[GBL06] J.-F. Gerbeau, C. Le Bris, and T. Lelièvre, *Mathematical Methods for the Magnetohydrodynamics of Liquid Metals*, Numerical Mathematics and Scientific Computation, Oxford University Press, New York, 2006. MR2289481 (2008i:76205)

[GLSW10] C. Greif, D. Li, D. Schötzau, and X. Wei, *A mixed finite element method with exactly divergence-free velocities for incompressible magnetohydrodynamics*, Comput. Methods Appl. Mech. Engrg. **199** (2010), 2840–2855. MR2740762 (2011h:76078)

[GMP91] M.D. Gunzburger, A.J. Meir, and J.S. Peterson, *On the existence and uniqueness and finite element approximation of solutions of the equations of stationary incompressible magnetohydrodynamics*, Math. Comp. **56** (1991), 523–563. MR1066834 (91m:76127)

[Pro08] A. Prohl, *Convergent finite element discretizations of the nonstationary incompressible magnetohydrodynamic system*, Math. Model. Numer. Anal. **42** (2008), 1065–1087. MR2473320 (2010a:65189)

[Sch04] D. Schötzau, *Mixed finite element methods for incompressible magnetohydrodynamics*, Numer. Math. **96** (2004), 771–800. MR2036365 (2005b:76088)

[Wei11] X. Wei, *Mixed Discontinuous Galerkin Finite Element Methods for Incompressible Magnetohydrodynamics*, Ph.D. thesis, The University of British Columbia, 2011. Electronic version availabe at `https://circle.ubc.ca`.

MATHEMATICS DEPARTMENT, UNIVERSITY OF BRITISH COLUMBIA, 1984 MATHEMATICS ROAD, VANCOUVER, BC, V6T 1Z2, CANADA
E-mail address: `schoetzau@math.ubc.ca`

MATHEMATICS DEPARTMENT, UNIVERSITY OF BRITISH COLUMBIA, 1984 MATHEMATICS ROAD, VANCOUVER, BC, V6T 1Z2, CANADA
E-mail address: `weixiaoxi@math.ubc.ca`

A compact splitting scheme approach on nonuniform grids

Qin Sheng and Shekhar Guha

This paper is dedicated to Professor Graeme Fairweather on the occasion of his 70th Birthday.

ABSTRACT. The strategy of compact schemes has been extremely popular in recent years. Most of such approaches are carried out together with straightforward finite difference approximations on uniform grids without splitting or adaptations. This paper explores a second order compact splitting method for computing the numerical solution of a two-dimensional paraxial Helmholtz equation on nonuniform transverse grids. Discussions and analysis of the numerical stability are given. Simulated examples are presented for illustrating our study.

1. A contemporary approach

Consider a two-dimensional paraxial Helmholtz equation,

$$2i\kappa u_z = u_{xx} + u_{yy}, \quad -\infty < x, y < \infty, \quad z \geq z_0, \tag{1.1}$$

where $i = \sqrt{-1}$, $\kappa \gg 10^3$ is the optical wavenumber, and z is the light beam propagation direction. Equation (1.1) has been utilized intensively in optical and electro-optical modeling and computations [**C, D, E**]. Different types of numerical procedures, including the latest eikonal splitting method [**G**], have been developed and studied for approximating the solution of (1.1). Among them, straightforward finite difference schemes play a crucially important role due to their remarkable simplicity and feasibility in applications [**A, B, F**]. Naturally, higher order schemes are more favorable in circumstances where κ is large. This desire, however, is in general difficult to satisfy when irregular grids are deployed.

This paper concerns a balance between the simplicity and accuracy of an adaptive splitting method via the reinforcement of a compact approach. To this end, we denote $\alpha = -i/(2\kappa)$ and rewrite (1.1) as a pair of differential equations,

$$u_z = \alpha u_{xx} + \alpha f(u), \tag{1.2}$$

$$u_z = \alpha u_{yy} + \alpha g(u), \tag{1.3}$$

where $f(u) = u_{yy}$, $g(u) = u_{xx}$, $-\infty < x, y < \infty$. Consider the initial condition

$$u(x, y, z_0) = u_0(x, y), \quad -\infty < x, y < \infty. \tag{1.4}$$

1991 *Mathematics Subject Classification.* Primary 65M06, 65M50; Secondary 65Z05, 78A15.

Key words and phrases. Compact finite difference scheme, splitting method, nonuniform grids.

The first author was supported in part by an AFRL Grant #FA-8650-11-D-5400.

Let $G_x = \{x_j\}$, $G_y = \{y_k\}$ be nonuniform grids superimposed upon transverse intervals in $x-$ and $y-$directions, respectively. Denote mesh step sizes $0 < p_j = x_{j+1} - x_j \le h \ll 1$, $0 < q_k = y_{k+1} - y_k \le h \ll 1$, $j, k \in \mathbb{Z}$. Then a straightforward discretization of (1.2), (1.3) yields the following semidiscretized system on $G_x \times G_y$,

$$u_z = \alpha \mathcal{D}_{2,x} u + \alpha f(u) - \frac{p_j - p_{j-1}}{3} u_{xxx} - P, \tag{1.5}$$

$$u_z = \alpha \mathcal{D}_{2,y} u + \alpha g(u) - \frac{q_k - q_{k-1}}{3} u_{yyy} - Q, \tag{1.6}$$

where

$$\mathcal{D}_{2,x} u = \frac{2}{p_j + p_{j-1}} \left(\frac{u_{j+1,k} - u_{j,k}}{p_j} - \frac{u_{j,k} - u_{j-1,k}}{p_{j-1}} \right),$$
$$\mathcal{D}_{2,y} u = \frac{2}{q_k + q_{k-1}} \left(\frac{u_{j,k+1} - u_{j,k}}{q_k} - \frac{u_{j,k} - u_{j,k-1}}{q_{k-1}} \right),$$

and

$$P = \frac{p_j^2 - p_j p_{j-1} + p_{j-1}^2}{12} (u_{xxxx})_{j,k} + O(h^3),$$
$$Q = \frac{q_k^2 - q_k q_{k-1} + q_{k-1}^2}{12} (u_{yyyy})_{j,k} + O(h^3).$$

Differentiating (1.2), (1.3) once generates

$$u_{xxx} = \frac{1}{\alpha} u_{xz} - f_x(u), \; u_{yyy} = \frac{1}{\alpha} u_{yz} - g_y(u).$$

Substituting them into (1.5), (1.6), we acquire consequently that

$$u_z = \alpha \mathcal{D}_{2,x} u - \frac{p_j - p_{j-1}}{3\alpha} \mathcal{D}_{1,x} u_z + \alpha \left(f(u) + \frac{p_j - p_{j-1}}{3\alpha} f_x(u) \right) - \tilde{P},$$
$$u_z = \alpha \mathcal{D}_{2,y} u - \frac{q_k - q_{k-1}}{3\alpha} \mathcal{D}_{1,y} u_z + \alpha \left(g(u) + \frac{q_k - q_{k-1}}{3\alpha} g_y(u) \right) - \tilde{Q},$$

where

$$\mathcal{D}_{1,x} u_z = \frac{(u_z)_{j+1,k} - (u_z)_{j-1,k}}{p_j + p_{j-1}}, \; \mathcal{D}_{1,y} u_z = \frac{(u_z)_{j,k+1} - (u_z)_{j,k-1}}{q_k + q_{k-1}},$$

and

$$\tilde{P} = P - \frac{(p_j - p_{j-1})^2}{6\alpha} (u_{xxz})_{j,k}, \; \tilde{Q} = Q - \frac{(q_k - q_{k-1})^2}{6\alpha} (u_{yyz})_{j,k}.$$

Apparently, $\tilde{P}, \tilde{Q} = O(h^2)$ for any $z > z_0$. Drop all truncation errors. A modified local one-dimensional procedure [**F, G**] leads immediately to our adaptive splitting method

$$\left(1 - \frac{\alpha\tau_\ell}{2} \mathcal{D}_{2,x} + \frac{p_j - p_{j-1}}{3\alpha} \mathcal{D}_{1,x}\right) v_k^{\ell+1/2} = \tau_\ell \left(\alpha f^\ell + \frac{p_j - p_{j-1}}{3} f_x^\ell \right) + \left(1 + \frac{\alpha\tau_\ell}{2} \mathcal{D}_{2,x} + \frac{p_j - p_{j-1}}{3\alpha} \mathcal{D}_{1,x}\right) v_k^\ell, \text{ for each fixed } y_k \in G_y; \tag{1.7}$$

$$\left(1 - \frac{\alpha\tau_\ell}{2} \mathcal{D}_{2,y} + \frac{q_k - p_{k-1}}{3\alpha} \mathcal{D}_{1,y}\right) w_j^{\ell+1} = \tau_\ell \left(\alpha g^{\ell+1/2} + \frac{q_k - q_{k-1}}{3} g_y^{\ell+1/2} \right) + \left(1 + \frac{\alpha\tau_\ell}{2} \mathcal{D}_{2,y} + \frac{q_k - q_{k-1}}{3\alpha} \mathcal{D}_{1,y}\right) w_j^{\ell+1/2}, \text{ for each fixed } x_j \in G_x, \tag{1.8}$$

where $v_k^\sigma = \left(u_{1,k}^\sigma, u_{2,k}^\sigma, \ldots, u_{N,k}^\sigma\right)^\top$, $w_j^\sigma = \left(u_{j,1}^\sigma, u_{j,2}^\sigma, \ldots, u_{j,N}^\sigma\right)^\top$, and p_j, q_k, τ_ℓ can be determined via proper monitoring functions [**A, C**]. Further, any finite set of the finite difference equations with given ending u-values can be comprised as

$$Av_k^{\ell+1/2} = Bv_k^\ell + \phi^\ell,\ k = 1, 2, \ldots, N; \tag{1.9}$$

$$Cw_j^{\ell+1} = Dw_j^{\ell+1/2} + \psi^{\ell+1/2},\ j = 1, 2, \ldots, N, \tag{1.10}$$

where $A,\ B,\ C,\ D \in \mathbb{C}^{N\times N}$ are tridiagonal matrices with

$$\begin{aligned}
a_{n,n} &= 1 + \mu_{2,n},\ b_{n,n} = 1 - \mu_{2,n},\ n = 1, 2, \ldots, N;\\
a_{n,n+1} &= \mu_{3,n} + \eta_n,\ b_{n,n+1} = -\mu_{3,n} + \eta_n,\ n = 1, 2, \ldots, N-1;\\
a_{n+1,n} &= \mu_{1,n} + \eta_{n+1},\ b_{n+1,n} = -\mu_{1,n} + \eta_{n+1},\ n = 1, 2, \ldots, N-1;\\
c_{n,n} &= 1 + \tilde{\mu}_{2,n},\ d_{n,n} = 1 - \tilde{\mu}_{2,n},\ n = 1, 2, \ldots, N;\\
c_{n,n+1} &= \tilde{\mu}_{3,n} + \tilde{\eta}_n,\ d_{n,n+1} = -\tilde{\mu}_{3,n} + \tilde{\eta}_n,\ n = 1, 2, \ldots, N-1;\\
c_{n+1,n} &= \tilde{\mu}_{1,n} + \tilde{\eta}_{n+1},\ d_{n+1,n} = -\tilde{\mu}_{1,n} + \tilde{\eta}_{n+1},\ n = 1, 2, \ldots, N-1,
\end{aligned}$$

for which

$$\mu_{1,n} = -\frac{\alpha\tau_\ell}{(p_{n+1}+p_n)p_n},\ \mu_{2,n} = \frac{\alpha\tau_\ell}{p_n p_{n-1}},\ \mu_{3,n} = -\frac{\alpha\tau_\ell}{(p_n+p_{n-1})p_n},$$

$$\tilde{\mu}_{1,n} = -\frac{\alpha\tau_\ell}{(q_{n+1}+q_n)q_n},\ \tilde{\mu}_{2,n} = \frac{\alpha\tau_\ell}{q_n q_{n-1}},\ \tilde{\mu}_{3,n} = -\frac{\alpha\tau_\ell}{(q_n+q_{n-1})q_n},$$

$$\eta_n = \frac{1}{3\alpha}\left(\frac{p_n - p_{n-1}}{p_n + p_{n-1}}\right),\ \tilde{\eta}_n = \frac{1}{3\alpha}\left(\frac{q_n - q_{n-1}}{q_n + q_{n-1}}\right).$$

Vectors ϕ and ψ are generated from discretizations involving $f,\ f_x,\ g,\ g_y$ and boundary values, respectively. Since they do not require a compactness, therefore the splitting scheme (1.7), (1.8), or (1.9), (1.10), on nonuniform grids is of second order accuracy as far as the CFL condition $\tau_\ell/h^2 = O(1)$ is maintained [**B**].

REMARK 1.1. Differentiating (1.2), (1.3) twice yields

$$u_{xxxx} = \frac{1}{\alpha}u_{xxz} - f_{xx}(u),\ u_{yyyy} = \frac{1}{\alpha}u_{yyz} - g_{yy}(u).$$

Substituting the above into P and Q we may obtain a third order compact approximation of the differential equations (1.2), (1.3). Discussions of the resulted algorithm are straightforward, although they are technically more sophisticated. Compact splitting schemes of even higher orders can be derived in similar ways. However, their computational advantages are severely limited due to the order barrier of splitting methods [**F**].

2. Numerical stability

We consider a linear Fourier analysis for the Cauchy problem concerned. Since κ is particularly large for optical applications, similar to [**G**], we have

DEFINITION 2.1. Let γ be the amplification factor of a finite difference method for solving (1.1), or (1.2), (1.3), together with (1.4). We say that the scheme is asymptotically stable for solving the highly oscillatory problem if there exists $c > 0$ such that

$$|\gamma| = 1 + O\left(1/\kappa^c\right). \tag{2.1}$$

Further, c is called the asymptotical stability index of the scheme. If (2.1) holds for all mesh steps then the scheme is said to be unconditionally asymptotically stable.

THEOREM 2.2. *If $p_j = p$, $q_k = q$ for all subindexes j, k, then the compact splitting method* (1.7), (1.8), *or* (1.9), (1.10), *is unconditionally asymptotically stable with a asymptotical stability index one.*

PROOF. The proof is straightforward. Note that the perturbation system corresponding to (1.7), (1.8), or (1.9), (1.10), can be written as

$$A\rho_k^{\ell+1/2} = B\rho_k^{\ell},\ k = 1, 2, \ldots, N; \tag{2.2}$$

$$C\varrho_j^{\ell+1} = D\varrho_j^{\ell+1/2},\ j = 1, 2, \ldots, N. \tag{2.3}$$

We only need to show the case with (2.2) since the other proof is similar. To this end, for any fixed k and $m \in \mathbb{Z}$, we have the amplification factor

$$\gamma = \frac{(-\mu_{1,m-1} + \eta_m)\, e^{-i\theta} + 1 - \mu_{2,m} + (-\mu_{3,m} + \eta_m)e^{i\theta}}{(\mu_{1,m-1} + \eta_m)e^{-i\theta} + 1 + \mu_{2,m} + (\mu_{3,m} + \eta_m)e^{i\theta}} = \frac{a_m}{b_m},\quad \theta \in \mathbb{R}, \tag{2.4}$$

where

$$a_m = \frac{\alpha\tau_\ell}{p}e^{-i\theta} + 2p - \frac{2\alpha\tau_\ell}{p} + \frac{\alpha\tau_\ell}{p}e^{i\theta},\quad b_m = -\frac{\alpha\tau_\ell}{p}e^{-i\theta} + 2p + \frac{2\alpha\tau_\ell}{p} - \frac{\alpha\tau_\ell}{p}e^{i\theta}.$$

It follows therefore

$$\begin{aligned} \frac{a_m}{b_m} &= \frac{p^2 + \alpha\tau_\ell\cos\theta}{p^2 - \alpha\tau_\ell\cos\theta} = \frac{\left(p^2 + \alpha\tau_\ell\cos\theta\right)^2}{p^4 - \alpha^2\tau_\ell^2\cos^2\theta} = \frac{p^4 + \beta^2\cos^2\theta - i2\beta\cos\theta}{p^4 + \beta^2\cos^2\theta} \\ &= 1 - \frac{i\tau_\ell\cos\theta}{\kappa(p^4 + \beta^2\cos^2\theta)} = \gamma, \end{aligned}$$

in which $\beta = \tau_\ell/(2\kappa) > 0$. Hence $|\gamma| = 1 + O(1/\kappa)$ and this completes the proof. □

THEOREM 2.3. *The compact splitting method* (1.7), (1.8), *or* (1.9), (1.10), *on nonuniform grids is unconditionally asymptotically stable with a asymptotical stability index one.*

PROOF. Again, we only need to consider (2.2). Recall (2.4). We have

$$\begin{aligned} a_m &= \left(\frac{\alpha\tau_\ell}{p_{m-1}} + \frac{p_m - p_{m-1}}{3\alpha}\right)e^{-i\theta} + (p_m + p_{m-1})\left(1 - \frac{\alpha\tau_\ell}{p_m p_{m-1}}\right) \\ &\quad + \left(\frac{\alpha\tau_\ell}{p_m} + \frac{p_m - p_{m-1}}{3\alpha}\right)e^{i\theta} = \frac{\alpha\tau_\ell}{p_{m-1}}e^{-i\theta} + \frac{\alpha\tau_\ell}{p_m}e^{i\theta} \\ &\quad + \frac{2(p_m - p_{m-1})}{3\alpha}\cos\theta + (p_m + p_{m-1})\left(1 - \frac{\alpha\tau_\ell}{p_m p_{m-1}}\right); \\ b_m &= \left(-\frac{\alpha\tau_\ell}{p_{m-1}} + \frac{p_m - p_{m-1}}{3\alpha}\right)e^{-i\theta} + (p_m + p_{m-1})\left(1 + \frac{\alpha\tau_\ell}{p_m p_{m-1}}\right) \\ &\quad + \left(-\frac{\alpha\tau_\ell}{p_m} + \frac{p_m - p_{m-1}}{3\alpha}\right)e^{i\theta} = -\frac{\alpha\tau_\ell}{p_{m-1}}e^{-i\theta} - \frac{\alpha\tau_\ell}{p_m}e^{i\theta} \\ &\quad + \frac{2(p_m - p_{m-1})}{3\alpha}\cos\theta + (p_m + p_{m-1})\left(1 + \frac{\alpha\tau_\ell}{p_m p_{m-1}}\right). \end{aligned}$$

Thus,

$$\begin{aligned}
\frac{a_m}{\alpha} &= \frac{\tau_\ell(p_m+p_{m-1})}{p_m p_{m-1}}(\cos\theta-1) - \frac{8\kappa^2(p_m-p_{m-1})}{3}\cos\theta \\
&\quad + i\left[2\kappa(p_m+p_{m-1}) + \frac{\tau_\ell(p_m-p_{m-1})}{p_m p_{m-1}}\sin\theta\right] \\
&= -\frac{2\tau_\ell(p_m+p_{m-1})}{p_m p_{m-1}}\sin^2\frac{\theta}{2} - \frac{8\kappa^2(p_m-p_{m-1})}{3}\cos\theta \\
&\quad + i\left[2\kappa(p_m+p_{m-1}) + \frac{\tau_\ell(p_m-p_{m-1})}{p_m p_{m-1}}\sin\theta\right] \\
&= (p_m+p_{m-1})\left\{-\frac{2\tau_\ell}{p_m p_{m-1}}\sin^2\frac{\theta}{2} - \frac{8\kappa^2(p_m^2-p_{m-1}^2)}{3}\cos\theta\right. \\
&\quad \left. + i\left[2\kappa + \frac{\tau_\ell(p_m^2-p_{m-1}^2)}{p_m p_{m-1}}\sin\theta\right]\right\}, \\
\frac{b_m}{\alpha} &= -\frac{\tau_\ell(p_m+p_{m-1})}{p_m p_{m-1}}(\cos\theta-1) - \frac{8\kappa^2(p_m-p_{m-1})}{3}\cos\theta \\
&\quad + i\left[2\kappa(p_m+p_{m-1}) - \frac{\tau_\ell(p_m-p_{m-1})}{p_m p_{m-1}}\sin\theta\right] \\
&= \frac{2\tau_\ell(p_m+p_{m-1})}{p_m p_{m-1}}\sin^2\frac{\theta}{2} - \frac{8\kappa^2(p_m-p_{m-1})}{3}\cos\theta \\
&\quad + i\left[2\kappa(p_m+p_{m-1}) - \frac{\tau_\ell(p_m-p_{m-1})}{p_m p_{m-1}}\sin\theta\right] \\
&= (p_m+p_{m-1})\left\{\frac{2\tau_\ell}{p_m p_{m-1}}\sin^2\frac{\theta}{2} - \frac{8\kappa^2(p_m^2-p_{m-1}^2)}{3}\cos\theta\right. \\
&\quad \left. + i\left[2\kappa - \frac{\tau_\ell(p_m^2-p_{m-1}^2)}{p_m p_{m-1}}\sin\theta\right]\right\}.
\end{aligned}$$

Consequently,

$$\frac{a_m}{b_m} = \frac{-a-b+i(2\kappa+c)}{a-b+i(2\kappa-c)} = 1 - \frac{1}{\kappa} \times \frac{2a-2ic}{(a-b)/\kappa + i\,(2-c/\kappa)} = \gamma,$$

where

$$a = \frac{2\tau_\ell}{p_m p_{m-1}}\sin^2\frac{\theta}{2},\quad b = \frac{8\kappa^2(p_m^2-p_{m-1}^2)}{3}\cos\theta,\quad c = \frac{\tau_\ell(p_m^2-p_{m-1}^2)}{p_m p_{m-1}}\sin\theta.$$

The above indicates that $|\gamma| = 1 + O(1/\kappa)$ which ensures our theorem. □

REMARK 2.4. The Fourier analysis utilized for Cauchy problems may not be suitable for boundary value problems unless proper restructures of underlying difference schemes can be formulated.

3. Computational examples

Denote $\kappa = z_1^2 + 1$, $z_1 \gg 10$. We consider equation (1.1) together with a Gaussian beam type initial function in small-f-number lens modeling computations [**D, G**],

$$u_0(x,y) = \frac{1-iz_1}{\kappa}e^{-(x^2+y^2)(1+iz_1)},\quad -\infty < x,y < \infty. \tag{3.1}$$

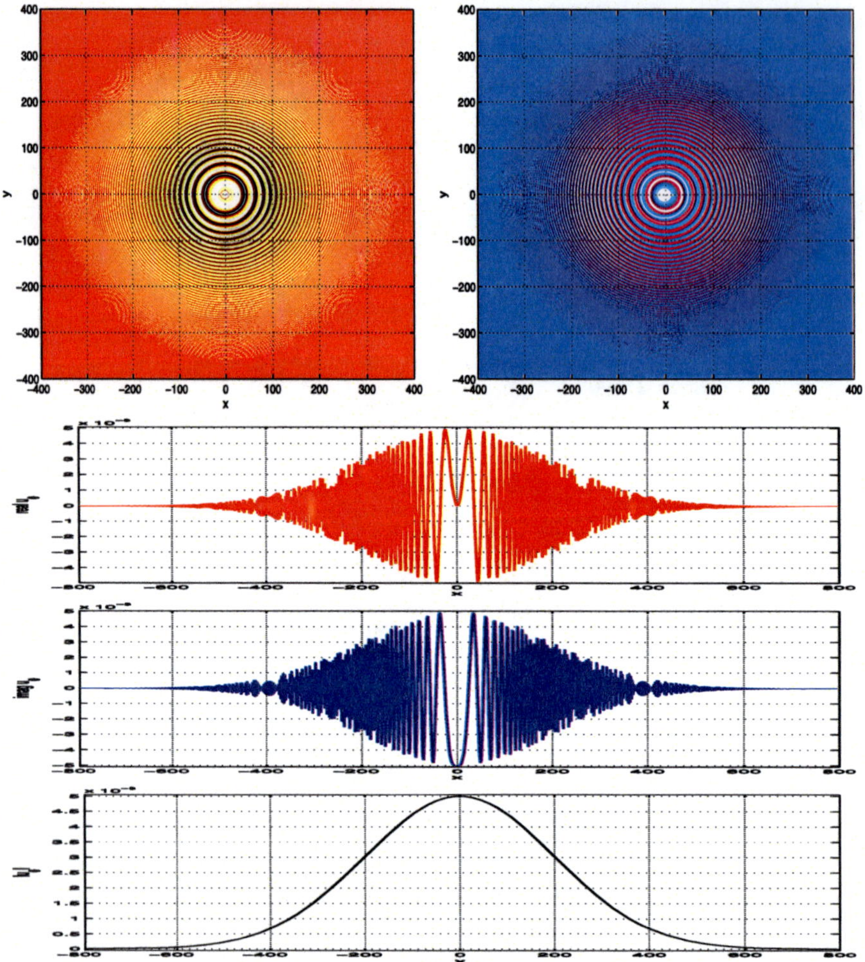

FIGURE 1. Contour maps of the real (top-left) and imaginary (top-right) parts of the oscillatory initial function $u = u_0(x, y)$ on the xy-plane; intersections of the real (red), imaginary (blue) parts and modulus (black) of u with corresponding xu-planes.

Due to the symmetry of (1.1), (3.1), asymptotic energy conservations may be observed through a much simplified integral along the x-axis, that is,

$$E(z) = \int_{-\infty}^{\infty} |u(x, 0, z)| dx. \tag{3.2}$$

On the other hand, because of the rapid decay of (3.1) in transverse directions, we are able to choose an actual computational domain for $-s \leq x, y \leq s,\ s > 0$, in space and import following axillary homogeneous Dirichlet boundary conditions [**B**]

$$u(x, y, z) = 0 \text{ whenever } |x| \text{ or } |y| = s. \tag{3.3}$$

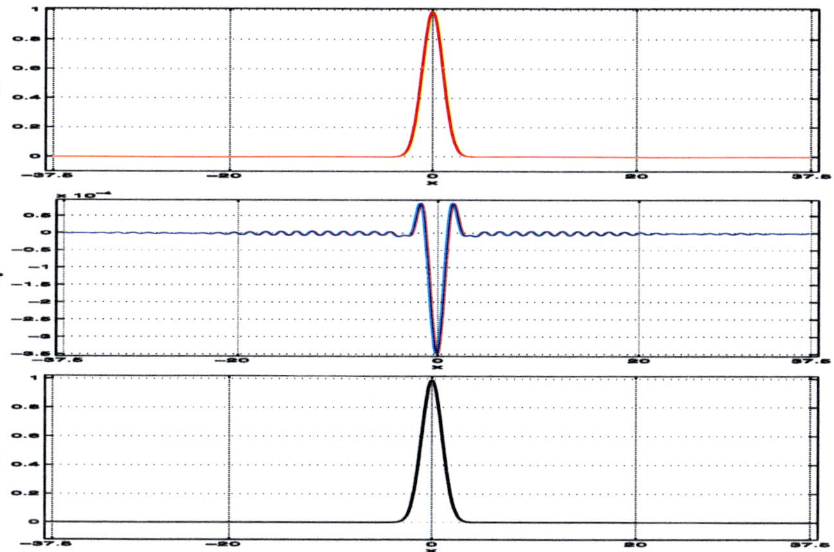

FIGURE 2. Intersections of the real (red), imaginary (blue) parts and modulus (black) of the function u with corresponding xu-planes at focusing. The numerical results are precise as validated by experiments. A typical optical parameter, $z_1 = 100$, is utilized.

EXAMPLE 3.1. For the simplicity of calculations, we set $s = 1$. Hence (1.1), (3.1) and (3.3) can be formulated to

$$u_z = -iu_{xx} - iu_{yy}, \quad -\sqrt{2\kappa} < x, y < \sqrt{2\kappa}, \quad z \geq z_0, \tag{3.4}$$

$$u(x, y, z_0) = \frac{1}{1 + iz_1} e^{-(x^2+y^2)/[2(1+iz_1)]}, \quad -\sqrt{2\kappa} < x, y < \sqrt{2\kappa}, \tag{3.5}$$

$$u(x, y, z) = 0 \text{ whenever } |x| \text{ or } |y| = \sqrt{2\kappa}. \tag{3.6}$$

It is known that the solution of (3.4)-(3.6) focuses at the distance $z = \tilde{z} - z_0 = z_1$. We let $1/4000 \leq p_j,\ q_k \leq 1/400$, and corresponding CFL numbers be bounded above by two on an exponentially graded symmetric mesh [**A**] in the space. Recall (3.2). An application of the composite trapezoidal rule leads to

$$E(z_0) \approx 2.50646936434319, \quad z_1 = 100. \tag{3.7}$$

In Figure 1, we show contour maps as well as surface projections of the initial function (3.5). High oscillations are clearly observed. A practically important parameter $z_1 = 200$ ($\kappa = 4.0001 \times 10^4$) [**D, E**] is employed.

Figures 2 and 3 are devoted to illustrations of the focusing phenomenon of numerical solution u at $z = z_1$. While the former figure demonstrates oscillatory features of the real, imaginary parts and modulus of u within the reduced interval $x \in [-37.6, 37.6]$ as $y = 0$, the latter picture offers more insights of 3D surfaces of the numerical solution near the focusing location. The peak value of $|u(x, y, z_1)|$ acquired is approximately $\zeta_1 = 0.98958512468675$ under a fast solution procedure enabled by the splitting method (1.9), (1.10). For given $z_1 = 100$, a straightforward numerical evaluation of the asymptotic energy via (3.2) indicates that

$$E(z_1) \approx 2.49240597631224,$$

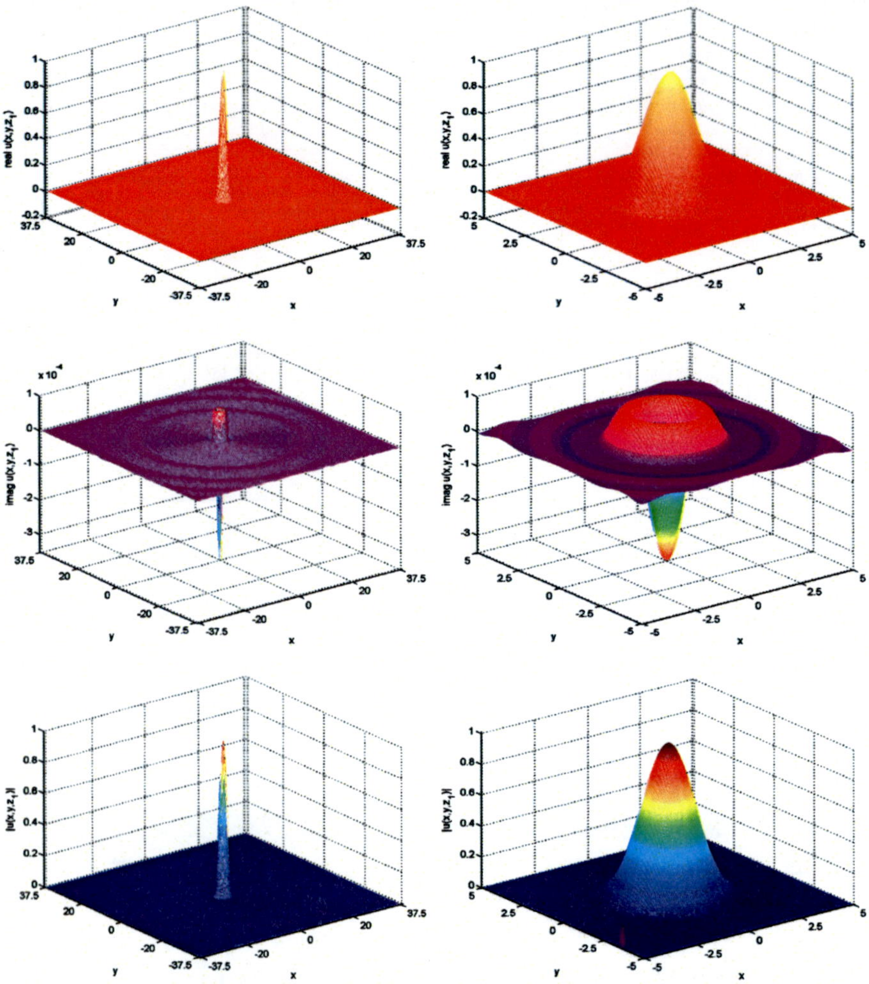

FIGURE 3. 3D plots of the numerical solution at focusing: real (top), imaginary (middle) parts and the modulus (bottom) of u. Figures on the right-hand-side are transversely enlarged. A typical optical parameter $z_1 = 100$ ($\kappa = 1.0001 \times 10^4$) is employed.

which is in a good agreement with (3.7). The numerical approximations obtained are thus highly satisfactory.

References

[A] M. A. Beauregard and Q. Sheng, *A semi-adaptive compact splitting method for the numerical solution of two-dimensional quenching problems*, Appl. Math. Comptat. **218** (2012), 1240–1254. MR2942406

[B] G. Fairweather, *Finite Element Galerkin Methods for Differential Equations*, CRC Press, London, 1978. MR0495013 (58:13781)

[C] L. P. Gonzalez, S. Guha, J. W. Rogers and Q. Sheng, *An effective z-stretching method for paraxial light beam propagation simulations*, J. Comptat. Phys. **227** (2008), 7264–7278. MR2433971 (2009f:78001)

[D] S. Guha, L. P. Gonzalez and Q. Sheng, *Description of light focusing by a spherical lens using diffraction integral method*, Proc. Appl. Math. Mech. **7** (2008), 1023301–1023302.

[E] J. V. Moloney and A. C. Newell, *Nonlinear Optics*, Westview Press, Boulder, CO, 2004. MR1163192 (93i:78010)

[F] Q. Sheng, *Solving linear partial differential equations by exponential splitting*, IMA J. Numer. Anal. **9** (1989), 199–212. MR1000457 (90g:65163)

[G] Q. Sheng, S. Guha and L. P. Gonzalez, *An exponential transformation based splitting method for fast computations of highly oscillatory solutions*, J. Comptat. Appl. Math. **235** (2011), 4452–4463. MR2802016 (2012f:65136)

[H] T. W. H. Sheu, L. W. Hsieh and C. F. Chen, *Development of a three-point sixth-order Helmholtz scheme*, J. Comptat. Acoustics, **16** (2008), 343–359. MR2524257 (2010f:65225)

Department of Mathematics and Center for Astrophysics, Space Physics and Engineering Research, Baylor University, Waco, Texas 76798-7328

E-mail address: qin_sheng@baylor.edu

Air Force Research Laboratory, Materials and Manufacturing Directorate, Wright-Patterson Air Force Base, Ohio 45433

E-mail address: shekhar.guha@wpafb.af.mil

Survey on discontinuous Galerkin methods for Hamilton-Jacobi equations

Chi-Wang Shu

ABSTRACT. In this paper we give a brief survey on the development, analysis and application of discontinuous Galerkin methods for solving Hamilton-Jacobi equations.

1. Introduction

We are interested in solving the following time dependent Hamilton-Jacobi equations

$$\varphi_t + H(\varphi_{x_1}, ..., \varphi_{x_d}) = 0, \qquad \varphi(x, 0) = \varphi^0(x), \tag{1.1}$$

where H is a Lipschitz continuous function. H could also depend on φ, x and t in some applications.

Hamilton-Jacobi equations appear often in many applications. Examples include front propagation, level set methods, image processing and computer vision, control and differential games.

The main difficulty for the design of efficient and reliable algorithms for solving the Hamilton-Jacobi equations (1.1) is that global C^1 solution does not in general exist for (1.1) regardless of the smoothness of the initial condition $\varphi^0(x)$. Singularities in the form of discontinuities in the derivatives of φ would appear at a finite time in most situations, thus the solutions would only be Lipschitz continuous but no longer C^1. Such solutions are extensively studied in the literature as the viscosity solutions [**8**], including their existence, uniqueness and well-posedness in the L^∞ norm. Numerical solutions for Hamilton-Jacobi equations have been pursued extensively in the literature, with finite difference schemes [**12**], finite element schemes, etc. A review can be found in [**19**].

At least in the one dimensional case, there is a strong relationship between the Hamilton-Jacobi equation

$$\varphi_t + H(\varphi_x) = 0, \qquad \varphi(x, 0) = \varphi^0(x) \tag{1.2}$$

1991 *Mathematics Subject Classification.* Primary 65M60, 65N30.

Key words and phrases. Hamilton-Jacobi equations, discontinuous Galerkin method, fast sweeping method, optimal control, error estimate.

The author was supported in part by NSF Grant DMS-1112700 and ARO grant W911NF-11-1-0091.

and the hyperbolic conservation law

$$u_t + H(u)_x = 0, \qquad u(x,0) = u^0(x). \tag{1.3}$$

In fact, if we identify $u = \varphi_x$, the two equations (1.2) and (1.3) are equivalent. This equivalency provides motivations for designing algorithms for one equation based on the success for another equation. In this article we survey the design of discontinuous Galerkin (DG) methods, which were first successfully designed to solve hyperbolic conservation laws (1.3), to solve the Hamilton-Jacobi equations (1.2) in one-dimension and then (1.1) in multi-dimensions.

Let us finish this introduction section by briefly describing the discontinuous Galerkin method for solving the one-dimensional hyperbolic conservation law (1.3). More details can be found in, e.g. [**5, 7, 10, 20**]. Suppose the partial differential equation (PDE) (1.3) is defined on the interval [0,1] and suppose, for simplicity of presentation, we have periodic boundary conditions. First, we divide [0,1] into N cells

$$0 = x_{\frac{1}{2}} < x_{\frac{3}{2}} < \cdots < x_{N+\frac{1}{2}} = 1,$$

and denote

$$I_j = \left(x_{j-\frac{1}{2}}, x_{j+\frac{1}{2}}\right), \quad x_j = \frac{1}{2}\left(x_{j-\frac{1}{2}} + x_{j+\frac{1}{2}}\right), \quad h_j = x_{j+\frac{1}{2}} - x_{j-\frac{1}{2}}$$

as the cells, cell centers and cell lengths respectively. We also define $h = h_{\max} = \max_j h_j$ and $h_{\min} = \min_j h_j$, and we consider only regular meshes, that is $h_{\max} \le \lambda h_{\min}$ where $\lambda \ge 1$ is a constant during mesh refinement. If $\lambda = 1$, then the mesh is uniformly distributed. Define the discontinuous Galerkin finite element space as

$$V_h^k = \{v : v|_{I_j} \in \mathcal{P}^k(I_j), j = 1, \cdots, N\}, \tag{1.4}$$

where $\mathcal{P}^k(I_j)$ denotes the space of polynomials in I_j of degree at most k. The DG scheme for solving (1.3) is: find $u_h \in V_h^k$, such that for any $v_h \in V_h^k$

$$\int_{I_j} (u_h)_t v_h dx - \int_{I_j} H(u_h)(v_h)_x dx + \hat{H}_{j+\frac{1}{2}}(v_h)^-_{j+\frac{1}{2}} - \hat{H}_{j-\frac{1}{2}}(v_h)^+_{j-\frac{1}{2}} = 0 \tag{1.5}$$

where $\hat{H}_{j+\frac{1}{2}} = \hat{H}((u_h)^-_{j+\frac{1}{2}}, (u_h)^+_{j+\frac{1}{2}})$ is a monotone numerical flux, which is an increasing function of the first argument and a decreasing function of the second argument, is Lipschitz continuous, and is consistent with the function H in the sense that $\hat{H}(u,u) = H(u)$. In the simple situation that $H'(u) \ge 0$, the monotone numerical flux can be taken as the simple upwind flux $\hat{H}(u^-, u^+) = H(u^-)$. A physical explanation is that, in this situation, the wind blows from the left to the right, hence at the cell interface $x_{j+\frac{1}{2}}$ we take the information from the upwind side $(u_h)^-_{j+\frac{1}{2}}$. The time discretization of (1.5) is typically given by the total-variation-diminishing (TVD), or strong-stability-preserving (SSP) Runge-Kutta methods [**9, 21**].

The DG scheme (1.5) can also be written in the equivalent form

$$\begin{aligned} &\int_{I_j} ((u_h)_t + H(u_h)_x)\, v_h dx + \left(\hat{H}_{j+\frac{1}{2}} - H((u_h)^-_{j+\frac{1}{2}})\right)(v_h)^-_{j+\frac{1}{2}} \\ &\quad - \left(\hat{H}_{j-\frac{1}{2}} - H((u_h)^+_{j-\frac{1}{2}})\right)(v_h)^+_{j-\frac{1}{2}} = 0 \end{aligned} \tag{1.6}$$

by performing an integration by parts for the second integral in (1.5). This equivalent form will be useful in Section 3 when we define our DG scheme for solving the Hamilton-Jacobi equations (1.2).

2. Discontinuous Galerkin method I

The first attempt to design a DG method was based on the observation that, at least in one dimension, the viscosity solution of the Hamilton-Jacobi equation (1.2) is equivalent to the entropy solution of the conservation law (1.3), when we identify $\varphi_x = u$. Therefore, a DG scheme for solving the conservation law (1.3), as given by (1.5), can be directly used to approximate the derivative of the viscosity solution of the Hamilton-Jacobi equation (1.2). This leads to the following DG algorithm of Hu and Shu [**11**]: Find $\varphi_h \in V_h^{k+1}$, such that

- $u_h = (\varphi_h)_x \in V_h^k$ is determined by the DG scheme (1.5);
- The missing degree of freedom is determined by

$$\int_{I_j} ((\varphi_h)_t + H(u_h))\,dx = 0.$$

This algorithm is well defined for one dimension. For multi-dimensional cases, e.g. in two space dimensions, the Hamilton-Jacobi equation

$$\varphi_t + H(\varphi_x, \varphi_y) = 0 \tag{2.1}$$

is (in some sense) equivalent to the following system of conservation laws

$$u_t + H(u,v)_x = 0, \qquad v_t + H(u,v)_y = 0 \tag{2.2}$$

when we identify $u = \varphi_x$ and $v = \varphi_y$. For example, a vanishing viscosity solution of (2.1) corresponds to a vanishing viscosity solution of (2.2), and vice versa [**13**]. However, (2.2) is not a strictly hyperbolic system, which may cause problems in its numerical solution if we treat u and v as independent variables. Therefore, we would like to still use a piecewise polynomial φ_h as our solution variable and take its derivatives to approximate u and v. The DG algorithm of Hu and Shu [**11**], as re-interpreted by Li and Shu [**16**], can be formulated as follows: Find $\varphi_h \in V_h^{k+1}$, such that

- $(u_h, v_h) = ((\varphi_h)_x, (\varphi_h)_y) \in W_h^k$ is determined by the standard DG scheme solving the conservation laws (2.2);
- The missing degree of freedom is determined by

$$\int_{I_j} ((\varphi_h)_t + H(u_h, v_h))\,dxdy = 0.$$

Here, I_j denotes two-dimensional elements (triangles or rectangles), and W_h^k is the locally curl-free subspace of $V_h^k \times V_h^k$:

$$W_h^k = \{(u,v) \in V_h^k \times V_h^k : u_y - v_x = 0 \quad \forall (x,y)\ in I_j\}.$$

Notice that the implementation of the DG method in locally curl-free subspace W_h^k is straightforward and it is less expensive than the standard DG method defined on $V_h^k \times V_h^k$, since W_h^k involves fewer degrees of freedom.

Some analysis for this DG method (including L^2 stability for a specific class of the Hamiltonian H) is given in [**15**].

3. Discontinuous Galerkin method II

The DG method in [11, 16], as described in the previous section, works well numerically, with provable stability results for certain special cases [11, 15]. However, since this method is based on the conservation law system satisfied by the derivatives of φ, a scalar problem (2.1) is converted to a system (2.2) for the multi-dimensional case, which is moreover only weakly hyperbolic at $(u, v) = (0, 0)$. This seems to have made the algorithm indirect and somewhat complicated. It is therefore desirable to design a DG method which solves directly the solution φ to the Hamilton-Jacobi equation (1.1). The scheme of Cheng and Shu [4] serves this purpose.

To motivate the method in [4], we first consider the special case for which the Hamiltonian is linear, namely $H(u) = au$ where a is a constant. The Hamilton-Jacobi equation (1.2) becomes

$$\varphi_t + a\varphi_x = 0 \tag{3.1}$$

which is also a conservation law (1.3). Therefore, we can write down the standard DG scheme for this conservation law, written in the form of (1.6). If the constant $a > 0$ and we take the upwind flux $\hat{H}(u^-, u^+) = H(u^-)$, this DG scheme becomes: find $\varphi_h \in V_h^k$, such that for any $v_h \in V_h^k$

$$\int_{I_j} \left((\varphi_h)_t + a(\varphi_h)_x \right) v_h dx + a[\varphi_h]_{j-\frac{1}{2}} (v_h)^+_{j-\frac{1}{2}} = 0 \tag{3.2}$$

where $[\varphi_h]_{j-\frac{1}{2}} = (\varphi_h)^+_{j-\frac{1}{2}} - (\varphi_h)^-_{j-\frac{1}{2}}$ denotes the jump of φ_h at the cell interface $x_{j-\frac{1}{2}}$. Likewise, if the constant $a < 0$ and we take the upwind flux $\hat{H}(u^-, u^+) = H(u^+)$, this DG scheme becomes: find $\varphi_h \in V_h^k$, such that for any $v_h \in V_h^k$

$$\int_{I_j} \left((\varphi_h)_t + a(\varphi_h)_x \right) v_h dx + a[\varphi_h]_{j+\frac{1}{2}} (v_h)^-_{j+\frac{1}{2}} = 0. \tag{3.3}$$

This motivates the definition of the DG scheme in [4] for solving the general nonlinear Hamilton-Jacobi equation (1.2): find $\varphi_h \in V_h^k$, such that

$$\begin{aligned} \int_{I_j} \left((\varphi_h)_t + H((\varphi_h)_x \right) v_h(x) dx + \left(\min_{x \in I_{j+1/2}} H'((\tilde{\varphi}_h)_x) \right)_- [\varphi_h]_{j+\frac{1}{2}} (v_h)^-_{j+\frac{1}{2}} \\ + \left(\max_{x \in I_{j-1/2}} H'((\tilde{\varphi}_h)_x) \right)_+ [\varphi_h]_{j-\frac{1}{2}} (v_h)^+_{j-\frac{1}{2}} = 0 \end{aligned} \tag{3.4}$$

holds for any $v_h \in V_h^k$. Here $a_- = \min(a, 0)$, $a_+ = \max(a, 0)$, and $H'(u)$ denotes the derivative of $H(u)$ with respect to u. The interval $I_{j+1/2} = [x_j, x_{j+1}]$, and the function $\tilde{\varphi}_h$ is the L^2 projection of φ_h (which is discontinuous at the interface point $x_{j+\frac{1}{2}}$) into $\mathcal{P}^{2k+1}(I_j \cup I_{j+1})$. Clearly, when the Hamiltonian $H(u) = au$ is linear, the scheme (3.4) becomes (3.2) or (3.3) depending on the sign of a. Therefore, the DG scheme (3.4) is a direct generalization of the standard DG scheme for conservation laws. The choice of the min and max in the scheme (3.4) is to make the scheme more stable, or more diffusive, within the scope of the designed high order of accuracy.

Extension of this method to multi-dimensions is straightforward, since it applies directly on the scalar Hamilton-Jacobi equation (2.1) rather than on the system (2.2) satisfied by the derivatives of the solution. For linear Hamilton-Jacobi equations (including those with smooth variable coefficients), this DG method can be proved to be L^2 stable with optimal $(k+1)$-th order convergence rate in L^2 for smooth solutions [**4**]. The same optimal $(k+1)$-th order convergence rate in L^2 for smooth solutions is also proved for nonlinear Hamilton-Jacobi equations in [**22**]. In the numerical experiments in [**4**], it is observed that the method converges to the viscosity solution, with the help of an entropy correction in some non-convex cases. In [**1, 2**], this DG method is applied to front propagation problems, without and with obstacles, producing excellent results. In [**18**], a central DG scheme based on overlapping staggered meshes is designed, which avoids the usage of the projected $\tilde{\varphi}_h$ in the definition of the scheme (3.4) at the cell interfaces.

4. Discontinuous Galerkin method III

Another DG method which solves directly the Hamilton-Jacobi equations (2.1) is that of Yan and Osher [**24**]. This method is motivated by the local discontinuous Galerkin (LDG) method for solving second order partial differential equations [**6**]. First, two "upwind-biased" piecewise polynomials for approximating φ_x are obtained by

- Find $u_h^- \in V_h^k$, such that for any $v_h \in V_h^k$

$$\int_{I_j} (u_h^-)_t v_h dx + \int_{I_j} \varphi_h (v_h)_x dx - (\varphi_h^-)_{j+\frac{1}{2}} (v_h)^-_{j+\frac{1}{2}} + (\varphi_h^-)_{j-\frac{1}{2}} (v_h)^+_{j-\frac{1}{2}} = 0; \tag{4.1}$$

- Find $u_h^+ \in V_h^k$, such that for any $v_h \in V_h^k$

$$\int_{I_j} (u_h^+)_t v_h dx + \int_{I_j} \varphi_h (v_h)_x dx - (\varphi_h^+)_{j+\frac{1}{2}} (v_h)^-_{j+\frac{1}{2}} + (\varphi_h^+)_{j-\frac{1}{2}} (v_h)^+_{j-\frac{1}{2}} = 0. \tag{4.2}$$

The DG method is then defined by: Find $\varphi_h \in V_h^k$, such that for any $v_h \in V_h^k$

$$\int_{I_j} \left((\varphi_h)_t + \hat{H}(u_h^-, u_h^+) \right) v_h dx = 0. \tag{4.3}$$

Here, $\hat{H}(u^-, u^+)$ is a monotone flux, for example the Lax-Friedrichs monotone flux.

Clearly, when the Hamiltonian $H(u) = au$ is linear, the scheme (4.3) becomes (3.2) or (3.3) depending on the sign of a. Therefore, the DG scheme (4.3) is again a direct generalization of the standard DG scheme for conservation laws. Generalization to multi-dimensions is again straightforward. Numerical results demonstrating the good performance of this DG method are given in [**24**]. The optimal $(k+1)$-th order convergence rate in L^2 for smooth solutions is proved for nonlinear Hamilton-Jacobi equations in [**22**].

5. Fast sweeping methods

If we are interested in solving the steady state solution of the Hamilton-Jacobi equation

$$\varphi_t + H(\varphi_{x_1}, ..., \varphi_{x_d}) = f(x), \tag{5.1}$$

i.e. we would like to obtain a time-independent solution $\varphi(x)$ of (5.1), we could just use the time-dependent algorithm with Runge-Kutta time stepping and run it for a very long time until steady state is reached. However, this will be computationally very expensive.

Alternatively, we could attempt to solve the discretization of the steady state equation

$$H(\varphi_{x_1}, ..., \varphi_{x_d}) = f(x), \tag{5.2}$$

directly. This discretization is however a large system of nonlinear algebraic equations, most likely with multiple solutions. It is a challenge to design an iterative solver for this nonlinear system which can converge to the desired approximation of the viscosity solutions of (5.2) as fast as possible.

Fast sweeping method is one of such efficient iterative methods for solving the discretization of the steady state equation (5.2). It uses Gauss-Seidel iterations and a local solver (that is, writing the local variable in cell I_j as explicit functions of the neighboring variables) to update the information. Upwinding and "causality" are used in an essential way to prove (fast) convergence of the fast sweeping method for first order monotone schemes [**3, 27**]. The generalization of the fast sweeping method to higher order finite difference schemes [**14, 23, 26**] suffers from the fact that downwind information must be used in the local solver because of the wide stencil of the scheme, hence convergence speed is compromised and it is no longer possible to have the number of iterations independent of the mesh size. On the other hand, the DG schemes are local (using information of only immediate neighbors) regardless of order of accuracy, hence they are naturally suited for efficient fast sweeping methods. In [**17, 25**], very efficient fast sweeping methods, which can converge to the steady state solutions of the Hamilton-Jacobi equations in very few iterations (independent of the mesh sizes) are designed for the piecewise linear (second order) DG method of Cheng and Shu [**4**], described in Section 3. Generalizations to higher order DG schemes constitute current research. The main challenge is how to obtain efficient local solvers for higher order DG schemes.

6. Concluding remarks

This short survey paper summarizes several types of discontinuous Galerkin (DG) schemes for solving time-dependent and steady state Hamilton-Jacobi equations. Because of their advantage in handling complicated geometry, flexibility in adaptivity, and fast convergence when applied in the fast sweeping framework, DG methods have a good potential in applications.

References

[1] Olivier Bokanowski, Yingda Cheng, and Chi-Wang Shu, *A discontinuous Galerkin solver for front propagation*, SIAM J. Sci. Comput. **33** (2011), no. 2, 923–938, DOI 10.1137/090771909. MR2801195 (2012j:65314)

[2] O. Bokanowski, Y. Cheng and C.-W. Shu, *A discontinuous Galerkin scheme for front propagation with obstacles*, submitted to Numerische Mathematik.

[3] Michelle Boué and Paul Dupuis, *Markov chain approximations for deterministic control problems with affine dynamics and quadratic cost in the control*, SIAM J. Numer. Anal. **36** (1999), no. 3, 667–695, DOI 10.1137/S0036142997323521. MR1681057 (2000a:49054)

[4] Yingda Cheng and Chi-Wang Shu, *A discontinuous Galerkin finite element method for directly solving the Hamilton-Jacobi equations*, J. Comput. Phys. **223** (2007), no. 1, 398–415, DOI 10.1016/j.jcp.2006.09.012. MR2314396 (2008c:65251)

[5] Bernardo Cockburn and Chi-Wang Shu, *TVB Runge-Kutta local projection discontinuous Galerkin finite element method for conservation laws. II. General framework*, Math. Comp. **52** (1989), no. 186, 411–435, DOI 10.2307/2008474. MR983311 (90k:65160)

[6] Bernardo Cockburn and Chi-Wang Shu, *The local discontinuous Galerkin method for time-dependent convection-diffusion systems*, SIAM J. Numer. Anal. **35** (1998), no. 6, 2440–2463 (electronic), DOI 10.1137/S0036142997316712. MR1655854 (99j:65163)

[7] Bernardo Cockburn and Chi-Wang Shu, *Runge-Kutta discontinuous Galerkin methods for convection-dominated problems*, J. Sci. Comput. **16** (2001), no. 3, 173–261, DOI 10.1023/A:1012873910884. MR1873283 (2002i:65099)

[8] Michael G. Crandall and Pierre-Louis Lions, *Viscosity solutions of Hamilton-Jacobi equations*, Trans. Amer. Math. Soc. **277** (1983), no. 1, 1–42, DOI 10.2307/1999343. MR690039 (85g:35029)

[9] Sigal Gottlieb, David Ketcheson, and Chi-Wang Shu, *Strong stability preserving Runge-Kutta and multistep time discretizations*, World Scientific Publishing Co. Pte. Ltd., Hackensack, NJ, 2011. MR2789749 (2012f:65107)

[10] Jan S. Hesthaven and Tim Warburton, *Nodal discontinuous Galerkin methods*, Texts in Applied Mathematics, vol. 54, Springer, New York, 2008. Algorithms, analysis, and applications. MR2372235 (2008k:65002)

[11] Changqing Hu and Chi-Wang Shu, *A discontinuous Galerkin finite element method for Hamilton-Jacobi equations*, SIAM J. Sci. Comput. **21** (1999), no. 2, 666–690 (electronic), DOI 10.1137/S1064827598337282. MR1718679 (2000g:65095)

[12] Guang-Shan Jiang and Danping Peng, *Weighted ENO schemes for Hamilton-Jacobi equations*, SIAM J. Sci. Comput. **21** (2000), no. 6, 2126–2143 (electronic), DOI 10.1137/S106482759732455X. MR1762034 (2001e:65124)

[13] Shi Jin and Zhouping Xin, *Numerical passage from systems of conservation laws to Hamilton-Jacobi equations, relaxation schemes*, SIAM J. Numer. Anal. **35** (1998), no. 6, 2385–2404, DOI 10.1137/S0036142996314366. MR1655852 (2000k:65147)

[14] Chiu-Yen Kao, Stanley Osher, and Yen-Hsi Tsai, *Fast sweeping methods for static Hamilton-Jacobi equations*, SIAM J. Numer. Anal. **42** (2005), no. 6, 2612–2632, DOI 10.1137/S0036142902419600. MR2139407 (2006c:65091)

[15] Olga Lepsky, Changqing Hu, and Chi-Wang Shu, *Analysis of the discontinuous Galerkin method for Hamilton-Jacobi equations*, Proceedings of the Fourth International Conference on Spectral and High Order Methods (ICOSAHOM 1998) (Herzliya), 2000, pp. 423–434, DOI 10.1016/S0168-9274(99)00109-9. MR1772920

[16] Fengyan Li and Chi-Wang Shu, *Reinterpretation and simplified implementation of a discontinuous Galerkin method for Hamilton-Jacobi equations*, Appl. Math. Lett. **18** (2005), no. 11, 1204–1209, DOI 10.1016/j.aml.2004.10.009. MR2170874 (2006g:65154)

[17] Fengyan Li, Chi-Wang Shu, Yong-Tao Zhang, and Hongkai Zhao, *A second order discontinuous Galerkin fast sweeping method for Eikonal equations*, J. Comput. Phys. **227** (2008), no. 17, 8191–8208, DOI 10.1016/j.jcp.2008.05.018. MR2442450 (2009e:65178)

[18] Fengyan Li and Sergey Yakovlev, *A central discontinuous Galerkin method for Hamilton-Jacobi equations*, J. Sci. Comput. **45** (2010), no. 1-3, 404–428, DOI 10.1007/s10915-009-9340-y. MR2679806 (2011i:65163)

[19] Chi-Wang Shu, *High order numerical methods for time dependent Hamilton-Jacobi equations*, Mathematics and computation in imaging science and information processing, Lect. Notes Ser. Inst. Math. Sci. Natl. Univ. Singap., vol. 11, World Sci. Publ., Hackensack, NJ, 2007, pp. 47–91, DOI 10.1142/9789812709066_0002. MR2395744 (2009b:65224)

[20] Chi-Wang Shu, *Discontinuous Galerkin methods: general approach and stability*, Numerical solutions of partial differential equations, Adv. Courses Math. CRM Barcelona, Birkhäuser, Basel, 2009, pp. 149–201. MR2531713

[21] Chi-Wang Shu and Stanley Osher, *Efficient implementation of essentially nonoscillatory shock-capturing schemes*, J. Comput. Phys. **77** (1988), no. 2, 439–471, DOI 10.1016/0021-9991(88)90177-5. MR954915 (89g:65113)

[22] T. Xiong, C.-W. Shu and M. Zhang, *A priori error estimates for semi-discrete discontinuous Galerkin methods solving nonlinear Hamilton-Jacobi equations with smooth solutions*, to appear in International Journal of Numerical Analysis and Modeling.

[23] Tao Xiong, Mengping Zhang, Yong-Tao Zhang, and Chi-Wang Shu, *Fast sweeping fifth order WENO scheme for static Hamilton-Jacobi equations with accurate boundary treatment*, J. Sci. Comput. **45** (2010), no. 1-3, 514–536, DOI 10.1007/s10915-010-9345-6. MR2679811 (2011g:65153)

[24] Jue Yan and Stanley Osher, *A local discontinuous Galerkin method for directly solving Hamilton-Jacobi equations*, J. Comput. Phys. **230** (2011), no. 1, 232–244, DOI 10.1016/j.jcp.2010.09.022. MR2734289 (2012a:65269)

[25] Yong-Tao Zhang, Shanqin Chen, Fengyan Li, Hongkai Zhao, and Chi-Wang Shu, *Uniformly accurate discontinuous Galerkin fast sweeping methods for Eikonal equations*, SIAM J. Sci. Comput. **33** (2011), no. 4, 1873–1896, DOI 10.1137/090770291. MR2831038 (2012k:65153)

[26] Yong-Tao Zhang, Hong-Kai Zhao, and Jianliang Qian, *High order fast sweeping methods for static Hamilton-Jacobi equations*, J. Sci. Comput. **29** (2006), no. 1, 25–56, DOI 10.1007/s10915-005-9014-3. MR2266807 (2007m:65096)

[27] Hongkai Zhao, *A fast sweeping method for eikonal equations*, Math. Comp. **74** (2005), no. 250, 603–627, DOI 10.1090/S0025-5718-04-01678-3. MR2114640 (2005k:65222)

Division of Applied Mathematics, Brown University, Providence, Rhode Island 02912
E-mail address: shu@dam.brown.edu

Coupling surface and subsurface flows with curved interfaces

Pu Song and Ivan Yotov

ABSTRACT. A mortar multiscale method is developed for the coupled Stokes and Darcy flows with the Beavers–Joseph–Saffman interface condition in irregular domains. Conforming Stokes elements and multipoint flux mixed finite elements in Darcy are used to discretize the subdomains on the fine scale. A coarse scale mortar finite element space is used to approximate interface stresses and pressures and impose weakly continuity of velocities and fluxes. Matching conditions on curved interfaces are imposed by mapping the physical grids to reference grids with flat interfaces.

1. Introduction

Coupled Stokes and Darcy flows have many applications, including groundwater-surface water flows, flows through fractured and vuggy rocks, industrial filtration, and blood flows. Existence and uniqueness of a weak solution using the Beavers-Joseph-Saffman interface condition was investigated in [**16, 8**]. Numerical approximations with various Stokes and Darcy elements have been developed, see, e.g., [**16, 8, 18, 10, 15, 9, 17, 11, 12**]. A coupling of Stokes-Darcy flows with transport was studied in [**20**].

In this paper we focus on accurate and efficient approximations on irregular domains with curved interfaces and non-matching grids. We employ the multipoint flux mixed finite element (MFMFE) method in Darcy [**23, 14, 21**], which uses specially chosen mixed finite element spaces and quadrature rules to eliminate locally velocity degrees of freedom and obtain cell-centered pressure schemes on irregular grids. For the coupled Stokes-Darcy problem we use a mortar multiscale method with non-matching grids, where the subdomains are discretized locally on a fine scale and interface conditions are imposed on a coarse scale using mortar finite elements. This approach was originally developed for MFE discretizations of Darcy flows in [**1, 2, 22**]. In [**4, 5**], mortar discretizations are developed for Stokes flow, where the mortar gird is chosen to be the trace of one of the neighboring subdomain grids. In [**18, 10, 6**], non-matching grids on Stokes-Darcy interfaces are considered utilizing a mortar space that is the normal trace of the Darcy velocity space. Very general multiscale mortar discretizations for Stokes-Darcy flows with Stokes-Stokes, Darcy-Darcy, and Stokes-Darcy interfaces have been developed in

---

2000 *Mathematics Subject Classification.* Primary 65M12, 65M15, 65N30, 76D07, 76S05.

Key words and phrases. Stokes-Darcy flows, mortar finite element, mixed finite element, multiscale finite element, multipoint flux approximation, curved interface.

This work was supported by DOE grant DE-FG02-04ER25618 and NSF grant DMS 1115856.

[13]. There, the mortar grid does not need to be the trace of either subdomain grid and the mortar space needs to satisfy a suitable inf-sup condition. In fact, the mortar grid can be on a coarse scale, leading to multiscale discretizations. Here we extend the approach in [13] to MFMFE discretizations in the Darcy region and curved interfaces between subdomains. Matching conditions on curved interfaces are imposed by mapping the physical grids to reference grids with flat interfaces.

The rest of the paper is organized as follows. In section 2 we present the coupled problem. The multiscale mortar discretization based on a non-overlapping domain decomposition weak formulation is described in section 3. Numerical examples are presented in section 4.

2. Problem description

Let the domain $\Omega \subset \mathbf{R}^n$ consist of Stokes flow in fluid region Ω_s and Darcy flow in porous media region Ω_d. Let $\Gamma_{sd} = \partial\Omega_s \cap \partial\Omega_d, \Gamma_d = \partial\Omega_d \setminus \Gamma_{sd}, \Gamma_s = \partial\Omega_s \setminus \Gamma_{sd}$. We denote the outward unit normal vector on Ω_s and Ω_d by $\mathbf{n}_s$ and $\mathbf{n}_d$, respectively. Let $\mathbf{u}_s, \mathbf{u}_d$, respectively p_s, p_d, denote the velocity, respectively the pressure, in the Stokes and Darcy regions. Let $\mathbf{D}(\mathbf{u}_s)$ and $\mathbf{T}(\mathbf{u}_s, p_s)$ denote, respectively, the deformation rate tensor and the stress tensor: $\mathbf{D}(\mathbf{u}_s) := \frac{1}{2}(\nabla\mathbf{u}_s + (\nabla\mathbf{u}_s)^T)$, $\mathbf{T}(\mathbf{u}_s, p_s) := -p_s\mathbf{I} + 2\nu\mathbf{D}(\mathbf{u}_s)$, where ν is the fluid viscosity. Let $\mathbf{f}_s$ be the body force in Ω_s and let $\mathbf{f}_d$ be the gravity force in Ω_d. Let $\mathbf{K}$ denote the rock permeability in the porous medium, which is a bounded, symmetric, and uniformly positive definite tensor. Let q_d denote an external source or sink term in Ω_d. The flow equations in the Stokes region are:

$$-\nabla\cdot\mathbf{T} \equiv -2\nu\nabla\cdot\mathbf{D}(\mathbf{u_s}) + \nabla p_s = \mathbf{f}_s \quad \text{in } \Omega_s, \tag{2.1}$$

$$\nabla\cdot\mathbf{u_s} = 0 \quad \text{in } \Omega_s, \tag{2.2}$$

$$\mathbf{u_s} = \mathbf{0} \quad \text{on } \Gamma_s. \tag{2.3}$$

In the Darcy region, the flow equations are:

$$\nu\mathbf{K}^{-1}\mathbf{u}_d + \nabla p_d = \mathbf{f}_d \quad \text{in } \Omega_d, \tag{2.4}$$

$$\nabla\cdot\mathbf{u}_d = q_d \quad \text{in } \Omega_d, \tag{2.5}$$

$$\mathbf{u}_d\cdot\mathbf{n}_d = 0 \quad \text{on } \Gamma_d, \tag{2.6}$$

where q_d is assumed to satisfy $\int_{\Omega_d} q_d d\mathbf{x} = 0$. The interface conditions are:

$$\mathbf{u}_s\cdot\mathbf{n}_s + \mathbf{u}_d\cdot\mathbf{n}_d = 0 \quad \text{on } \Gamma_{sd}, \tag{2.7}$$

$$-(\mathbf{T}\mathbf{n}_s)\cdot\mathbf{n}_s \equiv p_s - 2\nu(\mathbf{D}(\mathbf{u}_s)\mathbf{n}_s)\cdot\mathbf{n}_s = p_d \quad \text{on } \Gamma_{sd}, \tag{2.8}$$

$$-(\mathbf{T}\mathbf{n}_s)\cdot\boldsymbol{\tau}_j \equiv -2\nu(\mathbf{D}(\mathbf{u}_s)\mathbf{n}_s)\cdot\boldsymbol{\tau}_j = \frac{\nu\alpha}{\sqrt{K_j}}\mathbf{u}_s\cdot\boldsymbol{\tau}_j, \quad j = 1, d-1, \quad \text{on} \quad \Gamma_{sd}, \tag{2.9}$$

where $\boldsymbol{\tau}_j$, $j = 1, d-1$ is an orthonormal system of unit tangent vectors on Γ_{sd} and $K_j = (\mathbf{K}\boldsymbol{\tau}_j)\cdot\boldsymbol{\tau}_j$. The coefficient $\alpha > 0$ is an experimentally determined dimensionless constant. Conditions (2.7) and (2.8) denote the mass conservation across and continuity of normal stress on Γ_{sd}, respectively. Condition (2.9) is the slip with friction interface condition known as the Beavers-Joseph-Saffman law [3, 19].

3. Discretization of the coupled model

3.1. Non-overlapping domain decomposition variational formulation. We decompose Ω into non-overlapping subdomains, $\overline{\Omega}_s = \cup_{i=1}^{N_s} \overline{\Omega}_{s,i}$, $\overline{\Omega}_d = \cup_{i=1}^{N_d} \overline{\Omega}_{d,i}$. We will also use a single index i to number the subdomains, where i ranges from from $N_s + 1$ to $N = N_s + N_d$ in the Darcy region. Let $\Gamma_{ij} = \partial\Omega_i \cap \partial\Omega_j$. Similarly to Γ_{sd}, we use Γ_{dd} and Γ_{ss} to denote the set of Darcy-Darcy and Stokes-Stokes interfaces. We have the following interface conditions:

$$[\mathbf{u}_d \cdot \mathbf{n}] = 0, \quad [p_d] = 0 \quad \text{on } \Gamma_{dd}, \quad [\mathbf{u}_s] = \mathbf{0}, \quad [\mathbf{T}(\mathbf{u}_s, p_s)\mathbf{n}] = \mathbf{0} \quad \text{on } \Gamma_{ss}, \tag{3.1}$$

where the jumps on an interface Γ_{ij}, for $1 \le i < j \le N$, are defined as: $[\mathbf{v} \cdot \mathbf{n}] = \mathbf{v}_i \cdot \mathbf{n}_i + \mathbf{v}_j \cdot \mathbf{n}_j$, $[\mathbf{T}\mathbf{n}] = \mathbf{T}_i\mathbf{n}_i + \mathbf{T}_j\mathbf{n}_j$, and $[\mathbf{v}] = (\mathbf{v}_i - \mathbf{v}_j)|_{\Gamma_{ij}}$, where $\mathbf{v}_i = \mathbf{v}|_{\Omega_i}$ and $\mathbf{n}_i$ denotes the outward unit normal vector on $\partial\Omega_i$.

The pressure space is defined as $W = L_0^2(\Omega)$. The velocity space is $X = \{\mathbf{v} \in L^2(\Omega)^n\,;\, \mathbf{v}|_{\Omega_{d,i}} \in X_d, \mathbf{v}|_{\Omega_{s,i}} \in X_s\}$, where

$$X_d = \{\mathbf{v} \in L^2(\Omega_d)^n; \mathbf{v}_{d,i} := \mathbf{v}|_{\Omega_{d,i}} \in H(\text{div}; \Omega_{d,i}), 1 \le i \le N_d, \mathbf{v} \cdot \mathbf{n} = 0 \text{ on } \Gamma_d\},$$
$$X_s = \{\mathbf{v} \in L^2(\Omega_s)^n; \mathbf{v}_{s,i} := \mathbf{v}|_{\Omega_{s,i}} \in H^1(\Omega_{s,i})^n, 1 \le i \le N_s, \mathbf{v} = \mathbf{0} \text{ on } \Gamma_s\},$$

and the mortar space Λ is the dual of the trace of $\mathbf{u}_s$ on Γ_{ss} or the normal trace $\mathbf{u}_d \cdot \mathbf{n}$ on Γ_{sd} and Γ_{dd}. We define the following bilinear forms:

$$\forall\, (\mathbf{u}_{s,i}, \mathbf{v}_{s,i}) \in X_s \times X_s\,,\, a_{s,i}(\mathbf{u}_{s,i}, \mathbf{v}_{s,i}) = 2\nu \int_{\Omega_{s,i}} \mathbf{D}(\mathbf{u}_{s,i}) : \mathbf{D}(\mathbf{v}_{s,i})$$
$$+ \sum_{j=1}^{d-1} \int_{\partial\Omega_{s,i}\cap\Gamma_{sd}} \frac{\nu\alpha}{\sqrt{K_j}} (\mathbf{u}_{s,i} \cdot \boldsymbol{\tau}_j)(\mathbf{v}_s \cdot \boldsymbol{\tau}_j)\,,\; 1 \le i \le N_s,$$
$$\forall\, (\mathbf{u}_{d,i}, \mathbf{v}_{d,i}) \in X_d \times X_d\,,\, a_{d,i}(\mathbf{u}_{d,i}, \mathbf{v}_{d,i}) = \nu \int_{\Omega_{d,i}} \mathbf{K}^{-1}\mathbf{u}_{d,i} \cdot \mathbf{v}_{d,i}\,,\; 1 \le i \le N_d,$$
$$\forall\, \mathbf{v}_i \in X, \forall\, w_i \in L^2(\Omega)\,,\, b_i(\mathbf{v}_i, w_i) = -\int_{\Omega_i} w_i \text{div } \mathbf{v}_i\,,\; 1 \le i \le N.$$
$$a(\mathbf{u}, \mathbf{v}) = \sum_{i=1}^{N_s} a_{s,i}(\mathbf{u}, \mathbf{v}) + \sum_{i=1}^{N_d} a_{d,i}(\mathbf{u}, \mathbf{v}), \quad b(\mathbf{v}, w) = \sum_{i=1}^{N} b_i(\mathbf{v}, w).$$

The interface bilinear form is :

$$\forall\, (\mathbf{v}, \mu) \in X \times \Lambda\,,\, b_\Gamma(\mathbf{v}, \mu) = \int_{\Gamma_{ss}} [\mathbf{v}]\,\mu + \int_{\Gamma_{dd}} [\mathbf{v} \cdot \mathbf{n}]\,\mu + \int_{\Gamma_{sd}} [\mathbf{v} \cdot \mathbf{n}]\,\mu.$$

The variational formulation is: Find $(\mathbf{u}, p, \lambda) \in X \times W \times \Lambda$ such that

$$\forall \mathbf{v} \in X,\; a(\mathbf{u}, \mathbf{v}) + b(\mathbf{v}, p) + b_\Gamma(\mathbf{v}, \lambda) = \int_\Omega \mathbf{f} \cdot \mathbf{v}, \tag{3.2}$$

$$\forall\, w \in W\,,\, b(\mathbf{u}, w) = -\int_{\Omega_d} w\, q_d, \tag{3.3}$$

$$\forall\, \mu \in \Lambda\,,\, b_\Gamma(\mathbf{u}, \mu) = 0, \tag{3.4}$$

where $\lambda = -\mathbf{T}\mathbf{n}$ on Γ_{ss} and $\lambda = p_d$ on $\Gamma_{sd} \cup \Gamma_{dd}$.

3.2. Finite element discretization. Let $h > 0$ denote a discretization parameter, and for each h, let $\mathcal{T}_i^h$ be a partition on Ω_i made of triangles or tetrahedra in the Stokes region and triangles, tetrahedra, quadrilaterals, or hexahedra in the Darcy region. Partitions $\mathcal{T}_i^h$ and $\mathcal{T}_j^h$ need not match on the interface Γ_{ij} Let $\mathcal{T}_j^H$ denote a partition for Γ_{ij} made of segments, triangles or parallelograms of diameter bounded by H, where $H > 0$ is another discretization parameter.

Next, we will define the finite element space for these meshes. In the Stokes region, for each $\Omega_{s,i}$, let $(X_{s,i}^h, W_{s,i}^h) \subset H^1(\Omega_{s,i})^n \times L^2(\Omega_{s,i})$ be a pair of stable finite element spaces, such as the MINI element, the Taylor-Hood element or the Bernardi-Raugel element [7]. Assume that they contain at least polynomials of degree r_s and $r_s - 1$ for velocity and pressure, respectively. In the Darcy region, let $(X_{d,i}^h, W_{d,i}^h) \subset H(\text{div}; \Omega_{d,i}) \times L^2(\Omega_{d,i})$ be a pair of stable mixed finite element spaces, such as the Raviart-Thomas (RT) elements, the Brezzi-Douglas-Marini (BDM) elements, the Brezzi-Douglas-Fortin-Marini (BDFM) elements or the Brezzi-Douglas-Duràn-Fortin (BDDF) elements [7]. Assume that they contain at least polynomials of degree r_d and l_d for velocity and pressure, respectively, where $l_d = r_d$ or $l_d = r_d - 1$. Here we emphasize the use of the MFMFE method to discretize Darcy flow on irregular domains. There are two key ingredients for this method: the choice of finite element spaces and a numerical quadrature rule for the velocity mass matrix that allows for local flux elimination. On each element edge of face with s vertices, the velocity space must have s normal degrees of freedom, one associated with each vertex. This motivates the use of the BDM_1 space on simpices or quadrilaterals or an enhanced BDDF_1 space on hexahedra [**23, 14, 22**]. Note that the spaces are defined on a reference element $\hat{E}$ and then, given a bijection mapping $F_E : \hat{E} \to E$, the spaces are defined on a physical element E as

$$\mathbf{V}_h(E) = \frac{1}{J_E} DF_E \hat{\mathbf{V}}(\hat{E}) \circ F_E^{-1}, \quad W_h(E) = \hat{W}(\hat{E}) \circ F_E^{-1},$$

where the Piola transformation is used for the velocity space. Here DF_E is the Jacobian matrix and $J_E = |\det(DF_E)|$. Second, the quadrature rule on an element E is defined as: $(K^{-1}\mathbf{q}, \mathbf{v})_{Q,E} \equiv (\mathcal{K}^{-1}\hat{\mathbf{q}}, \hat{\mathbf{v}})_{\hat{Q},\hat{E}} \equiv \frac{|\hat{E}|}{s} \sum_{i=1}^{s} \mathcal{K}^{-1}(\hat{\mathbf{r}}_i)\hat{\mathbf{q}}(\hat{\mathbf{r}}_i) \cdot \hat{\mathbf{v}}(\hat{\mathbf{r}}_i)$, where $\mathcal{K} = JDF^{-1}\hat{K}(DF^{-1})^T$. This trapezoidal quadrature allows for eliminating locally velocity degrees of freedom associated with a vertex and the method can be reduced to a cell-centered pressure stencil.

The global finite element space are defined as: $X^h := \bigoplus X_i^h$, $W^h := \bigoplus W_i^h \cap L_0^2(\Omega)$. On each interface, let Λ_{ij}^H be a finite element space of continuous or discontinuous piecewise polynomials of degree at least r_{ss} on Γ_{ss}, r_{dd} on Γ_{dd}, and r_{sd} on Γ_{sd}. The global the mortar finite element space is $\Lambda^H := \bigoplus \Lambda_{ij}^H$.

The multiscale mortar FEM for Stokes-Darcy is given by: Find $(\mathbf{u}^h, p^h, \lambda^H) \in X^h \times W^h \times \Lambda^H$ such that

$$\forall \mathbf{v}^h \in X^h,\ a_h(\mathbf{u}^h, \mathbf{v}^h) + b(\mathbf{v}^h, p^h) + b_\Gamma(\mathbf{v}^h, \lambda^H) = \int_\Omega \mathbf{f} \cdot \mathbf{v}^h, \tag{3.5}$$

$$\forall w^h \in W^h,\ b(\mathbf{u}^h, w^h) = -\int_{\Omega_d} w^h q_d, \tag{3.6}$$

$$\forall \mu^H \in \Lambda^H,\ b_\Gamma(\mathbf{u}^h, \mu^H) = 0, \tag{3.7}$$

where $a_h(\cdot,\cdot) = a(\cdot,\cdot)$ in Ω_s and it is an approximation to $a(\cdot,\cdot)$ in Ω_d based on the MFMFE quadrature rule. A non-overlapping domain decomposition algorithm is used to reduce the algebraic system arising from (3.5)–(3.7) to an interface problem for λ^H, which is solved by a Krylov space solver. Each iteration requires solving subdomain problems of Stokes or Darcy types.

The analysis of the method will be presented in a forthcoming paper.

3.3. Implementation of curved interfaces. Here we describe how we handle domains with curved non-matching grid interfaces. The two grids on each side of the interface represent two different piecewise (bi)linear approximations of the interface curve or surface. The continuity condition (3.7) is imposed by mapping the subdomain and mortar grids to reference grids with flat interfaces. On Stokes-Darcy and Darcy-Darcy interfaces, we employ the Piola transformation for the velocity, which preserves the normal component of the vector: $\mathbf{u}\cdot\mathbf{n}|e| = \hat{\mathbf{u}}\cdot\hat{\mathbf{n}}|\hat{e}|$. The matching condition $\langle[\mathbf{u}^h\cdot\mathbf{n}],\mu^H\rangle = 0$ is imposed on the reference grid configuration by projecting the normal component on each side onto the reference mortar grid. On Stokes-Stokes interfaces, the grids are similarly mapped to reference grids to impose $\langle[\mathbf{u}^h],\mu^H\rangle = 0$, except that a standard change of variables is used to map the velocity space.

4. Numerical results

In this section we present some numerical examples to test the convergence of the method. In the first two examples the computational domain $\overline{\Omega} = \overline{\Omega}_s \cup \overline{\Omega}_d$ is a map of the unit square, where $\hat{\Omega}_s = (0,1)\times(\frac{1}{2},1)$ and $\hat{\Omega}_d = (0,1)\times(0,\frac{1}{2})$. For simplicity we set $\mathbf{T}(\mathbf{u}_s,p_s) = -p_s\mathbf{I} + \mu\nabla\mathbf{u}_s$ in Ω_s, and $\mathbf{K} = K\mathbf{I}$ in in Ω_d, where K is a positive constant. In Example 1, we test convergence on a domain with a curved Stokes-Darcy interface, see Figure 1. The analytical solution was specially constructed to satisfy the interface conditions (2.7)–(2.9) taking into account the geometry of the interface. In Example 2, the Stokes-Darcy interface is straight, but the Stokes-Stokes and Darcy-Darcy interfaces are curved, see Figure 2. The analytical solution is the same as in [**13**]. In both examples, for the Darcy region we specify the pressure on the left, right, and bottom boundaries and for the Stokes region we specify velocity on the left and top boundaries, and the normal and tangential stress on the right boundary. Each region is divided into two subdomains, giving a total of fours subdomains. The subdomain grids do not match across the interfaces. The Stokes subdomains are discretized by the lowest order Taylor–Hood triangular finite elements ($r_s = 2$). The Darcy subdomains are discretized by the MFMFE method on quadrilaterals ($r_d = l_d = 0$). We use discontinuous piecewise linear mortars on all interfaces ($r_{ss} = r_{dd} = r_{sd} = 1$). To test convergence, we solve the problem on a sequence of grid refinements. On the coarsest level, the subdomain grids are 6×8 in the lower left and upper right subdomains and 4×6 in the other two subdomains. We take $H = 2h$ with two mortar elements per interface on the coarsest level. The subdomain and mortar grids are refined by two each time. The numerical errors and convergence rates are reported in Tables 1–2, where l denotes the grid level, the errors are $E_s^u = \|\mathbf{u}_s - \mathbf{u}_s^h\|_{H^1(\Omega_s)}$, $E_s^p = \|p_s - p_s^h\|_{L^2(\Omega_s)}$, $E_d^u = \|\mathbf{u}_d - \mathbf{u}_d^h\|_{H(\mathrm{div};\Omega_d)}$, and $E_d^p = \|p_d - p_d^h\|_{L^2(\Omega_d)}$, and the respective convergence rates are R_s^u, R_s^p, R_d^u, and R_d^p. In both cases we observe convergence for the Stokes velocity and pressure of order between $h^{3/2}$ and h^2, as well as first order convergence for the Darcy velocity and pressure. We note that the optimal convergence rates for stand alone discretizations are second order for Stokes and first order for Darcy.

The reduction in the Stokes convergence in the coupled case is expected, due to the coupling with the lower order Darcy discretization and the effect of the non-matching mortar error. Figures 1–2 show the vertical velocity and errors for each example. In Example 1, the vertical velocity is not continuous across the Stokes-Darcy interface, as the latter is not flat, but the normal velocity is continuous. In Example 2, the vertical velocity is normal to the Stokes-Darcy interface and it is continuous. In both cases the velocity is continuous across the Stokes-Stokes and Darcy-Darcy interfaces. The depicted errors show that the error is generally smaller in the Stokes region, while the larger error in the Darcy region is associated with the non-matching interface or the corners of the domain.

We also present an example with a more realistic geometry simulating coupled surface water - groundwater flow, see Figure 3. In the top Stokes region, inflow and outflow conditions are specified on the left and right boundaries, respectively, as well as a combination of horizontal velocity and zero normal stress on the top surface boundary. In the bottom Darcy region, no flow is specified on the left and right boundaries and pressure on the bottom boundary. As expected, the horizontal velocity, which is close to tangential to the interface, is larger in the surface flow region, see Figure 3 left. The vertical velocity (right) is also discontinuous, due to the curvature in the interface. The velocity vector represented with the arrows has a continuous normal component, with fluid percolating into the groundwater region.

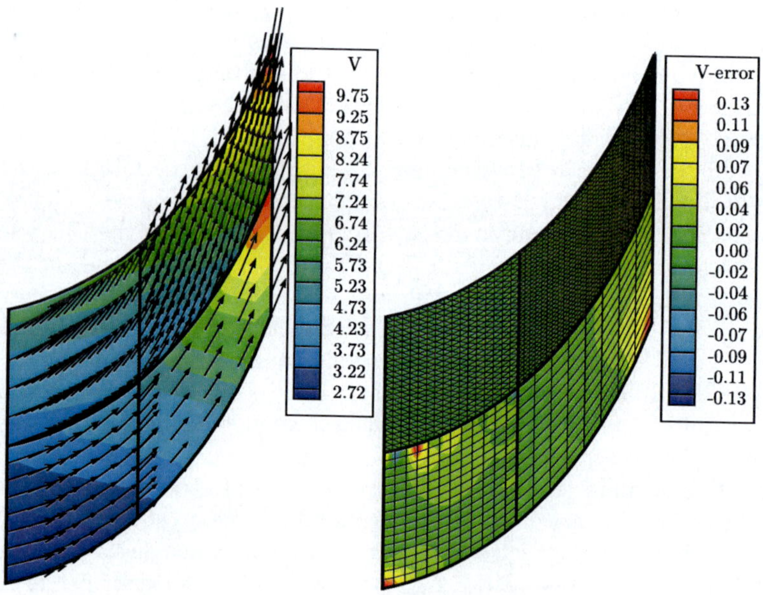

FIGURE 1. Computed vertical velocity (left) and error (right) on subdomain meshes 8×12 and 12×16 for Example 1.

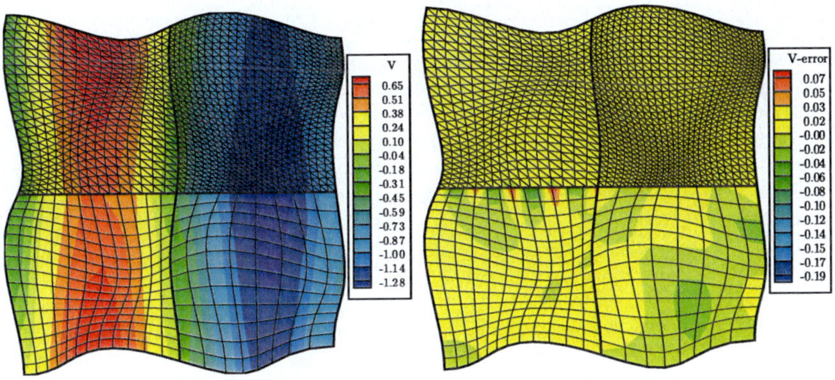

FIGURE 2. Computed vertical velocity (left) and error (right) on subdomain meshes 8×12 and 12×16 for Example 2.

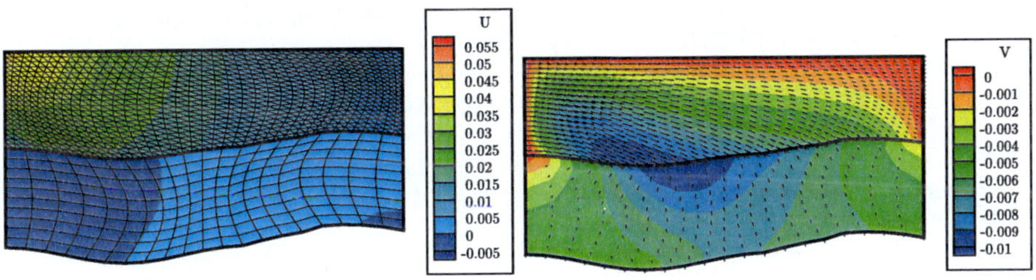

FIGURE 3. Computed horizontal and vertical velocity in the coupled surface-groundwater flow example.

TABLE 1. Numerical errors and convergence rates for Example 1.

l	E_s^u	R_s^u	E_s^p	R_s^p	E_d^u	R_d^u	E_d^p	R_d^p
1	4.08e-01		8.12e-02		4.27e-01		4.60e-01	
2	1.21e-01	1.76	1.72e-02	2.24	2.00e-01	1.09	2.31e-01	1.00
3	3.20e-02	1.92	4.00e-03	2.11	9.84e-02	1.02	1.16e-01	0.99
4	8.12e-03	1.98	1.41e-03	1.50	4.90e-02	1.01	5.78e-02	1.01

TABLE 2. Numerical errors and convergence rates for Example 2.

l	E_s^u	R_s^u	E_s^p	R_s^p	E_d^u	R_d^u	E_d^p	R_d^p
1	9.27e-02		7.56e-03		2.46e-01		5.41e-02	
2	2.17e-02	2.09	1.86e-03	2.02	1.23e-01	1.00	2.71e-02	1.00
3	5.47e-03	1.99	5.13e-04	1.86	6.14e-02	1.00	1.36e-02	0.99
4	1.58e-03	1.79	1.68e-04	1.61	3.06e-02	1.00	6.79e-03	1.00

References

1. T. Arbogast, L. C. Cowsar, M. F. Wheeler, and I. Yotov, *Mixed finite element methods on non-matching multiblock grids*, SIAM J. Numer. Anal. **37** (2000), 1295–1315. MR1756426 (2001h:65140)
2. T. Arbogast, G. Pencheva, M. F. Wheeler, and I. Yotov, *A multiscale mortar mixed finite element method*, Multiscale Model. Simul. **6** (2007), no. 1, 319–346. MR2306414 (2008k:65234)
3. G. S. Beavers and D. D. Joseph, *Boundary conditions at a naturally impermeable wall*, J. Fluid. Mech **30** (1967), 197–207.
4. Faker Ben Belgacem, *The mixed mortar finite element method for the incompressible Stokes problem: convergence analysis*, SIAM J. Numer. Anal. **37** (2000), no. 4, 1085–1100 (electronic). MR1756416 (2001b:65101)
5. ———, *A stabilized domain decomposition method with nonmatching grids for the Stokes problem in three dimensions*, SIAM J. Numer. Anal. **42** (2004), no. 2, 667–685 (electronic). MR2084231 (2005i:65176)
6. Christine Bernardi, Tomás Chacón Rebollo, Frédéric Hecht, and Zoubida Mghazli, *Mortar finite element discretization of a model coupling Darcy and Stokes equations*, M2AN Math. Model. Numer. Anal. **42** (2008), no. 3, 375–410. MR2423791 (2009d:65162)
7. F. Brezzi and M. Fortin, *Mixed and hybrid finite element methods*, Springer-Verlag, New York, 1991. MR1115205 (92d:65187)
8. Marco Discacciati, Edie Miglio, and Alfio Quarteroni, *Mathematical and numerical models for coupling surface and groundwater flows*, Appl. Numer. Math. **43** (2002), no. 1-2, 57–74, 19th Dundee Biennial Conference on Numerical Analysis (2001). MR1936 102
9. V. J. Ervin, E. W. Jenkins, and S. Sun, *Coupled generalized nonlinear Stokes flow with flow through a porous medium*, SIAM J. Numer. Anal. **47** (2009), no. 2, 929–952. MR2485439 (2010b:65254)
10. Juan Galvis and Marcus Sarkis, *Non-matching mortar discretization analysis for the coupling Stokes-Darcy equations*, Electron. Trans. Numer. Anal. **26** (2007), 350–384. MR2391227 (2009a:76120)
11. Gabriel N. Gatica, Salim Meddahi, and Ricardo Oyarzúa, *A conforming mixed finite-element method for the coupling of fluid flow with porous media flow*, IMA J. Numer. Anal. **29** (2009), no. 1, 86–108. MR2470941 (2010b:76118)
12. Gabriel N. Gatica, Ricardo Oyarzúa, and Francisco-Javier Sayas, *Analysis of fully-mixed finite element methods for the Stokes-Darcy coupled problem*, Math. Comp. **80** (2011), no. 276, 1911–1948. MR2813344 (2012i:65259)
13. V. Girault, D. Vassilev, and I. Yotov, *Mortar multiscale finite element method for Stokes and Darcy flows*, Tech. Report TR-MATH 12-06, Department of Mathematics, University of Pittsburgh, 2012.
14. R. Ingram, M. F. Wheeler, and I. Yotov, *A multipoint flux mixed finite element method on hexahedra*, SIAM J. Numer. Anal. **48** (2010), 1281–1312. MR2684336 (2012a:65333)
15. G. Kanschat and B. Rivière, *A strongly conservative finite element method for the coupling of Stokes and Darcy flow*, J. Comput. Phys. **229** (2010), no. 17, 5933–5943. MR2657851 (2011h:76081)
16. W. J. Layton, F. Schieweck, and I. Yotov, *Coupling fluid flow with porous media flow*, SIAM J. Numer. Anal. **40** (2003), no. 6, 2195–2218. MR1974181 (2004c:76048)
17. Mo Mu and Jinchao Xu, *A two-grid method of a mixed Stokes-Darcy model for coupling fluid flow with porous media flow*, SIAM J. Numer. Anal. **45** (2007), no. 5, 1801–1813. MR2346360 (2008i:65264)
18. B. Rivière and I. Yotov, *Locally conservative coupling of Stokes and Darcy flows*, SIAM J. Numer. Anal. **42** (2005), no. 5, 1959–1977. MR2139232 (2006a:76035)
19. P. G. Saffman, *On the boundary condition at the surface of a porous media*, Stud. Appl. Math., L (1971), no. 2, 93–101.
20. D. Vassilev and I. Yotov, *Coupling Stokes-Darcy flow with transport*, SIAM J. Sci. Comput. **31** (2009), no. 5, 3661–3684. MR2556557 (2011c:76105)
21. M. F. Wheeler, G. Xue, and I. Yotov, *A multipoint flux mixed finite element method on distorted quadrilaterals and hexahedra*, Numerische Mathematik **121** (2012), 165–204. MR2909918

22. M. F. Wheeler, G. Xue, and I. Yotov, *A multiscale mortar multipoint flux mixed finite element method*, ESAIM: Mathematical Modelling and Numerical Analysis (M2AN) **46** (2012), no. 4, 759–796. MR2891469
23. M. F. Wheeler and I. Yotov, *A multipoint flux mixed finite element method*, SIAM J. Numer. Anal. **44** (2006), no. 5, 2082–2106. MR2263041 (2008f:65228)

DEPARTMENT OF MATHEMATICS, UNIVERSITY OF PITTSBURGH, 301 THACKERAY, PITTSBURGH, PENNSYLVANIA 15260
E-mail address: pus6@pitt.edu

DEPARTMENT OF MATHEMATICS, UNIVERSITY OF PITTSBURGH, 301 THACKERAY, PITTSBURGH, PENNSYLVANIA 15260
E-mail address: yotov@math.pitt.edu

Reconstruction of obstacles embedded in waveguides

Jiguang Sun and Chunxiong Zheng

ABSTRACT. Reconstruction of obstacles embedded in a periodic waveguide with arbitrary geometry is considered in this paper. The measurement is on a line segment of the scattered field due to point sources inside the waveguide. A linear sampling type method is proposed to characterize the obstacles using the solutions of the near field ill-posed linear integral equations. Due to the fact that we consider waveguides with arbitrary geometry, to compute the background Green's function, we employ a method based on the limiting absorption principle and the recursive doubling technique. Furthermore, an algorithm is proposed to speed up the sampling procedure. Numerical examples are presented to demonstrate the performance of the proposed method.

1. Introduction

Inverse problems for periodic structures have a long history and have been investigated by many researchers, for example, see [**1–4**, **14**, **16**, **17**, **20**]. Most of these works address the reconstruction of structure profiles. In this paper, we consider the reconstruction of obstacles embedded in waveguides. Recently this issue has drawn a lot of attention. For example, in [**19**] Xu et al. applied a method using generalized dual space indicator for imaging an obstacle in a shallow water waveguide. They used the scattered field on a straight line and solved some ill-posed integral equations. Dediu and McLaughlin [**12**] proposed an eigensystem decomposition to recover weak inhomogeneities in a waveguide from far-field data. In [**5**], Bourgeois and Lunèville use far field data to reconstruct the inhomogeneity based on the linear sampling method and a factorization of the far field operator.

A major difficulty of the reconstruction of obstacles embedded in the waveguide is the computation of Green's functions. In contrast to the homogeneous medium, there is no closed form of the Green's functions for waveguides with arbitrary geometry. This is analogous to the case when applying linear sampling method to characterize objects embedded in an inhomogeneous complex background in open space. Methods of avoiding, at least partially, the computation of Green's function for inhomogeneous background in free space case have been studied in [**6**, **10**, **18**].

1991 *Mathematics Subject Classification.* Primary 78A46, 35B27; Secondary 78A45, 35R30.

Key words and phrases. Inverse scattering problems, periodic structures, waveguide, linear sampling method, limiting absorption principle.

This material is based upon work supported in part by NSF Grant #1016092 and the US Army Research Laboratory and the US Army Research Office under cooperative agreement number W911NF-11-2-0046.

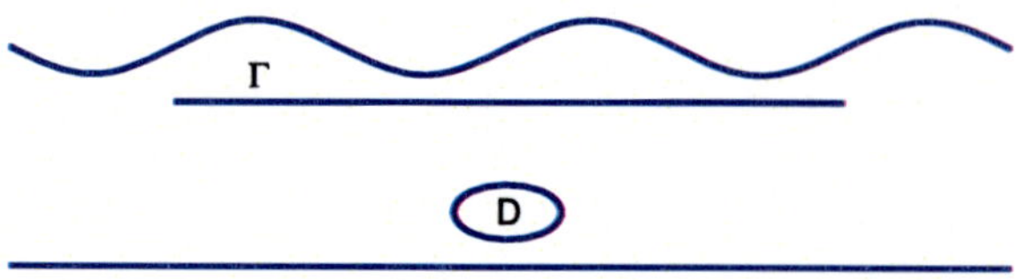

FIGURE 1. Schematic picture: scattering by an obstacle in a waveguide. The waveguide has a complicate geometry. The obstacle is D. The point sources and measurement of the scattered field are on Γ.

This difficulty accounts for why most existing works [**5**, **12**, **19**] only consider the homogeneous waveguide. In this paper, we consider waveguides with arbitrary geometry and employ the method developed in [**15**] to compute the background Green's function.

We assume that the incident field is due to a point source on a line Γ and the scattered field is measured on the same line (see Fig. 1). For the inverse problem, we will use the near field version of the linear sampling method, i.e., we solve linear integral equations for near field data and plot the solution norms in a sampling domain. The major ingredient of the linear sampling method is the computation of the Green's function. Since we need the Green's functions for many sampling points, we propose a Fourier expansion method to significantly reduce the computational cost.

The rest of the paper is organized as follows. Section 2 discusses the scattering of inhomogeneities due to a point source incident field in the waveguide. In Section 3, we present the linear sampling method to characterize the inhomogeneities using the near field equations. In addition, we propose a fast Fourier expansion method to compute the Green's functions. In Section 4, we present some preliminary numerical examples. We make some conclusions and discuss future works in Section 5.

2. The direct problem

The inverse problem we consider is to image an object D embedded in a two-dimensional periodic waveguide Ω (see Fig. 1). We denote the top boundary of Ω as $\partial\Omega^+$ and the bottom boundary of Ω as $\partial\Omega^-$. For simplicity, we assume that $\partial\Omega^-$ is a straight line and $\partial\Omega^+$ is a periodic function of x. The top boundary $\partial\Omega^+$ is supposed sound-soft and the bottom boundary $\partial\Omega^-$ sound-hard. Note that other combinations of boundary conditions are also possible. The point sources are located on a curve Γ above D. This implies that the incident field is simply the background Green's function at $\mathbf{r}_s \in \Gamma$ which solves

$$\Delta_{\mathbf{r}} G(\mathbf{r}, \mathbf{r}_s) + k^2 G(\mathbf{r}, \mathbf{r}_s) = \delta(\mathbf{r} - \mathbf{r}_s), \quad \text{in } \Omega, \tag{2.1a}$$

$$G(\mathbf{r}, \mathbf{r}_s) = 0, \quad \text{on } \partial\Omega^+, \tag{2.1b}$$

$$\partial_\nu G(\mathbf{r}, \mathbf{r}_s) = 0, \quad \text{on } \partial\Omega^-. \tag{2.1c}$$

Let $\Phi = \frac{i}{4}H_0^{(1)}(k|\mathbf{r}-\mathbf{r}_s|)$ be the Green's function in $\mathbb{R}^2$. It is known [5] that

$$G(\mathbf{r},\mathbf{r}_s) \sim \Phi(\mathbf{r},\mathbf{r}_s) \sim \frac{1}{2\pi}\log(k|\mathbf{r}-\mathbf{r}_s|), \quad \mathbf{r}\to\mathbf{r}_s.$$

Note that the above asymptotic relation also holds for any complex k with $\Re k > 0$, $\Im k > 0$ and $\Im k < \Re k$. For a homogeneous waveguide, it is possible to derive an analytical series expansion for G (see [5]). For a waveguide with arbitrary geometry, it is not possible in general.

We suppose the object D is sound-soft. This means that the scattered field $u^s(\cdot,\mathbf{r}_s)$ due to the appearance of D solves the following equation

(2.2a) $$\Delta_{\mathbf{r}} u^s(\mathbf{r},\mathbf{r}_s) + k^2 u^s(\mathbf{r},\mathbf{r}_s) = 0, \qquad \mathbf{r}\in\Omega\setminus\bar{D},$$

(2.2b) $$u^s(\mathbf{r},\mathbf{r}_s) = 0, \qquad \mathbf{r}\in\partial\Omega^+,$$

(2.2c) $$\partial_\nu u^s(\mathbf{r},\mathbf{r}_s) = 0, \qquad \mathbf{r}\in\partial\Omega^-,$$

(2.2d) $$u^s(\mathbf{r},\mathbf{r}_s) = -G(\mathbf{r},\mathbf{r}_s), \qquad \mathbf{r}\in\partial D.$$

To ensure the well-posedness, both G and u^s should also satisfy appropriate condition along the longitudinal axis of the waveguide as $|x|\to\infty$. However, how to define the boundary condition which ensures the well-posedness and has reasonable physical meaning is still an open problem. Following [15], we assume that the *limiting absorption principle* (LAP) holds for the waveguide Ω and the specific wave number k. The precise meaning of this principle is as follows. For any $\delta > 0$, replacing k^2 with $k^2 + i\delta$ in (2.2a) we obtain a well-posed problem with damping. Denote by $u_\delta^s \in H^1(\Omega\setminus D)$ the solution of the damped equation. The LAP implies that u_δ^s converges in $H^1_{loc}(\Omega\setminus D)$ as $\delta\to 0^+$. Though the physical meaning of LAP is rather natural, the rigorous mathematical justification of this principle for the considered problem, to the authors' knowledge, is still an open problem.

Using the LAP, we have the following reciprocity property of the Green's function $G(\mathbf{r},\mathbf{r}_s)$.

THEOREM 2.1. *The Green's function has the following reciprocity relation:*

$$G(\mathbf{r}_f,\mathbf{r}_s) = G(\mathbf{r}_s,\mathbf{r}_f),\ \forall \mathbf{r}_f,\mathbf{r}_s\in\Omega.$$

PROOF. Let $k_\delta = \sqrt{k^2+i\delta}$ with $\delta > 0$ and $G_\delta(\mathbf{r},\mathbf{r}_s)$ be the Green's function of (2.2a) by replacing k^2 with $k^2+i\delta$, i.e.,

$$\Delta_{\mathbf{r}} G_\delta(\mathbf{r},\mathbf{r}_s) + (k^2+i\delta)G_\delta(\mathbf{r},\mathbf{r}_s) = \delta(\mathbf{r}-\mathbf{r}_s),\ \forall \mathbf{r}\in\Omega.$$

For any fixed $\mathbf{r}_f,\mathbf{r}_s\in\Omega$, let $\epsilon > 0$ be small enough and we define

$$D_{\mathbf{r}_s,\epsilon} = \{\mathbf{r}\in\Omega : |\mathbf{r}-\mathbf{r}_s| < \epsilon\},$$
$$D_{\mathbf{r}_f,\epsilon} = \{\mathbf{r}\in\Omega : |\mathbf{r}-\mathbf{r}_f| < \epsilon\},$$

such that $\overline{D}_{\mathbf{r}_s,\epsilon}\cap\overline{D}_{\mathbf{r}_f,\epsilon} = \emptyset$ (see Fig. 2). Set

$$\Omega_\epsilon = \Omega\setminus(\overline{D}_{\mathbf{r}_s,\epsilon}\cap\overline{D}_{\mathbf{r}_f,\epsilon}).$$

Let us denote

$$(*) := G_\delta(\mathbf{r},\mathbf{r}_f)\frac{\partial G_\delta(\mathbf{r},\mathbf{r}_s)}{\partial n} - G_\delta(\mathbf{r},\mathbf{r}_s)\frac{\partial G_\delta(\mathbf{r},\mathbf{r}_f)}{\partial n}.$$

Applying Green formula on Ω_ϵ, we obtain

$$\int_{\partial\Omega_\epsilon} (*)\, ds = \int_{\Omega_\epsilon} [G_\delta(\mathbf{r},\mathbf{r}_f)\Delta G_\delta(\mathbf{r},\mathbf{r}_s) - G_\delta(\mathbf{r},\mathbf{r}_s)\Delta G_\delta(\mathbf{r},\mathbf{r}_f)]\, d\mathbf{r} = 0.$$

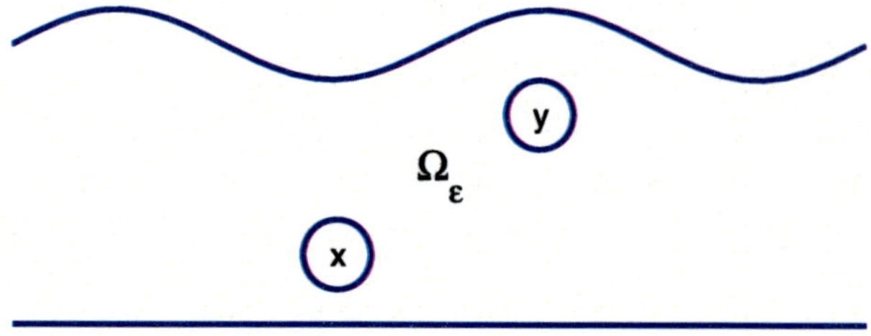

FIGURE 2. Schematic picture for the reciprocity property of the Green's function.

Thus we have

$$\int_{\partial D_{\mathbf{r}_s,\epsilon}} (*)\, ds + \int_{\partial D_{\mathbf{r}_f,\epsilon}} (*)\, ds = \int_{\partial\Omega} (*)\, ds.$$

The right hand side vanishes due to the boundary conditions on $\partial\Omega^{\pm}$, which implies

$$\int_{\partial D_{\mathbf{r}_s,\epsilon}} (*)\, ds + \int_{\partial D_{\mathbf{r}_f,\epsilon}} (*)\, ds = 0. \tag{2.3}$$

On $\partial D_{\mathbf{r}_s,\epsilon}$, there exists a constant M_1 depending on $\mathbf{r}_s$ and δ such that

$$\left|\frac{\partial G_\delta(\mathbf{r},\mathbf{r}_f)}{\partial n}\right| = |\nabla_{\mathbf{r}} G_\delta(\mathbf{r},\mathbf{r}_f)\cdot n| \le M_1.$$

Besides, by setting

$$h = G_\delta(\mathbf{r},\mathbf{r}_s) - \frac{1}{2\pi}\log(k_\delta|\mathbf{r}-\mathbf{r}_s|),$$

there exists a constant M_2 depending on $\mathbf{r}_s$ and δ such that

$$\|h\|_{C^1(\partial D_{\mathbf{r}_s,\epsilon})} \le M_2.$$

Hence

$$\left|\int_{\partial D_{\mathbf{r}_s,\epsilon}} \left(G_\delta(\mathbf{r},\mathbf{r}_s)\frac{\partial G_\delta(\mathbf{r},\mathbf{r}_f)}{\partial n}\right) ds\right| \le \left(\frac{1}{2\pi}|\log k_\delta\epsilon| + M_2\right) M_1 2\pi\epsilon \to 0,$$

and

$$\int_{\partial D_{\mathbf{r}_s,\epsilon}} G_\delta(\mathbf{r},\mathbf{r}_f)\frac{\partial G_\delta(\mathbf{r},\mathbf{r}_s)}{\partial n}\, ds = \epsilon\int_0^{2\pi} G_\delta(\mathbf{r}_s+\epsilon e^{i\theta},\mathbf{r}_f)\left(\frac{1}{2\pi\epsilon} + \nabla_{\mathbf{r}} h\cdot n\right) d\theta.$$

Since $|\nabla h\cdot n| \le 2M_2$, we have

$$\lim_{\epsilon\to 0}\epsilon\int_0^{2\pi} G_\delta(\mathbf{r}_s+\epsilon e^{i\theta},\mathbf{r}_f)\,(\nabla_{\mathbf{r}} h\cdot n)\, d\theta = 0.$$

Finally

$$\lim_{\epsilon\to 0}\int_{\partial D_{\mathbf{r}_s,\epsilon}} G_\delta(\mathbf{r},\mathbf{r}_f)\frac{\partial G_\delta(\mathbf{r},\mathbf{r}_s)}{\partial n}\, ds = \frac{1}{2\pi}\lim_{\epsilon\to 0}\int_0^{2\pi} G_\delta(\mathbf{r}_s+\epsilon e^{i\theta},\mathbf{r}_f)\, d\theta = G_\delta(\mathbf{r}_s,\mathbf{r}_f).$$

Similarly, we can show that the second term of the left hand side of (2.3) is $-G_\delta(\mathbf{r}_f, \mathbf{r}_s)$. Thus we have

$$G_\delta(\mathbf{r}_f, \mathbf{r}_s) = G_\delta(\mathbf{r}_s, \mathbf{r}_f).$$

Applying the LAP, we have $G(\mathbf{r}_f, \mathbf{r}_s) = G(\mathbf{r}_s, \mathbf{r}_f)$ by taking $\delta \to 0^+$. □

Consequently, we have the following results whose proofs are exactly the same as the free space case (see [**11**]).

COROLLARY 2.2. *Let u be the solution of the forward problem defined above. For all $\mathbf{r} \in \Omega \setminus \overline{D}$,*

$$u(\mathbf{r}) = \int_{\partial D} \left(u(\mathbf{t}) \frac{\partial G(\mathbf{r}, \mathbf{t})}{\partial \nu_{\mathbf{t}}} - \frac{\partial u(\mathbf{t})}{\partial \nu_{\mathbf{t}}} G(\mathbf{r}, \mathbf{t}) \right) ds(\mathbf{t}),$$

where $\nu_{\mathbf{t}}$ is the outward unit normal.

COROLLARY 2.3. *Let $u^s(\cdot, \mathbf{t})$ be the scattered field of D due to a point source at $\mathbf{t}$. Then*

$$u^s(\mathbf{r}, \mathbf{t}) = u^s(\mathbf{t}, \mathbf{r}), \quad \forall \mathbf{r}, \mathbf{t} \in \Omega \setminus \bar{D}.$$

To compute the Green's function G and the scattered field u^s numerically, we employ the fast algorithm based on the recursive doubling procedure which we refer the readers to [**15**] for details.

3. The inverse problem

The inverse problem is to reconstruct the support of the obstacle D, if we have the scattered field information on Γ. With the reciprocity property of G, we can prove the following uniqueness theorem similar to the one in [**5**] and thus we omit its proof here.

THEOREM 3.1. *Denote by D_1 and D_2 two sound soft obstacles with Lipschitz continuous boundaries. If for all incident waves $G(\cdot, \mathbf{r}_s)$ with $\mathbf{r}_s \in \Gamma$, the corresponding scattered fields $u_1^s(\cdot, \mathbf{r}_s)$ and $u_2^s(\cdot, \mathbf{r}_s)$ coincide on Γ, then $D_1 = D_2$.*

Since we measure the near field data, to apply the linear sampling method, we will need the following near field operator $\mathcal{N} : L^2(\Gamma) \to L^2(\Gamma)$ such that for any $g \in L^2(\Gamma)$,

$$(\mathcal{N} g)(\mathbf{r}) = \int_\Gamma u^s(\mathbf{r}, \mathbf{r}_s) g(\mathbf{r}_s)\, ds(\mathbf{r}_s), \ \forall \mathbf{r} \in \Gamma. \tag{3.1}$$

Let S be a region containing the object D. For any $\mathbf{r}_s \in S$, we consider the following integral equation

$$\mathcal{N} g = G(\cdot, \mathbf{r}_s). \tag{3.2}$$

It is well-known that the above equation does not have a solution in general. However, it is possible to find an approximate solution to (3.2) except a discrete set of wavenumber k's.

THEOREM 3.2. *Suppose k^2 is not a Dirichlet eigenvalue for D. Let $\mathcal{N}$ be the near-field operator defined by* (3.1).

1. *If* $\mathbf{r}_s \in D$, *then for any* $\epsilon > 0$ *there exists an approximate solution* $h^\epsilon(\cdot, \mathbf{r}_s)$ *of* (3.2) *such that*

$$\|(\mathcal{N}h^\epsilon)(\cdot, \mathbf{r}_s) - G(\cdot, \mathbf{r}_s)\|_{L^2(\Gamma)} \leq \epsilon.$$

In addition, $h^\epsilon(\cdot, \mathbf{r}_s)$ *converges as* $\epsilon \to 0$.

2. *If* $\mathbf{r}_s \to \partial D$ *and* $h^\epsilon(\cdot, \mathbf{r}_s)$ *satisfies*

$$\|(\mathcal{N}h^\epsilon)(\cdot, \mathbf{r}_s) - G(\cdot, \mathbf{r}_s)\|_{L^2(\Gamma)} \leq \epsilon,$$

then the norm of $h^\epsilon(\cdot, \mathbf{r}_s)$ *tends to infinity as* $\epsilon \to 0$.

PROOF. We first assume $D \subset C$. Similar to [**7**], we consider the linear equation

$$(\mathcal{N}\phi)(\mathbf{r}) = G(\mathbf{r}, \mathbf{r}_s), \quad \forall \mathbf{r} \in \partial C. \tag{3.3}$$

Note that $(\mathcal{N}\phi)(\mathbf{r})$ is the scattered field of D due to the incident field

$$(S\phi)(\mathbf{r}) = \int_{\partial C} G(\mathbf{r}, \mathbf{r}_s)\phi(\mathbf{r}_s)\, ds(\mathbf{r}_s). \tag{3.4}$$

If $\mathbf{r}_s \in D$, ϕ is a solution to the near field equation (3.3) if $S\phi$ solves

$$\Delta u + k^2 u = 0 \qquad \text{in } D, \tag{3.5a}$$

$$u = -G(\cdot, \mathbf{r}_s), \qquad \text{on } \partial D. \tag{3.5b}$$

It is known that $S\phi$ with $\phi \in L^2(\partial C)$ is dense in

$$\mathbb{H}(D) = \{u \in H^1(D); \Delta u + k^2 u = 0\}.$$

Hence there exists a ϕ such that $S\phi$ approximates the solution of (3.5) for $\mathbf{r}_s \in D$. As $\mathbf{r}_s$ approaches ∂D, the norm of $G(\cdot, \mathbf{r}_s)$ on ∂D blows up, so does the norm of the approxiamae solution of (3.5).

If Γ is a line segment, then Γ is a subset of ∂C for some domain C with D inside. Following [**8**], it is suffices to show that the set of functions

$$(S\phi)(\mathbf{r}) = \int_{\partial C} \Phi(\mathbf{r}, \mathbf{r}')\phi(\mathbf{r}')\, ds(\mathbf{r}'), \quad \phi \in L^2(\partial C) \quad \text{with support in} \quad \Gamma \subset \partial C$$

is complete in $L^2(\partial C)$. Then the approximation property follows Theorem 5.4 of [**11**]. Let $\varphi \in L^2(\partial C)$ and suppose that for a fixed $\Gamma \subset \partial C$

$$\int_{\partial C} \varphi(\mathbf{r}) \left[\int_\Gamma \phi(\mathbf{r}')G(\mathbf{r}, \mathbf{r}')\, ds(\mathbf{r}')\right] ds(\mathbf{r}) = 0$$

for every $\phi \in L^2(\Gamma)$. Interchanging the order of integration, we obain

$$\int_\Gamma \phi(\mathbf{r}') \left[\int_{\partial C} \varphi(\mathbf{r})\Phi(\mathbf{r}, \mathbf{r}')\, ds(\mathbf{r})\right] ds(\mathbf{r}') = 0$$

for every $\phi \in L^2(\partial C)$. This implies the single-layer potential

$$(S\varphi)(\mathbf{r}') = \int_{\partial C} \varphi(\mathbf{r})G(\mathbf{r}, \mathbf{r}')\, ds(\mathbf{r})$$

is zero on ∂C. By analyticity we have $(S\varphi)(\mathbf{r}') = 0$ on ∂C. Since the single-layer potential is a solution to the Helmholtz equation and k^2 is not a Dirichlet eigenvalue for Ω, we obtain $\varphi = 0$. □

The implementation of the above method requires us to choose a sampling domain S containing D. Then for each point $\mathbf{r}_s \in S$, we need to find an approximate solution $g_{\mathbf{r}_s}$ of (3.2). According to Theorem 3.2, the norm of $g_{\mathbf{r}_s}$ is relative small if $\mathbf{r}_s \in D$ and becomes larger as $\mathbf{r}_s$ approaching ∂D. The procedure is exactly the same as the linear sampling method or the reciprocity gap method in free space (see [**10, 13**]).

A major numerical difficulty we are facing is the evaluation of many Green's functions. Since we need $G(\mathbf{r}, \mathbf{r}_s)$ for each $\mathbf{r} \in \Gamma$ and $\mathbf{r}_s \in S$, the computation cost is prohibitive. To this end, we propose a Fourier expansion method to reduce the computation of $G(\mathbf{r}, \mathbf{r}_s)$. Let $\{\mathbf{r}_i\}, i = 1, 2, \dots, N_\Gamma$ be a discrete set of points on Γ and $\{\mathbf{r}_s^j\}, j = 1, 2, \dots, N_S$ be a discrete set of points on S. We need $G(\mathbf{r}_i, \mathbf{r}_s^j)$, $i = 1, 2, \dots, N_\Gamma$, $j = 1, 2, \dots, N_S$ to implement the inverse scheme.

Suppose $\Gamma_\delta \subset \Omega$ is a neighborhood of Γ satisfying $\Gamma_\delta \cap S = \emptyset$. Since

$$\Delta_{\mathbf{r}} G(\mathbf{r}, \mathbf{r}_s) + k^2 G(\mathbf{r}, \mathbf{r}_s) = 0, \ \forall \mathbf{r}, \mathbf{r}_s \in \Gamma_\delta \cup S,$$

by reciprocity we have

$$\Delta_{\mathbf{r}_s} G(\mathbf{r}, \mathbf{r}_s) = \Delta_{\mathbf{r}_s} G(\mathbf{r}_s, \mathbf{r}) = -k^2 G(\mathbf{r}_s, \mathbf{r}) = -k^2 G(\mathbf{r}, \mathbf{r}_s),$$

which implies that

$$\Delta_{\mathbf{r}_s} G(\mathbf{r}, \mathbf{r}_s) + k^2 G(\mathbf{r}, \mathbf{r}_s) = 0.$$

Note that we need $G(\mathbf{r}_i, \mathbf{r}_s)$ for all $\mathbf{r}_s \in S$ for a fixed point $\mathbf{r}_i \in \Gamma$. By reciprocity, we need $G(\mathbf{r}_s, \mathbf{r}_i)$ for all $\mathbf{r}_s \in S$. It is easy to see that $G(\mathbf{r}_s, \mathbf{r}_i)$ satisfies the following Helmholtz equation:

$$\Delta_{\mathbf{r}_s} G(\mathbf{r}_s, \mathbf{r}_i) + k^2 G(\mathbf{r}_s, \mathbf{r}_i) = 0, \quad \text{in } S. \tag{3.6}$$

Thus if we know the value of $G(\mathbf{r}_s, \mathbf{r}_i)$ for $\mathbf{r}_s \in \partial S$, we will have $G(\mathbf{r}_s, \mathbf{r}_i)$ for all $\mathbf{r}_s \in S$ by solving a Dirichlet boundary problem of the above Helmholtz equation. To further simplify the computation, we first choose S as a disk containing D. Let $\mathbf{r}_s^k, k = 1, 2, \dots, N$, be a discrete set of points S. For a fixed point $\mathbf{r}_i$ on Γ, we first compute $G(\mathbf{r}_i, \mathbf{r}_s^k), k = 1, 2, \dots, N$. Since the solution of the Helmholtz equation on a disk can be represented by a series of products of Hankel functions and trigonometric functions [**9**], $G(\mathbf{r}_i, \mathbf{r}_s)$ can be obtained after a Fourier expansion of the Dirichlet data on ∂S. The algorithm only computes the Green's functions on the boundary of S ($\sim O(N)$) instead of all the sampling points in S ($\sim O(N^2)$).

It is obvious that the algorithm can be applied to the free space problem. It is especially useful for three dimensional case since it reduces the computation of Green's functions in a three dimension sampling domain to a two dimension surface.

4. Numerical Examples

We consider a waveguide with period 1. The lower boundary of the waveguide is the x-axis. The upper boundary is given by $f(x) = 1 + 0.05\sin(\pi x)$. In Fig. 3, we show the Green's functions at $(0.3, 0.3)$ for $k = 1$, $k = 3$, and $k = 9$, respectively.

Now we turn to the inverse problems. We set $k = 2$. The first example is a circular obstacle located at $(0.4, 0.4)$ with radius of 0.1. We compute the scattered field on Γ which is used as input of the near field linear sampling method. For the ill-posed integral equations, we use Tikhonov regularization with parameter $\alpha = 10^{-6}$. Note that the choice of the regularization parameter is ad-hoc, i.e., trial and error. We show the reciprocal of the norm of the solutions for (3.2) on the

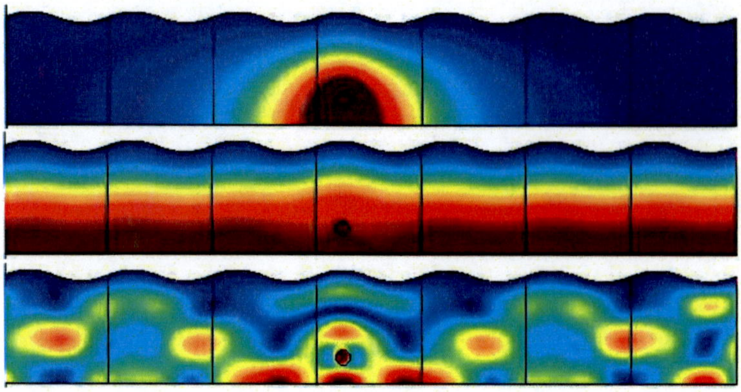

FIGURE 3. The background Green's functions for $k = 1$, $k = 3$, and $k = 9$ (from top to bottom).

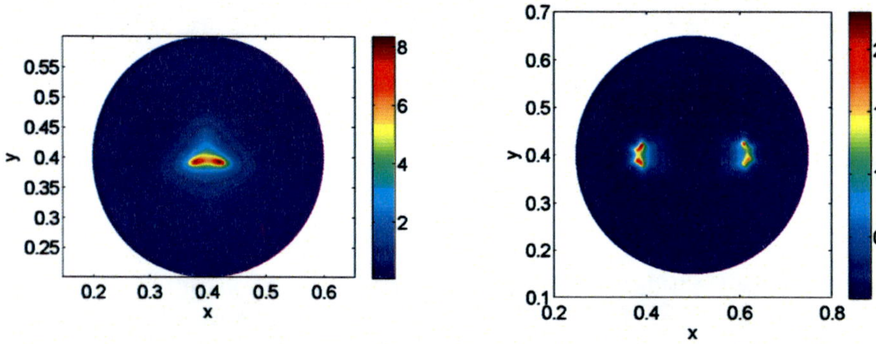

FIGURE 4. The plot of the reciprocal of the norm of the solution. Left: One circular target located at $(0.4, 0.4)$ with radius of 0.1. Right: Two circular obstacles are located at $(0.4, 0.4), (0.6, 0.4)$ with radius of 0.05.

left of Fig. 4. It can be seen that the location and the size of the obstacle can be obtained. However, it seems difficult to recover the exact shape of the obstacle.

For the second example, we choose two circular obstacles located at $(0.4, 0.4)$, $(0.6, 0.4)$, respectively, with radius of 0.05. Tikhonov regularization parameter is still $\alpha = 10^{-6}$. We show the reciprocal of the norm of the solutions for (3.2) on the right of Fig. 4. The construction is similar to the single obstacle case.

5. Conclusions and future work

We consider the inverse scattering problem of imaging objects embedded in periodic structures. This type of problems has important applications in industry, for example, defect detection. We develop a linear sampling method and show some preliminary examples.

The fundamental assumption is the limiting absorption principle for the periodic structure. It is desirable to give a rigorous proof for it. In future, we would like to consider the far field data, i.e., the measurement is at the end(s) of the waveguide which is far away from the objects. The extension of the method to diffraction grating would be another interesting project.

References

[1] Tilo Arens, Drossos Gintides, and Armin Lechleiter, *Direct and inverse medium scattering in a three-dimensional homogeneous planar waveguide*, SIAM J. Appl. Math. **71** (2011), no. 3, 753–772, DOI 10.1137/100806333. MR2796088 (2012j:35295)

[2] T. Arens and N. Grinberg, *A complete factorization method for scattering by periodic surfaces*, Computing **75** (2005), no. 2-3, 111–132, DOI 10.1007/s00607-004-0092-0. MR2163662 (2006d:78011)

[3] Gang Bao, *A uniqueness theorem for an inverse problem in periodic diffractive optics*, Inverse Problems **10** (1994), no. 2, 335–340. MR1269011 (95c:35263)

[4] Gang Bao and Avner Friedman, *Inverse problems for scattering by periodic structures*, Arch. Rational Mech. Anal. **132** (1995), no. 1, 49–72, DOI 10.1007/BF00390349. MR1360080 (96i:35131)

[5] L. Bourgeois and E. Lunéville, *The linear sampling method in a waveguide: a modal formulation*, Inverse Problems **24** (2008), no. 1, 015018, 20, DOI 10.1088/0266-5611/24/1/015018. MR2384777 (2009b:76147)

[6] Fioralba Cakoni, Michele Di Cristo, and Jiguang Sun, *A multistep reciprocity gap functional method for the inverse problem in a multilayered medium*, Complex Var. Elliptic Equ. **57** (2012), no. 2-4, 261–276, DOI 10.1080/17476933.2011.625089. MR2886741

[7] Fioralba Cakoni, M'Barek Fares, and Houssem Haddar, *Analysis of two linear sampling methods applied to electromagnetic imaging of buried objects*, Inverse Problems **22** (2006), no. 3, 845–867, DOI 10.1088/0266-5611/22/3/007. MR2235641 (2007e:35286)

[8] Fioralba Cakoni and David Colton, *Combined far-field operators in electromagnetic inverse scattering theory*, Math. Methods Appl. Sci. **26** (2003), no. 5, 413–429, DOI 10.1002/mma.360. MR1960013 (2003k:78017)

[9] David Colton, *Partial differential equations*, The Random House/Birkhäuser Mathematics Series, Random House Inc., New York, 1988. An introduction. MR943346 (89j:35002)

[10] David Colton and Houssem Haddar, *An application of the reciprocity gap functional to inverse scattering theory*, Inverse Problems **21** (2005), no. 1, 383–398, DOI 10.1088/0266-5611/21/1/023. MR2146182 (2006h:35275)

[11] David Colton and Rainer Kress, *Inverse acoustic and electromagnetic scattering theory*, 2nd ed., Applied Mathematical Sciences, vol. 93, Springer-Verlag, Berlin, 1998. MR1635980 (99c:35181)

[12] Sava Dediu and Joyce R. McLaughlin, *Recovering inhomogeneities in a waveguide using eigensystem decomposition*, Inverse Problems **22** (2006), no. 4, 1227–1246, DOI 10.1088/0266-5611/22/4/007. MR2249462 (2007f:76171)

[13] Michele Di Cristo and Jiguang Sun, *An inverse scattering problem for a partially coated buried obstacle*, Inverse Problems **22** (2006), no. 6, 2331–2350, DOI 10.1088/0266-5611/22/6/025. MR2277545 (2007j:35233)

[14] David C. Dobson, *Optimal design of periodic antireflective structures for the Helmholtz equation*, European J. Appl. Math. **4** (1993), no. 4, 321–339, DOI 10.1017/S0956792500001169. MR1251818 (94j:78021)

[15] Matthias Ehrhardt, Jiguang Sun, and Chunxiong Zheng, *Evaluation of scattering operators for semi-infinite periodic arrays*, Commun. Math. Sci. **7** (2009), no. 2, 347–364. MR2536442 (2010f:35019)

[16] Gregory Eskin, James Ralston, and Masahiro Yamamoto, *Inverse scattering for gratings and wave guides*, Inverse Problems **24** (2008), no. 2, 025008, 12, DOI 10.1088/0266-5611/24/2/025008. MR2408545 (2009i:35332)

[17] F. Hettlich and A. Kirsch, *Schiffer's theorem in inverse scattering theory for periodic structures*, Inverse Problems **13** (1997), no. 2, 351–361, DOI 10.1088/0266-5611/13/2/010. MR1445923 (98d:35034)

[18] Peter Monk and Jiguang Sun, *Inverse scattering using finite elements and gap reciprocity*, Inverse Probl. Imaging **1** (2007), no. 4, 643–660, DOI 10.3934/ipi.2007.1.643. MR2350219 (2009j:35385)

[19] Yongzhi Xu, Christopher Mawata, and Wei Lin, *Generalized dual space indicator method for underwater imaging*, Inverse Problems **16** (2000), no. 6, 1761–1776, DOI 10.1088/0266-5611/16/6/311. MR1801569 (2001j:86008)

[20] Jiaqing Yang, Bo Zhang, and Ruming Zhang, *A sampling method for the inverse transmission problem for periodic media*, Inverse Problems **28** (2012), no. 3, 035004, 17, DOI 10.1088/0266-5611/28/3/035004. MR2888529 (2012k:65141)

DEPARTMENT OF MATHEMATICAL SCIENCES, MICHIGAN TECHNOLOGICAL UNIVERSITY, HOUGHTON, MICHIGAN 49931

E-mail address: jiguangs@mtu.edu

DEPARTMENT OF MATHEMATICAL SCIENCES, TSINGHUA UNIVERSITY, BEIJING 100084, PEOPLE'S REPUBLIC OF CHINA

E-mail address: czheng@math.tsinghua.edu.cn

Full Eulerian modeling and effective numerical studies for the dynamic fluid-structure interaction problem

Pengtao Sun, Lixiang Zhang, Chun Liu, and Jinchao Xu

ABSTRACT. In this paper we present a full Eulerian model for a dynamic fluid-structure interaction (FSI) problem in terms of phase field approach, and design its effective numerical discretizations and iterative schemes based on the second-order backward difference formula (BDF), Galerkin/least-square (GLS) stabilization scheme and/or streamline-upwind/Petrov-Galerkin (SUPG) method. The presented full Eulerian FSI model effectively demonstrates the interaction between fluid flow and solid structure in terms of an uniform system of governing equations defined in a single domain. Numerical experiments are carried out for a cross rotating about its central axis due to the passing flow field, and the numerical results dramatically show the rotation of the cross while interacting with the fluid that is flowing around, showing that our model and numerical methods are effective to simulate the dynamic fluid-structure interaction phenomena.

1. Introduction

To define a fluid-structure interaction (FSI) model fully in Eulerian coordinates system, we first introduce a bounded domain $\Omega \subset \mathbb{R}^d$, $d = 2, 3$, such that $\Omega = \Omega_f \bigcup \Omega_s$, where Ω_f denotes the subdomain occupied by fluid flow and Ω_s the subdomain occupied by solid structure, and let $\Gamma_{\text{int}} = \partial\Omega_s$, representing the interface of fluid and structure. It is well known that the kinematic and dynamic interactions between fluid and structure are determined by continuity in the velocity and in the traction force at Γ_{int}, namely,

$$\mathbf{u}_f = \mathbf{u}_s, \quad \mathbf{x} \in \Gamma_{\text{int}}, \tag{1.1}$$

$$\boldsymbol{\sigma}_f \cdot \mathbf{n} = \boldsymbol{\sigma}_s \cdot \mathbf{n}, \quad \mathbf{x} \in \Gamma_{\text{int}}, \tag{1.2}$$

where $\mathbf{n}$ denotes the unit normal vector at the interface.

In a full Eulerian perspective of fluid-structure interaction problem, instead of separately partitioned two velocity fields $\mathbf{u}_f$ and $\mathbf{u}_s$ in Ω_f and Ω_s, respectively, it is

---

1991 *Mathematics Subject Classification.* Primary 65N30, 76D05; Secondary 76B15, 65Z05.

Key words and phrases. Fluid-structure interaction, full Eulerian model, phase field, Galerkin/least-square (GLS) stabilized finite element.

The first author was supported by NSF Grant DMS-0913757.

The second author was supported by National Natural Science Foundation of China (NSFC) [Grant no. 50839003].

The third author was supported by NSF Grant DMS-1109107.

The fourth author was supported by NSF Grant DMS-0915153.

convenient to introduce a monolithic velocity vector $\mathbf{u}$ applied to the entire domain Ω, that is here referred to as a *one-continuum* formulation [**LW01, LS02, SLX09**]. The one-continuum formulation would immediately satisfy (1.1) and (1.2) because $\mathbf{u}$ is supposed to be continuous across the interface Γ_{int}. Due to the inherent multiphase characteristic of FSI problem, in this paper, we employ phase field approach to define such monolithic velocity vector, and further, a full Eulerian FSI model.

Phase field models are an increasingly popular choice for modeling the motion of multiphase fluids (see [**AMW98**] for a comprehensive review). In the phase field model, sharp fluid interfaces are replaced by thin but nonzero thickness transition regions where the interfacial forces are smoothly distributed [**CC92**]. The basic idea is to introduce a conserved order parameter (e.g., mass concentration) that varies continuously over thin interfacial layers and is mostly uniform in the bulk phases. These models allow topological changes of the interface [**Boy99, GPn96, Jac99, LT98**] and have many advantages in numerical simulations of the interfacial motion [**ChH96**]. Thus, it is also known as the diffuse-interface model. More precisely, in this approach, a phase field variable ϕ_s is introduced, which can be thought of as the volume fraction, to demarcate the two species and indicate the location of the interface. A mixing energy is defined based on ϕ_s which, through a convection-diffusion equation, governs the evolution of the interfacial profile. The phase field method can be viewed as a physically motivated level-set method, and Lowengrub and Truskinovsky [**LT98**] have argued for the advantage of using a physically determined ϕ_s profile instead of an artificial smoothing function for the interface. When the thickness of the interface approaches zero, the diffuse-interface model becomes asymptotically identical to a sharp-interface level-set formulation. It also reduces properly to the classical sharp-interface model in general. Based on an energetic variational formulation, Liu and Shen [**LS02**] employed a phase field model to describe the mixture of two incompressible Newtonian fluids. The mixing energy studied is related to the usual Ginzburg-Laudau model for phase evolutions.

The statistical (or phase field approach) point of view represents the interface as a continuous, but steep, change of properties (density, viscosity, etc.) of the two phases (fluid-fluid or fluid-structure). Within a "thin" transition region, the phases are mixed and have to store certain amount of "mixing energy". Such an approach coincides with the usual phase field models in the theory of phase transition (see, e.g., [**CA51, CH58, MWB$^+$93, MS85, TC94**] and many others). Numerical simulations also demonstrate that the method captures many interesting physical phenomena and at the same time is very robust [**LS02, SLX09**]. The key observation is that in the phase field model, the surface force (such as surface tension) can be viewed as the limit of the bulk body force as the thickness of the interface approaches zero [**LW01**].

The rest of this paper is organized as follows. In Section 2, we introduce a full Eulerian model for fluid-structure interaction problem. The effective numerical techniques are presented in Section 3. In Section 4, a finite element approach is presented for streamline simulation. We carry out a 3D numerical simulation for a FSI model problem in Section 5, where the numerical results dramatically show that our model and numerical methods are effective to simulate the dynamic fluid-structure interaction phenomena.

2. A full Eulerian model of FSI problem

We introduce a phase field function ϕ_s such that

$$\phi_s(\mathbf{x}) = \begin{cases} 1, & \text{for } \mathbf{x} \in \Omega_s \\ 0, & \text{for } \mathbf{x} \in \Omega_f \end{cases} \tag{2.1}$$

where, ϕ_s takes two distinct values (1 and 0) in each of the phases, with a smooth change between both values in the zone around the interface, which is then diffuse with a finite width ε. ϕ_s is actually the volume fraction of solid inside a computational cell. Thus, a discrete location of the interface can be defined as the collection of all points where the phase field takes a certain value 0.5, i.e. $\Gamma_{\text{int}} = \{\mathbf{x}|\phi_s(\mathbf{x}) = 0.5\}$.

In terms of the phase field variable ϕ_s, we define the mixture velocity field $\mathbf{u}$ as

$$\mathbf{u} = (1-\phi_s)\mathbf{u}_f + \phi_s\mathbf{u}_s. \tag{2.2}$$

Note that such definition for $\mathbf{u}$ preserves the divergence free property, i.e., $\nabla\cdot\mathbf{u} = 0$ if $\nabla\cdot\mathbf{u}_f = \nabla\cdot\mathbf{u}_s = 0$, for an incompressible flow assumption in which all the densities are constant and identical. Similarly, we define the mixture density ρ as

$$\rho = (1-\phi_s)\rho_f + \phi_s\rho_s = \rho_f + (\rho_s - \rho_f)\phi_s \tag{2.3}$$

which is usually varied with ϕ_s if $\rho_f \neq \rho_s$, the mixture viscosity μ as

$$\mu = (1-\phi_s)\mu_f + \phi_s\mu_s, \tag{2.4}$$

and the mixture stress tensor $\boldsymbol{\sigma}$ as

$$\boldsymbol{\sigma}(\mathbf{u}, p) = -p\mathbb{I} + (1-\phi_s)\boldsymbol{\sigma}_f + \phi_s\boldsymbol{\sigma}_s, \tag{2.5}$$

where $\mathbb{I}$ is the unit tensor, $\boldsymbol{\sigma}_f = 2\mu_f\mathbb{D}$, $\boldsymbol{\sigma}_s = 2\mu_s\mathbb{D} + \boldsymbol{\sigma}_{sh}$, $\mathbb{D}$ is the velocity deformation tensor given by

$$\mathbb{D} = \frac{\nabla\mathbf{u} + \nabla\mathbf{u}^T}{2}.$$

$\boldsymbol{\sigma}_{sh}$ corresponds to the hyperelastic contribution, defined as [**SIT⁺11**]

$$\boldsymbol{\sigma}_{sh} = 2c_1\mathbb{B} + 2c_2(\text{tr}(\mathbb{B})\mathbb{B} - \mathbb{B}\cdot\mathbb{B}) + 4c_3(\text{tr}(\mathbb{B}) - 3)\mathbb{B}, \tag{2.6}$$

here $\mathbb{B} = \mathbb{F}\cdot\mathbb{F}^T$ represents the left Cauchy-Green deformation tensor, $\mathbb{F} = \partial\mathbf{x}/\partial\mathbb{X}$ the deformation gradient, $\mathbf{x}$ the current coordinates, and $\mathbb{X}$ the reference coordinates. For a two-dimensional case, the symmetry in the tensor $\mathbb{B}$ implies that $\mathbb{B}_{zz} = 1$, $\mathbb{B}_{xz} = \mathbb{B}_{zx} = \mathbb{B}_{yz} = \mathbb{B}_{zy} = 0$, and $\mathbb{B}_{xy} = \mathbb{B}_{yx}$. Thus, $\mathbb{B}$ owns three independent unknown components $\mathbb{B}_{xx}$, $\mathbb{B}_{yy}$ and $\mathbb{B}_{xy}$ in a two-dimensional domain.

The parameters in (2.6), c_1, c_2 and c_3, depend on what type of material we consider for this model. For instance, if we choose the neo-Hookean material, then $c_1 = \frac{\mu^s_{\text{Lamé}}}{2}$, $c_2 = c_3 = 0$, For another typical hyperelastic material, Saint Venant-Kirchhoff material, we take $c_1 = \mu^s_{\text{Lamé}}$, $c_2 = -\frac{\mu^s_{\text{Lamé}}}{2}$, $c_3 = \frac{\lambda^s_{\text{Lamé}} + 2\mu^s_{\text{Lamé}}}{8}$. Here, $\mu^s_{\text{Lamé}}$ and $\lambda^s_{\text{Lamé}}$ denote Lamé constants, given by $\mu^s_{\text{Lamé}} = \frac{E}{2(1+\nu)}$, and $\lambda^s_{\text{Lamé}} = \frac{E\nu}{(1+\nu)(1-2\nu)}$, where E is the Young's modulus and ν the Poisson's ratio.

We shall point out that, the stress, $\boldsymbol{\sigma}$, and the pressure, p, in (2.5) are not the actual stress tensor $\boldsymbol{\sigma}_{\text{actual}}$ and pressure p_{actual}. However, they hold the following relationships

$$\boldsymbol{\sigma}_{\text{actual}} = \boldsymbol{\sigma} - \frac{\text{tr}(\boldsymbol{\sigma})}{3}\mathbb{I}, \quad p_{\text{actual}} = p - \frac{\text{tr}(\boldsymbol{\sigma})}{3}, \tag{2.7}$$

which is because the actual stress tensor contains the deviatoric part $-\frac{\mathrm{tr}(\boldsymbol{\sigma})}{3}\mathbb{I}$. Thus, we also call $\boldsymbol{\sigma}$ and p as *pseudo* stress tensor and *pseudo* pressure, respectively.

From (2.3) we know that the mixture density ρ is no longer constant if $\rho_f \neq \rho_s$, resulting in a compressible flow assumption. Thus, with the use of phase field variable ϕ_s, the compressible full Eulerian fluid-structure interaction problem can be modeled as the following form

$$
\begin{aligned}
&(2.8) \quad & \rho\left(\frac{\partial \mathbf{u}}{\partial t} + \mathbf{u}\cdot\nabla\mathbf{u}\right) &= -\nabla p + \nabla\cdot(2\mu\mathbb{D} + \phi_s\boldsymbol{\sigma}_{sh}), \quad \text{in } \Omega\times(0,T],\\
&(2.9) \quad & \frac{\partial \rho}{\partial t} + \nabla\cdot(\rho\mathbf{u}) &= 0, \quad \text{in } \Omega\times(0,T],\\
&(2.10) \quad & \frac{\partial \phi_s}{\partial t} + \mathbf{u}\cdot\nabla\phi_s &= 0, \quad \text{in } \Omega\times(0,T],\\
&(2.11) \quad & \frac{\partial \mathbb{B}}{\partial t} + \mathbf{u}\cdot\nabla\mathbb{B} &= (\nabla\mathbf{u})^T\cdot\mathbb{B} + \mathbb{B}\cdot\nabla\mathbf{u}, \quad \text{in } \Omega\times(0,T].
\end{aligned}
$$

However, by means of (2.3) and (2.10), it is nice to see that, the continuity equation (2.9) can be actually reformulated to

$$\nabla\cdot\mathbf{u} = 0, \tag{2.12}$$

just the same with the continuity equation of an incompressible flow.

To improve the preservation of the structure's shape during a long term evolution, as well as to efficiently stabilize the numerical simulation for a transport equation, we update (2.10) to the following Allen-Cahn's type phase field equation [**LS02, SLX09**]

$$\frac{\partial \phi_s}{\partial t} + \mathbf{u}\cdot\nabla\phi_s = \gamma\left(\Delta\phi_s - \frac{1}{\varepsilon^2}\phi_s(\phi_s-1)(2\phi_s-1)\right), \tag{2.13}$$

where, ε represents the interfacial width, which is small compared to the characteristic length of the laboratory scale. In practice, ε is proportional to mesh size h. γ represents the elastic relaxation time. As $\gamma\to 0$, the limiting ϕ_s satisfies the transport equation (2.10), which is equivalent to the mass transport for incompressible fluids. In (2.13), if we let $f(\phi_s) = \phi_s(\phi_s-1)(2\phi_s-1)$, then $f(\phi_s) = F'(\phi_s)$ with $F(\phi_s)$ being a given energy potential, i.e., the Ginzburg-Landau double-well potential defined as $F(\phi_s) = \frac{1}{2}(\phi_s-1)^2\phi_s^2$. Note that this double-well potential function is not the same as defined in [**LW01, LS02, SLX09**] because the definition of the phase field variable ϕ_s in (2.1) is different, where the two distinct values in two phases are 1 and 0 instead of +1 and -1 therein.

Surface tension is an important factor for preserving the shape of the structure. To complete a phase field approach in fluid-structure interactional dynamics, we shall add the surface tension effect to (2.8). By means of phase field formulation [**LS02, SLX09**], we have the following modified momentum equation

$$\rho\left(\frac{\partial \mathbf{u}}{\partial t} + \mathbf{u}\cdot\nabla\mathbf{u}\right) + \nabla p - \nabla\cdot(2\mu\mathbb{D} + \phi_s\boldsymbol{\sigma}_{sh}) + \lambda\nabla\cdot(\nabla\phi_s\otimes\nabla\phi_s) = 0 \tag{2.14}$$

where $\lambda = \varepsilon\sigma$ is the surface tension parameter. A further calculation shows that

$$\lambda\nabla\cdot(\nabla\phi_s\otimes\nabla\phi_s) = \frac{\lambda}{2}\nabla|\nabla\phi_s|^2 + \lambda\Delta\phi_s\nabla\phi_s. \tag{2.15}$$

Let $\Omega\subset\mathbb{R}^2$ be a bounded domain with boundary $\partial\Omega = \Gamma_{\text{in}}\cup\Gamma_{\text{sf}}\cup\Gamma_{\text{out}}$ such that all three parts of the boundary, i.e., inlet Γ_{in}, surface Γ_{sf} and outlet Γ_{out}, are mutually disjoint. Based upon Allen-Cahn's type phase field formulation, therefore, a complete full Eulerian model of fluid-structure interaction problem can be further defined as follows

$$
\begin{aligned}
\rho\left(\frac{\partial \mathbf{u}}{\partial t}+\mathbf{u}\cdot\nabla\mathbf{u}\right) &= -\nabla p+\nabla\cdot(2\mu\mathbb{D}+\phi_s\boldsymbol{\sigma}_{sh}) \\
&\quad -\lambda\nabla\cdot(\nabla\phi_s\otimes\nabla\phi_s), \quad \text{in } \Omega\times(0,T], && (2.16)\\
\nabla\cdot\mathbf{u} &= 0, \quad \text{in } \Omega\times(0,T], && (2.17)\\
\mathbf{u}(0,\cdot) &= \mathbf{u}_0 \text{ in } \Omega,\\
\mathbf{u} &= \mathbf{g} \text{ on } \Gamma_{\text{in}}\times(0,T],\\
\boldsymbol{\sigma}(\mathbf{u},p)\cdot\mathbf{n} &= \mathbf{0} \text{ on } \Gamma_{\text{out}}\times(0,T],\\
\mathbf{u}\cdot\mathbf{n} &= 0 \text{ on } \Gamma_{\text{sf}}\times(0,T],\\
\mathbf{u}\cdot\boldsymbol{\tau}_k &= 0 \text{ on } \Gamma_{\text{sf}}\times(0,T],\\
\frac{\partial\phi_s}{\partial t}+\mathbf{u}\cdot\nabla\phi_s &= \gamma\left(\Delta\phi_s-\frac{1}{\varepsilon^2}\phi_s(\phi_s-1)(2\phi_s-1)\right), && (2.18)\\
&\quad \text{in } \Omega\times(0,T],\\
\phi_s(0,\cdot) &= \phi_{s0}=\begin{cases}1, & \text{in } \Omega_s\\ 0, & \text{in } \Omega_f\end{cases}\\
\phi_s &= 0 \text{ on } \Gamma_{\text{in}}\times(0,T],\\
\frac{\partial\mathbb{B}}{\partial t}+\mathbf{u}\cdot\nabla\mathbb{B} &= (\nabla\mathbf{u})^T\cdot\mathbb{B}+\mathbb{B}\cdot\nabla\mathbf{u}, \quad \text{in } \Omega\times(0,T], && (2.19)\\
\mathbb{B}(0,\cdot) &= \mathbb{I}\\
\mathbb{B} &= 0 \text{ on } \Gamma_{\text{in}}\times(0,T],
\end{aligned}
$$

where $\mathbf{n}$ denotes the outward pointing unit normal vector on $\partial\Omega$, the tangential vectors $\boldsymbol{\tau}_k$, $1\le k\le d-1$, are chosen such that $\mathbf{n},\boldsymbol{\tau}_1$ in $2D$ and $\mathbf{n},\boldsymbol{\tau}_1,\boldsymbol{\tau}_2$ in $3D$ build an orthonormal basis. The initial velocity $\mathbf{u}_0$, the final time T and the Dirichlet inflow boundary value $\mathbf{g}$ are prescribed. The do-nothing boundary condition is given at the outflow Γ_{out}. On Γ_{sf}, no slip and no penetration boundary conditions are applied. In order to prescribe the initial phase field ϕ_{s0}, we simply set ϕ_{s0} as unity in the computational cell whose barycenter falls into Ω_s and zero if the barycenter of a computational cell locates in Ω_f. The domain dimensions, physical and operating parameters are listed in Table 1.

3. Numerical methods

In this section, we numerically study the full Eulerian model of fluid-structure interaction problem (2.16)-(2.19) by means of a fully implicit Newton's method to linearize nonlinear governing equations, Taylor-Hood (Q2Q1) mixed finite element to discretize spatial parts of momentum equation, and the second-order backward difference formula (BDF) to discretize all the temporal derivatives, where Galerkin/least-square (GLS) scheme and streamline-upwind/Petrov-Galerkin (SUPG) method are appropriately applied to overcome the numerical instabilities induced by highly dominant convection due to the hyperelastic and transport effects arising from (2.16), (2.18) as well as (2.19).

To define the numerical discretizations of (2.16)-(2.19), we first introduce d-dimensional Sobolev space $[H_g^1(\Omega)]^d=\{\mathbf{v}\in[H^1(\Omega)]^d:\mathbf{v}|_{\Gamma_{\text{in}}}=\mathbf{g}\}$, and let $V_{h,g}^d=$

TABLE 1. Geometrical, physical and operating parameters

Parameters/properties	*Symbol*	*Value*	*Unit*
Modeling domain dimensions			
Length of channel	L_{channel}	1.452	m
Width of channel	W_{channel}	0.484	m
Length of cross	L_{cross}	0.121	m
Horizontal position of the base of cross	X_{cross}	0.13	m
Vertical position of the axis of rotation	Y_{axis}	0.242	m
Distance between base and axis of rotation	D_{axis}	0.0573	m
Initial position angle of cross	θ	30	degree
Physical and transport parameters			
Viscosity of fluid	μ_f	0.001	kg/(m·s)
Young's modulus	E	2.5×10^9	Pa
Poisson's ratio	ν	0.384	
Density of fluid	ρ_f	1000	kg/m^3
Density of structure	ρ_s	1280	kg/m^3
Operating parameters			
Maximum input velocity	$\mathbf{u}_{\text{in}}$	1.0	m/s
Tolerance of nonlinear iteration		10^{-6}	
Parameter for Galerkin-Least-Square scheme	δ_1	10^{-6}	
Parameter for streamline-diffusion scheme	δ_2	10^{-10}	

$\{\mathbf{v}_h|\mathbf{v}_h|_\tau \in [P^2(\tau)]^d, \mathbf{v}_h|_{\Gamma_{\text{in}}} = \mathbf{g}\}$ be the piecewise quadratic polynomial space of velocity vector function, thus $V^d_{h,g} \subset [H^1_g(\Omega)]^d$, and then $V_{h,g} \subset [H^1_g(\Omega)]$ is the piecewise quadratic polynomial space of scalar functions. Moveover, let $W_h = \{q_h|q_h|_\tau \in P^1(\tau)\} \subset H^1(\Omega)$ and $\Phi_h = \{\psi_h|\psi_h|_\tau \in P^1(\tau), \psi_h|_{\Gamma_{\text{in}}} = 0\} \subset H^1(\Omega)$ be the piecewise linear polynomial space of scalar functions. Then $\Phi_h^{\frac{d(d+1)}{2}}$ denotes the piecewise linear polynomial space of deformation tensor functions. On the other hand, let $\Delta t = \frac{T}{N}$, then $t_n = n\Delta t$, and denote $\phi(\mathbf{x}, t_n) = \phi^n$. $(p,q) = \int_\Omega pq\, d\mathbf{x}$ represents L^2 inner product.

Thus, we define the fully implicit finite element discretizations of (2.16)-(2.19) as follows by means of Newton's linearization scheme. Provided $(\mathbf{u}^n_h, p_h, \phi^n_h, \mathbb{B}^n_h) \in V^d_{h,g} \times W_h \times V_{h,0} \times \Phi_h^{\frac{d(d+1)}{2}}$ are given, let $(\mathbf{u}^{n+1}_{h,0}, p_{h,0}, \phi^{n+1}_{h,0}, \mathbb{B}^{n+1}_{h,0}) = (\mathbf{u}^n_h, p_h, \phi^n_h, \mathbb{B}^n_h)$, for $n = 0, 1, \cdots, N-1$ and $k = 0, 1, 2, \cdots$, find $(\mathbf{u}^{n+1}_{h,k+1}, p_{h,k+1}, \phi^{n+1}_{h,k+1}, \mathbb{B}^{n+1}_{h,k+1}) \in V^d_{h,g} \times W_h \times V_{h,0} \times \Phi_h^{\frac{d(d+1)}{2}}$, such that

$$
\begin{aligned}
&\left(\rho(\phi^{n+1}_{h,k})\frac{\partial \mathbf{u}^{n+1}_{h,k+1}}{\partial t}, \mathbf{v}_h\right) + \left(\rho(\phi^{n+1}_{h,k})\mathbf{u}^{n+1}_{h,k}\cdot\nabla\mathbf{u}^{n+1}_{h,k+1}, \mathbf{v}_h\right)\\
&+\left(2\mu(\phi^{n+1}_{h,k})\mathbb{D}(\mathbf{u}^{n+1}_{h,k+1}), \nabla\mathbf{v}_h\right) + \left(\rho(\phi^{n+1}_{h,k})\nabla\mathbf{u}^{n+1}_{h,k}\cdot\mathbf{u}^{n+1}_{h,k+1}, \mathbf{v}_h\right)\\
&+\delta_1(h^2)\left(\rho(\phi^{n+1}_{h,k})\frac{\partial \mathbf{u}^{n+1}_{h,k+1}}{\partial t} + L(\mathbf{u}^{n+1}_{h,k+1}, p_h), L(\mathbf{v}_h, q_h)\right)\\
&-(p_h, \nabla\cdot\mathbf{v}_h) + \left((\rho_s-\rho_f)\left(\frac{\partial \mathbf{u}^{n+1}_{h,k}}{\partial t} + \mathbf{u}^{n+1}_{h,k}\cdot\nabla\mathbf{u}^{n+1}_{h,k}\right)\phi^{n+1}_{h,k+1}, \mathbf{v}_h\right)
\end{aligned}
$$

$$
\begin{aligned}
&+\Big(2(\mu_s-\mu_f)\mathbb{D}(\mathbf{u}_{h,k}^{n+1})\phi_{h,k+1}^{n+1},\nabla\mathbf{v}_h\Big)\\
&+\Big(\sigma_{sh}(B_{h,k}^{n+1})\phi_{h,k+1}^{n+1},\nabla\mathbf{v}_h\Big)+\Big((2c_1-12c_3)\phi_{h,k}^{n+1}B_{h,k+1}^{n+1},\nabla\mathbf{v}_h\Big)\\
&+\Big((2c_2+4c_3)\phi_{h,k}^{n+1}(\mathrm{tr}(B_{h,k}^{n+1})B_{h,k+1}^{n+1}+B_{h,k}^{n+1}\mathrm{tr}(B_{h,k+1}^{n+1})),\nabla\mathbf{v}_h\Big)\\
&-\Big(2c_2\phi_{h,k}^{n+1}(B_{h,k}^{n+1}\cdot B_{h,k+1}^{n+1}+B_{h,k+1}^{n+1}\cdot B_{h,k}^{n+1}),\nabla\mathbf{v}_h\Big)\\
&-\Big(\lambda\nabla\phi_{h,k}^{n+1}\cdot\nabla\phi_{h,k+1}^{n+1},\nabla\cdot\mathbf{v}_h\Big)+\Big(\lambda\Delta\phi_{h,k}^{n+1}\nabla\phi_{h,k+1}^{n+1},\mathbf{v}_h\Big)\\
&-\Big(\lambda\nabla^2\phi_{h,k}^{n+1}\nabla\phi_{h,k+1}^{n+1},\mathbf{v}_h\Big)-\Big(\lambda\nabla\phi_{h,k}^{n+1}\nabla\phi_{h,k+1}^{n+1},\nabla\mathbf{v}_h\Big)\\
=&\Big(\rho(\phi_{h,k}^{n+1})\mathbf{u}_{h,k}^{n+1}\cdot\nabla\mathbf{u}_{h,k}^{n+1},\mathbf{v}_h\Big)+\Big(2(\mu_s-\mu_f)\mathbb{D}(\mathbf{u}_{h,k}^{n+1})\phi_{h,k}^{n+1},\nabla\mathbf{v}_h\Big)\\
&+\left((\rho_s-\rho_f)(\frac{\partial\mathbf{u}_{h,k}^{n+1}}{\partial t}+\mathbf{u}_{h,k}^{n+1}\cdot\nabla\mathbf{u}_{h,k}^{n+1})\phi_{h,k}^{n+1},\mathbf{v}_h\right)\\
&+\Big((2c_1-12c_3)\phi_{h,k}^{n+1}B_{h,k}^{n+1},\nabla\mathbf{v}_h\Big)-\Big(4c_2\phi_{h,k}^{n+1}B_{h,k}^{n+1}\cdot B_{h,k}^{n+1},\nabla\mathbf{v}_h\Big)\\
&+\Big((4c_2+8c_3)\phi_{h,k}^{n+1}\mathrm{tr}(B_{h,k}^{n+1})B_{h,k}^{n+1},\nabla\mathbf{v}_h\Big)
\end{aligned}
$$

$$
-\Big(\frac{\lambda}{2}|\nabla\phi_{h,k}^{n+1}|^2,\nabla\cdot\mathbf{v}_h\Big)+\Big(\lambda\Delta\phi_{h,k}^{n+1}\nabla\phi_{h,k}^{n+1},\mathbf{v}_h\Big), \tag{3.1}
$$

$$
(\nabla\cdot\mathbf{u}_h^{n+1},q_h)=0, \tag{3.2}
$$

$$
\begin{aligned}
&\left(\frac{\partial\phi_{h,k+1}^{n+1}}{\partial t},\psi_h\right)+\Big(\mathbf{u}_{h,k}^{n+1}\cdot\nabla\phi_{h,k+1}^{n+1},\psi_h\Big)\\
&+\Big(\nabla\phi_{h,k}^{n+1}\cdot\mathbf{u}_{h,k+1}^{n+1},\psi_h\Big)+\gamma(h^3)\Big(\mathbf{u}_{h,k}^{n+1}\cdot\nabla\phi_{h,k+1}^{n+1},\mathbf{u}_{h,k}^{n+1}\cdot\nabla\psi_h\Big)\\
&+\frac{\gamma(h^3)}{\varepsilon^2(h)}\Big([6(\phi_{h,k}^{n+1})^2-6\phi_{h,k}^{n+1}+1]\phi_{h,k+1}^{n+1},\psi_h\Big)
\end{aligned}
$$

$$
=\frac{\gamma(h^3)}{\varepsilon^2(h)}\Big(4(\phi_{h,k}^{n+1})^3-3(\phi_{h,k}^{n+1})^2,\psi_h\Big)+\Big(\nabla\phi_{h,k}^{n+1}\cdot\mathbf{u}_{h,k}^{n+1},\psi_h\Big), \tag{3.3}
$$

$$
\begin{aligned}
&\left(\frac{\partial\mathbb{B}_{h,k+1}^{n+1}}{\partial t},\mathbb{C}_h\right)+\Big(\mathbf{u}_{h,k}^{n+1}\cdot\nabla\mathbb{B}_{h,k+1}^{n+1},\mathbb{C}_h\Big)\\
&+\Big(\mathbf{u}_{h,k+1}^{n+1}\cdot\nabla\mathbb{B}_{h,k}^{n+1},\mathbb{C}_h\Big)+\delta_2(h)\Big(\mathbf{u}_{h,k}^{n+1}\cdot\nabla\mathbb{B}_{h,k+1}^{n+1},\mathbf{u}_{h,k}^{n+1}\cdot\nabla\mathbb{C}_h\Big)\\
&-\Big((\nabla\cdot\mathbf{u}_{h,k}^{n+1})^T\cdot\mathbb{B}_{h,k+1}^{n+1},\mathbb{C}_h\Big)-\Big(\mathbb{B}_{h,k+1}^{n+1}\cdot\nabla\cdot\mathbf{u}_{h,k}^{n+1},\mathbb{C}_h\Big)\\
&-\Big((\nabla\cdot\mathbf{u}_{h,k+1}^{n+1})^T\cdot\mathbb{B}_{h,k}^{n+1},\mathbb{C}_h\Big)-\Big(\mathbb{B}_{h,k}^{n+1}\cdot\nabla\cdot\mathbf{u}_{h,k+1}^{n+1},\mathbb{C}_h\Big)\\
=&-\Big((\nabla\cdot\mathbf{u}_{h,k}^{n+1})^T\cdot\mathbb{B}_{h,k}^{n+1},\mathbb{C}_h\Big)-\Big(\mathbb{B}_{h,k}^{n+1}\cdot\nabla\cdot\mathbf{u}_{h,k}^{n+1},\mathbb{C}_h\Big)
\end{aligned}
$$

$$
+\Big(\mathbf{u}_{h,k}^{n+1}\cdot\nabla\mathbb{B}_{h,k}^{n+1},\mathbb{C}_h\Big), \tag{3.4}
$$

$$
\forall(\mathbf{v}_h,q_h,\psi_h,\mathbb{C}_h)\in V_{h,g}^d\times W_h\times V_{h,0}\times\Phi_h^{\frac{d(d+1)}{2}},
$$

where $\mathbf{u}_h^0 = \mathbf{u}_0, \phi_h^0 = \phi_{s0}, \mathbb{B}_h^0 = \mathbb{I}$. The operator L on the left hand side of (3.1), defined as $L(\mathbf{u}, p) = \rho\mathbf{u} \cdot \nabla\mathbf{u} + \nabla p - \nabla \cdot (2\mu\mathbb{D})$, denotes the spatial differential operator of (2.8) with respect to $\mathbf{u}$. The discrete temporal derivatives in (3.1), (3.3) and (3.4) are defined as follows by the second-order BDF

$$\frac{\partial w_{h,k+1}^{n+1}}{\partial t} = \frac{1}{2\Delta t}(3w_{h,k+1}^{n+1} - 4w_{h,k+1}^{n} + w_{h,k+1}^{n-1}). \tag{3.5}$$

The algorithm of implementing the discretization schemes (3.1)-(3.4) is described as follows

Algorithm A1. Let $n = 0$.

(1) Let $k = 0$, $(\mathbf{u}_{h,0}^{n+1}, p_{h,0}, \phi_{h,0}^{n+1}, \mathbb{B}_{h,0}^{n+1})$=$(\mathbf{u}_h^n, p_h, \phi_h^n, \mathbb{B}_h^n)$.
(2) Solve (3.1)-(3.4) for $(\mathbf{u}_{h,k+1}^{n+1}, p_{h,k+1}, \phi_{h,k+1}^{n+1}, \mathbb{B}_{h,k+1}^{n+1})$.
(3) Determine if the following stopping criterion holds:

$$\begin{aligned} &\|\mathbf{u}_{h,k+1}^{n+1} - \mathbf{u}_{h,k}^{n+1}\|_{L^2(\Omega)} + \|p_{h,k+1} - p_{h,k}\|_{L^2(\Omega)} \\ &+\|\phi_{h,k+1}^{n+1} - \phi_{h,k}^{n+1}\|_{L^2(\Omega)} + \|\mathbb{B}_{h,k+1}^{n+1} - \mathbb{B}_{h,k}^{n+1}\|_{L^2(\Omega)} < \text{tolerance}. \end{aligned} \tag{3.6}$$

If (3.6) is false, then let $k \leftarrow k+1$, go to step 2. Otherwise, let $(\mathbf{u}_h^{n+1}, p_h, \phi_h^{n+1}, \mathbb{B}_h^{n+1})$=$(\mathbf{u}_{h,k+1}^{n+1}, p_{h,k+1}, \phi_{h,k+1}^{n+1}, \mathbb{B}_{h,k+1}^{n+1})$.
(4) Compute the streamlines field ψ_h^{n+1} with $\mathbf{u}_h^{n+1}$.
(5) If $n = N$, then stop the entire computation, otherwise, let $n \leftarrow n+1$, go to step 1.

4. Streamlines simulation

The stream function is defined for two-dimensional flows. It can be used to plot streamlines, which represent the trajectories of particles in a steady flow. The difference between the stream function values at any two points gives the volumetric flow rate (or volumetric flux) through a line connecting the two points. Since streamlines are tangent to the velocity vector of the flow, the value of the stream function must be constant along a streamline. The usefulness of the stream function lies in the fact that the velocity components in the x- and y- directions at a given point are given by the partial derivatives of the stream function at that point.

One way to define the stream function ψ for a two-dimensional flow is to use the following expression in terms of the flow velocity

$$\nabla \times \boldsymbol{\psi} = \mathbf{u}, \tag{4.1}$$

where $\boldsymbol{\psi} = (0, 0, \psi)^T$ if the velocity vector $\mathbf{u} = (u, v, 0)^T$. In Cartesian coordinate system, (4.1) is equivalent to

$$u = \frac{\partial\psi}{\partial y}, \quad v = -\frac{\partial\psi}{\partial x}. \tag{4.2}$$

A better approach to compute the stream function ψ is as follows

$$-\Delta\psi = \nabla \times \mathbf{u} \tag{4.3}$$

For two-dimensional flow, (4.3) can be further simplified as

$$-\Delta\psi = \frac{\partial v}{\partial x} - \frac{\partial u}{\partial y} \tag{4.4}$$

with Neumann boundary condition $\frac{\partial\psi}{\partial n}|_{\partial\Omega} = (-v, u)^T \cdot \mathbf{n}$, according to (4.2).

Thus, the numerical solution of streamlines field by means of linear finite element can be computed as follows: find $\psi_h \in W_h$ such that

$$(4.5)\quad (\nabla\psi_h, \nabla\phi) = (\nabla \times \mathbf{u}_h, \phi) + \int_{\Gamma_{sf}\bigcup\Gamma_{out}} \frac{\partial\psi_h}{\partial n}\phi ds + \int_{\Gamma_{in}} \frac{\partial\psi_h}{\partial n}\phi ds, \quad \forall\phi \in \Phi_h$$

from which, the coefficient matrix arising in the derived linear algebraic system is singular because no Dirichlet boundary condition is reinforced for (4.4), and the concurrent natural Neumann boundary condition in (4.5) produces infinite solutions differing by a constant. Considering that we only need streamlines to show the trajectory of particles in a steady flow, (4.4) does not have to produce a unique solution for the stream function. However, to directly solve the linear algebraic system arising from (4.5) by Gauss elimination solver, we better make its coefficient matrix nonsingular, namely, assign a Dirichlet boundary condition for (4.4). Since the velocity $\mathbf{u}_h$ is assigned to equal $\mathbf{g}$ on the inlet Γ_{in}, without loss of generality, we set $\psi\big|_{\Gamma_{in}} = yu_x - xu_y = yg_x - xg_y$, which satisfies (4.2), then the last boundary integral on Γ_{in} on the right hand side of (4.5) vanishes. Hence, we choose the linear finite element space $\Psi_h \subset H^1_\psi(\Omega) = \{\phi \in H^1(\Omega) : \phi|_{\Gamma_{in}} = yg_x - xg_y\}$, and (4.5) can be rewritten as, find $\psi_h \in \Psi_h$ such that

$$(4.6)\quad (\nabla\psi_h, \nabla\phi) = (\frac{\partial v_h}{\partial x} - \frac{\partial u_h}{\partial y}, \phi) + \int_{\Gamma_{sf}\bigcup\Gamma_{out}} (-v_h, u_h)^T \cdot \mathbf{n}\phi ds, \quad \forall\phi \in \Phi_h$$

Thus, for a given numerical solution of velocity field $\mathbf{u}_h$, we can solve (4.6) for a uniquely discrete streamlines field ψ_h, and further, illustrate the velocity field in terms of the attained streamlines.

5. Numerical results

Now we are ready to carry out a numerical experiment for the full Eulerian model of fluid-structure interaction problem (2.16)-(2.19). We consider a two-dimensional flow passes a cross in a certain time interval. The cross locates in a flow channel near its inlet with an angle θ from the horizontal axis in the counter-clockwise direction. When an incoming flow passes it from the left inlet to the right outlet, the cross is forced to rotate about its central axis, as shown in Fig. 1. To simulate such rotation of a cross about its central axis, we simply assume the

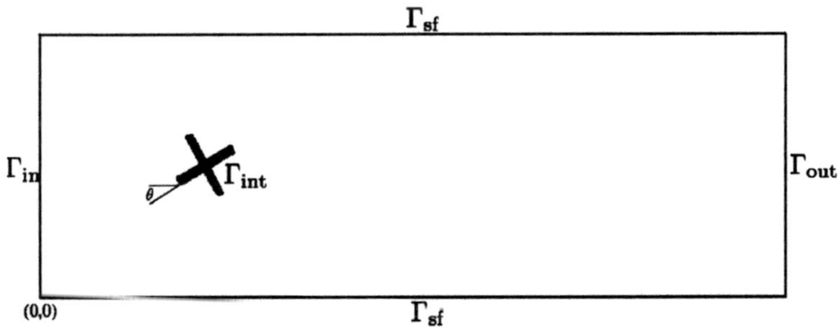

FIGURE 1. Two-dimensional channel with a cross whose central axis is fixed.

velocity field $\mathbf{u} = 0$ on the central axis of the cross all the time. The time interval is discretized with equidistant time step size 0.005s. As for the initial conditions, an impulsive start is used for velocity field, i.e. the initial discrete velocity is chosen to be zero on all degrees of freedom which are not on the inflow boundary. The degrees of freedom on Γ_{in} are given by the inflow boundary condition $\mathbf{u}|_{\Gamma_{\text{in}}} = \mathbf{g}$, which is set in our simulations as a parabolic profile, given by

$$\mathbf{g} = \begin{pmatrix} 1.5\mathbf{u}_{\text{in}}(1 - (\frac{y - W_{\text{channel}}/2}{W_{\text{channel}}/2})^2) \\ 0 \end{pmatrix}.$$

Therefore, the average speed of the background inflow fluids is $\mathbf{u}_{\text{in}}$ in the horizontal direction and zero in the vertical direction. The initial profile of phase field ϕ_{s0} is shown in Fig. 2.

FIGURE 2. Initial profiles of phase field ϕ_{s0}.

By employing the numerical discretizations (3.1)-(3.4) and Algorithm A1, we obtain the phase field and streamlines contours at different time levels, as displayed in Fig. 3 and Fig. 4. It shows that the phase field of structure, i.e., the cross, keep rotating forward (in clockwise direction) about its central axis in the first place for the angle of 1.25π after 1.5s (300 time marching steps) or so, then rotating backward (in counter-clockwise direction) for the angle of around π after about 1s. Afterwards, the cross rotates back and forth along with the surrounding flow in a random pattern. During the rotation, no any shift occurs for the cross, its central axis is strictly preserved at the original position. It is seen that the bars of cross turn out to be a bit curly after a certain time marching, which is because of the change of deformation tensor field $\mathbb{B}$, introducing the corresponding deformation of phase field of solid structure. Such deformation can be controlled by the parameters of viscosity and Young's modulus of the structure. The evolution of streamlines field shown in Fig. 4 shall correspond with that of the phase field, and many eddies are produced along the channel while the cross rotates itself due to the passing flow, presenting a dramatic change of flow field induced by the interaction between fluid and structure.

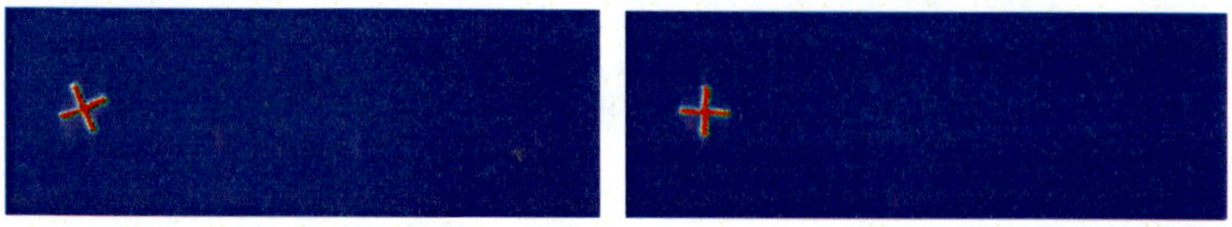

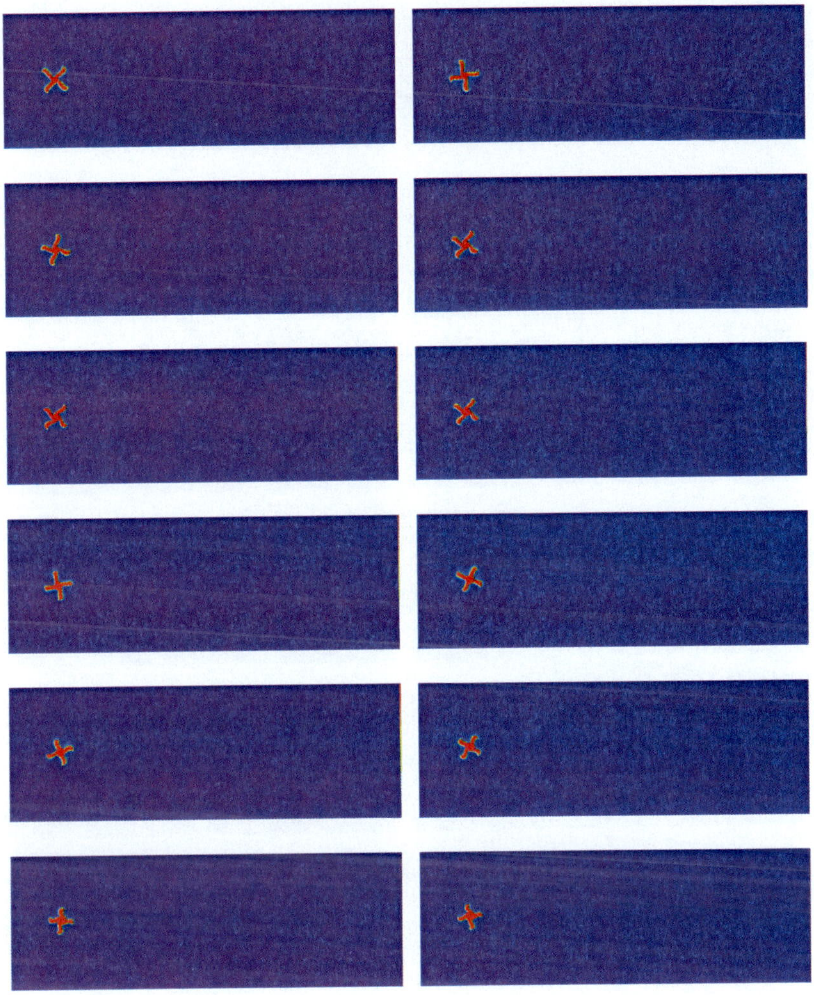

FIGURE 3. Evolution of phase fields from t=0.05s to 3.25s with time increment 0.25s.

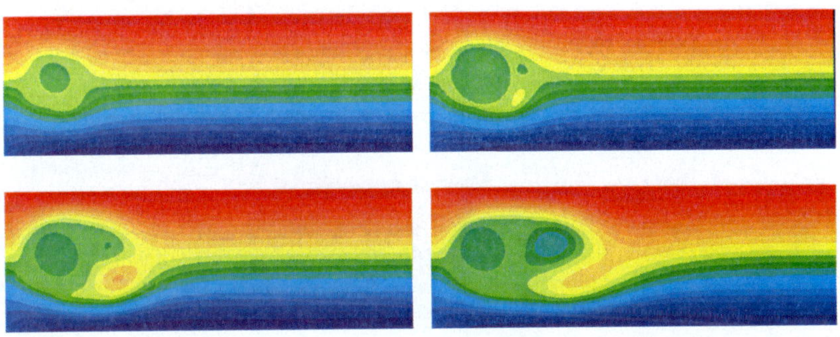

6. Conclusion

We build a full Eulerian model for fluid-structure interaction problem based on a complete phase field approach, including both diffuse interface and surface tension

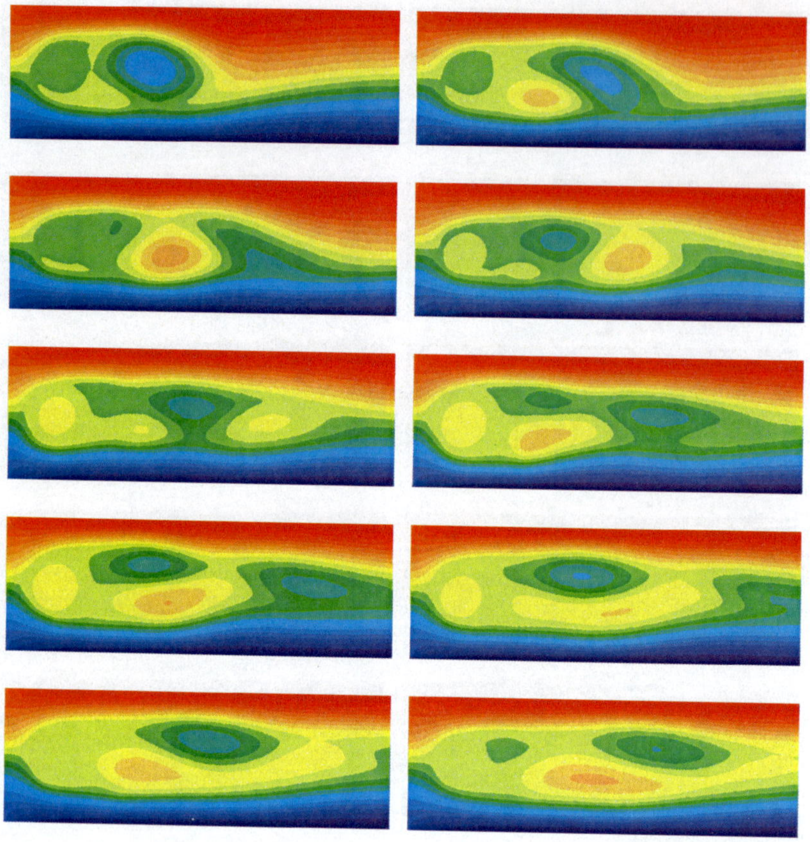

FIGURE 4. Evolution of streamline fields from t=0.05s to 3.25s with time increment 0.25s.

terms. In view of its highly dominant convection effect and/or even distinct hyperbolic properties, we design Galerkin/least-square and streamline-upwind/Petrov-Galerkin schemes to discretize the governing equations, and use fully implicit Newton's method to linearize the fluid-phase field-structure coupling system. Numerical experiments demonstrate that our model and numerical techniques are effective to explore the interaction between fluid and structure. Further investigations are underway to examine the in-depth mechanism that is involved in a real-life three-dimensional fluid-structure interaction problem.

References

[AMW98] D. M. Anderson, G. B. McFadden, and A. A. Wheeler, *Diffuse-interface methods in fluid mechanics*, Appl. Math. Lett. **30** (1998), 139–165. MR1609626 (99a:76048)

[Boy99] F. Boyer, *Mathematical study of multi-phase flow under shear through order parameter formulation*, Asymptot. Anal. **20** (1999), 175–212. MR1700669 (2000g:35166)

[CA51] J. W. Cahn and S. M. Allen, *A microscopic theory for domain wall motion and its experimental varification in fe-al alloy domain growth kinetics*, J. Phys. Colloque **38** (1977), pages=C7-51,).

[CC92] G. Caginalp and X. F. Chen, *Phase field equations in the singular limit of sharp interface problems*, On the evolution of phase boundaries (Minneapolis, MN, 1990–91), Springer, New York, 1992, pp. 1–27. MR1226912 (94d:35166)

[CH58] J. W. Cahn and J. E. Hillard, *Free energy of a nonuniform system. I. Interfacial free energy*, J. Chem. Phys. **28** (1958), 258–267.

[ChH96] *A level set formulation of eulerian interface capturing methods for incompressible fluid flows*, J. Comput. Phys. **124** (1996), 449–464. MR1383769 (97a:76087)

[GPn96] M. E. Gurtin, D. Polignone, and J. Vi nals, *Two-phase binary fluids and immiscible fluids described by an order parameter*, Math. Models Methods Appl. Sci. **6** (1996), 815–831. MR1404829 (99e:76123)

[Jac99] D. Jacqmin, *Calculation of two-phase Navier-Stokes flows using phase-field modeling*, J. Comput. Phys. **155** (1999), 96–127. MR1716497 (2000g:76088)

[LS02] C. Liu and J. Shen, *A phase field model for the mixture of two incompressible fluids and its approximation by a fourier-spectral method*, Physica D **179** (2002), 211–228. MR1984386 (2004j:35233)

[LT98] J. Lowengrub and L. Truskinovsky, *Quasi-incompressible Cahn-Hilliard fluids and topological transitions*, R. Soc. Lond. Proc. Ser. A Math. Phys. Eng. Sci. **454** (1998), 2617–2654. MR1650795 (2000e:82022)

[LW01] C. Liu and N.J. Walkington, *An eulerian description of fluids containing visco-elastic particles*, Arch. Ration. Mech. Anal. **159** (2001), 229–252. MR1857673 (2003d:76161)

[MS85] W. W. Mullins and R. F. Sekerka, *On the thermodynamics of crystalline solids*, J. Chem. Phys. **82** (1985), 5192–5202. MR789911 (86e:73008)

[MWB+93] G. B. McFadden, A. A. Wheeler, R. J. Braun, S. R. Coriell, and R. F. Sekerka, *Phase-field models for anisotropic interfaces*, Phys. Rev. E (3) **48** (1993), 2016–2024. MR1377920

[SIT+11] Kazuyasu Sugiyama, Satoshi Ii, Shintaro Takeuchi, Shu Takagi, and Yoichiro Matsumoto, *A full eulerian finite difference approach for solving fluid-structure coupling problems*, J. Comput. Phys. **230** (2011), 596–627. MR2745446 (2011k:65112)

[SLX09] P. Sun, C. Liu, and J. Xu, *Phase field model of thermo-induced marangoni effects in the mixtures and its numerical simulations with mixed finite element method*, Commun. Comput. Phys. **6** (2009), 1095–1117.

[TC94] J. E. Taylor and J. W. Cahn, *Linking anisotropic sharp and diffuse surface motion laws via gradient flows*, J. Statist. Phys. **77** (1994), 183–197. MR1300532 (95j:58029)

Department of Mathematical Sciences, University of Nevada, Las Vegas, 4505 Maryland Parkway, Las Vegas, Nevada 89154
E-mail address: pengtao.sun@unlv.edu

Department of Mechanical Engineering, Kunming University of Science and technology, Kunming, 68 Wenchang Road, Yunnan, China
E-mail address: zlxzcc@126.com

Department of Mathematics, Pennsylvania State University, University Park, Pennsylvania 16802
E-mail address: liu@math.psu.edu

Department of Mathematics, Pennsylvania State University, University Park, Pennsylvania 16802
E-mail address: xu@math.psu.edu

A multiple-endpoints Chebyshev collocation method for high order differential equations

Shan Wang and Zhiping Li

ABSTRACT. A new collocation method with multiple-endpoints and a new boundary condition technique is established for high order differential equations. Numerical examples on 1D sixth order and 2D fourth order linear problems show that the new method efficiently improves the condition numbers and the convergence rates. An example on nonlinear elastic thin film buckling shows the advantage of the new method for high order nonlinear partial differential equations with complex boundary conditions.

1. Introduction

Pseudospectral methods as meshless methods are successfully used for widely diverse applications [**1**]. The Chebyshev type collocation methods are among the most popular spectral methods because of computational convenience [**2**]. A typical choice of collocation points for solving boundary value problems of second order differential equations with a Chebyshev method is to use the Chebyshev-Gauss-Lobatto collocation points, which include certain inner collocation points and two end points. Chebyshev-Gauss collocation method, which has no endpoints, is also a popular choice. However, difficulties arise when pseudospectral method is applied to higher order differential equations, especially in high dimensions [**2**], for example it usually leads to an over-determined system. There are two standard ways to deal with the problem (Chapter 6 in [**2**]). One is to use either more base functions [**2**] or less inner collocation points [**3**] with a shortcoming that the condition number is typically very large. The other is to introduce proper variable substitutions so that part of the boundary conditions are satisfied naturally by the new unknown functions (e.g. [**4**]). However, for complicated boundary conditions, such as reciprocally periodic connection boundary conditions [**5**] and nonlinear boundary conditions [**6**], it can hardly work.

In this paper, we designed a new multiple-endpoints collocation points for high order differential equations. Numerical examples on 1D 6th-order and 2D 4th-order linear differential equations are presented to show the improved condition numbers of the differential matrices and better accuracy of the new method. In particular,

---

2010 *Mathematics Subject Classification.* Primary 65N35, 65L60.

Key words and phrases. Chebyshev collocation method, multiple-endpoints, high order partial differential equations, complex boundary conditions.

Corresponding author: Zhiping Li, email address: lizp@math.pku.edu.cn.

This research was supported by the NSFC project 11171008 and RFDP of China.

we present an example on an elastic thin film buckling problem governed by a nonlinear von Kárman equation, for which the standard Chebyshev methods failed to produce physically consistent solutions.

The rest of the paper is organized as follows. The new multiple-endpoints collocation points and the new boundary condition technique are introduced in Section 2. In Section 3, the numerical results are presented, discussed, and comparisons are made. The paper ends with some concluding remarks in Section 4.

2. Multiple-endpoints Collocation Points

2.1. The base functions of multiple-endpoints collocation method and the corresponding collocation points. Let T_n be the Chebychev polynomial with degree n and $\omega(x)$ be the corresponding weight function [**1**]. Follow the theory of Gauss-Lobatto collocation points [**1**], we consider polynomials

$$Q_{N+1,m}(x) = T_{N+1}(x) + \sum_{k=1}^{2m} a_{N+1,k} T_{N+1-k}(x), \quad N+1 \geq 2m, \tag{2.1}$$

where $\{a_{N+1,k}\}_{k=1}^{2m}$ are so taken that $Q_{N+1,m}^{(s)}(\pm 1) = 0$, $s = 0, 1, \cdots, m-1$.

Two of the most important properties of the polynomials $\{Q_{N,m}\}_{N=2m}^{\infty}$ are revealed by the following two theorems, i.e. for certain properly chosen weight, they are a sequence of orthogonal polynomials, and their zeros in $(-1,1)$ are simple and are the corresponding Gauss quadrature points.

THEOREM 2.1. *The polynomials* $\{\frac{Q_{N,m}}{(1-x^2)^m}\}_{N=2m}^{\infty}$ *are orthogonal in* $L^2_{\omega_m}(-1,1)$ *with the weight* $\omega_m(x) = (1-x^2)^m \omega(x) = (1-x^2)^{m-\frac{1}{2}}$.

As a consequence of Theorem 2.1, it can be shown that the polynomial $Q_{N+1,m}(x)$ has $N+1-2m$ separated simple zeros in $(-1,1)$ (see Lemma 1.2.2 in [**1**] page 10) and two m-zeros at each of the boundary points.

Let $x_0 = x_1 = \cdots = x_{m-1} = -1, x_{N+1-m} = \cdots = x_{N-1} = x_N = 1$ and $-1 < x_m < \cdots < x_{N-m} < 1$ be the $N+1$ roots of the polynomial $Q_{N+1,m}$ given by (2.1), we call $\{x_i\}_{i=0}^{N}$ the multiple-endpoints Chebychev collocation points.

THEOREM 2.2. *Let* $-1 < x_m < x_{m+1} < \cdots < x_{N-m} < 1$ *be the* $N+1-2m$ *zeros of the polynomial* $\frac{Q_{N+1,m}}{(1-x^2)^m}$. *Let* $\{w_m^{N+1,m}, w_{m+1}^{N+1,m}, \cdots, w_{N-m}^{N+1,m}\}$ *be the corresponding quadrature weights for the integral* $\int_{-1}^{1} p(x)\omega_m(x)dx$ *with weight* $\omega_m(x) = (1-x^2)^{m-\frac{1}{2}}$. *Then, we have* $w_j^{N+1,m} > 0$, $j = m, \cdots, N-m$, *and*

$$\sum_{j=m}^{N-m} p(x_j) w_j^{N+1,m} = \int_{-1}^{1} p(x)\omega_m(x)dx, \qquad \forall p \in \mathbb{P}_{2N-4m+1}.$$

By the expression (2.1), if the parameters $\{a_k\}_{k=1}^{2m}$ are given, the values of the polynomials $Q_{N+1,m}$ and their derivatives $Q_{N+1,m}^{(s)}$ can be easily calculated by working with the corresponding results of the Chebyshev polynomials. Substitute the boundary values of the Chebyshev polynomials [**1**] into the expression (2.1) of the polynomials $Q_{N+1,m}$, then the $2m$ boundary conditions $Q_{N+1,m}^{(s)}(\pm 1) = 0$, $s = 0, 1, \cdots, m-1$ give a linear system that the parameters $\{a_k\}_{k=1}^{2m}$ must satisfy. After some elementary manipulations with the equations, we are lead to the following linear systems $A\vec{a}_{odd} = \mathbf{0}, A\vec{a}_{even} = -(1, (N+1)^2, \cdots, (N+1)^{2(m-1)})^T$, where $\vec{a}_{odd} = (a_1, a_3, \cdots, a_{2m-1})^T$, $\vec{a}_{even} = (a_2, a_4, \cdots, a_{2m})^T$, and where the matrix $A := (A_{ij})_{i,j=1}^m$ with $A_{ij} = (N+1-2j)^{2i-2}$. Obviously, for $N > 2m$, A is a nonsingular Vandermonde matrix. Thus the parameteres $\{a_k\}_{k=1}^{2m}$ are uniquely solvable, and in particular, $\vec{a}_{odd} = \mathbf{0}$.

2.2. Discretization with Chebyshev Collocation Methods. Consider boundary value problems of differential equations of the form

$$\mathcal{L}u(x) = f(x) \quad x \in \Omega, \tag{2.2}$$

$$\mathcal{B}u(x) = g(x) \quad x \in \partial\Omega. \tag{2.3}$$

In 1D, for simplicity, let $\Omega = (-1, 1)$. Let k be the number of boundary conditions, let $\{x_i\}_{i=1}^{N+1-k} \subset (-1, 1)$ be the inner collocation points, and $\bar{x}_i \in \{-1, 1\}$, $1 \le i \le k$ be the multiple boundary collocation points corresponding to the boundary conditions. Substituting $u_N(x) = \mathbf{T}_N(x)\hat{\mathbf{U}}_N$ into (2.2) and (2.3), and evaluating at the collocation points, we obtain a system of $N+1$ equations for $N+1$ unknowns $\hat{\mathbf{U}}_N = (\hat{u}_0, \hat{u}_1, \cdots, \hat{u}_N)^T$:

$$\begin{pmatrix} \mathbf{T}_{in}^N L_N \\ \mathbf{T}_{br}^N \odot B_N \end{pmatrix} \hat{\mathbf{U}}_N = \begin{pmatrix} \mathbf{f}_{in} \\ \mathbf{g}_{br} \end{pmatrix}, \tag{2.4}$$

where $\mathbf{T}_{in}^N$ is a $(N+1-k) \times (N+1)$ matrix whose i-th row is $\mathbf{T}_N(x_i)$, $\mathbf{T}_{br}^N$ is a $k \times (N+1)$ matrix whose i-th row is $\mathbf{T}_N(\bar{x}_i)$, L_N is the differential matrix of the differential operators $\mathcal{L}$, $\mathbf{T}_{br}^N \odot B_N$ is a $k \times (N+1)$ matrix whose i-th row is given by $\mathbf{T}_N(\bar{x}_i)B_N^i$ with B_N^i being the differential operator of the differential operator $\mathcal{B}$ at $\bar{x}_i$, $\mathbf{f}_{in} = (f(x_1), \cdots, f(x_{N+1-k}))^T$ and $\mathbf{g}_{br} = (g(\bar{x}_1), \cdots, g(\bar{x}_k))^T$. The method works in a similar way for higher dimensions (see for example [**1**]).

Notice that, if $\mathcal{B}$ and $\mathcal{L}$ are linear, then instead of solving the system (2.4) for $\hat{\mathbf{U}}_N$, it could be more convenient to solve the following smaller system for $\bar{\mathbf{U}}_{in}$:

$$\mathbf{T}_{in}^N L_N S_1 \bar{\mathbf{U}}_{in} = \mathbf{f}_{in} - \mathbf{T}_{in}^N L_N S_2 \mathbf{g}_{br}, \tag{2.5}$$

where (S_1, S_2) is the inverse matrix of $\begin{pmatrix} \mathbf{T}_{in}^N \\ \mathbf{T}_{br}^N \odot B_N \end{pmatrix}$, *i.e.* $(S_1, S_2) \begin{pmatrix} \mathbf{T}_{in}^N \\ \mathbf{T}_{br}^N \odot B_N \end{pmatrix} = I$. The approach can also be extended to the case where $\mathcal{L}$ is nonlinear.

2.3. Solving problems with homogeneous boundary conditions. For a standard homogeneous boundary value problem of a $2k$th-order partial differential equation defined on $(-1, 1)$, we can use the $N+1-2k$ separated simple zeros of $Q_{N+1,k}$ as the inner collocation points, and find numerical solutions of the problem by solving the equations (2.4) or (2.5) with $\mathbf{g}_{br} = 0$.

Noticing that the polynomials $\{Q_{K,k}\}_{K=2k}^N$ are a set of base functions for the space of polynomials $\mathbb{Q}_N = \{p \in \mathbb{P}_N : p^{(j)}(\pm 1) = 0, \ j = 0, \cdots, k\}$, we may as well directly express the approximation solution as $u_N(x) = \mathbf{Q}_N(x)\tilde{\mathbf{U}}_N$, where $\mathbf{Q}_N(x) = (Q_{2k,k}, Q_{2k+1,k}, \cdots, Q_{N,k})$, and solve for $\tilde{\mathbf{U}}_N = (\tilde{u}_{2k}, \cdots, \tilde{u}_N)^T$. By (2.1), we have $\mathbf{Q}_N(x) = \mathbf{T}_N(x)\mathbf{A}^{N,k}$, where $\mathbf{A}^{N,k}$ is a $(N+1) \times (N+1-2k)$ matrix. This implies that $\hat{\mathbf{U}}_N = \mathbf{A}^{N,k}\tilde{\mathbf{U}}_N$. Thus, the discrete system (2.4) is transformed to an equivalent reduced system

$$\mathbf{T}_{in}^N L_N \mathbf{A}^{N,k} \tilde{\mathbf{U}}_N = \mathbf{f}_{in}. \tag{2.6}$$

The method can be naturally extended to solve homogeneous boundary value problems defined on the domain $(-1, 1)^d$ in d-dimensions.

2.4. Solving problems with inhomogeneous boundary conditions. For an inhomogeneous boundary value problem of a $2k$th-order linear partial differential equation defined on $(-1,1)$, we decompose the approximation solution $u_N(x) = \mathbf{T}_N(x)\hat{\mathbf{U}}_N$ into two parts, that is $u_N(x) = \mathbf{T}_N(x)(\hat{\mathbf{U}}_N^0 + \hat{\mathbf{U}}_N^1)$, where $\mathbf{T}_N(x)\hat{\mathbf{U}}_N^1$ satisfies the inhomogeneous boundary condition (see (2.4))

$$(\mathbf{T}_{br}^N \odot B_N)\hat{\mathbf{U}}_N^1 = \mathbf{g}_{br}, \tag{2.7}$$

and $\mathbf{T}_N(x)\hat{\mathbf{U}}_N^0$ satisfies the homogenous boundary condition and the equation

$$\mathbf{T}_{in}^N L_N \hat{\mathbf{U}}_N^0 = \mathbf{f}_{in} - \mathbf{T}_{in}^N L_N \hat{\mathbf{U}}_N^1. \tag{2.8}$$

Obviously, the above decomposition is not unique. For the sake of simplicity and stability, we determine $\hat{\mathbf{U}}_N^1$ by solving the equation (2.7) with the least square method, in which we require that the L^2 norm of $L_N\hat{\mathbf{U}}_N^1$ is minimized. Instead of solving (2.8), we solve, by the method given in §2.3 (see (2.6)), the equation

$$\mathbf{T}_{in}^N L_N \mathbf{A}^{N,k}\tilde{\mathbf{U}}_N^0 = \mathbf{f}_{in} - \mathbf{T}_{in}^N L_N \hat{\mathbf{U}}_N^1 \tag{2.9}$$

for the homogeneous part of the solution which is now given in the form $\mathbf{T}_N(x)\hat{\mathbf{U}}_N^0 = \mathbf{Q}_N(x)\tilde{\mathbf{U}}_N^0$. This approach can also be naturally extended to solve inhomogeneous boundary value problems defined on the domain $(-1,1)^d$ in d-dimensions.

Notice that, in the standard approaches for inhomogeneous boundary value problems in higher dimensions, the discrete equation, which is in general a underdetermined system because of the multi-counted boundary corner collocation points, is solved either by the least square method or by eliminating certain numbers of the highest order base polynomials. In contrast, our approach here produces in theory an exact solution which generally consists of all admissible base polynomials. In addition, our numerical experiments show that our method is more robust and can produce more accurate numerical solutions.

3. Numerical examples

3.1. 1D 6th-order linear problem. First, we consider a 6th-order linear problem with homogeneous boundary conditions:

$$\begin{cases} u^{(6)}(x) - u(x) = f(x), & -1 < x < 1, \\ u(\pm 1) = 0, u'(\pm 1) = 0, u''(\pm 1) = 0. \end{cases} \tag{3.1}$$

For simplicity of notations, we rewrite the linear equations (2.5) (for $\mathbf{g}_{br} = 0$) and (2.6) in the form $\mathbf{L}_N^1\bar{\mathbf{U}}_{in} = \mathbf{f}_{in}$ and $\mathbf{L}_N^0\tilde{\mathbf{U}}_N = \mathbf{f}_{in}$ respectively, where $\mathbf{L}_N^1 = \mathbf{T}_{in}^N L_N S_1$ and $\mathbf{L}_N^0 = \mathbf{T}_{in}^N L_N \mathbf{A}^{N,k}$. The condition numbers cond$(\mathbf{L}_N^i)$ of the matrixes $\mathbf{L}_N^i$ corresponding to the Chebyshev-Gauss, Chebyshev-Gauss-Lobatto, and Chebyshev-Multiple-Endpoints collocation points are compared in the left sub-figure of Figure 1, where it is seen that the Chebyshev-Multiple-Endpoints method has obviously much smaller condition numbers cond$(\mathbf{L}_N)$.

To compare the accuracy of numerical solutions, let the exact solution be given by $u(x) = e^x + p(x)$ with $p(x) = f(x) \in \mathbb{P}_5$, and let $u_N^1(x) = \mathbf{T}_N(x)\hat{\mathbf{U}}_N = \mathbf{T}_N(x)S_1\bar{\mathbf{U}}_{in} \in \mathbb{P}_N(-1,1)$ be the numerical solutions obtained by solving $\mathbf{L}_N^1\bar{\mathbf{U}}_{in} = \mathbf{f}_{in}$ and $u_N^0(x) = \mathbf{Q}_N(x)\tilde{\mathbf{U}}_N \in \mathbb{P}_N(-1,1)$ be the numerical solutions obtained by solving $\mathbf{L}_N^0\tilde{\mathbf{U}}_N = \mathbf{f}_{in}$ respectively. In the right sub-figure of Figure 1, the error $e_u = \|u - u_N^i\|_{L^2(-1,1)}$ of the numerical solutions are compared, where we see that

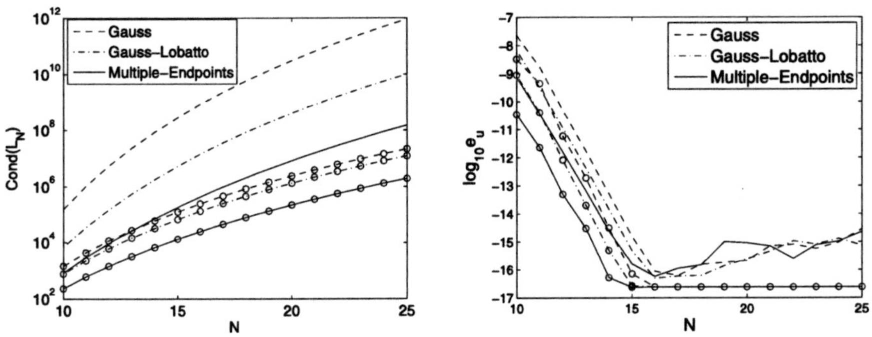

FIGURE 1. The numerical results of $\text{cond}(\mathbf{L}_N^i)$ and e_u^i, $i = 0, 1$, where the results with respect to $i = 0$ are marked by $\circ$.

the new method has smaller error and reaches the machine accuracy faster than the other methods. It is also seen that u_N^0 reaches higher accuracy than u_N^1.

3.2. Biharmonic equation. Next, we consider the 2D biharmonic equation with clamped boundary conditions:

$$\begin{cases} \Delta^2 u(x,y) = f(x,y), & (x,y) \in \Omega, \\ u(x,y) = g_0, \ \dfrac{\partial u}{\partial n}(x,y) = g_1, & (x,y) \in \partial\Omega. \end{cases} \tag{3.2}$$

where Δ denotes the Laplace operator, and $\Omega = (-1,1)^2$.

We choose the same polynomial base functions for the two dimensions, and write the discrete solution in the form $u_N(x,y) = \sum_{i=0}^{N}\sum_{j=0}^{N} \hat{u}_{ij}T_i(x)T_j(y)$.

We consider homogenous and inhomogeneous boundary conditions seperately. We compare three types of collocation points, which are the Chebyshev-Gauss, Chebyshev-Gauss-Lobatto, and Chebyshev-Multiple-Endpoints collocation points. At the same time, the new boundary technique introduced in §2.3 and §2.4 (referred to as the first method in Figures 2 and 3) is compared with a standard method (referred to as the second method in Figures 2 and 3), in which the four highest order terms are omitted and the discrete solution is written as

$$u_N(x,y) = \sum_{i=0}^{N}\sum_{j=0}^{N} \hat{u}_{ij}T_i(x)T_j(y) - \sum_{i=N-1}^{N}\sum_{j=N-1}^{N} \hat{u}_{ij}T_i(x)T_j(y). \tag{3.3}$$

The numerical results of homogenous case is shown in Figure 2, in which the exact solution is given by $u(x,y) = \pi^{-4}(1+\cos(\pi x))(1+\cos(\pi y))$. In the left sub-figure of Figure 2, the comparison of the error $e_u = \|u_N(x) - u(x)\|_{L^2(-1,1)^2}$ shows clearly that our boundary technique leads to higher accuracy. The numerical results of Multiple-Endpoints-Chebyshev collocation points are also seen to perform slightly better than the others. The numerical results on the condition number $\text{Cond}(L_N)$ is shown in the right sub-figure of Figure 2, in which L_N corresponding to $\mathbf{T}_{in}^N L_N \mathbf{A}^{N,k}$in (2.6) for the first method and to $\mathbf{T}_{in}^N L_N S_1$ in (2.5) for second method. It is clearly seen that the Multiple-Endpoints-Chebyshev collocation points and the new boundary technique result in significantly lower condition numbers.

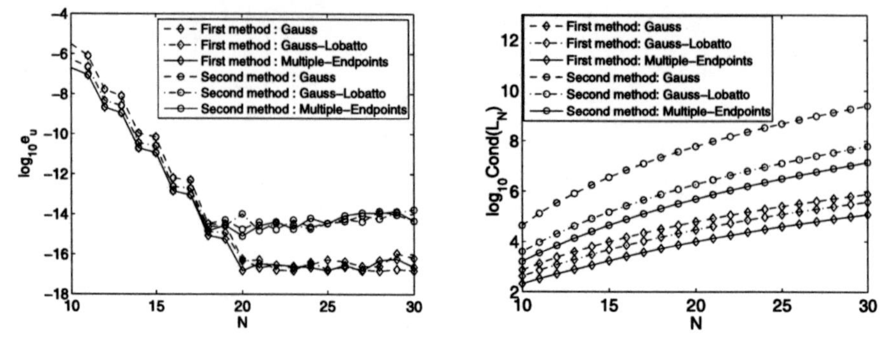

FIGURE 2. Numerical results for homogeneous problem.

For the inhomogeneous case, we consider two exact solutions $u^1(x, y) = \pi^{-4}(1+\cos(\pi x))(1 + \cos(\pi y)) + 0.1(y + 1)$ and $u^2(x, y) = \ln(2 + xy) + 1$. The errors are shown in Figure 3. Obviously, the error of latter decrease much slower, this is not a surprise since the Fourier expansion of the latter converges much slower. The first method is still seen to work better than the second method, and the difference can be crucial sometimes. The advantage of Multiple-Endpoints-Chebyshev collocation points on the accuracy also appears to be more significant.

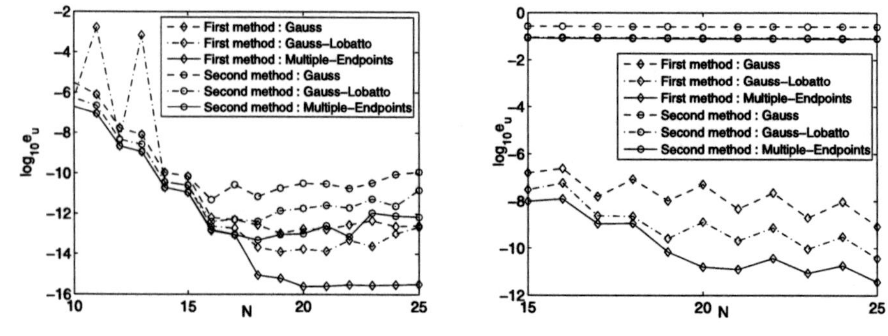

FIGURE 3. Numerical results for u^1 (on the left) and u^2 (on the right).

3.3. An example in thin film delamination. In this sunsection, we consider an example of telephone-cord buckling of elastic thin film [**7, 8**], which can be modeled by a nonlinear von Kárman plate equations [**5, 7, 8, 9, 10, 11**].

In the annular sector model [**5**], an equilibrium state of a buckle can be obtained by solving the following nonlinear dynamic system

$$
\begin{cases} w_{,tt} + cw_{,t} + \Delta^2 w - \dfrac{4\pi^2}{(1-r_0)^2}\sigma_0 \Delta w - NLT_W(w,u,v) = 0, \\ LT_U(u,v) + NLT_U(w) = 0, \\ LT_V(u,v) + NLT_V(w) = 0, \end{cases} \tag{3.4}
$$

defined on an annular sector region $\Omega = \{(r,\theta)|\, r_0 < r < 1, -\theta_0 < \theta < \theta_0\}$, where LT, NLT stand for the corresponding linear and nonlinear terms (see [**5**] for details). The system is coupled with the clamped boundary conditions on $r = r_0$ and $r = 1$:

$$
\begin{cases} w(r_0,\theta) = 0, \; w_{,r}(r_0,\theta) = 0, \; u(r_0,\theta) = 0, \; v(r_0,\theta) = 0, \\ w(1,\theta) = 0, \quad w_{,r}(1,\theta) = 0, \quad u(1,\theta) = 0, \quad v(1,\theta) = 0, \end{cases} \tag{3.5}
$$

and the reciprocally periodic connection boundary conditions on $\theta = -\theta_0, \theta_0$:

$$
\begin{cases} \dfrac{1}{r^k}\dfrac{\partial^k w}{\partial \theta^k}\Big|_{(r,-\theta_0)} = \dfrac{1}{(1+r_0-r)^k}\dfrac{\partial^k w}{\partial \theta^k}\Big|_{(1+r_0-r,\,\theta_0)}, \qquad k = 0,1,2,3, \\ u(r,-\theta_0) = -u(1+r_0-r,\theta_0), \; \dfrac{u_{,\theta}(r,-\theta_0)}{r} = -\dfrac{u_{,\theta}(1+r_0-r,\theta_0)}{1+r_0-r}, \\ v(r,-\theta_0) = v(1+r_0-r,\theta_0), \quad \dfrac{v_{,\theta}(r,-\theta_0)}{r} = \dfrac{v_{,\theta}(1+r_0-r,\theta_0)}{1+r_0-r}. \end{cases} \tag{3.6}
$$

Map the normalized annular sector domain $\Omega = \{(r,\theta)\big| r_0 \le r \le 1, \; -\theta_0 \le \theta \le \theta_0\}$ onto the standard computational domain $\hat{\Omega} = \{(x,y)\big| -1 \le x \le 1, \; -1 \le y \le 1\}$ with $x = (2r - 1 - r_0)/(1 - r_0)$, $y = \theta/\theta_0$. Rewrite the out-of-plane dimensionless displacement $w(x,y)$ in the form $w(x,y) = (1-x^2)q(x,y)$, so that the clamped boundary conditions (3.5) are naturally satisfied [**5**].

We use Chebyshev-Gauss-Lobatto collocation points in the last two equations of (3.4) for x and y and in the first equation of (3.4) for x, and test the three types of collocation points in the first equation of (3.4) for y. For time discretization, the Newmark-β method is used for the first equation in (3.4).

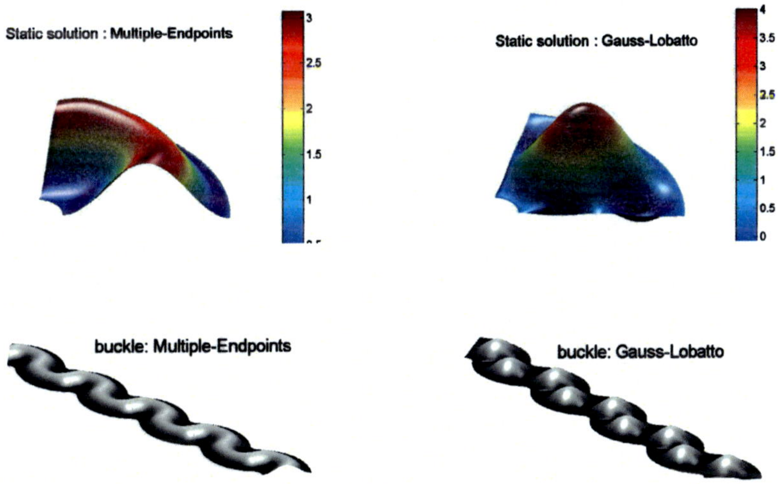

FIGURE 4. Numerical solutions and corresponding buckle morphologies.

For a set of physically relevant data $\sigma_0 = 10$, $r_0 = 0.15$, $\theta_0 = 0.7$, take small smooth functions $w_0 = 0.01(1-r)^2(r_0-r)^2$, $u_0 = 0$, $v_0 = 0$ as the initial state. Our numerical experiments showed that, while the scheme using the Chebyshev-Gauss collocation points always leads to blow up, the schemes using the other two types of collocation points can converge and produce static numerical solutions. The numerical solutions, with respect to $M = 20$, $N = 10$ and $c = 1000$, obtained by using the Multiple-Endpoints and Chebyshev-Gauss-Lobatto collocation points are shown in Figure 4.

Compare the numerical solutions with the physical experiments (see figures in [**9, 10, 11**], e.g. figure 1(c) in [**10**]), we see that the numerical results produced by using the Multiple-Endpoints fit the physical experiment well, while the numerical result produced by using the Chebyshev-Gauss-Lobatto collocation points is nonphysical with obvious negative values in the out-of-plane displacement.

4. Conclusions

The Chebyshev-Multiple-Endpoints collocation method and a new boundary condition technique are established in this paper for boundary value problems of high order differential equations. The new collocation points can be easily obtained by working with the Chebyshev polynomials. For linear problems, our numerical results showed that the new collocation points and the new boundary condition technique really helped to improve the condition numbers of differentiation matrices and the approximation accuracy of the numerical solutions. In particular, an example on nonlinear elastic thin film buckling showed that the improvement in the condition number can be crucial to the success of solving nonlinear problems.

References

[1] J. Shen, T. Tang, 2006, Spectral and high-order methods with applications, Science press, Beijing. MR2723481 (2012b:65001)

[2] J. P. Boyd, 1989, Chebyshev & Fourier spectral methods, Springer-Verlag Berlin, Heidelberg.

[3] C. Bernardi, G. Coppoletta, Y. Maday, 1992, Some spectral approximations of two-dimensional fourth-order problems, mathematics of computation 59, 63-67. MR1134714 (92k:65173)

[4] R. M. Kirby and Z. Yosibash, 2004, Solution of von-Karman dynamic non-linear plate equations using a pseudo-spectral method, Computer Methods in Applied Mechanics and Engineering 193, 575-599. MR2033964

[5] S. Wang and Z. P. Li, Mathematical modeling and numerical simulation of telephone cord buckles of elastic films, 2001, Science China Mathematics 54,1063-1076. MR2800927 (2012f:74080)

[6] J.F. Bonder, J.D. RossiExistence results for the p-Laplacian with nonlinear boundary conditions, 2001, Journal of Mathematical Analysis and Applications, 195-223. MR1864315 (2002h:35063)

[7] G. Gioia, M. Ortiz, 1997, Delamination of compressed thin films, Advances in Applied Mechanics 33, 119-192.

[8] L. B. Rreund and S. Suresh, 2003, Thin film materials: stress, defect formation, and surface evolution, Cambridge University Press.

[9] J. Colin, F. Cleymand, C. Coupean, J. Grilhe, 2000, Worm-like delamination patterns of thin stainless steel films on polycarbonate substrates. Philos. Mag. A 80 (11), 2559-2568.

[10] M. W. Moon, H. M. Jensen, J. W. Hutchinson, K. H. Oh, A. G. Evans, 2002, The characterization of tepephone cord buckling of compressed thin films on substrates, Journal of the Mechanics and Physics of Solids 50, 2355-2377.

[11] M. Cordill, D. Bahr, N. Moody, W. Gerberich, 2007, Adhesion measurements using telephone cord buckles, Materials Science and Engineering: A Volume 443, 150-155.

LMAM & School of Mathematical Sciences, Peking University, Beijing 100871, People's Republic of China

E-mail address: shasti0@pku.edu.cn

LMAM & School of Mathematical Sciences, Peking University, Beijing 100871, People's Republic of China

E-mail address: lizp@math.pku.edu.cn

Discontinuous residual distribution schemes for time-dependent problems

Andrzej Warzyński, Matthew E. Hubbard, and Mario Ricchiuto

ABSTRACT. This paper is concerned with the design of a new approach to numerical simulation of time-dependent hyperbolic PDEs within the residual distribution framework. The underlying representation of the solution is discontinuous-in-space which introduces extra flexibility into the design of new schemes. This is mainly due to edge-based residuals, which are necessary to impose communication between the cells and which are in addition to their cell-based counterparts. We show how to construct a scheme free of spurious oscillations and present numerical results for 2d scalar nonlinear equations on unstructured triangular grids to validate the method. The proposed framework leads to explicit time-stepping which is a promising alternative to more frequently applied implicit approaches.

1. Introduction

In this paper we consider a class of new numerical schemes for scalar hyperbolic conservation laws governing the evolution of an unknown quantity $u(\mathbf{x}, t)$, given by

$$\partial_t u + \nabla \cdot \mathbf{f}(u) = 0 \qquad \text{in } \Omega \times [0, T], \tag{1.1}$$

where Ω is a closed subset of $\mathbb{R}^2$ and $\mathbf{f}$ is the flux of u. Equation (1.1) is equipped with appropriate boundary conditions and the initial conditions:

$$u(\mathbf{x}, 0) = u_0(\mathbf{x}).$$

The framework we shall design our schemes in is that of Residual Distribution ($\mathcal{RD}$). Extension of the work presented here to systems, although possible with the aid of well-understood methods like the one in [11], is beyond the scope of this paper and will not be considered here.

Since their inception in [15], $\mathcal{RD}$ methods have proven to be reliable and accurate ways of discretizing hyperbolic equations. Their ability to carry out genuinely multidimensional upwinding enables them to capture more accurately the properties of the underlying physical problem which are incorporated within the mathematical model. This, at least in theory [9], means that they are capable of producing solutions of higher resolution then those produced by other known algorithms. For promising experimental observations on this matter see [12]. It was also demonstrated (see, for instance, [3] and [4]) that residual distribution methods are very

2000 *Mathematics Subject Classification.* 76M25, 65M99.
The first and the second author were supported by EPSRC grant EP/G003645/1.

robust and perform well when applied to complex problems arising in engineering and other applications.

In the case of steady state problems, at least for scalar equations, the $\mathcal{RD}$ framework has already reached a high level of sophistication and understanding (this was summarized in [**2**]). Even though further research is still being carried out, the emphasis is now mainly laid on the development of residual distribution methods for time-dependent problems. The main challenge is to design a scheme which retains all the properties of its steady counterpart(s), in particular positivity and linearity preservation [**10**] at the same time, and which is relatively efficient.

The space-time framework investigated in [**7**] (see also [**10**] and references therein) allows construction of discretizations with *all* the desired properties. Unfortunately, those methods are subject to a CFL-type restriction on the time-step, which is particularly disappointing when taking into account that they are, by construction, implicit. In the *two layer* variant [**8**] one couples two space-time slabs at a time and solves the equations simultaneously in both. On one hand the resulting system to be solved at each step is larger, but on the other the construction removes from one of the layers the restriction on the time-step. In theory this means that an arbitrarily large time-step can be used. For a full discussion see [**7**]. Recently, Hubbard and Ricchiuto [**14**] proposed to drive the height of one of the space-time slabs (and hence its associated time-step) to zero so that the scheme becomes discontinuous-in-time. The resulting formulation is simpler than the original whereas all of the properties are retained.

In this contribution we investigate a somewhat similar in spirit approach. We also assume that the discrete representation of the data is discontinuous, however we will consider discontinuities in space instead of time. Such an approach will enable us to construct explicit schemes with a localised system and a simple framework within which $h-$ and $p-$ adaptivity can be incorporated. This work extends previous work on discontinuous residual distribution schemes initiated by Hubbard [**13**] and Abgrall [**1**] in which only steady state problems were considered. We first apply those schemes to time-dependent equations and then show how to design positive schemes. Only first order methods will be considered. This paper also extends work of Abgrall and Shu [**5**] in the sense that it shows how ideas from the discontinuous Galerkin framework can be incorporated within residual distribution methods. This is, briefly speaking, done by considering flux differences (*edge residuals* in the RD nomenclature) instead of the fluxes themselves.

This paper is organised as follows. In the next section the residual distribution and discontinuous residual distribution methods are presented. A quick overview of available distribution strategies for the edges is given and one new distribution method (based on the discontinuous Galerkin method) is then introduced. Next, in Section 4, we present discontinuous-in-space residual distribution methods for time-dependent problems. This is followed by numerical results and conclusions.

2. Continuous Residual Distribution Schemes for Steady State Problems

The residual distribution framework (both in the continuous and discontinuous setting) was originally introduced for steady state problems:

$$\nabla \cdot \mathbf{f}(u) = 0 \quad \text{or} \quad \mathbf{a}(u) \cdot \nabla u = 0 \qquad \text{in } \Omega, \tag{2.1}$$

in which $\mathbf{a}(u) = \frac{\partial \mathbf{f}}{\partial u}$. Throughout the text we will assume that a triangulation of Ω is given and will denote it by $\mathcal{T}_h$ (h being the mesh parameter). The numerical solution u_h is assumed to be of the same form as in the case of finite-element-type approximations, i.e. given its nodal values $u_i = u_h(\mathbf{x}_i)(\mathbf{x}_i \in \mathcal{T}_h)$, it reads

$$u_h(\mathbf{x}) = \sum_{i \in \mathcal{T}_h} \psi_i(\mathbf{x}) u_i, \tag{2.2}$$

in which ψ_i is the standard Lagrange basis function associated with $\mathbf{x}_i$. Cell interfaces will be denoted by e and $\mathcal{D}_i$ will stand for the subset of triangles containing node i. The median dual cell, obtained by joining the gravity centres of triangles in $\mathcal{D}_i$ with the midpoints of the edges meeting at i, will be denoted by S_i (illustrated in Figure 1).

To construct a set of linear equations for the nodal values of u_h one first, for each cell $E \in \mathcal{T}_h$, calculates the cell residual

$$\phi^E = \int_E \nabla \cdot \mathbf{f}(u)\, d\Omega.$$

Next, fractions of this residual are distributed among the vertices of E so that the resulting linear system reads:

$$\sum_{E \in \mathcal{D}_i} \beta_i \phi^E = 0 \qquad \forall i. \tag{2.3}$$

The distribution coefficients β_i signify the fraction of the cell residual that is assigned to the node i and are used to impose various properties of the scheme.

Here we will consider only **the N** (for Narrow) **scheme [2]** - the most successful linear positive distribution strategy. In order to introduce it, for each vertex $i \in E$, we define the so-called flow sensors:

$$k_i = -\frac{\mathbf{a}(u) \cdot \vec{\mathbf{n}}_i}{2} |e_i|, \quad k_i^+ = \max(0, k_i), \quad k_i^- = \min(0, k_i),$$

in which $\vec{\mathbf{n}}_i$ is the outward pointing unit normal vector to edge e_i (opposite the ith vertex). We will also need the following quantities:

$$N = \left(\sum_{j \in E} k_j^+ \right)^{-1}, \qquad u_{in} = -\sum_{j \in E} N k_j^- u_j.$$

Finally, the distribution itself reads:

$$\beta_i \phi^E = k_i^+ (u_i - u_{in}). \tag{2.4}$$

In practice, system (2.3) is solved with the aid of pseudo time-stepping:

$$u_i^{n+1} = u_i^n - \frac{\Delta t}{|S_i|} \sum_{E \in \mathcal{D}_i} \beta_i \phi^E \qquad \forall i,$$

which is used to iterate to the steady state.

3. Discontinuous Residual Distribution Schemes For Steady State Problems

Discontinuous-in-space residual distribution schemes were introduced simultaneously by Abgrall [**1**] (based on a report written in French in 2007) and Hubbard [**13**]. In this framework the numerical solution is no longer assumed to be globally continuous which means that, at least in each cell $E \in \mathcal{T}_h$, its discrete representation remains unchanged (cf. Equation (2.2)):

$$\forall \mathbf{x} \in E \qquad u_h(\mathbf{x}) = \sum_{i \in \mathcal{T}_h} \psi_i(\mathbf{x}) u_i.$$

Relaxing the continuity constraint means that one needs to impose some sort of communication between the cells. In both cases the authors suggest using the edge residuals:

$$\phi^e = \int_e [\mathbf{f}(u_h) \cdot \vec{\mathbf{n}}] \, d\Gamma$$

where $[\mathbf{f}(u_h) \cdot \vec{\mathbf{n}}]$ represents the jump of the function $\mathbf{f}(u_h) \cdot \vec{\mathbf{n}}$ across the edge, the sign of the difference being dictated by the direction chosen for $\vec{\mathbf{n}}$. To be more precise

$$\begin{aligned} \phi^e(u_h) = \int_e [\mathbf{f}(u_h) \cdot \vec{\mathbf{n}}] \, d\Gamma &= \int_e (\mathbf{f}_L \cdot \vec{\mathbf{n}}_L + \mathbf{f}_R \cdot \vec{\mathbf{n}}_R) \, d\Gamma \\ &= \int_e (\mathbf{f}_L - \mathbf{f}_R) \cdot \vec{\mathbf{n}}_L \, d\Gamma = \int_e (\mathbf{f}_R - \mathbf{f}_L) \cdot \vec{\mathbf{n}}_R \, d\Gamma, \end{aligned}$$

where the subscripts $_L$ and $_R$ mean that we take the value of a quantity in E_L and E_R respectively, the cells associated with e (see Figure 1). Obviously ϕ^e is zero if

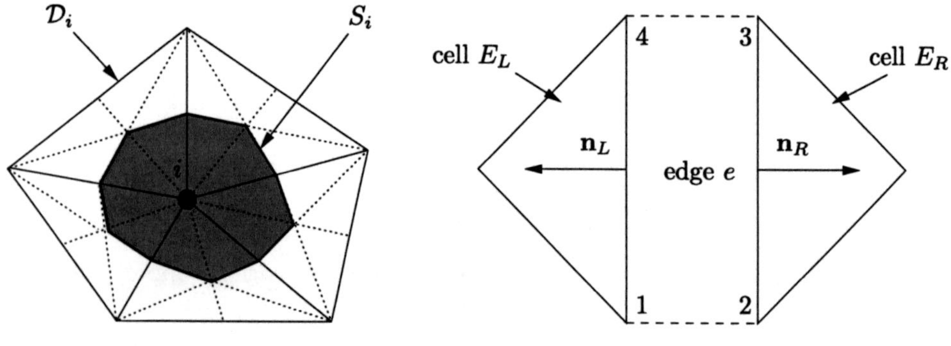

FIGURE 1. Edge e and the two cells associated with it: E_L and E_R.

u_h is assumed to be globally continuous. As in the case of cell residuals, fractions of ϕ^e are *sent* to the vertices of e. Similar to continuous residual distribution, to find the numerical solution u_h one first assembles signals sent to each degree of freedom i and then solves the resulting linear system with the aid of pseudo time-stepping:

$$u_i^{n+1} = u_i^n - \frac{3\Delta t}{|E|} \left(\beta_i^E \phi^E + \alpha_i^{e_1} \phi^{e_1} + \alpha_i^{e_2} \phi^{e_2} \right) \qquad \forall i. \tag{3.1}$$

In analogy to the cell residual, $\alpha_i^{e_1}$ and $\alpha_i^{e_2}$ are the distribution coefficients for the degree of freedom $i \in E$ corresponding to the edges $e_1 \in E$ and $e_2 \in E$, respectively, adjacent to vertex i. Note that in the discontinuous setting each degree of freedom belongs to only one cell and two of its edges.

As far as cell residuals are concerned we will utilize only the aforementioned N scheme (2.4). It is the distribution strategy for the edges that is the main novelty within the discontinuous framework and we shall look into it in more detail. We point out that it gives us extra freedom in the design of new schemes compared to the continuous case where only the cell distribution can be altered. Three natural choices that will be applied in this work are as follows.

The mED scheme of Hubbard [**13**] is linear, positive and linearity preserving. For a generic edge e and its vertices $1, 2, 3$ and 4 (see Figure 1) it is given by

$$\phi_1^e = \frac{1}{2}\left[\mathbf{a}_{12}\cdot\vec{\mathbf{n}}\right]^-(u_1-u_2) = \alpha_1\,\phi^e,$$
$$\phi_2^e = \frac{1}{2}\left[\mathbf{a}_{12}\cdot\vec{\mathbf{n}}\right]^+(u_1-u_2) = \alpha_2\,\phi^e,$$
$$\phi_3^e = \frac{1}{2}\left[\mathbf{a}_{43}\cdot\vec{\mathbf{n}}\right]^+(u_4-u_3) = \alpha_3\,\phi^e,$$
$$\phi_4^e = \frac{1}{2}\left[\mathbf{a}_{43}\cdot\vec{\mathbf{n}}\right]^-(u_4-u_3) = \alpha_4\,\phi^e.$$

The $\mathbf{a}_{ij}$ are conservatively averaged values of the flux Jacobian defined as:

$$\mathbf{a}_{12} = \frac{1}{3}\left(\mathbf{a}_1+\mathbf{a}_2+\frac{\mathbf{a}_3+\mathbf{a}_4}{2}\right),\qquad \mathbf{a}_{43} = \frac{1}{3}\left(\mathbf{a}_3+\mathbf{a}_4+\frac{\mathbf{a}_1+\mathbf{a}_2}{2}\right)$$

in which $\mathbf{a}_i$ $(i=1,\ldots,4)$ are the values of $\mathbf{a}$ at the vertices of e and $\vec{\mathbf{n}} = |e|\vec{\mathbf{n}}_R$.

The (local) **Lax-Friedrichs distribution** for edges was proposed by Abgrall in [**1**] and is based on its counterpart for cells. It is defined as

$$\alpha_i\,\phi^e = \frac{\phi^e}{4} + \alpha^e(u_i-\bar{u}),\qquad i=1,\ldots,4,$$

with

$$\bar{u} = \frac{u_1+u_2+u_3+u_4}{4},$$

where $u_1, u_2, u_3.u_4$ are the values of u_h at the vertices of e (cf. Figure 1) and

$$\alpha^e \geq \max_{x\in e}\|\mathbf{f}'(u_h(x))\|.$$

It is also positive and linear, but not linearity preserving.

The DG distribution is simply the edge integral appearing in the strong formulation of the discontinuous Galerkin approximation of (2.1) (see [**6**] for a review of discontinuous Galerkin methods):

$$\alpha_1^{DG}\,\phi^e = \int_e (\mathbf{f}^*-\mathbf{f})\,(u_h)\cdot\vec{\mathbf{n}}_L\,\psi_1^{E_L}\,d\Gamma,$$
$$\alpha_2^{DG}\,\phi^e = \int_e (\mathbf{f}^*-\mathbf{f})\,(u_h)\cdot\vec{\mathbf{n}}_R\,\psi_2^{E_R}\,d\Gamma,$$
$$\alpha_3^{DG}\,\phi^e = \int_e (\mathbf{f}^*-\mathbf{f})\,(u_h)\cdot\vec{\mathbf{n}}_R\,\psi_3^{E_R}\,d\Gamma,$$
$$\alpha_4^{DG}\,\phi^e = \int_e (\mathbf{f}^*-\mathbf{f})\,(u_h)\cdot\vec{\mathbf{n}}_L\,\psi_4^{E_L}\,d\Gamma,$$

in which the numerical flux $\mathbf{f}^*$ is defined as usual and ψ_i^E is the Lagrange basis function in cell E associated with edge vertex i. Now, since $\vec{\mathbf{n}}_R = -\vec{\mathbf{n}}_L$ it follows immediately that:

$$\sum_{i\in e}\alpha_i^{DG} = 1$$

which makes the above a valid distribution. It has yet to be tested on steady state problems, but in this paper we concentrate on transient problems and will not address that.

4. Discontinuous Residual Distribution Schemes for Transient Problems

To the authors' best knowledge there have been no (successful) attempts to apply discontinuous residual distribution methods to time-dependent problems. We start by assuming that the temporal domain $[0,T]$ is discretized into a set of N discrete levels $\{t^n\}_{n=0,1\dots,N-1}$ such that

$$t^0 = 0, \quad t^{N-1} = T, \quad t^n < t^{n+1} \quad \text{and} \quad \Delta t^n = t^{n+1} - t^n.$$

Given an appropriate initial condition $u_0(\mathbf{x})$, the method has the following form:

$$u_i^{n+1} = u_i^n - \frac{3\Delta t^n}{|E|}\left(\beta_i \phi^E(u_h^n) + \alpha_i\, \phi^{e_1}(u_h^n) + \alpha_i\, \phi^{e_2}(u_h^n)\right) \quad \forall E \in \mathcal{T}_h \; \forall i \in E \tag{4.1}$$

in which $u_h^n = u_h(\mathbf{x}, t^n)$ is the known numerical solution at time t^n. The numerical solution $u_h(\mathbf{x}, 0)$ is taken as the interpolation of the analytical initial condition u_0. The above is simply (3.1) applied to time-dependent problems which means that now Δt has physical meaning. The limit on the time-step guaranteeing positivity is given by (cf. Equation (40) in [**13**]):

$$\Delta t \le \frac{|E|/3}{(k_i^E)^+ + (k_i^{e_1})^+ + (k_i^{e_2})^+} \qquad \forall E \in \mathcal{T}_h \; \forall i \in E,$$

in which cell E and edges e_1 and e_2 are the only ones i receives signals from.

We note that, since the time derivative was discretized with the aid of a first order approximation, the overall order of the approximation error should be expected to be no higher than one. Extension to higher order schemes is a subject of ongoing work.

5. Numerical Results

The two dimensional inviscid Burgers' equation,

$$\partial_t u + \nabla \cdot \mathbf{f}(u) = 0 \qquad \text{in } \Omega \times [0,T],$$

with $\mathbf{f} = \left(\frac{u^2}{2}, \frac{u^2}{2}\right)$ is approximated over the square $\Omega = [-1,1]^2$ with the discontinuous initial condition:

$$u(\mathbf{x}, 0) = \begin{cases} 1 & \text{if } \mathbf{x} \in (-0.5, 0) \times (-0.5, 0), \\ 0 & \text{otherwise.} \end{cases}$$

The problem is solved up to the final time $T = 1$. In all the experiments the N scheme was used to distribute the cell residuals and the three methods outlined in Section 3 were applied to the edges. In Figure 2 we present cross sections of the solutions and in Table 1 we show the corresponding minimal and maximal values.

The results show that both mED and LF lead to numerical approximations free of unphysical oscillations whereas the solution obtained with the aid of the DG-type approach exhibits some overshoots. This was expected as both mED and LF distributions were designed so that the resulting scheme is positive and in the case of discontinuous Galerkin methods extra limiting is usually applied to achieve similar results. We also note that the LF method is very diffusive and of all the considered approaches the mED algorithm performed the best.

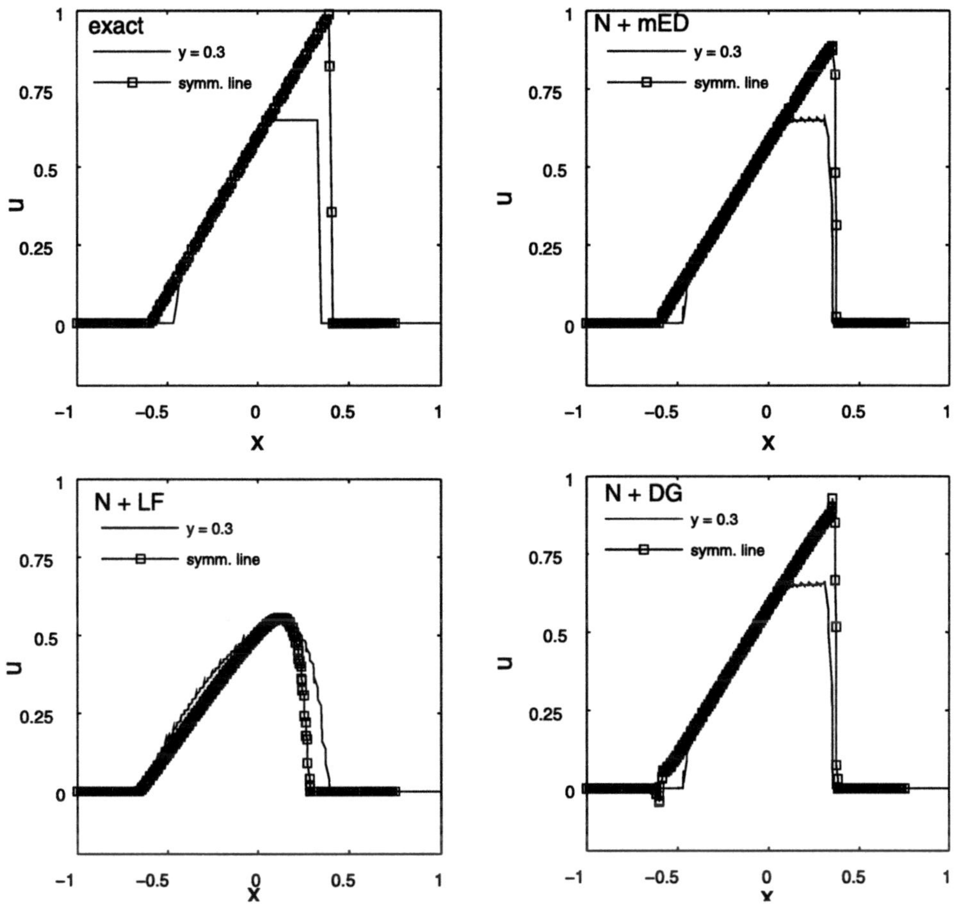

FIGURE 2. Solution along the line $y = 0.3$ and the symmetry line $y = x$ for the 2d Burgers' equation at time $t = 1$. Top left: the exact solution. Top right: N + mED. Bottom left: N + LF. Bottom right: N + DG.

TABLE 1. Mininum and maximum values of the exact and numerical solution to the 2d Burgers' equation at time $t = 1$.

	exact	N + mED	N + LF	N + DG
u_{min}	0	0	0	-1.026e-01
u_{max}	1	8.833e-01	5.567e-01	9.319e-01

6. Conclusions

We proposed a new class of numerical approximations to time-dependent hyperbolic conservation laws. This new formulation was successfully applied to the inviscid Burgers' equation and three different methods for distributing the edge residuals were compared. The numerical results confirmed that as long as the cell and edge residuals are distributed in a positive manner the resulting method is positive. All the presented schemes are first order accurate.

Our future work will be focused on the design of second order schemes. This can be achieved by, for instance, applying second order time-stepping and consistent mass matrices to discretize the time derivative. Incorporating discontinuities in time into our framework is also being investigated.

References

[1] Rémi Abgrall, *A residual distribution method using discontinuous elements for the computation of possibly non smooth flows*, Adv. Appl. Math. Mech. **2** (2010), no. 1, 32–44. MR2576745

[2] Remi Abgrall, *A review of residual distribution schemes for hyperbolic and parabolic problems: the July 2010 state of the art*, Commun. Comput. Phys. **11** (2012), no. 4, 1043–1080, DOI 10.4208/cicp.270710.130711s. MR2864076 (2012j:65276)

[3] R. Abgrall, G. Baurin, P. Jacq, and M. Ricchiuto, *Some examples of high order simulations parallel of inviscid flows on unstructured and hybrid meshes by residual distribution schemes*, ComputFluids **61** (2012), no. 0, 6 – 13.

[4] R. Abgrall, H. Deconinck, and K. Sermeus, *Status of multidimensional upwind residual distribution schemes and applications in aeronautics*, AIAA Paper 2000-2328, Fluids 2000/Denver, June 2000.

[5] Rémi Abgrall and Chi-Wang Shu, *Development of residual distribution schemes for the discontinuous Galerkin method: the scalar case with linear elements*, Commun. Comput. Phys. **5** (2009), no. 2-4, 376–390. MR2513691 (2010g:65140)

[6] Bernardo Cockburn, *Discontinuous Galerkin methods for convection-dominated problems*, High-order methods for computational physics, Lect. Notes Comput. Sci. Eng., vol. 9, Springer, Berlin, 1999, pp. 69–224. MR1712278 (2000f:76095)

[7] Á. Csík and H. Deconinck, *Space-time residual distribution schemes for hyperbolic conservation laws on unstructured linear finite elements*, Internat. J. Numer. Methods Fluids **40** (2002), no. 3-4, 573–581, DOI 10.1002/fld.315. ICFD Conference on Numerical Methods for Fluid Dynamics, Part II (Oxford, 2001). MR1932998 (2003m:65174)

[8] Á. Csík, H. Deconinck, M. Ricchiuto, and S. Poedts, *Space-time residual distribution schemes for hyperbolic conservation laws*, 15th AIAA Computational Fluid Dynamics Conference, Anaheim, CA, USA, June 2001.

[9] H. Deconinck, *Upwind methods and multidimensional splittings for the Euler equations*, Computational Fluid Dynamics, VKI LS 1991-01, von Karman Institute for Fluid Dynamics, 1991.

[10] H. Deconinck and M. Ricchiuto, *Residual distribution schemes: foundations and analysis*, Encyclopedia of Computational Mechanics, vol. 3, John Wiley and Sons, Ltd., 2007.

[11] H. Deconinck, P. L. Roe, and R. Struijs, *A multidimensional generalization of Roe's flux difference splitter for the Euler equations*, Comput. & Fluids **22** (1993), no. 2-3, 215–222, DOI 10.1016/0045-7930(93)90053-C. MR1231645 (94e:76060)

[12] S. M. J. Guzik and C. P. T. Groth, *Comparison of solution accuracy of multidimensional residual distribution and Godunov-type finite-volume methods*, Int. J. Comput. Fluid Dyn. **22** (2008), no. 1-2, 61–83, DOI 10.1080/10618560701748176. MR2405126 (2009c:76048)

[13] Matthew Hubbard, *Discontinuous fluctuation distribution*, J. Comput. Phys. **227** (2008), no. 24, 10125–10147, DOI 10.1016/j.jcp.2008.08.017. MR2467946 (2009m:65191)

[14] M.E. Hubbard and M. Ricchiuto, *Discontinuous upwind residual distribution: A route to unconditional positivity and high order accuracy*, ComputFluids **46** (2011), no. 1, 263 – 269.

[15] P. L. Roe, *Fluctuations and signals - a framework for numerical evolution problems*, Numerical Methods for Fluid Dynamics, Academic Press, 1982, pp. 219–257.

School of Computing, University of Leeds, Leeds, LS2 9JT, United Kingdom
E-mail address: `scaw@leeds.ac.uk`

School of Computing, University of Leeds, Leeds, LS2 9JT, United Kingdom
E-mail address: `m.e.hubbard@comp.leeds.ac.uk`

351 cours de la Liberation, Bat. A29 bis 33405 Talence Cedex - France
E-mail address: `Mario.Ricchiuto@inria.fr`